Crisis on the Coast and Hinterland

Ashis Kumar Paul • Anurupa Paul
Editors

Crisis on the Coast and Hinterland

Assessing India's East Coast with Geomorphological, Environmental and Remote Sensing and GIS Approaches

Springer

Editors
Ashis Kumar Paul
Department of Geography and Environment Management
Vidyasagar University
Midnapore, West Bengal, India

Anurupa Paul
Department of Remote Sensing and GIS
Vidyasagar University
Midnapore, West Bengal, India

ISBN 978-3-031-42230-0 ISBN 978-3-031-42231-7 (eBook)
https://doi.org/10.1007/978-3-031-42231-7

This Springer imprint is published by the registered company Springer Nature Switzerland AG
The registered company address is: Gewerbestrasse 11, 6330 Cham, Switzerland

Paper in this product is recyclable.

The volume is dedicated to my Mother
who protected me from all odds in the Earth

Preface

The volume of *Crisis on the Coast and Hinterland: Assessing India's East Coast with Geomorphological, Environmental and Remote Sensing and GIS Approaches*, represents 32 book chapters distributed in 3 parts with focal themes of contemporary geo-environmental issues. All the book chapters are written by a group of experts who are involved in coastal research under a school of thought developed by the chief editor of the book volume for the tropical coast of India.

The east coast of India comprises four administrative states and one union territory along the coastal fringes of the Bay of Bengal, and one island group under the status of a union territory in the Andaman Sea and Bay of Bengal of the North Indian Ocean. The volume demonstrates, in the first part, geomorphological diversity at the east coast, coastal vulnerability, sedimentary depositional landforms, geoarchaeological remains, beach stage modelling, Chilika lagoon hydro-geomorphic process, barrier spit morphology, and beach ridge chenier deposits with geomorphological approaches. However, the second part of the volume highlights habitat degradations, mangrove sensitivity to the ephemeral rise of sea waters, land degradation in the coastal river basin, riparian corridors of the hinterland drainage basin, beach tourism potentials, climate variability and agricultural modifications in the coastal hinterland, changing tidal flat environments, estuarine environments, hazards and flood risks, premature land reclamations, geomorphosites for the promotion of geotourism, the historical and geomorphological background of ancient ports, impacts of marine litter in the coastal regions, and coastal groundwater contamination problems with environmental approaches. The third part of the volume represents assessments of the coral reef geomorphology of the Andaman-Nicobar Islands, the cyclone battered Sundarban, multivariate analysis of coastal vulnerabilities, managing coastal squeeze and wetland loss, assessment of tourism carrying capacity, tourism climate index, diversity in landscape ecology in the coastal tracts, overwash vulnerabilities in the Chilika lagoon, changing livelihood security index along the coastal belt, and coastal urbanization with emerging consequences in response to climate variability using remote sensing techniques.

The structure of the volume includes 18 issues from West Bengal, 10 issues from Odisha, 3 issues from Andhra Pradesh, 3 issues from the Tamil Nadu and Pondicherry coasts, and 4 issues from the Andaman and Nicobar group of islands, particularly to represent the crisis on the coast and hinterland in the tropical environment of eastern India along the Bay of Bengal. Though the number of coastal issues differs from place to place, the importance of climate change-induced reworking of the coastal landscapes and anthropogenic-controlled fluvial discharges in the regional settings of India's East Coast, West Bengal, and Odisha demonstrated multiple coastal issues in the volume. However, the uniqueness of the volume lies in the representation of coastal issues at the local scale, regional scale, and national scale with in-depth field studies and the application of remote sensing techniques and spatial modelling techniques for the east coast of India.

The issues of the coastal crisis represented in this volume will provide an attractive database on India's East Coast for understanding the major geo-environmental problems in coastal management and in the sustainable development of coastal resources in response to future vulnerabilities. All the research contributions included in the volume will provide a new avenue of research in the coastal studies, particularly for the upcoming researchers in the fields of coastal geomorphology, coastal environment, and the application of remote sensing techniques in the coastal studies.

Finally, the initiative taken by the chief editor in the volume will reflect the impact of 40 years of professional experiences in the field of coastal research in India, with a concentration of extensive field survey methods at both temporal and spatial scales. I hope it will be a very good reference book for the students and researchers of various disciplines in coastal studies.

Midnapore, West Bengal, India

Ashis Kumar Paul
Anurupa Paul

Acknowledgements

The idea for the volume *Crisis on the Coast and Hinterland* originated in 2021, when I was addressing the researchers of coastal studies and sharing my field experiences on the climate change reworking of the tropical coastal landscapes of India. I congratulate Springer Nature for accepting my book proposal on such attractive coastal issues in the Indian tropics during 2022. I convey my thanks to Zachary Romano, Senior Publishing Editor of Earth Sciences, Geography, and Environment at Springer, New York, and other associate publishing editors Herbert Moses, Henry Rodgers, Evangeline Christoper, and Marion Schneider of Springer for reviewing the proposal and for materializing the publication processes of the volume.

All the contributors of the volume are specially acknowledged for their active participation in the initiatives as well as for keeping their patience in waiting for the outcome of the publication. I also appreciate the mental boost provided by my wife, Mrs. Mousumi Paul, and my daughter, Dr. Anurupa Paul, to fulfill my dream project.

I extend my gratitude to Mr. Kshitish Karak, tourism agent for the Andaman and Nicobar group of islands, for his help in conducting field work in different destinations of the Emerald Islands from 2005 to 2018. Dr. D. Haldar, of GSI, is acknowledged as the Deputy Director for his association with the Andaman field visits during 2003 and 2005. Dr. Sujit Dasgupta, Deputy Director, GSI, for his co-operation in the field work (2018) in the Andaman Islands, is also acknowledged. I also remember the role of various NGOs in the Indian Sundarban and my Ph.D. research scholars in organizing the field studies and extending their co-operation in different stages of the field works in different coastal belts of the country. The assistance of Sri. Joydeb Sardar, Centre for Environmental Studies, Vidyasagar University, in all stages of the work helped me a lot in finishing the volume successfully. He is specially acknowledged for extending his co-operation in typing the manuscripts and computing the data sets for conducting the publication processes for the volume. At last, but not least, the students of the coastal management special paper of Vidyasagar University (2005–2022) are also acknowledged for their co-operation in the organization of field works in the coastal parts of the Bay of Bengal from time to time in the validation of coastal studies using remote sensing techniques and the application of spatial modelling techniques.

Ashis Kumar Paul
Anurupa Paul

Contents

List of Figures

List of Tables

Abbreviations

ASCII	American Standard Code for Information Interchange
AD	Anno Domini
ANI	Andaman Nicobar Island
AHP	Analytic Hierarchy Process
AUC	Area Under Curve
ASTER	Advanced Spaceborne Thermal Emission and Reflection Radiometer
ACI	Adaptive Capacity Index
BCE	Before Common Era
BPL	Below Poverty Line
CVI	Coastal Vulnerability Index
CB	Chandrabhaga Beach
CE	Common Era
CDA	Chilika Development Authority
CWR	Crop Water Requirement
CRZ	Coastal Regulation Zone
CCI	Clean Coast Index
CGWB	Central Ground Water Board
CWC	Central Water Commission
CBO	Complete Barrier Overwash
CS	Cyclonic Storm
DoD	Dem of Difference
DEM	Digital Elevation Model
DF	Deligent Fault
DGPS	Differential Global Positioning System
DDPR	District Disaster Plan Report
DSDA	Digha Shankarpur Development Authority
EGM	Earth Gravity Model
EPA	Environmental Protection Act
EIA	Environmental Impact Assessment
EMF	Eastern Margin Fault

EVI	Enhanced Vegetation Index
EI	Exposure Index
ECC	Effective Carrying Capacity
GSI	Geological Survey of India
GCP	Ground Control Point
GIS	Geographic Information System
GPS	Global Positioning System
GAP	Ganga Action Plan
GEBCO	General Bathymetric Chart of the Oceans
GDP	Gross Domestic Product
GDEM	Global Digital Elevation Model
HLSI	Household Livelihood Security Index
HCA	Hierarchical Cluster Analysis
HHT	Highest High Tide
HAT	Highest Astronomic Tide
HTL	High Tide Line
HH	Horizontal and Horizontal
INCOIS	Indian National Centre for Ocean Information Services
IMD	India Meteorological Department
IWR	Irrigation Water Requirement
IPCC	Intergovernmental Panel on Climate Change
ICZM	Integrated Coastal Zone Management
LSI	Livelihood Security Index
LAI	Leaf Area Index
LTL	Low Tide Line
LULC	Land Use and Land Cover
LiDAR	Light Detection and Ranging
MOIR	Maximum Overwash Intrusion Recurrence
MSI	Multispectral Instrument
MC	Management Capacity
MoEF	Ministry of Environment and Forests
MHWL	Mean High Water Level
MMTL	Mean Monthly Tidal Level
MCM	Million Cubic Metres
MT	Metric Tonne
MF	Mentawei Fault
MSL	Mean Sea Level
NDVI	Normalized Difference Vegetation Index
NOAA	National Oceanic and Atmospheric Administration
NCAER	National Council of Applied Economic Research
NBSS & LUP	National Bureau of Soil Survey and Land Use Planning
OLI	Operational Land Imager
OSL	Optically Stimulated Luminescence
OSR	Overwashed Shoreline Ratio
PSMSL	Permanent Service for Mean Sea Level

PET	Potential Evapotranspiration
PAR	Photosynthetically Active Radiation
PCC	Physical Carrying Capacity
PADI	Professional Association of Diving Instructors
RCC	Real Carrying Capacity
ROC	Receiver Operating Characteristic Curve
RMSE	Root Mean Square Error
SST	Sea Surface Temperature
SAM	Spectral Angle Mapper
SVM	Support Vector Machine
SLR	Sea Level Rise
SCS	Severe Cyclonic Storm
SI	Sensitivity Index
SD	Standard Deviation
SDDI	Stress Degree Day Index
SBR	Sundarban Biosphere Reserve
SPI	Standardized Precipitation Index
SOI	Survey of India
SAC	Space Application Centre
S	Swale
SRTM	Shuttle Radar Topography Mission
SMC	Coastal Modelling System
TE	Tourism Expenditure
TDR	Transfer Development Right
TCC	Tourism Carrying Capacity
TCI	Tourism Climate Index
TS	Total Station
USGS	United States Geological Survey
YBP	Year Before Present

About the Editors

Ashis Kumar Paul is teaching Geography and Environment Management, Earth System Sciences (ESS), and Remote Sensing and GIS as a Professor in the Department of Geography, Vidyasagar University, West Bengal, India. He is a reputed Coastal Geomorphologist in India. He has guided 31 Ph.D. Research Scholars and 3 M.Phil. theses in Geography and ESS, Geomorphology, Geo-archaeology, Remote Sensing and GIS, and Environmental Studies. His research interests cover studies on Coastal Geomorphology, Coastal Sediments, Coastal Environment, Coastal Tourism, Bio-Geomorphology, and Coastal Management on the Eastern and Western shores of India, including the Andaman Group of Islands. He has over 150 research papers in international and national journals, 31 book chapters, 4 reference books, and other edited volumes. He has authored the books *Coastal Geomorphology and Environment*, *Tsunami: An Assessment of Disasters, Tectonics and Landforms*, and *Physical Environment*. He was the President of the Indian Institute of Geomorphologists in 2020 and was awarded membership as an authorized CERF Society Member of JCR, USA, in 2017. He is also known as a field geomorphologist and field ecologist.

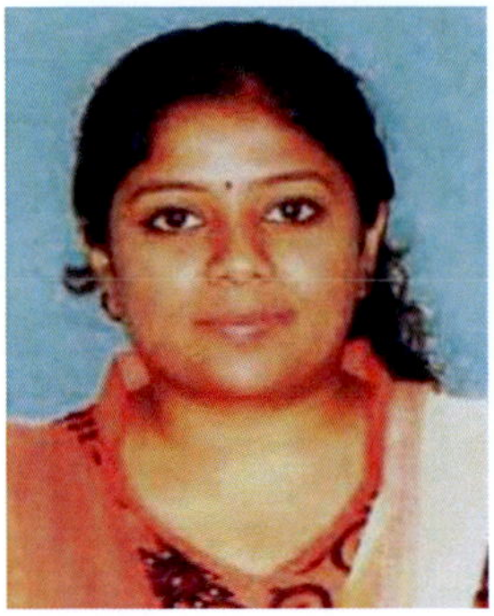

Anurupa Paul has completed her M.Sc. in Geography from Savitribai Phule Pune University, Maharashtra, and obtained her second M.Sc. degree in Environmental Science from Vidyasagar University, West Bengal. She has been awarded the Ph.D. degree from the Department of Remote Sensing and GIS of Vidyasagar University. She was awarded the title of Young Scientist in 2013 from Savitribai Phule Pune University, India. She has published 26 articles in reputed international and national journals and two book chapters in Springer Nature and Taylor & Francis. Anurupa has presented her research work in Kuala Lumpur, Malaysia (the 38th ACRS Conference), other international conferences (12), and 34 national conferences. She has also participated in 36 workshops, international training programmes, and 121 webinars on the application of remote sensing techniques in coastal research and disaster management. She has received seven best paper presentation awards in seminars. Her area of interest in research is mainly focused on Karst Morphology, Coastal Geomorphology, Limestone Geomorphology, Environmental Geoscience, Earth Science, and the application of Remote Sensing and GIS. She has guided 18 M.Sc. dissertation in coastal geomorphology and environment with the application of geospatial techniques in the Department of Remote Sensing and GIS, Vidyasagar University.

Contributors

Sk Saharukh Ali Department of Geography, Vidyasagar University, Midnapore, West Bengal, India

Rashmi Rani Anand Saheed Bhagat Singh College, University of Delhi, New Delhi, Delhi, India

Abhinanda Bal K.D. College of Commerce and General Studies, Midnapore, West Bengal, India

Jatisankar Bandyopadhyay Department of Remote Sensing & GIS, Vidyasagar University, Midnapore, West Bengal, India

Suparna Banerjee Department of Remote Sensing & GIS, Vidyasagar University, Midnapore, West Bengal, India

Nilay Kanti Barman Department of Geography, Midnapore College (Autonomous), Midnapore, West Bengal, India

Swagata Bera Department of Geography, Dum Dum Motijheel College, Kolkata, West Bengal, India

Phalguni Bhattacharyya Balarampur M N Vidyamandir, Bonhooghly, Narendrapur, Kolkata, West Bengal, India

Bubay Bid Department of Remote Sensing & GIS, Vidyasagar University, Midnapore, West Bengal, India

Sukumar Chand Department of Remote Sensing and GIS, Vidyasagar University, Midnapore, West Bengal, India

Sudip Dasgupta Department of Geography, Asutosh College, Kolkata, West Bengal, India

Kanailal Das Parameswar Mahavidyalaya, Namkhana, West Bengal, India

Karabi Das Department of Geography, Dr. Kanailal Bhattacharyya College, Howrah, West Bengal, India

Punam Debnath Department of Geography, Vidyasagar University, Midnapore, West Bengal, India

M. Devi Department of Physics, KiiT University, Bhubaneswar, Odisha, India

Abantika Dey Department of Remote Sensing & GIS, Vidyasagar University, Midnapore, West Bengal, India

Soumen Dey Department of Remote Sensing & GIS, Vidyasagar University, Midnapore, West Bengal, India

Satyajit Dhara Acharya Jagadish Chandra Bose College, Kolkata, West Bengal, India

Debabrata Ghorai Department of Research and Development, Dvara E-Registry, DBS House, Hyderabad, India

Swati Ghosh Department of Geography, Dum Dum Motijheel College, Kolkata, West Bengal, India

Asutosh Goswami Department of Earth Sciences and Remote Sensing, JIS University, Agarpara, Kolkata, West Bengal, India

Soumita Guha Sishutirtha Sukanta Vidyaniketan, Kolkata, West Bengal, India

Sk. Majharul Islam Bangabasi College, Rajkumar Chakraborty Sarani, Kolkata, West Bengal, India

Subrata Jana Department of Geography, Belda College, Paschim Medinipur, West Bengal, India

Amrit Kamila Department of Remote Sensing & GIS, Vidyasagar University, Midnapore, West Bengal, India

Khadija Khatun Department of Remote Sensing and GIS, Vidyasagar University, Midnapore, West Bengal, India

Lila Mahato Department of Geography, Krishnagar Government College, Nadia, West Bengal, India

Rimpa Maji Department of Remote Sensing and GIS, Vidyasagar University, Midnapore, West Bengal, India

Dipanjan Das Majumdar Department of Geography, Netaji Satabarshiki Mahavidyalaya, Ashoknagar, West Bengal, India

Ipsita Mallick Department of Remote Sensing & GIS, Vidyasagar University, Midnapore, West Bengal, India

Subhajit Mandal Department of Remote Sensing & GIS, Vidyasagar University, Midnapore, West Bengal, India

Dipankar Mondal Department of Geography, Vidyasagar University, Midnapore, West Bengal, India

Debasmrity Mukherjee Department of Geography, Asutosh College, Kolkata, West Bengal, India

Shrabani Mukherjee Department of Remote Sensing & GIS, Vidyasagar University, Midnapore, West Bengal, India

Subhankar Naskar Department of Remote Sensing & GIS, Vidyasagar University, Midnapore, West Bengal, India

Kushal Nayak Department of Remote Sensing & GIS, Vidyasagar University, Midnapore, West Bengal, India

Damodar Panda Department of Geography, Utkal University, Bhubaneswar, Odisha, India

Jasmin Parvin Department of Geography, Vidyasagar University, Midnapore, West Bengal, India

Anurupa Paul Department of Remote Sensing & GIS, Vidyasagar University, Midnapore, West Bengal, India

Ashis Kumar Paul Department of Geography, Vidyasagar University, Midnapore, West Bengal, India

Mihir Pradhan Kalagachia Jagadish Vidyapity, Khejuri, West Bengal, India

Barendra Purkait Department of Geology, University of Calcutta, Kolkata, West Bengal, India

Ratnadeep Ray Department of Earth Sciences and Remote Sensing, JIS University, Kolkata, India

Avik Saha Department of Remote Sensing & GIS, Vidyasagar University, Midnapore, West Bengal, India

Ratan Kumar Samanta Swarnamoyee Jogendranath Mahavidyalaya, Amdabad, West Bengal, India

Joydeb Sardar Department of Remote Sensing & GIS, Vidyasagar University, Midnapore, West Bengal, India

Centre for Environmental Studies, Vidyasagar University, Midnapore, West Bengal, India

Suman Saren Department of Geography, Swami Niswambalananda Girls' College, Uttarpara, Hooghly, West Bengal, India

Tuli Sen Sagar Mahavidyalaya, Sagar Island, West Bengal, India

Farhin Sultana Department of Geography, Vidyasagar University, Midnapore, West Bengal, India

Part I
Assessment Through Geomorphological Approaches

Chapter 1
Geomorphological Diversity and Sea Level Rise Vulnerabilities on India's East Coast

Ashis Kumar Paul and Anurupa Paul

1.1 Introduction

The east coast of India is characterized by multiple bands of shoreline parallel landforms that extend from the current shoreline to landward parts of the eastern ghat hills. Considering the contour height, morphological features, denudational activities, lithological diversities, and location of coastal plain rivers, the east coast is divided into five different stages of landforms. The angle of the coast line differs from place to place in three different directions from Kanyakumari to Masulipatnam Bay (northward), from Masulipatnam Bay to Chandipur Bay (north-northeastward), and from Chandipur Bay to the border of the Indian Sundarban (eastward). The longshore current cells are guided by such configurations of the shoreline, particularly during summer months with south-west monsoon currents and also during winter months with northeast monsoon currents. The river systems of the east coast contributed discharges and sediment to feed the regional and local cells in the nearshore for their activation by generating strength from the landward sides, and seasonal sea conditions with energy input systems guided them to transport sediment in a rhythmic path along the shore in parallel directions. Shoreline morphology of the alluvium coast evolves through the adjustment between energy and sediments inputs into the cell circulation systems from the land and sea over a time period in a cyclic development.

The younger coastal plains have preserved the signatures of marine depositional features, deltaic advancement, configurations of lagoons, backwaters, coastal embayments, estuaries, barrier bars, tombolo's, series of beach ridges, transgressive

A. K. Paul
Department of Geography, Vidyasagar University, Midnapore, West Bengal, India

A. Paul (✉)
Department of Remote Sensing & GIS, Vidyasagar University,
Midnapore, West Bengal, India
e-mail: anurupashis@gmail.com

A. K. Paul, A. Paul (eds.), *Crisis on the Coast and Hinterland*,
https://doi.org/10.1007/978-3-031-42231-7_1

sandy tracts, distal islands, drainage abandonment, isostatic upliftments, subsidence, basin deposition, marine terraces, coastal sand dunes, Teri sands, sea level variations, dune calcarenites, coral fringe islands, notches, cliffs, shore platforms, as well as coastal promontories, headlands, bays and coves along the shorelines and shore parallel hinterlands (Paul, 2002; Paul et al., 2017, 2023a, b; Kamila et al., 2021). Further, they may be categorized as the features of alluvium coast, rocky coast, carbonate coast, and coral fringed coast. Emergent and submergent features with active neo-tectonics, isostatic changes, and transgressive and regressive stages of the dynamic sea level variation demonstrated diversity in the younger coastal features of the Bay of Bengal fringe areas in India. The younger coastal features were formed by fluvio-marine deposits since the last glacial low stands (18000–21000 YBP) and were modified by the Holocene marine transgressive and regressive stages (9000–1000 YBP) due to the changing wetness and dryness of the past environment. A significant break in slope marks the band of coastal features behind the younger coastal plain, with surface elevation ranging from 30 to 100 m contour height along the present-day shore parallel locations towards the hinterlands of the shore fringe coastal plains. This part of the elevated tract with ferricrete surface is known as the mature coastal plain, in which the fluvial deposits and estuarine flood plain deposits are lateritized by Pleistocene–Holocene climate change activities and weathering phenomena.

The lateritized tract has two different surface elevations: an upper (50–80 m) and lower (30–40 m) surface dipping towards the east with detrital nodule laterites of pebble conglomerates and mottled clay and silt residual deposits. This formation also contains silicified wood and highly oxidized gravels at different depths (Roy & Chattopadhyay, 1997; Niyogi, 1970; Niyogi et al., 1970; Monsur et al., 2001; Ahmed et al., 2004; Ravenscroft et al., 2005), and the frontal surface is separated by valley cuts with significant valley flats in the region. Another set of geomorphic features extends landward sides behind the lateritic high land tract in a shore parallel location, with a denuded surface and exposed residual hills or outliers. The area is known as 'the plain of marine denudation with outliers', and it represents the paleoenvironment of sufficient erosional activities of surrounding rock masses by marine processes. The dots of isolated hills over the denuded surface represented islands of an ancient geological period when marine etch-planation was taking place with high sea levels (Pliocene period, 3 million years ago) in the region (Cushing, 1913).

The eastern ghat hills to the west of the denuded surface, on the other hand, are depicted as summit plains of the same altitude ridges with elevated peneplain and valley slopes. All the consequent streams of the east coast have entrenched meandering courses across the elevated peneplain, and gradual development of subsequent streams has dissected the ridges of the eastern ghat in the geological past. The physical vulnerabilities are estimated as following by threats arising from sea level rise and other emerging adverse physical factors for the east coast of India in this section. The younger alluvium coast with their natural habitats will be affected by erosion, inundations, and repeated land fall of cyclonic storms in the discourse of climate change impacts along the low-lying tracts fringed with the Bay of Bengal.

1.2 Study Area

The study area includes the mainland and island coasts of eastern India, fringed with the Bay of Bengal and Andaman Sea. India's east coast comprises the categories of alluvial coast, rocky coast, coral fringe carbonate coast, and coastal hinterlands (Fig. 1.1). The younger coast is geomorphologically more diverse than the ancient coasts and elevated peneplains of the eastern ghat hills. However, the peninsular and extra peninsular drainage systems of the subcontinent are drained towards the Bay of Bengal across the coastal gradients. Tectonically, the mainland coast is located in the passive margines and the island coasts of Andaman and Nicobar group belong to the active margin setup in the region. The east coast of India is occupied by the states of West Bengal, Odisha, Andhra Pradesh, Tamil Nadu, Pondicherry, and the Andaman and Nicobar Islands. The North Circar coast extends from West Bengal to Andhra Pradesh along the Bay of Bengal, and the Coromandel coast, on the other hand, extends from Masulipatnam to Kanyakumari in Tamil Nadu. The island coasts of the Andaman and Nicobar groups are influenced by the Andaman Sea and Bay of Bengal. India's east coast also demonstrates as an ideal section of Tropical maritime zones extended from 6°45′N latitude to 21°45′N latitude in the northern Indian ocean region. The total length of the Indian mainland coast is about 1800 km, and the archipelago of the Andaman and Nicobar group of islands is about 766 km long and extends from Indira Point to Landfall Island in the Bay of Bengal. Such a long coast is strongly influenced by significant fetch distances and prevailing swell waves in the region of seasonal monsoon winds.

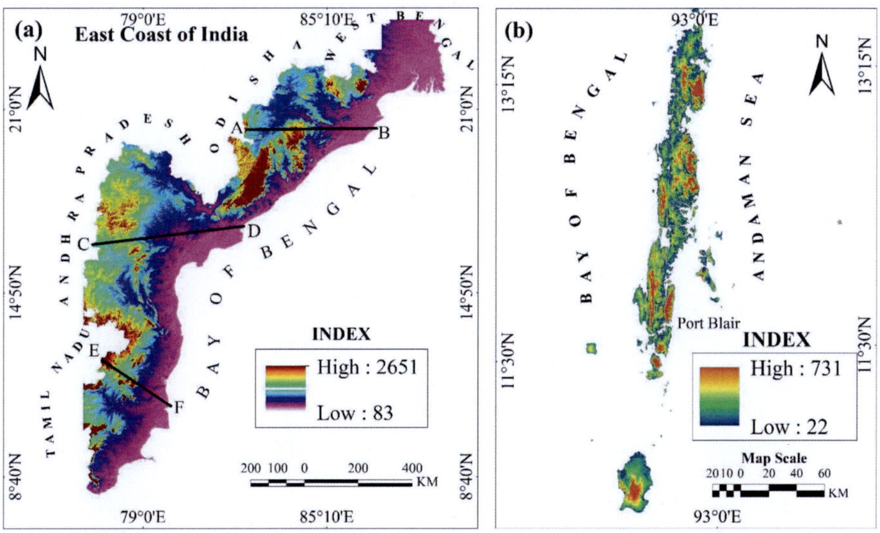

Fig. 1.1 The diversity in elevation zones of the major geomorphic units: (**a**) India's East Coast along the Bay of Bengal shoreline; (**b**) Andaman group of islands in between Andaman Sea and Bay of Bengal

1.3 Methods of the Study

The geomorphological diversity of the east coast is examined using SRTM DEM (spatial resolution 30 m) and ALOS PALSAR DEM (spatial resolution 12.5 m) to generate contour maps and morphogenetic regions. Surface expressions of coastal zones are investigated using cross profiles or coastal transects drawn from seaward to landward sides. The field survey at different regional sectors reveals spatial diversity of the landforms and process variation. Other attribute data (sea level data from PSMSL, tide gauge stations, INCOIS data on ocean information, cyclone and tsunami data from IMD, and seismo-tectonic zones data from the GSI) are used for assessing coastal vulnerability in the region. The USGS method of assessing physical vulnerability includes six major factors (i.e., geomorphology, coastal slopes, relative sea level changes, shoreline erosion accretion rates, mean tide range, and mean wave height) of coastal risks. However, the method is modified after adding four major local risk factors (i.e., frequency of cyclone landfalls, storm surge elevations, tsunami wave heights, and seismo-tectonic events) in the present study to estimate the actual vulnerability of the coastal regions more accurately. To represent the spatial diversity of vulnerabilities, 32 coastal stations are considered along long stretches of tropical coasts.

1.4 Results and Discussion

The section demonstrates the nature and diversity of geology, tectonics, physiography, geomorphology, drainage conditions, hazards, and vulnerabilities of the coastal zones on a regional scale. Future management plans may be formulated after considering the nature of elevations, geomorphology, and available risk factors along the tropical coast in the discourse of climate change. The genesis of coastal landforms is also highlighted in the section to identify the geomorphological diversity of ancient and modern coasts.

1.4.1 Geology and Tectonic Setup

Geologically, the eastern coast of India is covered mostly by Quaternary alluvium, in which the deltaic and non-deltaic coastal plains, basin fill sediments, estuarine and lagoonal deposits, and early Holocene marine transgressive deposits occur. The east coast laterites occupy mature coastal plain over high land surface behind the younger deposits. At Visakhapatnam and Srikakulam region of Andhra Pradesh coast, Granulite and Gneiss with supracrustal rocks are projected towards the sea to develop the rocky highland coast in the region. The green stone and allied supracrustal rocks and platform cover sediments are found to be exposed at the sections of Nellore in Tamil Nadu and Keonjhar in Odisha. Acid volcanic suites in the form of hills are exposed at Tiruchirappalli and Madurai towards the west of coastal quaternaries in Tamil Nadu. Gondwana and related sediments are found along the Godavari and Mahanadi rift valleys in sections towards the west of Quaternary deposits (Sastri et al., 1981). Proterozoic sediments are well preserved in the Cuddapah basin in the east central part of the Dharwar craton, which is parallel to the Nellore schist belt, thc Eastern Ghat Mobile Belt, and the east coast.

However, Proterozoic crystalline rocks of the unclassified group containing charnockite and khondalite can be found in the eastern ghat hills of Tamil Nadu, Andhra Pradesh, and Odisha. The Andaman and Nicobar Islands, on the other hand, have ophiolites and pillow lava at the basement, which are overlain by Andaman flysch sediments, Pleistocene limestones, and a younger group of archipelago formations with calcareous sandstone, mudstone, white claystone, and chalks or marls. On the shore cliff stratigraphy of the South Andaman and Nicobar group of islands, fossiliferous groups of limestone, sandstone, grit, and conglomerates have been discovered. The cemented carbonates, coral rubble, reef materials, beach rocks, sandstones, claystones, shales, and limestone are found on the wave-abrasive shore platforms of the island fringes.

The structural trends of the east coast from West Bengal to Tamil Nadu can be categorized as Bengal Basin, basin margin faults, hinge lines, Mahanadi Basin with faulted basement structures, Odisha Highlands with cratonic belts, Mahanadi Rift, Baster craton, Godavari Rift, eastern boundary thrust of the Cuddapah Basin, Eastern Ghat Mobile Belt, Bhavani Palghat Mobile Belt, and southern granulite terrain. The passive margin coast, or the transition between oceanic and continental crust, under such structural trends reveals the sites of ancient and modern sediment deposition in fluvio-marine environments, and they are not active plate boundaries, though they were originally created by rifts. However, India's east coast overlies the two different passive margins as the north-east passive margin coast and the southern volcanic passive margin coast – along the western boundary of the Bay of Bengal. Initiation of igneous processes was associated with the rifting of the volcanic passive continental margin, but such activity was absent along the passive margin coast. Surface expressions in such coasts are clearly defined by sediment deposition over the basement structures of ancient rifts. The nature of the transitional crust of the continental and ocean plates along the passive margin coasts is revealed by various seismic data.

However, the island arc setup of the Andaman and Nicobar group islands along the subduction and obduction zones of the ocean plates represents an ideal example of active margin coasts, which contain accretionary prismes, forearc basins, and volcanic arc basin structures under the strong thrust belts with uplift and volcanic activities. Terraces, knick points, RIA coasts, and shoreline modifications have resulted from various neo-tectonic and seismotectonic activities in the Bay of Bengal and Andaman Sea group of islands.

1.4.2 Physiographic Diversity

Morphological expressions of the east coast can be grouped as (i) younger coastal plains, (ii) mature coastal plains, (iii) the Plain of Marine Denudation with outliers, (iv) elevated peneplains, the Eastern Ghat, (v) Andaman Island ridges, and (vi) the Archipelago group of islands. Physiographically, the diversity is also reflected in the coastal transects from east to west on the basis of satellite DEMs (SRTM DEM and ALOS PALSAR) and contour plans of the regional settings of the East Coast. The wide strand plains, delta plains, islands and bars, high lands, river valleys, basins, residual hills, outliers, denuded surfaces, elevated peneplains, eastern ghat hills, valley flats, and terraces are major physiographic features of the east coast mainland. On the other hand, the island groups of Andaman and Nicobar represent piedmont slopes, hill ridges, valley flats, shore cliffs, reef terraces, and shore platforms of different elevations, wetland swamps, and channel creeks in the oceanic environment (Fig. 1.2).

1.4.3 Geomorphology

(i) Sedimentary depositional landforms of the east coast are restricted to the younger coastal plains in the forms of river deltas, marine strand plain deposits, islands and bars, beach-dune complexes, flood plain deposits, tidal flats, lagoonal and estuarine deposits, river valley deposits, and basin fill deposits by marine and fluvio-marine processes. The coastal plain topography of uncon-

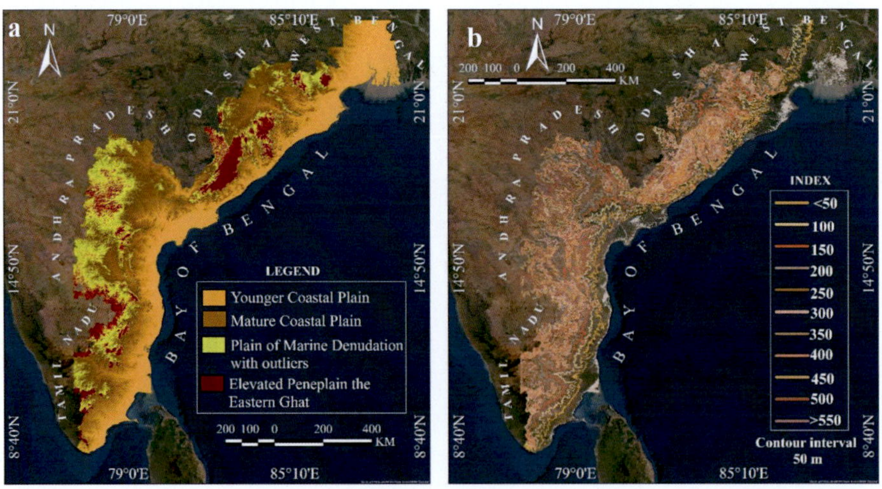

Fig. 1.2 (**a**) Spatial extension of the major coastal landforms, the east coast of India, and (**b**) The contour plan created to depict topographic diversity on the east coast of India is based on SRTM DEM

solidated alluviums above the mean sea level extends as 60–140 km wide and extensive surface from the shoreline to the inland areas and mostly bounded by 40 m contour lines towards west. The entire surface is fringed with the Bay of Bengal to the east and segmented into many local and regional units by the number of east and southeast flowing river systems. They are named as Sundarban delta plain, Midnapore littoral tract, Balasore strand plain with chenier delta, Delta Plains of Brahmani-Baitarni and Mahanadi, Utkal plain, Ganjam coast, Andhra plain, Waltair highland fringe platforms, Godavari-Krishna Delta Plains, Beach ridge chenier plains, Masulipatnam Bay fringe strand plains, Tamil Nadu Plains, Cuavery Delta Plain, marine transgressive cover tracts, coral fringe coasts with islands, shore platforms, beach rocks, and marine terraces. The studies on the Ganga-Brahmaputra delta (Goodbred & Kuehl, 2000a, b) reveal that delta building activities took place over time in six stages as (a) low-stand system with incised alluvial valleys, (b) initial phase of delta growth with rising sea level, (c) aggradation of the subaerial delta during rapid sea level rise, (d) decelerating rise of sea level, progradation of coastal plains, and initial growth of the subaqueous delta, (e) progradation of subaerial and subaqueous deltas, and (f) delta similar to the modern system. Thus, during the late Pleistocene epoch, subaqueous delta building advanced, but during the early to late Holocene stages, both aggradation and progradation were active in the final stage of delta growth to extend the coastal alluvial depositional surfaces. Similarly, other deltas built by the Subarnarekha, Brahmani-Baitarani, Mahanadi, Godavari, Krishna, Cauvery, and Pennar river along the east coast of India extended the areas under the younger coastal plain. The morphology of the depositional surfaces was gradually altered by the deltaic growth and evolution in the coastal plain fringed by the Bay of Bengal through lobe switching, channel switching, and channel extension.

The presence of paleo beach ridges, wide and extensive sandy tracts, and sand dunes in the inner parts of the delta plains represent the impacts of the Holocene transgressive and regressive phases of the coastal evolution. Other features like barriers, islands, embayments, lagoons, and estuaries were built up by sea level variations and associated coastal marine processes from the early Holocene to the late Holocene and modern periods. However, the coastal high stands in between inter deltaic coastal plains, with available geomorphic features, demonstrate the marine transgressive cover surfaces that are affected by neo-tectonics. They are also drained by several minor channels or river courses in the coastal plain. Several marine terraces with overlying sand dunes are visible in the region of Tamil Nadu and Andhra plains particularly along the coastal strand plains of the geomorphic unit. The lake Pullicat, Kolleru Lake, and Chilika lagoon are also restricted within the region, and presently they are reducing in size and depth due to sedimentation activity influenced by fluvio-marine processes in the regional setting.

(ii) The lateritic crust from 30 to 200 m contours in shore parallel location towards the west of younger coastal plain is known as mature coastal plain. The fluvial sediments, pedi-sediments, and estuarine sediments of the Pleistocene epoch were lateritized and duricrusted by the vertical and horizontal translocation of sesquioxide through pedological, hydrological, and geomorphological processes during the Pleistocene-Holocene climate changes in the region. They are terraced into two different surfaces of older and relatively younger origin and also known as Pleistocene terrace in the western margin of Bengal basin. Similarly, the concretionary laterite, boulder conglomeratic laterite, and duricrusted hardpan surface of laterites of reddish, yellow, pale brown, and dark color variations are found in the states of West Bengal, Odisha, Andhra Pradesh, Puducherry, and Tamil Nadu along the east coast of India. The perfect horizonations of such secondary laterites are visible in Garbeta laterite (Gangani) and Gopiballavpur laterite (Ghora Pincha) of West Bengal. The eastern margins of the lateritic crust (mature coastal plain) are marked by valley cuts, valley fills, terraced benches, and incised valleys. The high land laterites are also found at the pediment slopes of the residual hills. The mature coastal plain of the east coast contains oxidized and petrified wood fossils below the lateritic horizons. Terrace sands and gravels, as well as the piedmont sediments of Andhra Pradesh (Waltair highlands) and Tamil Nadu areas, are developed into ferricrete in the transported materials.

Budel's (1982) inselberg-studded plains, or etch plains, are found in the humid tropics of the Tamil Nadu Plain (east side of the Deccan), where the wash surface and the basal weathering surface between two inselbergs have distinct double planation surfaces. Laterites occur as cap rock on the recent alluviums (upper Gondwana Sandstone and Shales and Precambrian basement rock), and they extend in Tamil Nadu (around Madras) from 5 to 45 km inland from the present-day shoreline (Achyuthan, 1996). Her study reveals that the lateritic crust surface has been originated during the early to late Neogene and that neo-tectonics played a significant role in shaping the present landscape. However, in Odisha coast (southwestern fringe of Chilika lagoon), the studies on sedimentary history by (Ahmad et al., 2022; Paul et al., 2014) a group of geologists reveal that the lateritic soil crust of 1.30-m-long sediment core history ranges between 8.49 and 0.99 ka (8490 YBP–996 YBP) due to the impact of Holocene climate change and presence of iron oxide into the transported sediments of the region. The lateritic associated soils of east coast Andhra Pradesh on peneplains of Nellore district on Archean age schists dominated by silica to sesquioxide concentration under hot semi-arid climate through the pedogenic process (Bhaskar & Tiwari, 2017). Such lateritic soil crusts are also found in the Visakhapatnam, Vijaynagaram, and Srikakulam districts of coastal Andhra Pradesh. The fossil woods from the Cuddalore series near Puducherry are also excavated in the lateritic soil of Mio-Pliocene age and belong to pebbly, gravelly, ferricrete, and concretionary soils with reddish brown sands that occur as veneer over the sandstone towards the east (Fig. 1.3).

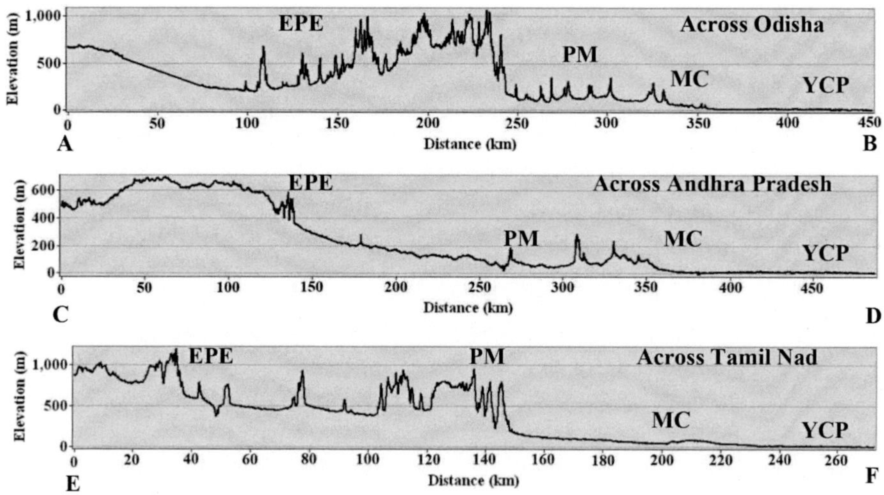

Fig. 1.3 The SRTM DEM was used to draw the major landform units of the East Coast of India (*YCP* - Younger Coastal Plain, *MCP* - Mature Coastal Palin, *PMD* - Plains of Marine Denudations with outliers, *EPEG* - Elevated Peneplain the Eastern Ghat)

(iii) Further inland (above the 200 m contour) towards the west of coastal laterite, the landform unit is categorized as plains of marine denudation with outliers and characterized by a peneplain surface resulting from the weathering of the basement rocks and active marine denudation processes during the Cretaceous phase. The shore parallel region is located over metamorphic terrain and is relatively less dissected due to its low altitude and semiarid environment. The nature of the surface and its continuity are actually influenced by the presence of extensive trunk streams along the region's east coast. However, the planated surface is thoroughly lateritized with ferruginous materials. There is a strong evenness to the surface over the distorted underlying structures and the location of steep-sided outliers on the plains of the regional unit. The resistant quartzite rocks of a few locations are not consumed by marine denudation and exist as outliers on the shallow surface. They are separated from the surrounding schists and gneisses of the denuded surface by marine transgression during the Cretaceous and Oligocene periods. Such residual hills are concentrated between Vijayawada to Vishakhapatnam over the flat plain surface in Andhra Pradesh. All the hill crests are lying at summit level and having equal heights in the region. A few portions of the outliers also indicated the location of elevated sea caves on their slopes. The denuded surface gently dips towards the eastern seaboard or in a southeast direction, followed by the courses of major consequent river systems. The summit plains of outliers reveal that they were part of the elevated peneplains of the eastern Ghat prior to marine erosion, for which the marine denudation plain develops towards the east of Ghat provinces in the region as a result of scarp retreat. Thus, the former extension of elevated peneplains is actually denuded, but not by the retreat of a fault scarp on the eastern margin of the Eastern Ghat Hills. The Odisha highlands to the west of the mature coastal plain represent parts of the denuded plain with outliers.

A series of outliers are also located over the denuded plains of Ganjam, Srikakulam, Vijaynagaram, Vishakhapatnam, Pallavaram, and Vijayawada to the eastern fringe of the Eastern Ghat Hills, parallel to the present-day shoreline. Similarly, the eastern fringe of the Nallamala Range and Tamil Nadu uplands reveal the extension of marine denuded plains with outliers.

(iv) The eastern ghat hills in the form of dissected units on the east coast of India represent an elevated peneplain, and the regional unit acts as a barrier wall towards the west of other physiographic units. The elevated peneplain is much higher in the southwest of Madurai and can be reached at three different elevations: 600 m, 1200 m, 1800 m, and above 1800 m. The resistant series of metamorphic rocks (Quartzites, Gneisses, and Schists) preserve the high-altitude hills of equal heights as well of the ancient sea before the marine invasion in the region of sub-parallel ridges. Major rivers are incised and meandered in places, but wide extensive river valleys appear in the areas of relatively weaker series of metamorphic rocks in the elevated peneplain.

In the northern sections up to Rajahmundry, where the peneplain is relatively low in elevation, the mature dissection has allowed the elevated ridge to develop a gentle slope (Cushing, 1913). Rivers are widely spaced in areas of less resistant rocks, and in spite of its ancient elevation, the region shows a youthful stage of dissection. The valleys of the consequent streams (i.e., Cauvery, Godavari, Palar, Krishna, Mahanadi, Brahmani, and Subarnarekha rivers) across the elevated peneplains demonstrated a youthful stage in many sections. Another striking feature is that the metamorphic terrains of the elevated peneplain are subparallel to one another and also to the nearby coast, but the river systems are extended in a transverse direction. In many sections, the eastern slope of the subparallel ridges abruptly rises from the local surface, and it all looks like sea cliffs from the ancient marine invasion of the region. The sea wall appears boldest to the south of Tamil Nadu high lands in Kodaikanal in which the eastern slope ranges from 30° to 35° to the basement surface of the elevated peneplain.

The major physiographic units with contrasting landforms are evolved over the geological times by ancient marine transgressive-regressive phases (Permian to Late Cretaceous, Late Oligocene to Middle Miocene and Early Holocene to Late Holocene), active fluvial phases from upper most Jurassic to Lower Cretaceous, Eocene-Paleocene, and Pleistocene to recent epochs. On the east coast of India, the formation of some regional geo-fracture and rift systems, as well as horst-graben type basins with minor folding and tilting, occurred during the Paleozoic era (251,902 m.y.–541.0 m.y.). Finally, the Pleistocene-Holocene modifications of land forms manifested the diversity in physiography of the coastal plain by developing sediment cover surfaces with active neo-tectonics (Fig. 1.4).

Fig. 1.4 Geomorphic features of the east coast: (**a**) Reef terrace on the basement of Charnockite outcrop with notching marks; (**b**) Sand dunes over the marine terrace fronted by sea beach with placer deposits at Cape Comorin; (**c**) Chilika wetland fringe with Palur hills; (**d**) Beach berm platform of Puri with erosive shore front area in Odisha; (**e**) Dolphin nose with Ramakrishna beach in Visakhapatnam; (**f**) Notching marks on the Khondalite outcrop along the shoreline of Thotlakonda in Andhra Pradesh; (**g**) Tidal flat wetland of the Sundarban with mangrove swamp; and (**h**) Diminishing sand dunes of Digha coast in West Bengal

(v) The Andaman ridge at the subduction zone reveals the island arc setup with accretionary prisms fore arc, volcanic arc, and back arc basins in the northern Indian Ocean and the volcanic arc basin with active volcano, dormant volcano, and a submarine lava plateau in the Andaman Sea towards the east of the island ridge. The three sections of the Andaman group of islands (North, Middle, and South Andaman) have an extended 362 km long ridge and are separated by creeks and channels. They appear as a series of longitudinal ridges and intermontane valley flats from south to north in the active thrust belt of compressive zones.

The island system is gradually elevated from south to north at an average elevation of 500 m surface above the sea level. The residual peaks of the hill ridges are Saddle Peak (732 m) in the North Andaman, Mt. Diavolo (512 m) in the Middle Andaman, and Mt. Harriet (365 m) in the South Andaman, and in width, the island system ranges from 30 to 50 km from the west to the east. Thus, the consequent longitudinal valley streams are relatively longer than the transverse channels across the hill slopes. All the streams of Andaman Island system are marked by several knick points along their courses with presence of water falls.

Cross profiles of the island sections from SRTM DEM and ALOS PALSAR DEM represent the nature of topography with variation of altitude and relief characters. In the cross profiles and highlands, parallel ridges, valleys, shore platforms, sea cliffs, marine terraces, peaks, and highlands are clearly visible as physiographic features of an active tectonic setup in the island arc system. The coastal terraces and their radiocarbon dates are considered by a group of researchers (Rajendran et al., 2008) to estimate the uplift rates, and the study reveals the variation in uplift rates from 1.33 mm/year^{-1} in the Little Andaman to 2.80 mm/year^{-1} in the South Andaman and 2.45 mm/year^{-1} in the North Andaman. The east ward dipping fault slices and folds contain the younger series of rocks belong to limestone, which are typically formed coral rags and shell sands. There is presence of six terraces on the sea ward sides of Middle Andaman within 300 m distance inland from the shoreline dated as 7000–27,500 YBP. The coral fringe coasts are marked by shore platforms, sea cliffs, terraces, promontories, coves, bays, headlands, and 'Ria' type shores. Subsidence was reported in the mouths of submerged estuaries during the event of 2004 earthquake event. Andaman Flysch sediments are found along the shore fringes of the fore arc basin and outer arc basin, consisting of sandstone, grit, siltstone, shales, and conglomerates. The basement rocks of older series consisting of Pre-Cretaceous meta sedimentary rocks and upper cretaceous ophiolites and Palaeogene-Neogene sedimentary formations (Bandyopadhyay & Carter, 2017a, b; Allen et al., 2008). Various limestone geomorphological features, including sea caves, island interior caves, tafoni marks, phyto karstic features, natural arches, springs, resulted from the dissolution of soluble rocks and demonstrated the presence of thick limestones, jointed rocks, and a high precipitation rate in the tropical oceanic island environment (Fig. 1.5).

(vi) The archipelago group of islands in Nicobar consist more younger series of rocks belonging to calcareous mudstones, sandstones, white claystone, chalk,

Fig. 1.5 Morphologic features of the Andaman group of Islands: (**a**) Andaman flysch exposed on the shore platform; (**b**) Plunging shore cliff in South Andaman; (**c**) Corbyn's Cove damaged after the tsunami hazard during the Great earthquake (2004); (**d**) Incised meander channel over the intermontane valley flat; (**e**) Ophiolite exposure on the shore face in South Andaman; (**f**) Boulder drift in the tsunami event (2004); (**g**) Stilt roots of mangroves (*Rhizophora* Sp.) standing over the beach rock (**h**) Munda Pahar area under dense evergreen forest in South Andaman

marl, and coralline limestones. There are 19 major islands in Nicobar group which extends 300 km long outer arc ridge to south southeast direction from 10-degree channel to Indira Point in the form of isolated land masses over the Indian Ocean separated by wide gap of sea stretches. Three major sections of the archipelago include Car Nicobar, Little Nicobar, and Great Nicobar Islands. Topographically, Car Nicobar is characterized by shore fringe flat plain (<10 m) surface with central high land region (74–80 m), Little Nicobar on the other hand is dominated by high land surface ranging from 80 m to over 400 m altitudes and drained by some significant river courses. Ria type of submerged estuaries is visible towards the north of Little Nicobar after the event of great earthquake and tsunami incidence, 2004, in the region. Mount Deoban (435 m) is the highest peak in Little Nicobar, but many channel beds are raised by the effect of tilting to the east during the co-seismic effects (2004) in the island (Table 1.1).

The two different sub-parallel ridges of 500–600 m altitude in SSE to NNW directions contain the skeleton of Great Nicobar Island. The shore fringe terraces and intermontane valley flats are lying at 20–40 m above sea level, and the valley flat is drained by Galathea river. On the basis of the post-tsunami outcrop, the geological map is prepared by Bandyopadhyay and Carter (2017a, b) for the island, which indicates that the western shore is fringed by Holocene clay, sand, and coral reefs, with the southern shores of the archipelago group (chalk, mudstone, and reworked tuff) of a younger series. However, the other parts of the entire island are consisting with Andaman flysch and Mithakari group of rocks. The shoreline is crenulated with bays, coves, headlands, promontories, and river mouths. The island was affected by subsidence ranging from −1 to −3 m in Indira Point during the event of Great earthquake (2004). The exposures of outcrops are visible along the Galathea bay and Campbell Bay fringe shores from the Google Earth Image (2023), which also showing the signatures of tsunami damages. Similarly, the submerged shores and valleys are found along the river mouths and shore platforms due to the effects of a seismo-tectonic event. Signatures of the 'Ria' type shores are visible in the sections of the NW shore, SW shore, and SE shoreline of Great Nicobar. A significant knick point is developed in the downstream section of the Galathea river (5 m upstream from the mouth) on the island after the event of the great earthquake (2004).

1.4.4 Drainage Characters

The river systems of the east coast of India represent elongated courses across the sub-continent and transported sediments into the structural basins over a geological time in the passive margin coast. The Cauvery River, Vaigai river, Penner River, and Krishna River of Tamil Nadu and Andhra Pradesh have extended their courses into the Bay of Bengal, cutting across the eastern ghat hills and Tamil Nadu highlands through the Palghat Gaps with incised and widely spaced valleys. The courses of the

Table 1.1 Geomorphological diversity in the East coast of India along the Bay of Bengal and Andaman Sea

Sl. No.	Geomorphic unit	Geomorphic features	Processes involved	Age of development
1	Younger coastal plain with alluvium cover surface	River delta with estuaries, lagoons, spits, barrier bars, beach ridges, bays, embayments, marine strand plains	Fluvial, coastal, tidal, marine, lagoonal, aeolian, subsidence, upliftment, neo-tectonics	Late Pleistocene to Holocene; Recent-Subrecent
2	Mature coastal plain with lateritic crust and highland surface	Older and younger lateritic surfaces with graveliferous, concretionary pisolithic grains, red soils and duricrust materials, cap rocks, pedi- sediments, and secondary laterites	Ancient river deposits, geochemical weathering processes, pedogenic process, sesquioxide concentration, case hardening by oxides	Late Pleistocene to Holocene and Recent
3	Plains of marine denudation with outliers and summit plains	Marine etch planation, ancient marine transgression, scarp retreats, pedi-planation, wide spaced river valleys around the less resistant rocks, Outliers, Questa's, and Hogbacks	Vertical etching, marine planation, geo-fracturing, and structural basins	Archaean to late Proterozoic, Cretaceous to Pleistocene and Holocene
4	Elevated peneplain the Eastern Ghat Hills	Dissected hills, uplifted peneplains, summit plains, incised river valleys, rifts, and grabens	Planation surfaces, upliftment of peneplains, tectonics, denudational processes	Archaean to late Proterozoic, Permian to Late Cretaceous, Late Oligocene to Middle Miocene and Early Holocene to Late Holocene
5	Andaman ridges and islands	Outer arc ridges, accretionary prisms, intermontane valley flats, piedmont terraces, shore fringe marine terraces, fringe coral reefs, broken shorelines, cliffs, and outcrops	Fracturing, thrust belt, faulting, seismo-tectonics, neo-tectonics, coastal and marine processes, biological processes, calcification processes, karstification processes, volcanism	Late Cretaceous to Eocene
6	Nicobar ridge with isolated land masses and archipelago	High lands, shore platforms, parallel to sub-parallel hill ridges, terraces, Ria shores, coves, bays, promontories, outcrop, coastal plain	Subsidence, upliftments, neo-tectonics, Ria type shores, karstification, calcification, and coral reefs evolution	Pliocene to Pleistocene and Recent

Godavari River and Mahanadi River have been guided by structural rifts or geo-fractures across the eastern ghat hills in Andhra Pradesh and Odisha states. The river Brahmani-Baitarani and Subarnarekha systems also extended their courses into the Bay of Bengal, cutting across the Odisha highlands and Ranchi Plateau. Similarly, the Ganga-Brahmaputra River system entered into the Bengal Basin cutting across the Garo-Rajmahal Gap and have been built up the largest delta in the east coast, and the fluvial sediments are deposited over the geological times (Permian to Quaternary) into the Cauvery basin, Palar basin, Godavari-Krishna basin, Pranhita-Godavari Graben, Son-Mahanadi Graben, Mahanadi basin, and Bengal basin to build up the deltas and shelf floors along the Bay of Bengal. The longitudinal courses of the east coast rivers show gradation and concavity in their profile shapes, which represent a quasi-quasi-equilibrium state of the system.

Seasonal discharge variations are a distinguishing feature of alluvial rivers in the peninsular and extrapeninsular regions where precipitation rates vary. Ramkumar et al. (2016) estimated the annual water and sediment discharges for the six major rivers of the east coast of India, among which the Ganga (493 $m^3 \times 10^8$ and 413 MT), Godavari (92 $m^3 \times 10^8$ and 187 MT), and Penner River (93 $m^3 \times 10^8$ and 6.9 MT) systems transport a significant amount of water and sediment discharges into the deltaic parts. Previous studies (Das, 2021; Das et al., 2022; Nageswara et al., 2010; Mukhopadhyay, 2007) demonstrated that there is a 40% reduction in sediment discharge and over 60% reduction in freshwater discharge in the river deltas of the east coast of India due to climate change impacts and anthropogenic impacts in the river basins of peninsular India. Holocene evolution of the deltas of the east coast and deltaic subsidence of Godavari system are also studies by the earlier workers (Nageswara & Sadakata, 1993; Nageswara et al., 2010; Rao et al., 2015), which showed the stages of deltaic evolution in the coastal parts of the Bay of Bengal. The drainage systems of all the six major basins on the east coast also reveal the dendritic pattern, which is the reflection of the homogenous character of rock types and thick deposits of floodplain alluvium. However, the delta plain rivers are characterized by the presence of a significant number of distributary channels (Fig. 1.6). The Holocene transgressive and regressive phases have produced submerged river valleys, beach ridge deposits, transgressive sand sheets, and meandering courses of the channel on the east coast of India. The lower courses of the river systems are influenced by tidal inflow and marine impacts. The drainage characters of Andaman and Nicobar Islands in the active plate margin coast tectonic control played significant role in modification of channels. Several knick points are recorded along the drainage courses in the island units. Ria-type estuaries are developed in the lower sections of the river courses due to subsidence and submergence, particularly during past seismo-tectonic events like the great earthquake event of 2004. Many rivers are also fault-controlled channels and extend in a rectangle pattern under the active thrust belt of the Andaman Island system.

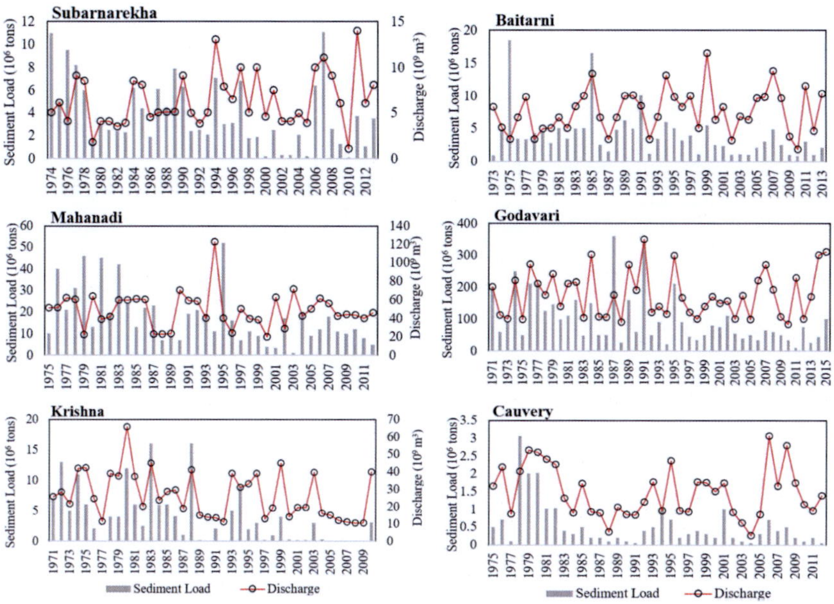

Fig. 1.6 The time series data depicts sediment and water discharges by six major river systems on the east coast of India. (Data Source: CWC, India)

1.4.5 Hazard and Vulnerabilities

India's east coast is affected by multiple hazards, and ten significant factors are considered to estimate the physical vulnerabilities in the present section of the study. As per the method of the United States Geological Survey, the vulnerability factors contain geomorphology, coastal slopes, relative sea level change, shoreline erosion accretion rate, mean tide range, and mean wave height for assessing the CVI of any coastal section. However, other factors such as frequency of cyclone landfall, storm surge elevation, tsunami wave height-2004, and seismo-tectonic zones were included in the study to assess the physical vulnerability of the 32 sections of the Bay of Bengal fringe coasts (Table 1.2).

Sea level rise vulnerabilities for the east coast of India will be triggered by other physical factors, which are mainly considered in the present study. The coastal plains touching with the sea at least bounded by 10 m contour lines will be affected by saltwater inundations during the events of coastal hazards in the Bay of Bengal. Considering the vulnerability classes (VH, H, M, L, and VL), the entire coastal stretches of the Bay of Bengal are classified into 5 categories of vulnerability, showing significant diversity in vulnerability zones. Among the 32 coastal stations, two stations from the Sundarbans and Great Nicobar Island represent very high vulnerability zones; three stations from the Hugli estuarine coast with Car Nicobar and Little Nicobar Islands of the Indian Ocean demonstrate high vulnerability zones on India's east coast. The moderate vulnerable one occupies 7 coastal stations of which

Table 1.2 Ranking for coastal vulnerability index variables (total 10 factors) for the East coast of India

Ranking of coastal vulnerability index variables					
Factors	Very low (1)	Low (2)	Moderate (3)	High (4)	Very high (5)
Geomorphology	Rocky, Cliff coast, Fiords, Fiards	Medium cliff, indented coast	Low cliffs, Glacial drift, Alluvial plains	Cobble beaches, Estuary, Lagoon	Barrier beaches, sand beaches, salt marsh, mudflats, deltas, mangroves, coral reefs
Coastal slope (%)	>0.2 (0.72°)	0.2–0.07 (0.26°)	0.07–0.04 (0.15°)	0.04–0.025 (0.09°)	<0.025 (<0.09°)
Relative sea level change (mm/year)	<1.8	1.8–2.5	2.5–2.95	2.95–3.16	>3.16
Shoreline erosion and accretion (m/year)	>2.0	1.0–2.0	−1.0 – +1.0	−1.1 – −2.0	<−2.0
Mean tide range (m)	>6.0	4.1–6.0	2.0–4.0	1.0–1.9	<1.0
Mean wave height (m)	<0.55	0.55–0.85	0.85–1.05	1.05–1.25	>1.25
Frequency of cyclone landfall (1942–2022)	<1	1–3	3–5	5–7	>7
Storm surge elevation (m)	<2	2–4	4–6	6–8	>8
Tsunami wave height-2004 (m)	<1	1–3	3–5	5–7	>7
Seismo tectonics (seismic zoning)	1	2	3	4	5

3 from Contai coastal plain of West Bengal, 1 from Gopalpur on sea of Odisha coast, 1 from Masulipatnam Bay fringe coast of Andhra Pradesh, and 2 from Middle Andaman and South Andaman districts of Andaman Group of islands (Figs. 1.7 and 1.8, Table 1.3). However, the current status of 11 coastal stations and 9 remaining coastal stations on the east coast belong to low to very-low vulnerable zones under the impact of physical vulnerable factors. The climate change and sea level rise threats in the northern Indian Ocean will trigger the coastal vulnerabilities in the Bay of Bengal.

1.5 Conclusion

The following conclusions may be drawn on the basis of the above study. The geomorphological diversities of the east coast of India are highlighted in six major regional settings by the study. The geomorphological significance of the ancient coast and modern coast is clearly identified based on the satellite DEMs, contour

Fig. 1.7 Shore cliffs and shore platforms in South Andaman: (**a**) A large coral boulder drifted on the shore platform from the coral bank after the impact of tsunami energy in 2004; (**b**) a wave cut notching at the base of the shore cliff in Neil Island; (**c**) landward dipping strata of alternate calcareous white claystone and shale limestone on the sea cliff at Havelock Island with fissure caves; and (**d**) a Natural Sea Arch after erosion of the Headland coast in Neil Island

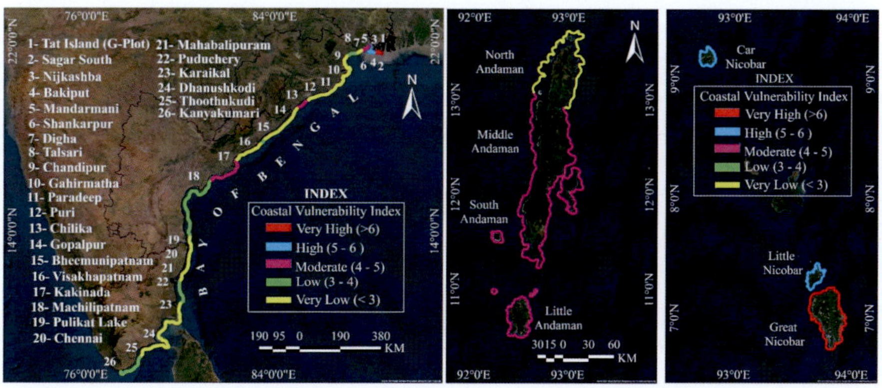

Fig. 1.8 Assessment of vulnerabilities for the East coast of India with Andaman and Nicobar Islands

Table 1.3 Integrated scores for the ten factors of vulnerability variables in the 32 coastal sections of India's east coast for assessment of the CVI

Sl. No	Location	Geomorphology	Coastal Slope (%)	Relative Sea level Change (mm/yr)	Shoreline erosion and accretion (m/yr)	Mean tide range (m)	Mean wave height (m)	Frequency of cyclone landfall (1942-2022)	Storm surge elevation (m)	Tsunami wave height-2004 (m)	Seismo Tectonics (Seismic Zoning)	CVI	CVI
1	Tat Island (G-Plot)	5	5	5	5	2	2	5	3	1	4	6.23	VH
2	Sagar South	4	5	5	5	2	2	5	3	1	4	5.89	H
3	Nijkashba	4	4	3	5	3	1	5	3	1	3	4.25	M
4	Bankiput	4	4	3	4	3	1	5	3	1	3	4.02	M
5	Mandarmani	5	4	3	5	3	2	4	2	1	3	4.56	M
6	Shankarpur	3	4	2	5	3	2	4	2	1	3	3.63	L
7	Digha	3	4	2	4	3	3	4	2	1	3	3.79	L
8	Talsari	3	4	2	4	2	2	3	1	1	3	2.43	VL
9	Chandipur	3	4	2	3	3	2	2	1	1	2	2.04	VL
10	Gahirmatha	5	5	3	4	2	3	3	2	1	2	3.84	L
11	Paradeep	5	5	3	2	3	3	4	2	1	2	3.84	L
12	Puri	4	4	2	2	3	4	4	2	1	2	3.33	L
13	Lake Chilika	4	4	3	4	4	4	4	1	1	2	3.96	L
14	Gopalpur on sea	4	4	2	3	4	5	3	3	1	2	4.32	M
15	Bheemunipatnam	4	5	1	2	3	5	2	1	2	2	2.64	VL
16	Visakhapatnam	1	1	1	1	3	5	2	1	2	2	1.05	VL
17	Kakinada	4	4	1	2	4	4	3	2	2	3	3.69	L
18	Machilipatnam	5	4	1	3	4	4	3	3	2	3	4.78	M
19	Lake Pulicat	4	4	1	2	4	4	2	2	3	3	3.69	L
20	Chennai	4	3	1	1	4	4	2	2	4	3	3.09	L
21	Puducherry	3	3	1	1	5	5	1	1	4	2	2.06	VL
22	Mahabalipuram	4	3	1	1	5	5	1	1	4	2	2.22	VL
23	Karaikal	5	5	1	2	4	4	1	1	4	2	2.83	VL
24	Dhanushkodi (Rameswaram)	5	5	1	2	5	5	1	1	4	2	3.17	L
25	Tuticorin	3	4	1	1	5	5	2	1	4	2	2.64	VL
26	Kanyakumari	2	2	1	1	5	5	1	1	4	2	1.69	VL
27	North Andaman	5	3	2	1	4	5	2	1	5	3	3.671	L
28	Middle Andaman	5	3	2	1	4	5	2	1	5	5	4.17	M
29	South Andaman	5	3	2	2	4	5	2	1	5	5	4.95	M
30	Car Nicobar	5	3	3	2	4	5	2	1	5	5	5.48	H
31	Little Nicobar	5	3	3	2	4	5	2	1	5	5	5.48	H
32	Great Nicobar	5	3	3	3	4	5	2	1	5	5	6.07	VH

plans, topographic features, geological features, structural features, and processes involved.

India's east coast is characterized by two different types of tectonic settings as passive margin coast and active margin coast. Thus, the continental margin coasts of peninsular India and the island arc setup of the Andaman and Nicobar coasts represent diverse geomorphological features with activity of processes. The basic structural features, tectonic, rock types, sedimentation rates, fluvio-marine deposition, sea level fluctuations, and denudational processes have played the significant roles in the evolution of the coast. The wide and extensive alluvium plains of the east coast are produced by basin filling since Permian to Quaternary period. Ancient marine transgressive phases and the Holocene marine transgressive-regressive phases have modified the depositional cover surfaces on the deltaic and non-deltaic coasts. The series of delta development has built up the younger alluvium coast through re-shaping the shoreline configurations in the east coast of India. However, the strand-plain surfaces of marine deposition are extended between the non-deltaic areas of the younger coast. The uplifted peneplains, marine denudations, or marine etch planation, weathering, and lateritization processes have left the features of the ancient coast up to the eastern ghat hills. The ancient sedimentary basins between the cratons also preserved marine and fluvial sediments on the east coast of India.

The cross sections of the east coast also reveal the geomorphological diversities with the signatures of elevated surfaces, valley flats, and coastal plains from west to east. However, several fractures, scarps, valleys, hill ridges, and marine terraces are visible from the cross sections of the Andaman and Nicobar group of islands. Different features are found in the outer arc, fore arc, volcanic arc, and back arc basins of the tectonic stress fields in the island arc setup. Various karstic features, coral reefs, and seismo-tectonic features are also reflected in the rock outcrops of the younger series and older series of the islands.

Finally, the assessment of sea level rise vulnerability along with other physical vulnerabilities of the east coast of India has produced a good data base by the study. The areas liable to very high, high, and moderate vulnerabilities need immediate attention for the restoration of habitats and sustainable management of the coastal zones. Marine influences have increased in the coastal zones of younger coastal plains as a result of a significant reduction in fresh water discharges by river systems into coastal fringes over the previous decades. The landward extent of tidal ingression through the river mouths, increased soil salinity into the coastal alluviums, encroachment of saltwater into the coastal aquifers, saltwater inundations into the low-lying delta plains, and estuarine sediment filling by marine processes are major environmental crises developed due to marine influences in the coastal plains of the region.

References

Achyuthan, H. (1996). Geomorphic evolution and genesis of laterites around the east coast of Madras, Tamil Nadu, India. *Geomorphology, 16*(1), 71–76. https://doi.org/10.1016/0169-555X(95)00085-J

Ahmad Shah, R., Khan, I., Rahman, A., Kumar, S., Achyuthan, H., Shukla, A. D., et al. (2022). Holocene climate events and associated land use changes in the eastern coast of India: Inferences from the Chilika Lagoon. *The Holocene, 32*(10), 1081–1090.

Ahmed, K. M., Bhattacharya, P., Hasan, M. A., Akhter, S. H., Alam, S. M. M., Bhuyian, M. A. H., Imam, M. B., Khan, A. A., & Sracek, O. (2004). Arsenic enrichment in groundwater of the alluvial aquifers in Bangladesh: An overview. *Applied Geochemistry, 19*, 181–200.

Allen, R., Carter, A., Najman, Y., Bandopadhyay, P. C., Chapman, H. J., Bickle, M. J., & Gerring, C. (2008). New constraints on the sedimentation and uplift history of the Andaman-Nicobar accretionary prism, South Andaman Island. *Geological Society of America Special Papers, 436*, 223–255. https://doi.org/10.1130/2008.2436(11)

Bandopadhyay, P. C., & Carter, A. (2017a). Chapter 2 Introduction to the geography and geomorphology of the Andaman–Nicobar Islands. *Geological Society, London, Memoirs, 47*(1), 9–18. https://doi.org/10.1144/M47.2

Bandopadhyay, P. C., & Carter, A. (Eds.). (2017b). *The Andaman–Nicobar accretionary ridge: Geology, tectonics and hazards*. Geological Society of London.

Bhaskar, B. P., & Tiwari, G. (2017). Geochemical interpretations of laterite associated soils of east coast Andhra Pradesh. *Chemical Science Reviews and Letters, 6*(21), 187–197.

Budel, J. (1982). *Climatic geomorphology*. Princeton University Press. ISBN-100691082952.

Cushing, S. W. (1913). The East Coast of India. *Bulletin in the American Geographical Society, 45*(2), 81–92. https://doi.org/10.2307/200621

Das, S. (2021). Dynamics of streamflow and sediment load in peninsular Indian rivers (1965–2015). *Science of the Total Environment, 799*, 149372. https://doi.org/10.1016/j.scitotenv.2021.149372

Das, S., Kandekar, A. M., & Sangode, S. J. (2022). Natural and anthropogenic effects on spatio-temporal variation in sediment load and yield in the Godavari basin, India. *Science of the Total Environment, 845*, 157213. https://doi.org/10.1016/j.scitotenv.2022.157213

Goodbred, S. L., Jr., & Kuehl, S. A. (2000a). The significance of large sediment supply, active tectonism, and eustasy on margin sequence development: Late Quaternary stratigraphy and evolution of the Ganges–Brahmaputra delta. *Sedimentary Geology, 133*(3–4), 227–248.

Goodbred, S. L., Jr., & Kuehl, S. A. (2000b). Enormous Ganges-Brahmaputra sediment discharge during strengthened early Holocene monsoon. *Geology, 28*(12), 1083–1086.

Kamila, A., Paul, A. K., & Bandyopadhyay, J. (2021). Exploration of chronological development of coastal landscape: A review on geological and geomorphological history of Subarnarekha Chenier delta region, West Bengal, India. *Regional Studies in Marine Science, 44*, 101726.

Monsur, M. H., Tooley, M. J., Ghatak, G. S., Chandra, P. R., Roy, R. K., Adhikari, P. C., & Akhter, S. H. (2001). A review and correlation of quaternary deposits exposed in the Bengal Basin and its surrounding areas. *Bangladesh Journal of Geology, 20*, 33–54.

Mukhopadhyay, S. K. (2007). *The Hooghly estuarine system, NE coast of Bay of Bengal, India*. Paper presented at the workshop on Indian estuaries, CSRI-National Institute of Oceanography, Goa, 25 July, 2007.

Nageswara Rao, K., & Sadakata, N. (1993). Holocene evolution of deltas on the east coast of India. In *Deltas of the world* (pp. 1–15). ASCE.

Nageswara Rao, K., Sadakata, N., Shinde, V., & Rajawat, A. S. (2010). Subsidence of Holocene sediments in the Godavari delta, India. *Frontiers of Earth Science in China, 4*, 410–416. https://doi.org/10.1007/s11707-010-0124-3

Niyogi, D. (1970). Morphology and evolution of the Balasore shoreline, Orissa. In *Studies in earth sciences, west commemoration* (pp. 289–304). Today and Tomorrow Publishers.

Niyogi, D., Mallick, S., & Sarkar, S. K. (1970). A preliminary study of laterites of West Bengal, India. In S. P. Chatterjee & S. P. Das Gupta (Eds.), *Selected papers physical geography (Vol. 1)* (21st International geographical congress) (pp. 443–449). National Committee for Geography.
Paul, A. K. (2002). *Coastal geomorphology and environment* (pp. 1–342). ACB Publication.
Paul, A. K., Islam, S. M., & Jana, S. (2014). An assessment of physiographic habitats, geomorphology and evolution of Chilika Lagoon (Odisha, India) using geospatial technology. In *Remote sensing and modeling: Advances in coastal and marine resources* (pp. 135–160). Springer.
Paul, A. K., Ray, R., Kamila, A., & Jana, S. (2017). *Mangrove degradation in the Sundarbans* (In coastal wetlands: Alteration and remediation) (pp. 357–392). Springer. https://doi.org/10.1007/978-3-319-56179-0
Paul, A., Paul, A. K., & Sardar, J. (2023a). Carbonate landforms of Rameswaram Island, Tamil Nadu coast, India: A review. *Arabian Journal of Geosciences, 16*(3), 176. https://doi.org/10.1007/s12517-023-11222-6
Paul, A. K., Paul, A., & Majumdar, D. D. (2023b). Coastal sand dunes along the western and eastern shores of India. In *Sand dunes of the northern hemisphere* (pp. 350–382). CRC Press.
Rajendran, K., Rajendran, C. P., Earnest, A., Prasad, G. R., Dutta, K., Ray, D. K., & Anu, R. (2008). Age estimates of coastal terraces in the Andaman and Nicobar Islands and their tectonic implications. *Tectonophysics, 455*(1–4), 53–60. https://doi.org/10.1016/j.tecto.2008.05.004
Ramkumar, M., Menier, D., Mathew, M., & Santosh, M. (2016). Geological, geophysical, and inherited tectonic imprints on the climate and contrasting coastal geomorphology of the Indian peninsula. *Gondwana Research, 36*, 65–93. https://doi.org/10.1016/j.gr.2016.04.008
Rao, K. N., Saito, Y., Nagakumar, K. C. V., Demudu, G., Rajawat, A. S., Kubo, S., & Li, Z. (2015). Palaeogeography and evolution of the Godavari delta, east coast of India during the Holocene: An example of wave-dominated and fan-delta settings. *Palaeogeography, Palaeoclimatology, Palaeoecology, 440*, 213–233. https://doi.org/10.1016/j.palaeo.2015.09.006
Ravenscroft, P., Burgess, W. G., Ahmed, K. M., Burren, M., & Perrin, J. (2005). Arsenic in groundwater of the Bengal Basin, Bangladesh: Distribution, field relations, and hydrogeological setting. *Hydrogeology Journal, 13*, 727–751.
Roy, R. K., & Chattopadhyay, G. S. (1997). Quaternary geology of the environs of Ganga Delta, West Bengal and Bihar. *Indian Journal of Geology, 69*(2), 177–209.
Sastri, V. V., Venkatachala, B. S., & Narayanan, V. (1981). The evolution of the east coast of India. *Palaeogeography, Palaeoclimatology, Palaeoecology, 36*(1–2), 23–54. https://doi.org/10.1016/0031-0182(81)90047-X

Chapter 2
Nearshore Morphological Alteration through Sediment Dynamics: An Integrated Software Based (Coastal Modelling System-SMC) Approach for Deltaic Balasore Coast, Odisha, India

Nilay Kanti Barman

2.1 Introduction

Littoral morphological change includes changes to the beach morphology, shoreline position, and near-shore bottom topography caused by a variety of hydrodynamic and geomorphic processes, including fluvial, aeolian, and terrestrial ones. Sediment mobility is related to changes in near-shore bathymetry and the resulting series of adjustment dynamics. Hydrodynamic processes, including marine waves, tides, and wind-induced currents, are what cause morphological changes in coastal environments. Thus, sediment rearrangement through morphodynamic processes and morphological changes that actually take place in the maritime environment are all covered under littoral morphological alteration. There must be room for hydrodynamic reworking since sediment loads take time to reorganize. Therefore, the sediment load can be viewed as a time-dependent amalgamation apparatus. As a result, the boundary between the limits of terrestrial and aquatic dynamics frequently deviates from the limit of hydrodynamics. This can be a sign that the littoral setting never achieves equilibrium. The coastal region's systems approach was created by Wright and Thorn (1977). The current focus is on identifying and classifying the many dynamic zones related to sediment movement in the littoral zone in and around the Subarnarekha delta plain. The necessary characteristics of coastal morphodynamism are predicated on the interaction of bathymetry and hydrodynamics, which propel sediment drift and result in morphological changes. Either a negative or positive response is possible. Negative reaction denotes self-regulation characteristics in relation to mild disturbances (Wright & Thorn, 1977). Positive results show

N. K. Barman (✉)
Department of Geography, Midnapore College (Autonomous),
Midnapore, West Bengal, India
e-mail: nilaycsws@gmail.com

A. K. Paul, A. Paul (eds.), *Crisis on the Coast and Hinterland*,
https://doi.org/10.1007/978-3-031-42231-7_2

an increase in variability and deliberate self-organization, which leads to the emergence of new operational paradigms (Philips, 1991). The onset of morphodynamic performance is marked by response reversal.

2.2 Study Area

Three diverse coastline segments of the low-lying Subarnarekha delta plain have been subjected to the coastal modelling system technique in order to comprehend the near-shore wave climate and ensuing sediment drifting processes of these regions. The three different studied points are Kirtaniya (21°32′20″N. −21°33′58″N. to 87°25′30″E. −87°27′33″E) Choumukh (21°32′07″N. −21°33′28″N. to 87°13′55″E. −87°16′19″E.), and Rasalpur (21°25′25″N. −21°26′10″N. to 87°02′58″E. −87°05′02″E.) respectively (Fig. 2.1).

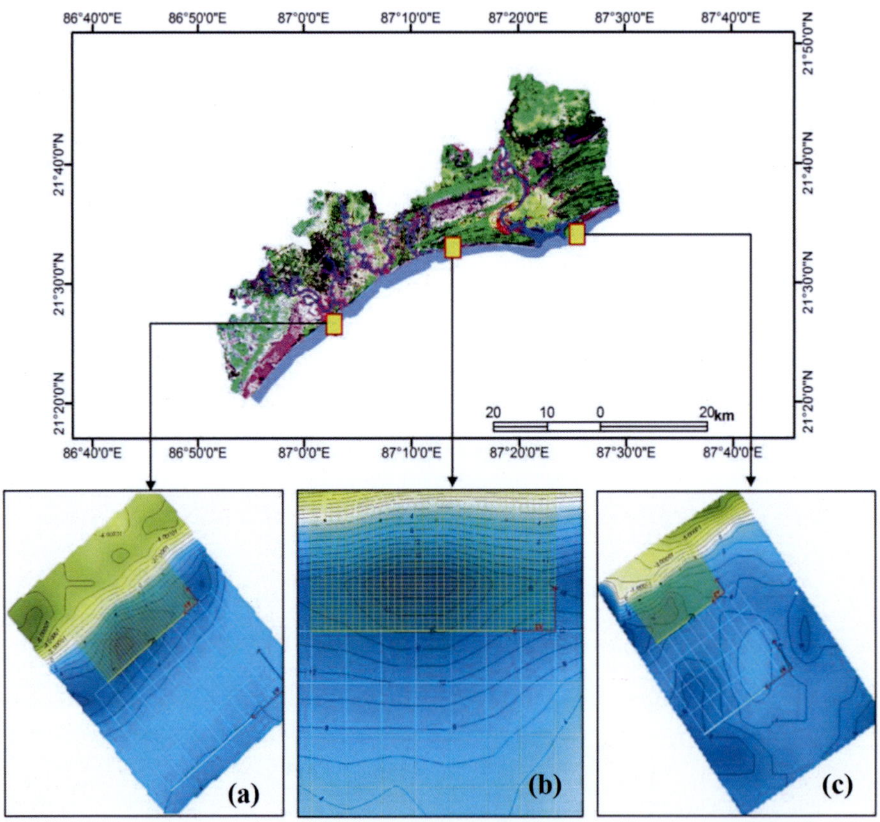

Fig. 2.1 Studying three separate coastal areas: (**a**) Kirtaniya, (**b**) Choumukh, and (**c**) Rasalpur

2.3 SMC for Coastal Modelling

The Spanish Beach Nourishment Manual contains a section called SMC, which stands for Coastal Modelling System (SBM). Between 1995 and 2003, the DGC collaborated with the University of Cantabria's Ocean and Coastal Research Group (GIOC) to create the SBM. The Coastal Modelling System (SMC) integrates a series of programs and numerical models created with the spatial and temporal measures of the many dynamic factors disturbing the morphology of the littoral and beaches in accordance with various themes and referred texts. The SMC incorporates the entire exposition of the Coastal Modelling System, but it also allows for unconventional approval right out of the box. The SMC can be seen as an ascendant tool with a clear permanence in this nous (Table 2.1).

2.3.1 Methodology

Numerous numerical models are included in the short-term module, such as the MOPLA module, to examine littoral dynamics on a short-time scale (hours–days). It consists of beach plan 2DH and cross-profile 2DV (PETRA) morphodynamic development models (MOPLA). MOPLA enables the following: (1) to show how native coastal dynamics (wave and current patterns) work, (2) to assess how the beach responds to transient events (storm and surge), and (3) to think about the approach and design for the coastal constructions. The MOPLA module, which consists of the wave transformation module (Oluca), the depth-averaged currents module (Copla), and the sediment transport and morphological development module, has been used to calculate current exertion (Eros). The input data for the SMC (Mopla model) must be translated into a particular format that the model could easily operate in order to be used (Table 2.1).

2.4 Results

2.4.1 Wave Propagation

In a surface wave, the saltwater particles move in ellipses, moving up and forward during the wave endorsement as the wave crest gets in front of it and down and backward as the wave trough exceeds it (Siddiqui & Maajid, 2004). In shallower water, the waves travel forward more slowly and collectively because the stirring velocity of each wave develops in exactly the same proportion to the square root of the water depth (De Vriend et al., 1993). Breaking waves contain a lot of dynamism, part of which is used to generate confined currents, mostly longshore currents, and

Table 2.1 Data input in SMC for the investigation of monochromatic wave parameters at Kirtaniya, Choumukh, and Rasalpur coastal sectors

Monochromatic wave	Monsoonal condition		
Study points	Kirtaniya	Choumukh	Rasalpur
1. In wave dynamics			
Wave height (H) m	1	0.79	0.85
Direction	S 14.3 W	S 50.0 E	S 10.0 W
Period (T) s	8.39	7.69	9.32
Tidal range (m)	2.46	2.11	2.25
2. In dynamics/currents			
Time range	Second to hour		
Total time (Min)	500	500	500
Chezy roughness	1	1	1
Eddy viscosity (m^2/s)	39	36	26
3. In dynamics/transport (Sediment characteristics)			
D_{50} (mm)	0.16	0.18	0.17
D_{90} (mm)	0.25	0.22	0.22
Friction angle (DMS)	30°	33°	31°
Density	2.65	2.65	2.65
Porosity	0.3	0.3	0.3
Standard deviation	1	1	1
4. In dynamics/transport (Water characteristics)			
Density (T/m^3)	1025		
Viscosity (m^2/s)	1.10–6		
5. In dynamics/transport (Simulation characteristics)			
Morphodynamic evolution			
Duration (h)	48	48	48
Maximum bottom variation (m)	0.85	0.54	0.69
6. In "dynamics/transport" change these values:			
Duration (h)	48	48	48
Irregular waves propagation			
7. Spectrum/parameters (Frequency spectrum)			
Depth (h) m	1.13	1.01	1.08
Significant wave height (Hs) m	1	1	1
Peak frequency (fp) Hz	0.18	0.13	0.15
Maximum frequency (fmax) Hz	0.28	0.22	0.25
Spectral width (m)	10	10	10
Number of components (Nf)	10	10	10
8. Spectrum/parameters (directional spectrum)			
Mean direction (m)	0°	0°	0°
Dispersion or shape parameter (m)	5°	5°	5°
Number of components (N)	15	15	15

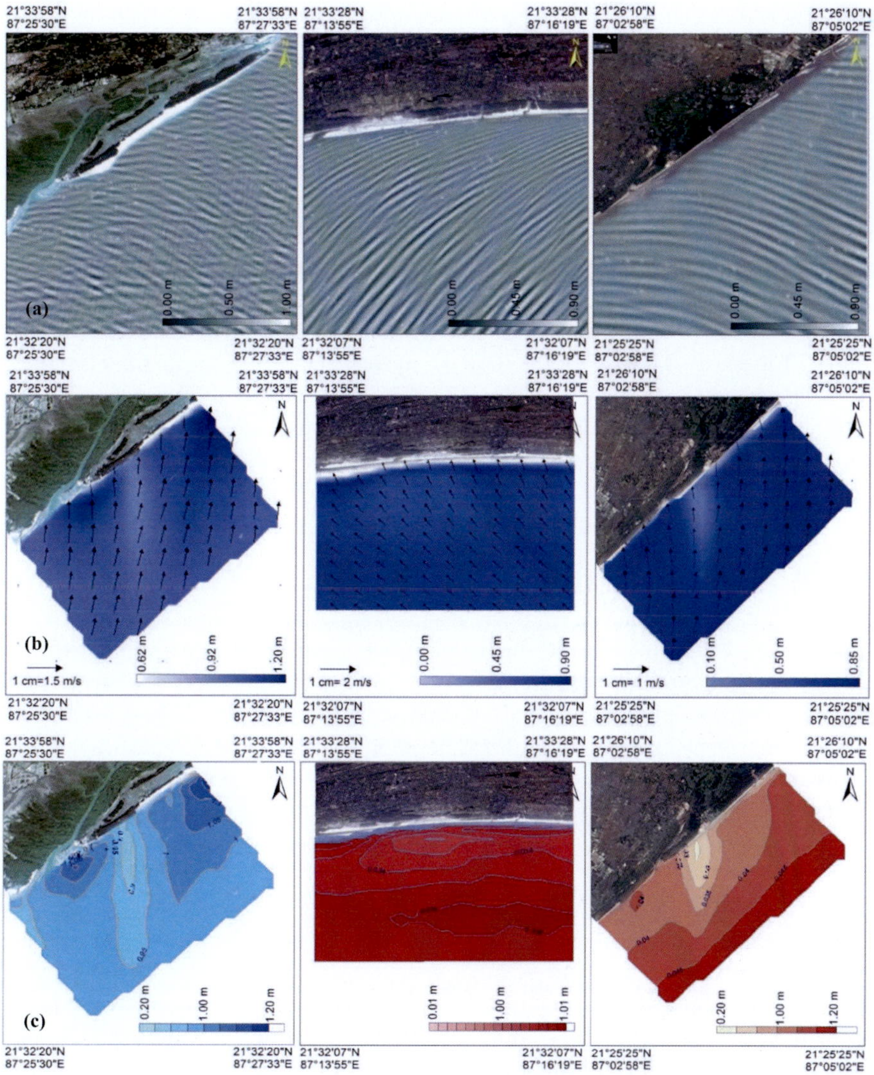

Fig. 2.2 (**a**) Bottom topography and the dominant wind both influence wave propagation and the resulting wave height bending; (**b**) significant wave height and rapidity of waves during the solitary transformation; and (**c**) change in wave height from offshore to nearshore

rip currents that return water to the ocean (Hsu et al., 1987). Figure 2.2a depicts the model-generated wave transmission at the three specific coastal locations, with wave altitudinal ranges of 0.00–0.10 m, 0.00–0.09 m, and 0.00–0.09 m, respectively, at Kirtaniya, Choumukh, and Rasalpur. The figure also shows the wave's configuration and direction. Therefore, the Kirtaniya sector has a greater wave altitudinal range than the other two chosen coastal sectors, which is particularly important for beach morphodynamics.

2.4.2 Significant Wave Heights

The significant wave height, which is typically utilised in visual assessments of sea state, is defined as the average height of the highest 1/3 of waves. The distinct wave could be up to twice the height of the significant wave (Dean, 1977). The results show that the significant wave heights in Kirtaniya, Choumukh, and Rasalpur are 0.62–1.20 m, 0.00–0.90 m, and 0.10–0.85 m, respectively (Fig. 2.2b). The three places under study had wave crest velocities of 1.5 m/s, 2 m/s, and 1 m/s, respectively.

2.4.3 Wave Height

Wave height is the perpendicular distance between the wave crest and trough. The relationship between wind speed, fetch distance, and wave height is straightforward (Larson et al., 1999). In contrast, the depth of the ocean and the height of its waves are inversely proportional to one another (González & Medina, 1999). The model's considered wave height at three coastal locations is 0.01–1.01 m, 0.20–1.20 m, and 0.20–1.20 m, respectively (Fig. 2.2c). The approaching energy of the wave crest rapidly increases in relation to the wave height, which is much more operational in nature to affect the morphodynamics of the beach and the movement of silt.

2.4.4 Phase Component

At a point where the wave has completed its last transition past the reference position, it takes a little while for the wave to complete its full cycle in horizontal travel (Battjes & Janssen, 1978). When the consequence points of two periodic motions simultaneously affect the maximum or minimum transpositions, they are said to be in phase. Alternately, phase difference is the difference between two waves with the same frequency that are related to the same instant in time, represented in electrical degrees or time (Liu & Mei, 1976). The amount by which these waves are out of phase with one another can be expressed in radians from 0 to 2 π or in degrees ranging from 0 to 360. The two waves are considered to be in anti-phase if the angle between their phases is 180 degrees, or π radians (Watanabe et al., 1984). At Kirtaniya, 0.00° and 180° of the wave phase components predicted by the model are observed to vary, as are 0.00° and 200° at Choumukh and 0.00° and 200° at Rasalpur (Fig. 2.3a).

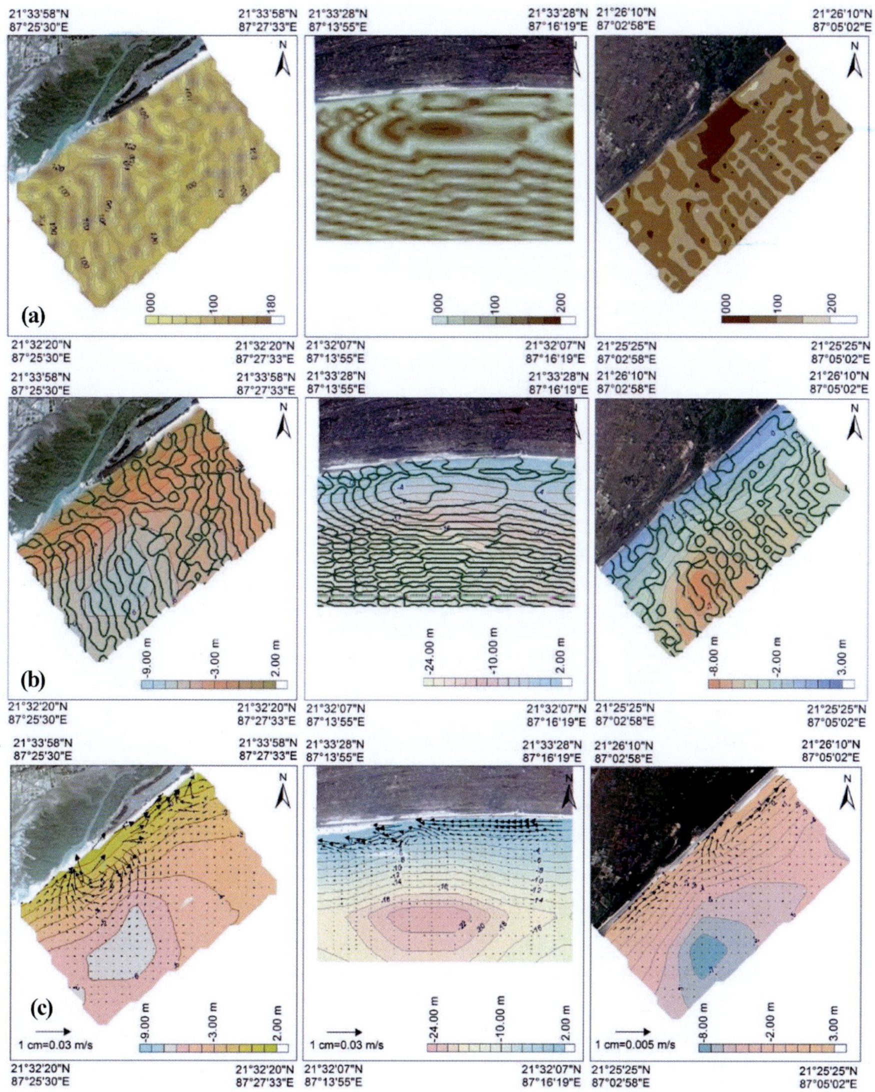

Fig. 2.3 (**a**) Wave phase component through the solitary transformation from offshore to near-shore positions; (**b**) phase direction and bottom topography modify the wave front component; and (**c**) relationship between current dynamics and bottom topography in a monsoonal wave climate

2.4.5 Wave Front

The wave front is defined as the fixed points in a given space that a wave reaches simultaneously while passing through a medium. A line, curve, or surface for a wave travelling in three dimensions are examples of points with the same phase that make up the wave front (Birkemeier, 1994). The graphic above accurately depicts the wave front scenario in the coastal areas of Kirtaniya, Choumukh, and Rasalpur (Fig. 2.3b).

2.4.6 Current and Topography

A momentum is developing as the wave travels through space. The momentum of currents is another transformation of this momentum (Méndez & Medina, 2001). Currents are intimidated by dissipating waves in the direction of their wave vector k. A long beach current is forced in the direction of the wave's k when it breaks. The surface zone's wave breaking always results in poorly adjusted currents. The long coast stream may become unstable and split into eddies even on a beach with a constant slope. It works best to think of the flow in the surf zone as consisting of two fields: a current field that, together with depth changes, refracts the incoming waves and a wave field that pushes currents when the waves break. The wave-induced current velocity is depicted in Fig. 2.3c in relation to the bathymetry of the nearby beach. At Rasalpur, the current is simply split into two opposing directions due to the topography of the nearby shore (Fig. 2.3c).

2.4.7 Sediment Transport Through the Current

The current project reveals potential sediment movement in the nearshore zone, with rates ranging from 0.25 to 0.1 m^3/h/ml in the Kirtaniya sector, from 0.01 to 0.04 m^3/h/ml in Choumukh, and from 0.015 to 0.06 m^3/h/ml in the Rasalpur sector (Fig. 2.4a). The model predicted that there would be larger transference expanses during extreme environmental conditions than under normal circumstances (Méndez & Medina, 2001; Bailard, 1984). Additionally, the model's predicted path for sediment transference in Kirtaniya and Rasalpur is south west to north east, and in Choumukh, it is north east to south west, which is consistent with field findings of sediment flow in these regions. According to the model, in Kirtaniya and Rasalpur, sand is moving from the south west to the north east, while in Choumukh, it is moving from the north east to the south west. According to the simulation of average monsoonal conditions, sediment movement in the Choumukh area continues even during monsoon seasons.

2.4.8 Sediment Transport and Morphological Alteration

The sediment mobility and topography change on beaches throughout a 48-h period. In light of the aforementioned circumstances, the model's calibrated result shows that erosion occurs primarily at the vortex's centre and in the near-shore region,

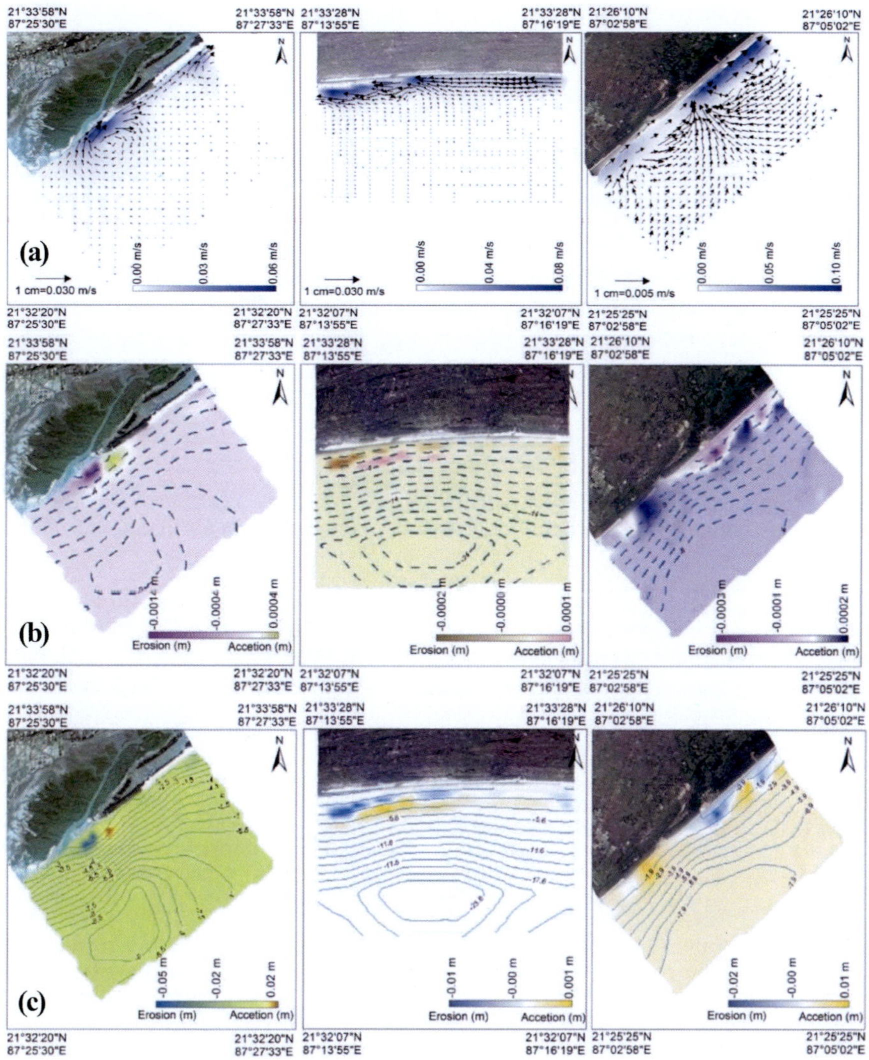

Fig. 2.4 (**a**) Potential for transporting sediment based on fetch openness and long-shore current intensity; (**b**) initial topography was created by the movement of sediment and morphological changes brought on by waves and wave-induced current patterns; and (**c**) after calibration over a 48-h period on the same space-time continuum, the final topography is created by sediment movement and morphological change caused by wave and wave-induced current patterns

whereas accretion occurs at the vortex's edge and in the off-shore zone (Fig. 2.4b). The likely bedload transportation of sediment in the near-shore region is limited to between 0.00 m^3 and 0.1 m^3/h/ml at Kirtaniya, 0.00 m^3 and 0.04 m^3/h/ml at Choumukh, and 0.00 m^3 and 0.06 m^3/h/ml at Rasalpur coastal sector. After the 48-h period of model calibration, the height of sediment from erosion and accretion is constrained between −0.05 m and 0.02 m at Kirtaniya, −0.01 m and 0.001 m at

Choumukh, and −0.02 m and 0.01 m at Rasalpur sector. At each location of investigation, the extent of erosion and accretion can also be given (Fig. 2.4c).

2.5 Discussion

For the three places in the area that were chosen, the model was run as a stationary model. These runs allow for some broad observations to be made about the wave climate for these regions, including wave propagation orientation, wave height, wave phase component, wave front crenulations, and significant wave height. In the Kirtaniya study location, the wave height, significant wave heights, current velocity, and potential transport of sediment are all high; at Choumukh, they are low; and at Rasalpur, they are of intermediate intensity (Fig. 2.2). As higher waves have a greater tendency to mobilize sediment, storm conditions can have a significant impact on the transport of sediment in the area (WMO, 2003). The first research location, Kirtaniya, is situated on the Subarnarekha delta Chenier plain, which is a zone with dynamic fluvial, marine, and terrestrial processes as well as frequent tropical cyclones and related natural disturbances. This region has a great deal of complexity, diversity, and fragility (Niyogi, 1970). The ability of coastal morphological units, such as wetlands, beaches, lagoons, estuaries, and sand dunes, to perform multiple parallel systems, including providing rich swamps and breeding grounds for marine creatures, feeding grounds for birds, recreational areas for tourists, and resource availability for the local communities, is crucial to changing the coastal morphology.

Regressive younger coastal ridges, mudflats, and floodplains that appear as depressed zones serve as a representation of the region. Balasore's coastal stretch's northernmost portion of the seafront is made up of a series of beach barriers and washover deposits (Barman et al., 2015). In general, the Kirtaniya coastline stretch is mostly a portion of the Subarnarekha flood plain, which was created when the river Subarnarekha avulsed to the west (Paul, 2002). This form of bed load transfer has been mindful of interactions among maritime transgression processes, abundant sediment supply, and dominant wave tidal dynamics, which have altered the coastal beach morphodynamics appropriately over time (Niyogi & Chakraborty, 1973). Geomorphological indicators imply that a phase of maritime incursion has likely already begun in this coastal area. There has been a slight increase in both cyclone frequency and intensity. Higher concentration of the sediment transportation at Kirtaniya has been found to be caused by cyclone-induced storm surges, torrential rain in the upper catchment of the Subarnarekha River, enormous discharge of the Subarnarekha River to the ocean at Kirtaniya, as well as vulnerable geomorphic location. Recent climate and environmental changes may be to blame for the increased intensity and severity of beach morphological transformation. Additionally, the river Subarnarekha discharges a significant amount of sediment in a huge quantity (Figs. 2.3 and 2.4). Due to the strong south-westerly monsoon wind, the resulting cross-shore current and waves, and the high magnitude of the tidal inflow, this discharge flow naturally causes conflict. At Kirtaniya areas, this results in huge water and silt accumulation.

Thin mangrove scraps that were once present in this region only a few years ago have vanished as a result of changes in the sedimentological makeup of the coastal deposits that serve as the substrate for mangrove swamps. The current research area's land use pattern has undergone such alterations that they have increased the likelihood that the area's beaches may change. On the other hand, Choumukh Beach is located on the right bank of the river Subarnarekha. Because the prevailing wind always comes from the south and south-east, this particular location is helping Choumukh protect the high intensity of the longshore current that may be amplified due to the enormous discharge of the river Subarnarekha (tropical monsoonal wind). The fact that the prevailing wind and wave are moving from east to north east in this location is quite fascinating because of the beach's 24° north-east orientation. Therefore, it is also a crucial requirement to lessen the transit of silt in this area. Less mobility of sediment as bedload is also a result of the lack of sediment in this area due to the absence of any estuaries. As a result, this site's results for the wave climatic parameters are lower than those of the other two research locations (Fig. 2.4). Because of this, it is entirely normal that the waves and wave-induced currents have less potential energy to move the sediment bed. Rasalpur Beach, on the other hand, sees intermediately intense sediment transport.

2.6 Conclusion

Any coastal design or management requires a thorough understanding of coastal dynamics. For Balasore in particular, this is accurate. To further understand near-shore wave and sediment dynamics, additional numerical modelling on fresh wave data should be done. Wave action and wave-induced currents are the primary mechanisms for sediment mobility in coastal environments. These primarily act from the south and south-west to the north-east during the monsoon season, but during the retreating monsoon phase, they tend to act in the other direction. Because of this, sediment transport patterns and rates often follow this seasonal pattern, with the majority of sediment migrating in the opposite direction during the receding monsoon and from the south and south-west to the north-east during the monsoon. SMC-based numerical wave modelling successfully predicted major wave heights in good agreement with observed wave data. While there was no evidence to corroborate the directed wave output from the model, SMC have a tendency to anticipate shorter wave periods than the observed wave data revealed. A number of immobile runs' findings indicate that the model can anticipate the estuarine environment's wave climate with a fair level of accuracy and that it should be utilized to generate a full year's worth of wave hind casts. Estimates for the pattern of sediment mobility in the region were then made using near-shore data that was taken from the model results.

It is advised that the area be subjected to at least a thorough full-year wave hind cast using better resolution bathymetry. In order to improve each hind cast's accuracy and describe the coastline's extreme wave climate in Balasore, 10 years of wave hindcasts should be conducted. The peculiarities of the sediment mobilization in the area may then be explained considerably more effectively using these hind casts. Any wave hindcast should be subjected to wave transformation modelling in order to provide wave parameters for a number of places in the nearshore zone. This information would allow for a more accurate prediction of the features of sediment movement in addition to providing intended wave data at various sites. It is advised that an integrated plan be implemented for the entire coastal region, from the Sundarbans to Goa. This would ensure future studies and integrated information on the coastal challenges in this region.

References

Bailard, J. A. (1984). A simplified model for longshore sediment transport. In *Proceedings of the 19th coastal engineering conference* (pp. 1454–1470).

Barman, N. K., Chatterjee, S., & Khan, A. (2015). Quantification of panchayat-level flood risks in the Bhograi Coastal Block, Odisha, India. *Journal of Indian Geophysics Union, 19*(3), 322–332.

Battjes, J. A., & Janssen, J. P. F. M. (1978). Energy loss and set-up due to breaking of random waves. In *Proceedings of 16th international conference on coastal engineering* (pp. 569–587). ASCE.

Birkemeier, W. A. (1994). *The DUCK94 near shore field experiment – Experiment overview*. U.S. Army Engineer Waterways Experiment Station, Vicksburg, MS.

De Vriend, H. J., Copabianco, M., Chesher, T., de Swart, H. W., Latteux, B., & Stive, M. J. F. (1993). Approaches to long-term modelling of coastal morphology: A review. *Coastal Engineering, 21*, 225–269.

Dean, R. G. (1977). *Equilibrium beach profiles, U.S. Atlantic and Gulf Coasts* (Tech. Rep. No. 12). U. Delaware, Newark.

González, M., & Medina, R. (1999). Equilibrium shoreline response behind a single offshore breakwater. In *Proceedings coastal sediments '99* (pp. 844–859). ASCE.

Hsu, J. R. C., Silvester, R., & Xia, Y. M., (1987). New characteristics of equilibrium shaped bays. In *Proceedings 25th coastal engineering conference* (pp. 3986–3999). ASCE.

Larson, J., Lynch, G., Games, D., & Seubert, P. (1999). Alterations in synaptic transmission and long-term potentiation in hippocampal slices from young and aged PDAPP mice. *Brain Research, 840*, 23–35.

Liu, P. L.-F., & Mei, C. C. (1976). Water motion on a beach in the presence of a breakwater, 1, waves. *Journal of Geophysical Research, 81*, 3079–3084.

Méndez, F. J., & Medina, R. (2001). A perturbation method for wave and wave-induced current computations in beach morphology models. In *Proceedings of coastal dynamics*. Lünd (Suecia). Enprensa.

Niyogi, D. (1970). Quaternary geology and geomorphology of the Kharagpur-Digha area W.B. Guide book for the field trips; Section of Geology and Geography. In *57th Session of Indian Science Congress* (pp. 1–18). Kharagpur.

Niyogi, D., & Chakraborty, A. (1973). Applied geomorphology along Digha beach, Midnapore district. In *W.B. seminar geomorphological studies in India. Proceedings* (pp. 205–210). Sagar.

Paul, A. K. (2002). *Coastal geography and environment: Sundarban Coastal Plain, Kanthi Coastal Plain and Subarnarekha Delta Plain* (p. 1). ACB Publications.

Philips, J. D. (1991). Nonlinear dynamical systems in geomorphology: Revolution or evolution? *Geomorphology, 5*, 219–229. [This article comfirms the applicability of nonlinear dynamical systems theory to the study of landforms].

Siddiqui, M., & Maajid, S. (2004). Monitoring of geomorphological changes for planning reclamation work in coastal area of Karachi, Pakistan. *Advances in Space Research, 33*, 1200–1205. https://doi.org/10.1016/S0273-1177(03)00373-9

Watanabe, A., Hara, T., & Horikawa, K. (1984). Study on breaking condition for compound wave trains. *Coastal Engineering. Japan, 27*, 71–82.

WMO. (2003). Tropical Meteorology Research Programme (TMRP), Commission for Atmospheric Sciences. In *Proceedings of the 5th WMO International Workshop on Tropical Cyclones (IWTCV)* (TMRP Rep No. 68, Chapter 8). Tropical Cyclone Impacts.

Wright, L. D., & Thorn, B. G. (1977). Coastal depositional landforms: A morphodynamic approach. *Progress in Physical Geography, 1*(3), 412–459. (Virginia Institute of Marine Science-Scopus 460).

Chapter 3
Shoreline Change and Associated Beach Ridge Chenier Formations in the Subarnarekha Delta Region, India

Subrata Jana

3.1 Introduction

The terrestrial sediments are downdrifted by the Subarnarekha River into the shallow sea region and redistributed along the shore by the marine agents of waves, tides, currents, and storm surges. In the time period between 7000 and 50 YBP, a series of Chenier sand ridges developed in front of stable shore positions (Paul, 1996a, b, 1997, 2002; Maiti, 2013; Kamila et al., 2021). In the initial phase of sand ridge formation, the coarser sands and marine shell fragments were deposited in the shallow mudflats by the strong storm surge from the shallow sea region to form the beach ridge (Jana & Paul, 2019; Sebastian et al., 2019). The shape, size, and extension of the respective sand ridges coexist with the direction of the longshore drift associated with the dominance of monsoon winds (southwestern and northeastern monsoons) (Kunte & Wagle, 2005; Kaliraj et al., 2014). The volume of terrestrial sediment supply into the shallow sea region is associated with the strengthening and weakening of the monsoon through the actions of weathering and erosion in the entire basin area (Banerji et al., 2020; Mandal et al., 2021). The repeated nature of sea level indicates the inter-dune spacing across the delta region (Jana & Paul, 2022). In these concerns, the barrier bars have been formed in the shallow sea region, backed by the shallow mudflats dominated by the tidal environment in a stillstand phase of the sea level (Paul, 1996a, 2002; Jana & Paul, 2019; Kamila et al., 2021). In the regressive phase, the sea level retreats and the barrier bar forms in a further seaward position, and again the shallow mudflats form in backwater regions dominated by the tidal environment (Paul, 2002; Barman et al., 2016). The previously known barrier bars were transformed into dune ridges as a result of sediment supplied from the wide beach zone. As a result of the regressive phases of sea

S. Jana (✉)
Department of Geography, Belda College, Belda, West Bengal, India
e-mail: subrata.vumid@gmail.com

A. K. Paul, A. Paul (eds.), *Crisis on the Coast and Hinterland*,
https://doi.org/10.1007/978-3-031-42231-7_3

level, shallow mudflats emerged and transformed into swale landforms. The width of the swale landscape indicates the rate of sea level retreat during the respective regressive phases of sea level. The presence and absence of dune ridges and swales on both sides of the Subarnarekha river indicate the strength of longshore drift, reflected in the strength of the northeastern or southwestern monsoon, along with the volume of terrestrial sediment supply into the shallow sea region.

Till now, lots of work has been done in this region regarding the landscape setup, geomorphology, and mechanism of chenier formations (Maiti, 2013; Jana et al., 2014; Jana & Paul, 2018, 2019, 2020, 2022; Kamila et al., 2021). The majority of these studies, however, have been conducted on the delta's eastern side, leaving the western side largely unexplored, despite being mentioned in a few works (Niyogi, 1968; Paul, 1996a, 2002). Although the actual mechanism of chenier formations cannot be explained properly without the study of the entire delta region in one work, there are lots of diversities in landscape and geomorphology on both sides of the delta region. Therefore, this study has been carried out over the entire Subarnarekha delta region on the aspects of landscape, geomorphology, and beach ridge-chenier formations in association with the concomitant nature of shoreline positions.

3.2 Material and Methods

3.2.1 Study Area

The study has been carried out in the deltaplain region on both sides of the Subarnarekha River course, which extends from the Rasulpur river in the east to the Budhabalanga river in the west, covering parts of the states of West Bengal and Odisha (Fig. 3.1). This study area is about 90 km long along the Bay of Bengal coastline and about 30 km wide in the interior part of the deltaplain. Although the present study has been done on the chenier mechanism of the Subarnarekha delta region, the study area has been selected far inland and extends beyond the chenier landforms to explain the process of delta formations as well as the mechanism of chenier formations over the deltaplain during the different time periods (Jana, 2017, 2021).

3.2.2 Database and Data Processing

In the study, the shuttle radar topography mission (SRTM) 1-arc second (30 m) resolution digital elevation model (DEM)-based data from 2015 was considered for terrain analysis and also to find out the different morphological features. The differential global positioning systems (DGPS) survey extracted elevation data from 1000 ground control points (GCPs) from various terrain units. The DGPS provides elevation data considering the earth as an ellipsoid surface, the World Geodetic System 1984 (WGS84), and the Earth Gravity Model (EGM96). Therefore, the

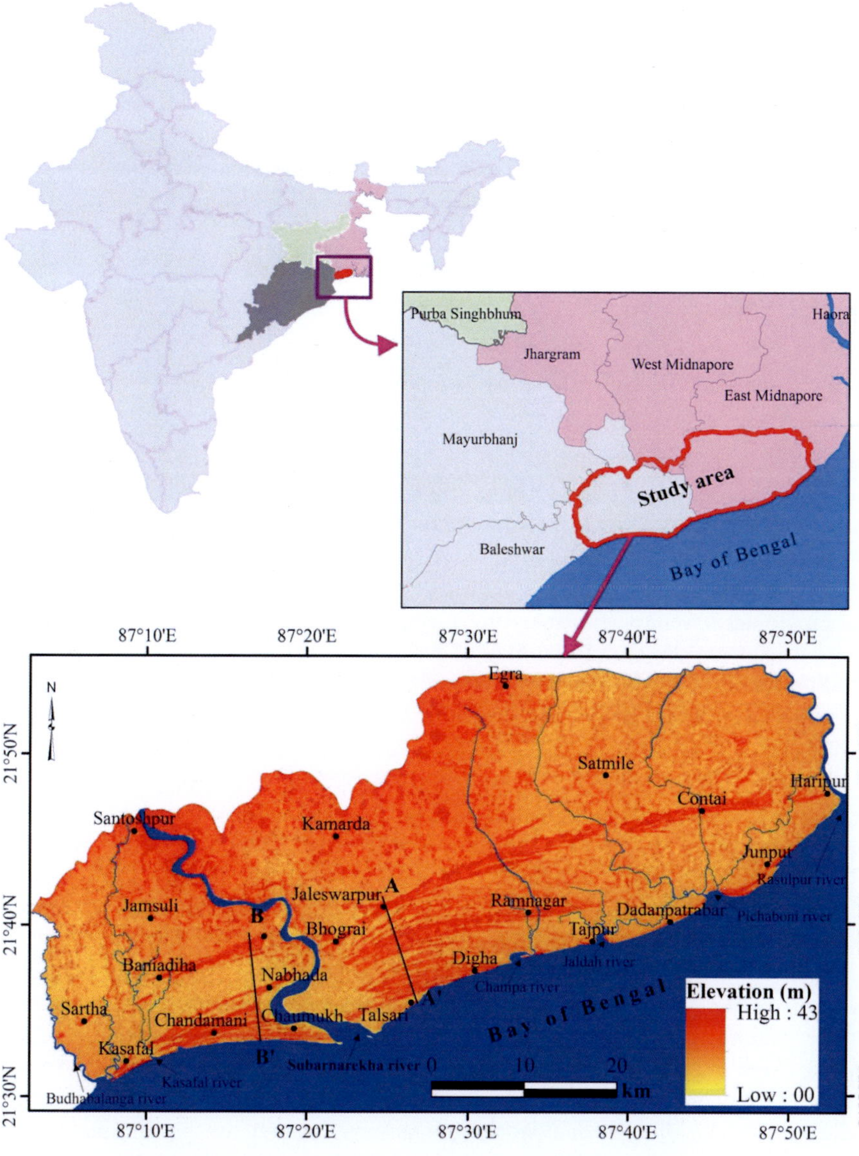

Fig. 3.1 The study area in the Subarnarekha deltaplain, selected from the Rasulpur river in the east to Budhabalanga river in the west. The DEM shows the sequential pattern of dune ridges

collected DGPS data were converted into an equipotential geoidal surface corresponding to the mean sea level (MSL) as a local vertical datum (Pavlis et al., 2012; Patel et al., 2016). The SRTM image was resampled to 10 m spatial resolution after subpixel analysis (Mokarrama & Hojati, 2018). The vertical accuracy of the resampled SRTM data was validated with randomly taken 300 DGPS-based elevations, which divulge ±0.12 m root mean square error (RMSE).

The landscape features of the Subarnarekha deltaplain were extracted from the finally processed 10 m-resolution SRTM data. The differential elevations of the various landscape features were defined by the contour patterns. Moreover, the Google Earth image was also considered for accurate detection of the landscape features. In a similar way, the geomorphological diversities of the entire study area were extracted and finally processed into ArcGIS 10.1 software to get the final output. Also, the cross-profiles were extracted from the SRTM data.

3.3 Results and Discussion

3.3.1 Landscape Evolution of the Subarnarekha Delta

The delta head of the Subarnarekha deltaplain exists near Keshiary, about 30 km north of the present study area. About 7000–6000 YBP, the eastern tectonic shelf zone of the Bengal Basin was subsided along the subsurface Damodar fault (DF) (Ghosh & Guchhait, 2015; Jana & Paul, 2018). Moreover, along the margin of the Medinipur-Farakka fault (MFF), a fault-line scarp has been formed due to tremendous erosional activities (Biswas, 1987). Simultaneously, a marine transgression event occurred on the current surface of the Subarnarekha deltaplain. During the early Pleistocene, the entire ancient deltaplain (between MFF and DF) subsided and came under the depositional environment of the marine aggradational process during highstands. Furthermore, the lateritic upland (between the Chotanagpur foothills and the MFF) was uplifted as a result of increased weathering activity for a prolonged arid climate and erosional activity, which was exacerbated by the strengthening of the monsoon in this region between 5800 and 4300 YBP (Biswas, 1993; Banerji et al., 2020; Jana & Paul, 2020). However, the shoreline was located at the margin of the ancient deltaplain during 80,000 YBP, indicating that the entire study area was under pro-deltaic conditions influenced by brackish water and a marine environment until the early Pliocene (Mallick et al., 1972). After that, the sea level regressed and the depositional environment was shifted to a shallow marine, estuarine, and fluvial-dominated environment (Mallick et al., 1972). Therefore, the ancient fluvio-marine delta fan terrace was formed in the northern part of the study area (Fig. 3.2) by the sequential phases of marine and fluvial environments, followed by an ancient marine terrace formed by the dominance of marine environments in the deltaplain (Niyogi, 1968; Paul, 1996a, b, 2002; Jana & Paul, 2019).

In the earlier period of 7000 YBP, the rivers Subarnarekha and Rasulpur were two distributaries that fell into the Bay of Bengal across the entire deltaplain (Paul, 2002; Maiti, 2013; Kamila et al., 2021). Furthermore, the Pichaboni, Jaldah, and Champa rivers of the delta region were linked to the main stream of the Subarnarekha River. However, the distributaries of Subarnarekha and Rasulpur carried huge sediments from the upper catchment areas and deposited them over the deltaplain along with sediment downdrift into the shallow sea region of the Bay of Bengal. During

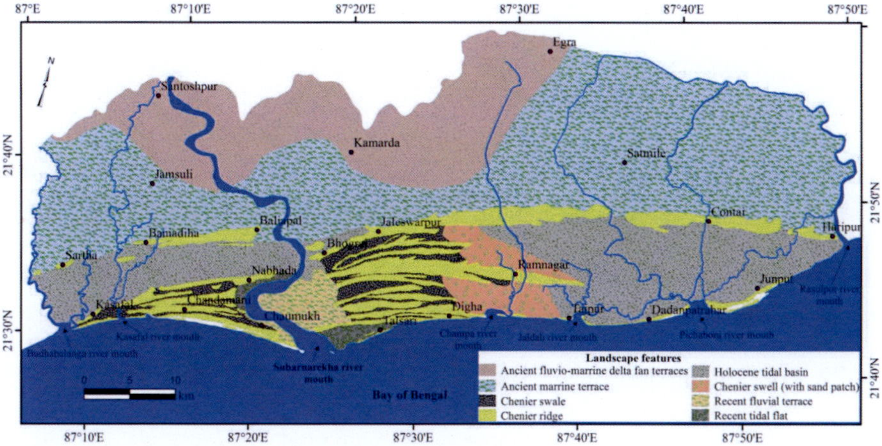

Fig. 3.2 Landscape features of the Subarnarekha deltaplain

5000 YBP, the river Hugli drifted more sediment into the northern head of the Bay of Bengal, around the Rasulpur river mouth region, with stronger sedimentary environmental focie (Allison et al., 2003; Kamila et al., 2021). The finer mud particles were thrown farther westward with the dominance of the northeastern monsoon. During 7000 YBP, the Rasulpur river was beheaded from the main course of the Subarnarekha River (Paul, 2002; Maiti, 2013). The downdrift sediments into the shallow sea region of the Bay of Bengal were redistributed along the shoreline by the longshore drift to form the beach ridge, which was turned into a dune ridge (Contai – Baliapal dune ridge) (Fig. 3.2). Afterward, the Subarnarekha River became a single channel system and threw its finer sediment towards the west in the Baliapal region. Therefore, a wide and elongated tidal basin was formed in the front side of the elongated Contai–Baliapal dune ridge (first stage) during the Holocene period (Figs. 3.2 and 3.3).

The other dune ridges in the deltaplain are not as long as the Contai–Baliapal dune ridge. During 5760–5000 YBP, the second and third stages of dune ridges (Figs. 3.2 and 3.3) were formed to a shorter extent in comparison to the Contai–Baliapal dune ridge. However, during 5800–4300 YBP, when the Indian summer monsoon was strengthened by nature (Banerji et al., 2020; Jana & Paul, 2020), the Subarnarekha River was the only channel that downdrifted sediments into the shallow sea region (the Rasulpur river had already been severed). As a result, the longshore current was unable to transport much sediment along the shoreline (Paul, 1996a). This situation is also applicable for the other conjugative stages (4th–9th) of the dune ridge formations in the deltaplain (Fig. 3.3). The inter-dune chenier swales were formed as a result of the concomitant nature of sea level caused by the nature and temporal extents of stillstand and the aggressive and regressive nature of the sea levels. In general, dune ridges formed during stationary sea level positions. But at the reiterating phase of sea level, the drifted sediments did not get much time to settle in a suitable place to form dune ridges. Therefore, the chenier swales were

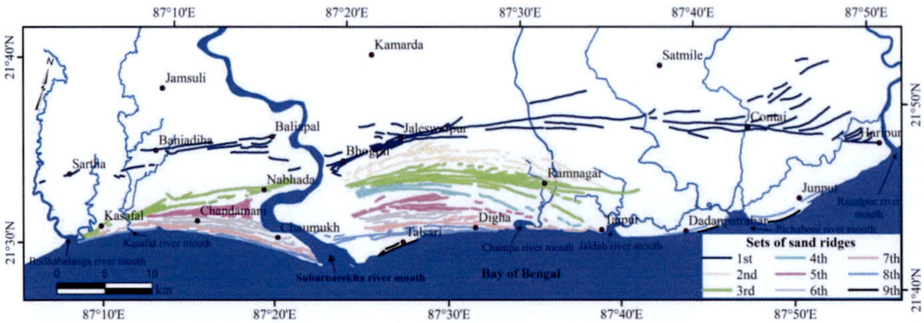

Fig. 3.3 Nine sats of dune ridges within the chenier landscape of the Subarnarekha deltaplain

formed in between the dune ridges as marine terraces composed of silt and clay materials (Figs. 3.2 and 3.3). The chenier swales with sand patches (Fig. 3.2) formed in the far western extent of the Holocene tidal basin where finer mud from the Hugli system could not reach and finer sands from the Subarnarekha system were thrown east by longshore drift after settling of relatively coarse sands into the beach ridge and dune ridge surface.

The recent fluvial terrace of the Subarnarekha River system was formed after the westward shifting of the Subarnarekha River course just after the Bhograi section (Fig. 3.2). This shifting of the river course probably happened during the 2900 ± 160 YBP (Maiti, 2013) guided by the westerly littoral drift (Paul, 1996a; Niyogi, 1968). The fifth to seventh stages of dune ridges were formed during the same period (Maiti, 2013). Moreover, the dune ridges of the fifth stage extended more westward (on the right side of the Subarnarekha River) than the eastern extent of the dune ridge (Figs. 3.2 and 3.3). But the seaward bulge of the deltaplain was more on the eastern side. This contradictory situation indicates that the southwestern monsoon was strong enough during this period (after a brief period of dominance by the northeastern monsoon) to drift the sediment towards the east, forcing the Subarnarekha course to turn again towards the east, and the sedimentary foci of the Subarnarekha river were also extended towards the east, causing more sediment to be thrown in the same direction. The recent tidal flat in the Talsari region was formed during the 200–50 YBP, as the dune ridge of the eighth stage was developed during the 200 YBP (Maiti, 2013), and the shorefront dune at Talsari was formed around 50 years ago (Paul, 2002). The marine sediment was deposited over the tidal flat dominated by the backwater marine environment between 200 and 50 YBP.

3.3.2 Changing Pattern of Shorelines and Associated Marine Environment

The seven sets of shoreline positions coexist with the geomorphic features of the beach ridges (Fig. 3.4). Beach ridges indicate shoreline positions because they are formed directly by the storm energy wave environment through the deposition of

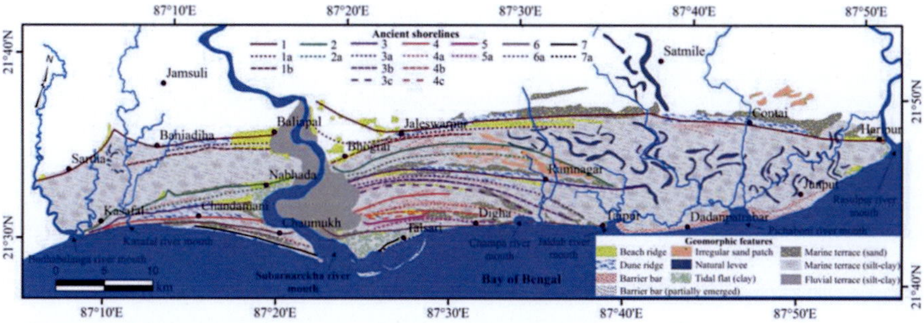

Fig. 3.4 Diversified geomorphological features of the Subarnarekha deltaplain. (Modified after Niyogi, 1968)

coarser sands and shell fragments in appropriate shorefront areas (Paul, 1996a). Moreover, the barrier bars have also been formed with the effects of longshore drift regulated by the shoreline positions. Therefore, the beach ridge and barrier bar give a clear idea about the configurations of the ancient shorelines (Paul, 2002). The first phase of the shoreline (shoreline 1) and adjacent shorelines (shorelines 1a and 1b) indicate the RIA type of submergent shoreline (Niyogi, 1968) near the Subarnarekha River course (around Baliapal) and Kasafal river course (around Baniadiha and Sartha section) (Fig. 3.4). The configuration of the shorelines 1a and 1b appears to extend to a shorter extent adjacent to the beach ridges in front of the estuary and is absent away from the estuary. When compared to the earlier RIA type shoreline (Niyogi, 1968), these two shorelines (1a and 1b) indicate that the estuary valley was gradually filled up and straightened. The second phase of shorelines (shorelines 2 and 2a) remain closer to shoreline 1b on the deltaplain's eastern side but further away from the 1b shoreline on the western side, separated by the wide marine terrace (Fig. 3.4). The process of delta formation (delta extension) moved more quickly on the east side of the deltaplain than on the west. Therefore, shoreline 3 exists on both sides of the delta region, and the other three conjugative shorelines (3a, 3b, and 3c) do not exist on the western side of the deltaplain (Fig. 3.4). Similar conditions have been found for shorelines 4, 4a, 4b, and 4c. The 5 and 5a shorelines exist along the barrier bars in the eastern part of the deltaplain, which are not found on the western side. This also shows the bulge of sediment on the deltaplain's eastern side, which is dominated by the backwater cum marine environment on the backside of the Digha dune (eighth phase). Very interestingly, shoreline 6a is only present in the western part of the deltaplain along the configuration of barrier bars, which indicates the delta extension rate was higher on the western side than the eastern side of the deltaplain. With the bulge out of the recent deltaplain, the current shorelines (shorelines 7 and 7a) were shifted further towards the sea. In addition, nine sets of dune ridges are found in the eastern deltaplain, whereas six sets of dune ridges are found in the western deltaplain (Fig. 3.3). The second, fourth, and ninth sets of dune ridges do not exist in the western deltaplain. However, all the dune ridges indicate the successive positions of retreating shorelines. The wide and extended Holocene tidal basin exists in the parallel position of the second set dune ridge on the western side.

3.3.3 Geomorphology of Subarnarekha Delta

Ten diverse geomorphological features have been discovered within a 12 km wide and 90 km lateral extension of the Subarnarekha basin's nearly symmetrical delta-plain (Fig. 3.4). In comparison to the straight course, three prominent river meanders extended the river course by about 10 km. At the swinging positions of the meandering course, the fluvial terrace has been formed and is dominated by the silt and clay material (Fig. 3.4) (Jana & Paul, 2022). Within this 12 km wide deltaplain, a series of beach ridges, dune ridges, barrier bars, and marine terraces have been formed under the different depositional environments controlled by the marine agents within an ancient lagoonal setting (Niyogi, 1968). The ancient beach has been converted into a marine terrace (sandy), which has been further modified into a dune ridge by the deposition of windblown sand (Paul, 2002). The beach ridges formed by the dominant wave actions (storms) with the deposition of coarser sand and marine shell fragments run parallel to present and ancient shorelines. The elevation of the ridges is controlled by the wave height corresponding to the wave energy to carry and deposit the materials. In most of the cases, the beach ridges are about 4–7 m elevated, and somewhere the older beach ridges are lowered down by the aeolian actions. In front of the estuary, these ridges are wider due to the effects of the high-energy wave action in the deeper water section, and they gradually narrow down when they form away from the estuary margins. The barrier bars are formed in the offshore region with sediment deposition under the dominance of longshore drift. The barrier bars are formed initially as submerged bars that remain underwater (as seen in the shallow sea region of the Subarnarekha mouth) (Fig. 3.4), then wave action acts on them when the bar becomes elevated and is only submerged by high tides (Jana et al., 2014). At the later stages of growth, the bars are subjected to aeolian action, with windblown sands being deposited on the bar surface. At the backside of the barrier bar, the lagoonal basins or tidal flats are formed under the backwater and tide-dominated marine environment with the deposition of finer materials (silt and clay).

The sand dunes or dune ridges are formed by the deposition of windblown sand from the beach sections. The dune elevation varies from 1–2 m to 15–18 m within the Subarnarekha deltaplain. The most prominent dune ridge (Contai–Baliapal dune ridge) is found at the northern (landward) extent of the Chenier landscape (Fig. 3.4). The shape, size, and extension of the present and ancient dune ridges indicate the shoreline configurations as well as the relative stability of the shorelines at certain positions. Most of the ancient beach ridges and some of the barrier bars have been entirely or partially converted into dune ridges. Therefore, in some places, the ridge collision or ridge bifurcation types of morphological features are found with the crenulated landward margins, regardless of whether the seaward margin is straight enough (Paul, 1996a). The beach ridges, barrier bars, and dune ridges are situated over the marine terraces, which are made up of sand, silt, or a mixture of both these materials. At the initial phase, this marine terrace remained a submerged coastal plain dominated by the marine and fluvio-marine environments. Then the

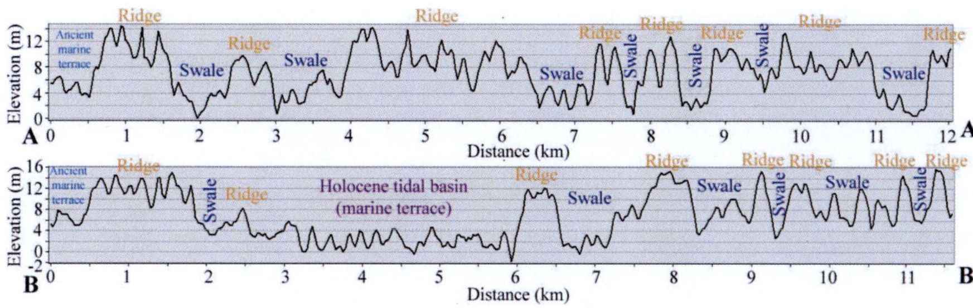

Fig. 3.5 Sequential dune ridge and swale landforms extracted by the cross-profiles

upliftment of the deltaplain (relative to the sea level) took place, and finally the marine terraces rose up in successive phases and appeared in their present form. The natural levees of fluvial deposits and irregular sandy patches of the marine deposits are also found over the marine terraces. The ancient beach that formed at the landward edge of the lagoonal settings is transformed into a marine terrace (sandy in nature), which remains neither on the dune ridge top nor at the edge of the lagoonal clays.

The fragmented natural levees represent the pattern of palaeo rivers and tidal channels over the deltaplain (Fig. 3.4). The natural levee was formed by the deposition of sand and silt-size sediments by the fluvial process beside the river or channel course. The river courses are filled up by sediments after the river flow is disconnected or dies from the main river, and the natural levees remain relatively elevated. The fluvial terraces remained on both sides of the Subarnarekha River (Fig. 3.4) in the avulsion position, composed of silt and clay materials. Both the cross-profiles (AA′ and BB′) provide ample evidence about the evolution scenario of the Subarnarekha deltaplain (Fig. 3.5). The AA′ cross-profile is extracted across the eastern part of the deltaplain, whereas the BB′ is extracted from the western side (Figs. 3.1 and 3.5). The profiles have been drawn from the landscape portion of the ancient marine terrace (landward margin) to the shorefront portion. Both cross-profiles show sequential ridge and swale topography (Fig. 3.5). However, the Holocene tidal basin as well as the marine terrace is dominated by the silt and clay types of materials within 3–6 km of the BB′ cross-profile, whereas dune ridges are present in the same position as the AA′ cross-profile (Fig. 3.5).

3.4 Conclusion

The Subarnarekha deltaplain's eight diverse landscapes and ten distinct geomorphological features formed and evolved over time under the distinct depositional environments controlled by the fluvial, marine, and fluvio-marine environments. The nine sets of Chenier dune ridges and swales were formed between 7000 and 50 YBP across the deltaplain. All these dune ridges are analogous to the existence of progressive shorelines towards the sea. The second, fourth, and ninth sets of dune ridges are absent in the western side of the deltaplain, as are the 5 and 5a shorelines, revealing the western deltaplain bulge at a minimum rate due to the lower volume of sediment deposited in that side in comparison to the eastern side of the deltaplain. The westward shifting of the Subarnarekha River near Chaumukh section was directed by the strong northeastern littoral drift, which again shifted the eastern side with the strengthening of the southwestern monsoon.

References

Allison, M. A., Khan, S. R., Goodbred, S. L., Jr., & Kuehl, S. A. (2003). Stratigraphic evolution of the late Holocene Ganges–Brahmaputra lower delta plain. *Sedimentary Geology, 155*(3–4), 317–342.

Banerji, U. S., Arulbalaji, P., & Padmalal, D. (2020). Holocene climate variability and Indian summer monsoon: An overview. *The Holocene*. https://doi.org/10.1177/0959683619895577

Barman, N. K., Chatterjee, S., & Paul, A. K. (2016). *Coastal morphodynamics: Integrated spatial modeling on the deltaic Balasore coast, India*. Springer.

Biswas, A. (1987). Laterites and lateritoids of Bengal. In V. S. Datye, J. Diddee, S. R. Jog, & C. Patil (Eds.), *Exploration in the tropics* (pp. 157–167). Prof. K.R. Dikshit Felicitation Committee.

Biswas, S. K. (1993). Major neotectonic events during the Quaternary in Krishna-Godavari basin. *Current Science India, 64*(11–12), 797–803.

Ghosh, S., & Guchhait, S. (2015). Characterization and evolution of primary and secondary laterites in northwestern Bengal Basin, West Bengal, India. *Journal of Palaeogeography, 4*(2), 96–123.

Jana, S. (2017). *Assessment of hydro-geomorphological aspects of middle and lower courses of Subarnarekha River and its Deltaic Part*. Unpublished PhD thesis submitted to Vidyasagar University.

Jana, S. (2021). An automated approach in estimation and prediction of riverbank shifting for flood-prone middle-lower course of the Subarnarekha River, India. *International Journal of River Basin Management, 19*(3), 359–377. https://doi.org/10.1080/15715124.2019.1695259

Jana, S., & Paul, A. K. (2018). Genetical classification of deltaic and non-deltaic sequences of landforms of Subarnarekha middle course and lower course sections in Odisha and parts of West Bengal with application of geospatial technology. *Journal of Coastal Science, 5*(1), 16–26.

Jana, S., & Paul, A. K. (2019). Assessment of morphogenetic sedimentary depositional environments of different morphological surfaces of middle-lower and deltaic courses of Subarnarekha River. *Journal of Coastal Science, 6*(1), 1–11.

Jana, S., & Paul, A. K. (2020). Chronological evolution of the channel functional units in association with palaeo-hydrogeomorphological environment in the ancient delta fan of Subarnarekha basin, India. *Environmental Earth Sciences, 9*, 331. https://doi.org/10.1007/s12665-020-09093-1

Jana, S., & Paul, A. K. (2022). Plan shape geomorphology of Alluvial Valley in the middle-lower and deltaic courses of the Subarnarekha River basin, India. In Shit et al. (Eds.), *Drainage basin dynamics* (pp. 63–87). Springer Nature. https://doi.org/10.1007/978-3-030-79634-1_3

Jana, S., Paul, A. K., & Islam, S. M. (2014). Morphodynamics of barrier spits and tidal inlets of Subarnarekha Delta: A study at Talsari-Subarnapur spit, Odisha, India. *Indian Journal of Geography and Environment, 13*, 23–32.

Kaliraj, S., Chandrasekar, N., & Magesh, N. S. (2014). Impacts of wave energy and littoral currents on shoreline erosion/accretion along the south-west coast of Kanyakumari, Tamil Nadu using DSAS and geospatial technology. *Environment and Earth Science, 71*(10), 4523–4542.

Kamila, A., Paul, A. K., & Bandyopadhyay, J. (2021). Exploration of chronological development of coastal landscape: A review on geological and geomorphological history of Subarnarekha chenier delta region, West Bengal, India. *Regional Studies in Marine Science, 1*(44), 101726.

Kunte, P. D., & Wagle, B. G. (2005). The beach ridges of India: A review. *Journal of Coastal Research, 1*, 174–183.

Maiti, S. (2013). Interpretation of coastal morphodynamics of Subarnarekha estuary using integrated cartographic and field techniques. *Current Science India, 104*(12), 1709–1714.

Mallick, S., Bhattacharya, A., & Niyogi, D. (1972). A comparative study of the Quaternary formations in the Baitarani valley, Orissa with those of the Damodar-Ajoy delta area, Lower Ganga Basin. In *Proceeding of seminar 'Geomorphology, geohydrology and geotectonic of the Lower Ganga Basin'* (A91–A104).

Mandal, U., Sena, D. R., Dhar, A., Panda, S. N., Adhikary, P. P., & Mishra, P. K. (2021). Assessment of climate change and its impact on hydrological regimes and biomass yield of a tropical river basin. *Ecological Indicators, 126*, 107646. https://doi.org/10.1016/j.ecolind.2021.107646

Mokarrama, M., & Hojati, M. (2018). Landform classification using a subpixel spatial attraction model to increase spatial resolution of digital elevation model (DEM). *Egyptian Journal of Remote Sensing and Space Science, 21*(1), 111–120.

Niyogi, D. (1968). Morphology and evolution of the Subarnarekha Delta, India. *Tridsskrift Saertrky af Geografisk, 67*, 230–241.

Patel, A., Katiyar, S. K., & Prasad, V. (2016). Performances evaluation of different open source DEM using Differential Global Positioning System (DGPS). *Egyptian Journal of Remote Sensing and Space Science, 19*(1), 7–16.

Paul, A. K. (1996a). Chenier beach ridge and chenier sand ridge formations around Subarnarekha estuary. *National Geographer, XXXi*(1&2), 143–153.

Paul, A. K. (1996b). Identification of coastal hazards in West Bengal and parts of Orissa. *Indian Journal of Geomorphology, I*(1) New Delhi, Academy & Law serials, 1–27.

Paul, A. K. (1997). Coastal erosion in West Bengal. *MAEER, MIT Pune Journal, IV*(15k &16k) (special issue on coastal environmental management), 66–84.

Paul, A. K. (2002). *Coastal geomorphology and environment* (pp. 1–342). ACB Publication.

Pavlis, N. K., Holmes, S. A., Kenyon, S. C., et al. (2012). The development and evaluation of the Earth Gravitational Model 2008 (EGM2008). *Journal of Geophysical Research: Solid Earth, 117*, B044406. https://doi.org/10.1029/2011JB008916

Sebastian, M., Behera, M. R., & Murty, P. L. N. (2019). Storm surge hydrodynamics at a concave coast due tovarying approach angles of cyclone. *Ocean Engineering, 191*. 106437 https://doi.org/10.1016/j.oceaneng.2019.106437

Chapter 4
Estimation of Seasonal Sediment Budget of Chandrabhaga Beach-Dune System, Bay of Bengal, India

Dipanjan Das Majumdar, Ashis Kumar Paul, and Barendra Purkait

4.1 Introduction

Sand dunes along the coast are sedimentary depositional features caused by wind activity moving sand inland from the shore. In terms of the sediment budget, the beach is the source and the dunes are the sink. The first and most important aspect in determining whether dunes can form is the availability of sand. Initial sediment supply has primarily been driven by sea level rise, and for many Holocene coastal dune fields throughout the world, post-glacial marine transgression has been the main driver. Sand supply was initially influenced by factors like shore face slope (Cowell et al., 1995), the availability of sediment to be reworked or redistributed on the shore face (Carter, 1988; Orford et al., 1991; Carter & Woodroffe, 1994; Orford et al., 2003), the local and regional supply from rivers, and wave energy once the sea levels reached the present (at different times on diverse coasts) (Short & Hesp, 1982). The second element that was essential for dune growth was wind energy over a minimum threshold velocity. The potential for dune growth increases with wind energy, especially along coasts where onshore winds are predominant, then, depending on the pioneer plants' capacity to colonize the backshore and sustain general stability even during times of storm erosion, rapid accretion/progradation, or not (Paul, 2002; Hesp &Walker, 2013).

In the backshore, foredunes are shore-parallel dune ridges developed by aeolian sand deposition within a vegetation colony (Hesp, 2002; Hesp &Walker, 2013).

D. D. Majumdar (✉)
Department of Geography, Netaji Satabarshiki Mahavidyalaya, Ashoknagar, West Bengal, India
e-mail: dipddm@gmail.com

A. K. Paul
Department of Geography, Vidyasagar University, Midnapore, West Bengal, India

B. Purkait
Department of Geology, University of Calcutta, Kolkata, West Bengal, India

A. K. Paul, A. Paul (eds.), *Crisis on the Coast and Hinterland*,
https://doi.org/10.1007/978-3-031-42231-7_4

They have been referred to as "incipient frontal dunes" and "embryo dunes" (while they are still developing). Sand may accumulate within distinct or relatively distinct vegetation clumps, or within individual plants, driftwood, etc., to build neo-dunes. In simple models of dune systems, the embryonic dunes with beach progradation grow in height collectively, eventually coalescing into a new foredune. The embryonic dunes are ephemeral landforms that develop over a period of months to years on stable and eroding shorelines before being swept away by wave action during strong storms. Embryo dunes typically range in height from a few centimetres to one or two meters and in width from one to two meters (Paul et al., 2023). In regions with strong onshore winds and abundant sand supplies, they may expand to over 100 m.

Foredunes store and cycle large amounts of sand in the backshore, making them important reservoirs of the coastal sediment budget in regions with abundant onshore sand sources (Hesp, 2002; Psuty, 2004; Walker & Hesp, 2013). Interactions between dunes and the beach are intricate, including a wide variety of processes and potential morphological responses across different timescales (Sherman & Bauer, 1993; Arens et al., 2001; Aagaard et al., 2004). The sediment budgets of both the beach and the frontal dunes have a significant impact on the frontal dune morphology of any given beach (Psuty, 1988, 1992). Therefore, if the beach has a negative sediment budget, embryonic dunes will not grow and the foredune will be eroded. In extreme cases, blowouts can cause the foredune to become disorganized, leading to the formation of transgression sand sheets, overwash fan deposits, and elongated dunes. In contrast, if the beach sediment budget is positive, new embryo dunes and foredune ridges will form, causing the frontal dunes to grow vertically or prograde seaward (Paul et al., 2023). Because of the complex interplay between topography, vegetation, and the aeolian processes that move the sand throughout the system, coastal dunes are among the most difficult landforms to study. Due to the system's complexity, the established process response models cannot foretell the integral pattern of change. In order to provide a full picture of the changes that have occurred, an alternate method is to take numerous measurements of the topography throughout time in relation to a fixed point or landmark. The coastal scientist adopting such an approach must decide whether to prioritize micro (metres to tens of meters)-, meso (tens of meters to hundreds of meters)-, or macro (hundreds of meters to thousands of meters)-scale methods (Andrews et al., 2002).

For instance, the study of beach-dune morphodynamics frequently involves the development of digital elevation models (DEMs) for sediment volumetric change research (Anthony et al., 2006; Gares et al., 2006). However, care must be taken in predicting and interpreting volumetric changes and related geomorphic responses using DEMs because of the uncertainties and inaccuracies associated with surveying and interpolation procedures that modify the precision and accuracy of DEMs. A variety of data acquisition methodologies, such as light detection and ranging (LiDAR), digital photogrammetry, topographic surveys using laser total stations, real-time kinematic GPS, are used in recent studies on dune morphological changes using DEMs (Woolard & Colby, 2002; Mitasova et al., 2005; Anthony et al., 2006; Reitz et al., 2010; Das, 2016). Repeat DEMs must be compared across time in order to quantitatively assess the seasonal influence on beach-dune volumetric change. The main objective of this chapter is to precisely measure and describe the sediment volume and resulting geomorphic responses within the Chandrabhaga beach-dune system (CB) in the Puri district of Odisha, Bay of Bengal, India. The work specifically uses statistical change detection methods to predict seasonal sand volume variations and characterize associated geomorphic reactions using high-resolution DEMs collected from seasonal laser total station surveys. As a result, this research offers a preliminary evaluation of the annual beach-dune sediment budget while taking into account the seasonal effects (pre-monsoon, monsoon, and post-monsoon) on the current process response system.

4.2 Study Area

Odisha's 480 km of coastline, which juts out into the northwest Bay of Bengal, is a prograding and depositional environment with six major estuaries, the second-largest mangrove belt in India, Asia's largest brackish water lagoon (Chilika Lake), enormous non-vegetated wetlands, and one of the world's longest sandy beaches rich in heavy minerals. Natural resources and ecosystems along the shore, particularly beaches and sand dunes, have been under more pressure over the past 20 years. For this study, Chandrabhaga Beach (CB) in Konark was chosen (Fig. 4.1). The location was chosen for its accessibility, low population density, and distinctive combination of beach, berm, spit, and foredune. Chandrabhaga Beach (CB) is situated near the abandoned confluence of the Chandrabhaga River. High beach gradients, a sizable berm surface with sporadic neo-dunes, and a large foredune body that runs parallel to the shoreline are all features of CB. The climate is moderate yet extremely humid and sticky in the seaside region of Puri. The average low temperature in the winter is 17 °C, while the summertime maximum temperature is 35 to 40 °C. The climate of Puri is classified as Aw by Köppen and Geiger (dry winter and wet summer). Approximately 78% of the annual rainfall in Odisha falls between

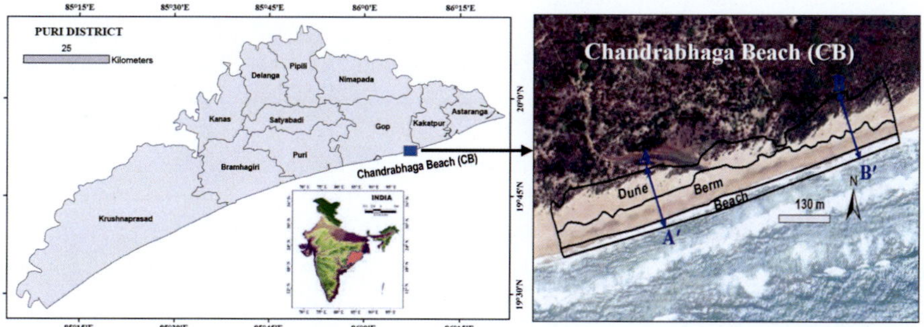

Fig. 4.1 Location of Chandrabhaga Beach (CB) segmented into distinct geomorphic units are shown on a map of the study area together with the beach-dune profile lines; A-A′ and B-B′

June and September, with the remaining 22% falling throughout the year. Monsoon depressions during the southwest monsoon season (June to September) and cyclonic storms primarily during the post-monsoon (October to November) and sporadically during the pre-monsoon (March to May) period are the main contributors to rainfall in coastal regions.

The tidal patterns along the Odisha coast are of the "mixed type," which is primarily semidiurnal in nature. The average spring tidal range is 2.39 m, and the neap tidal range is about 0.85 m. The Odisha coast is wave-dominated during the monsoon season, while it is mixed wave and tide dominated during the non-monsoon period. The average significant wave height along the Odisha coast is between 1.25 and 1.40 m, primarily falling between June and December and rising between January and May. However, during the southwest monsoon, winds produce strong waves of at least 3 m that impact the shore at an angle and cause the long shore drift, which yearly transports around 1.5 million cubic meters of sand from the southwest to the northeast along the nearshore region (Ramesh et al., 2011). Previous data indicates that the Odisha coast has been affected by a number of tropical cyclones. Cyclones that hit the Odisha coast between 1877 and 1987 had erratic tracks and passed between the mouth of the Dhamra and Paradip. The Very Severe Cyclonic Storm Phailin was the most powerful tropical cyclone to make landfall in Odisha since the 1999 "Orissa Cyclone."

4.3 Data Collection and Methodology

In order to define a mesoscale pattern of beach-dune sediment budgets and their resulting morphology, this work uses dense elevation measurements, geostatistics, and DEM analysis in a GIS environment. Between March 2013 and January 2014, data for this study were collected. A thorough nested sampling procedure was used for point collection. Each point's X (easting), Y (northing), and Z (elevation) values were tallied in meters. Each survey at CB yielded 1000–1500 points on average, spaced 0.02 m^2 on average, or around one point every 5 m. In the backdune and fixed crestal regions of the sand dunes, benchmarks were set up. To improve the accuracy of the survey, at least two benchmarks were fixed. To ensure that the surveys could be compared to one another and to check for instrument and other random errors, the benchmarks were measured and referenced during each survey (Table 4.1, Fig 4.2a).

The next phase includes converting raw data into readable ASCII format, fixing mistakes discovered in the field such as incorrect data entry, aligning the measured elevations with the standard datum, etc. At this stage, we may evaluate the sampling procedure, interpolation models, grid cell size, and distribution of the elevation spikes and other artefacts by analysing the spatial distribution of a continuous surface (Table 4.1). Kriging is the most suitable and appropriate method of interpolation, according to previous studies and the verification done in this research (Andrews et al., 2002; Walker et al., 2013; Das, 2016). In order to create a spatially

Table 4.1 The seasons with the dates of the surveys, the surveyed area, and the number of surveyed points are shown for the CB dune site in columns 3 and 4, while columns 5, 6, and 7 display the error statistics related to the interpolation method that was used to generate the DEM

Dune Site	Seasons	Surveyed area in km^2	Number of points	R^2	Root mean square error (RMSE)	Average standard error (ASE)
Chandrabhaga Beach (CB)	Pre-monsoon (09/03/2013)	0.112	1486	0.98	0.21	0.42
	Monsoon (16/08/2013)		1467	0.97	0.25	0.45
	Post-monsoon (21/12/2013)		1440	0.97	0.26	0.47

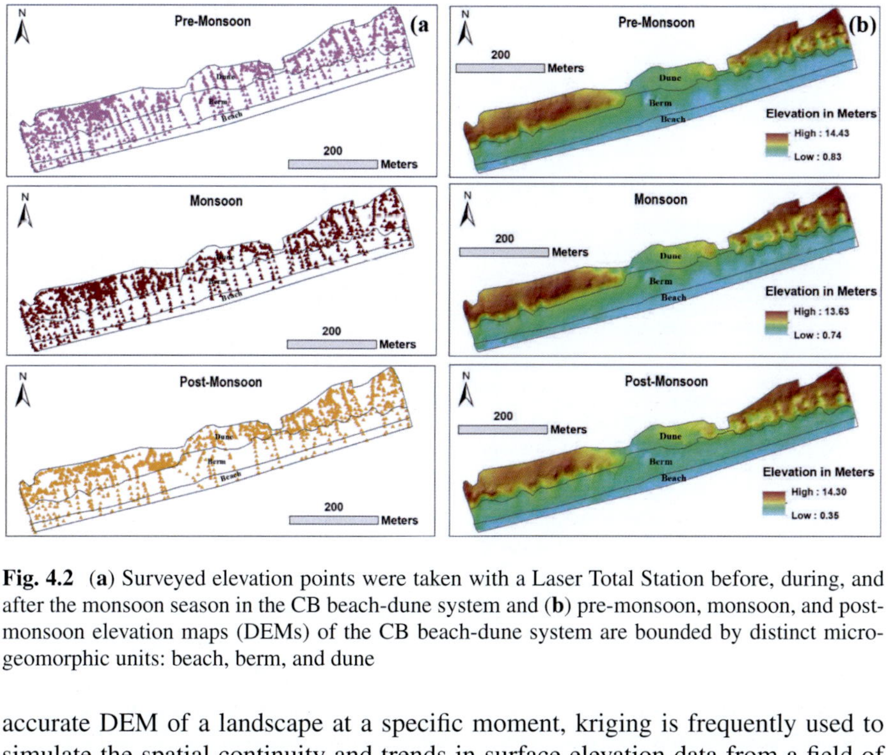

Fig. 4.2 (**a**) Surveyed elevation points were taken with a Laser Total Station before, during, and after the monsoon season in the CB beach-dune system and (**b**) pre-monsoon, monsoon, and post-monsoon elevation maps (DEMs) of the CB beach-dune system are bounded by distinct micro-geomorphic units: beach, berm, and dune

accurate DEM of a landscape at a specific moment, kriging is frequently used to simulate the spatial continuity and trends in surface elevation data from a field of spatially discontinuous point observations. Then, using a spherical model variogram that best captured the semi-variance in each geomorphic segment for each survey, DEMs were created at 1 m grid resolution (Fig. 4.2b). The Geostatistical Analyst toolkit in ArcGIS 10.1 was used for all geostatistical analyses and grid preparation.

In addition to the three-dimensional volumetric estimates and surface mapping from the DEMs, cross-shore profiles were derived from the interpolated DEMs along the fixed transects for the dune site (CB) to further visualize two-dimensional changes across the geomorphic units (beach, berm, and dune). During the initial field visit, various geomorphic units were primarily delimited using a hand-held GPS, and they were then further corrected with the aid of high-resolution Google Earth data. Finally, the boundaries of the geomorphic units were determined using the knowledge of the regional slope breaks that were discovered from the interpolated DEMs (Fig. 4.2a, b). To enable volumetric response comparisons within the same unit across time, these boundaries were kept constant throughout the study period (Das, 2016). The DoDs (DEM of Difference) were derived from the seasonal DEMs by simply subtracting, on a cell-by-cell basis, the elevations found in each DEM from the elevations found in the DEM derived from the survey conducted the following season (Fig. 4.3). The calculated elevation change was used as a depth measurement, and the surface area of each cell was multiplied by that value to arrive at an estimate of the net volumetric change (i.e., 1 m^2). Following the addition of these volumes, erosional and depositional categories were summed up to create a net volumetric budget for the area under investigation.

4.4 Results and Discussion

The annual survey data for CB show a dominant erosional regime in each section of the beach-dune profile (pre-monsoon, monsoon, and post-monsoon), as depicted by the net volume change in Table 4.2. The foredune ridge, which runs parallel to the beach, as well as any unplanted ramps or accompanying embryo dunes to its seaward, comprises the main dune system. A seaward-facing waterway that splits the foredune into two sections disturbs the continuity of the foredune (Figs. 4.1 and 4.3). Except during the highest high tide, the large berm that runs along the foredune prevents the tidal jet from connecting to the channel. The high amount of rainwater from the upstream portion during the monsoon is the only chance to become tethered to the sea. With the commencement of high energy events, the rate of wind deflation and sand transport to the inland increases since the crestal region of the foredune is sparsely vegetated. The moderate to high beach gradient, large berm with wave deposits (moderately coarse sand), and average 12 m high dune cliffs are all signs of an intermediate to reflective beach state. The elevation change map indicates that the beach, berm, and dune section experienced erosion during the monsoon, and the situation worsened during the post-monsoon, when the volume lost from the beach, berm, and dune was 5107 m^3, 10,005 m^3, and 16,700 m^3, respectively, in areas of 5015 m^2, 8330 m^2, and 12,463 m^2 (Fig. 4.3). The average depth of erosion and deposition in Table 4.2, where the average erosion depth is greater than the average deposition depth and indicates a net loss of sediment from the pre-monsoon to the monsoon to the post-monsoon, also shows the negative annual sediment budget in CB.

The profiles in Fig. 4.4 show two extreme situations occurring along the beach-dune portions. Along the transect A-A, the post-monsoon profile in the berm and dune segments experienced an average 1.5 m fall from the monsoon profile and an

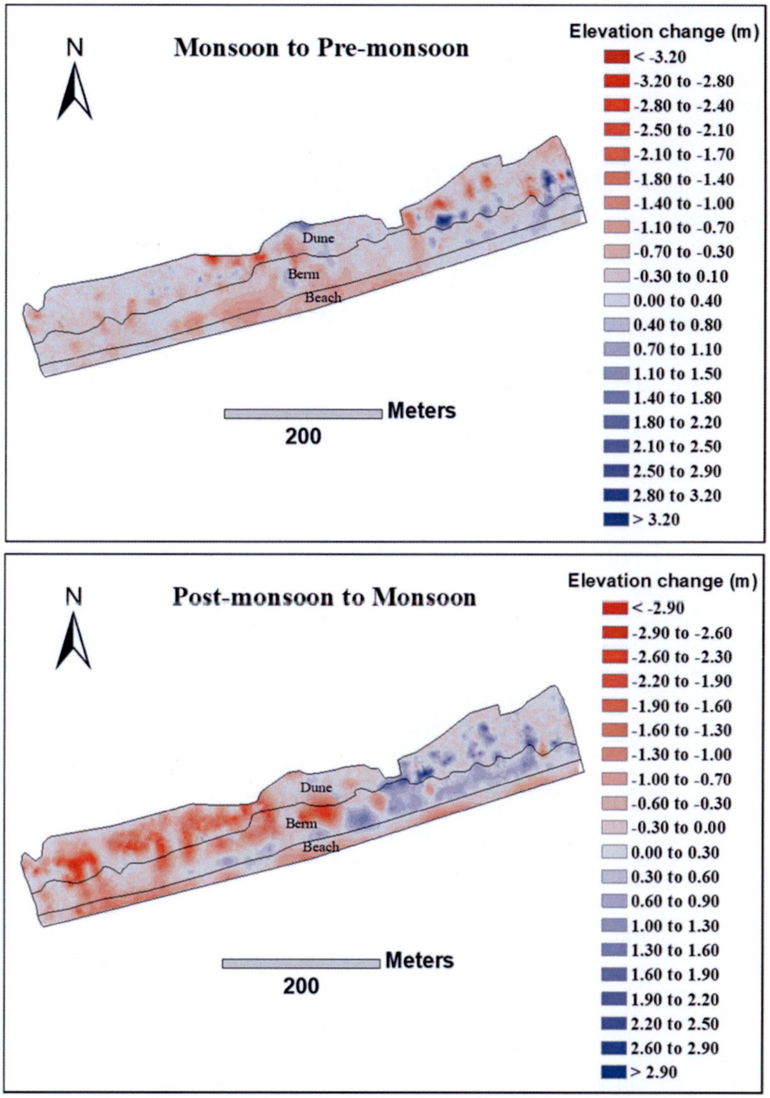

Fig. 4.3 The distinct micro-geomorphic units of beach, berm, and dune define the boundaries of DEM of Difference (DOD) maps from monsoon to pre-monsoon and post-monsoon to monsoon of the CB beach-dune system

average 2 m fall from the pre-monsoon profile, respectively, whereas in the transect B-B, the monsoon profile suffered from erosion and retreat across the dune segment, which was refilled during the post-monsoon, making the shape similar to that of the pre-monsoon stage. The landfall of the Phailin cyclone just as the post-monsoon season was about to begin disturbed the normal sediment circulation pattern seen in tropical coastal locations such as Puri. Each part suffered considerable

Table 4.2 For the CB beach-dune system, the DODs of monsoon to pre-monsoon and post-monsoon to monsoon were used to figure out the average summary statistics for volumetric, areal, and vertical averages

Dune site		Chandrabhaga Beach (CB)					
DEM of difference (DoDs)		Monsoon to pre-monsoon			Post-monsoon to monsoon		
Geomorphic units		Beach	Berm	Dune	Beach	Berm	Dune
Volumetric	Erosion in m^3	1242	2468	5875	5107	10,005	16,700
	Deposition in m^3	–	550	3848	18	4101	3489
	Net change in m^3	−1242	−1918	−2027	−5089	−5904	−13,211
Areal coverage	Erosion in m^2	1586	2827	4982	5015	8330	12,463
	Deposition in m^2	–	671	2801	24	4213	3040
Vertical average	Average depth of erosion (m)	0.78	0.87	1.18	1.02	1.2	1.34
	Average depth of deposition (m)	–	0.82	1.37	0.75	0.97	1.15
	Average net thickness difference (m) for area of interest	−0.07	−0.05	−0.04	−0.31	−0.14	−0.26
	Average net thickness difference (m) for area of detectable change	−0.78	−0.55	−0.26	−1.01	−0.47	−0.85

The volumetric metric shows erosion, deposition, and the overall change in volume. In the areal aspect, you can see the total area of erosion and the total area of deposition. The vertical average takes into account the average height of erosion and deposition as well as the average net thickness difference in terms of total area and area with detectable change

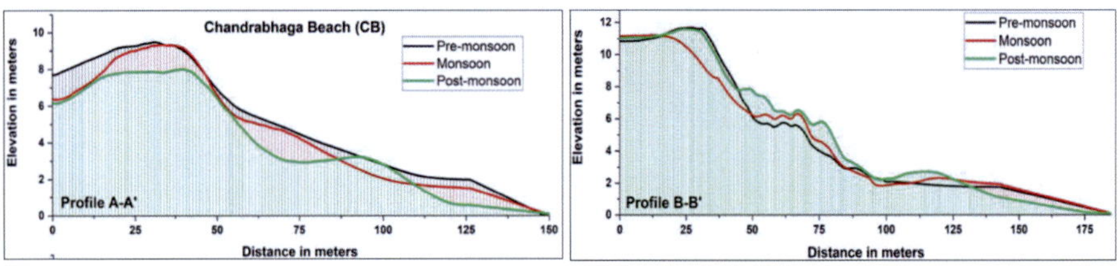

Fig. 4.4 The two-dimensional cross-shore beach-dune profile along the fixed transects (A-A′ and B-B′) shows the cycles of erosion and deposition in the studied area for the pre-monsoon, monsoon, and post-monsoon seasons

erosion as a result of the high intensity wave and wind regimes. The cyclonic wind threw a substantial amount of sand inland from the crestal part of the dune, particularly from the vegetation-free zones, and the sand produced through dune scarping was moved by the storm-induced longshore current, leaving the area appearing like a sediment-starved zone. The monsoon season exacerbated beach erosion, but this loss was offset by an increase in the average depth of sediment thickness on the dune surface as a result of processes that exchanged sand between the beach and the dunes (Fig. 4.4). During storms, narrow beaches can be replenished by foredune sand, in this way retaining their shape, whereas wider beaches can protect the

Fig. 4.5 (**a**) Formation of a foredune ridge parallel to the CB coastline. (**b**) Sand dunes migrating inland in the study area. (**c**) Numerous neo-dune formations on the berm surface, as well as blowouts in the foredune ridge running parallel to the beach. (**d**) The broad crestal section of the foredune demonstrates that the beach-dune sand-sharing system has a substantial sediment sink

foredunes at the expense of their sand budget. It is also critical that the beach and foredune budgets are not time-synchronized. Sand accretion on a beach occurs when a sandbar welds to the beach during calm weather conditions. Meanwhile, foredune accretion occurs during a storm season. Due to alongshore sediment transport, this process may also interact with nearby coastal sectors.

The foredunes in the study area range in height from very low, scattered dunes of less than a meter to large dune complexes of 12–15 m (Fig. 4.5). The low, scattered dunes occur on some barrier islands dominated by overwash and in areas of limited sediment supply. The highest dunes are rare and tend to occur on erosional coasts, where human interference acts to artificially induce greater foredune heights, for example, by maintaining absolute lee slope stability through plantation. Established foredunes are developed from incipient foredunes and are commonly distinguished by their greater morphological complexity, height, width, age, and geographical position. In the present study area, the species that initiate the formation of incipient foredunes also dominate the established foredunes (Fig. 4.5) (Hesp, 2002). Thus, morphological complexity, height, volume, and geographical position are the prime parameters to distinguish the incipient from the established foredune. Beach width, sediment supply, and wind velocity are the three main factors that control or influence foredune development (Davidson-Arnott & Law, 1996). The availability of sediment on the backshore is dependent on the deposition of beach or dune sediment during storms when water levels tend to be elevated above normal tidal heights (Houser & Ellis, 2013). The transport of this sediment is dependent on the ability of onshore winds to entrain and transport more sediment from the beach to the dune than is lost from the dune during storms that erode the dune. Ultimately, the amount of sediment transported across the backshore from beach to dunes depends on the adjustment of the wind to the beach morphology and size (fetch), the limits of transport, and the presence of vegetation to capture the sediment. As the vegetation density or cover decreases, sand is transported further up along the stoss faces, leading to an increase in the steepness of the CB dune system. High wind speeds combine with topographic acceleration over dunes to suspend grains and transport sediment across dunes to inland areas (Davidson-Arnott & Law, 1996; Gares, 1992). The stronger winds are accompanied by storm surges that not only limit the fetch length but also erode the beach face and scar the dune.

The shape of the beach-dune profile (Fig. 4.4) in the study area is strongly controlled by the magnitude and frequency of occurrence of high-water levels (Ruz & Meur-Ferec, 2004), the duration that storm waves impact the beach face (Van Gent et al., 2008), and the pre-storm beach width (Thornton et al., 2007). The sediment deposited on the beach and the berm can be quickly eroded during a subsequent storm despite a strong transport potential associated with the onshore winds (Nordstrom & Jackson, 1993). Even if the storm surge has just exceeded the berm elevation, dune recession can occur through the loss of stability at the base of the dune and mass failure (Erikson et al., 2007). In this respect, the storm duration may have a direct influence on the amount of dune erosion since a short storm does not provide sufficient time for the swash to impact the dune. As a result, the rate of erosion decays exponentially during a storm as the beach and nearshore establish a new equilibrium with the dune sediment. Intra-site variability in dune morphology along the studied dune sites is controlled by shoreline orientation relative to the prevailing wind, leading to the development of foredunes with varying sizes, the formation of blowouts, and storm-induced transgressive dunes along the coastlines of CB (Fig. 4.5).

Digital elevation models (DEMs) are used in this study to detect and measure geomorphic and volumetric alteration in beach-dune landscapes. However, due to variations in DEM creation and error handling, comparing data from such studies can be challenging (Andrews et al., 2002; Mitasova et al., 2005; Wheaton et al., 2010). This study offers a repeatable method for identifying significant geomorphic and sediment volume changes in dune environments using open-source software and standard surveying tools like the Laser Total Station (Fig. 4.2a). The directionality in the morphology and/or sand transport pathways that are frequently observed within distinct beach-dune landscape units is used in this study. This enables the creation of more accurate and repeatable DEMs from the floating-point cloud data that is frequently found in field surveys when used in conjunction with the Kriging interpolation method (Figs. 4.2b and 4.3). The present study highlights the seasonal variations of beach-dune sediment budgets and the resultant morphological responses within the broader context of the available theories on beach-dune interaction around the globe. The comparison of the data obtained from the seasonal field survey provides some insight into the behaviour of this particular dune system. These dune areas were predominantly eroding, with the total volumetric loss far exceeding the gain (Table 4.2). It is evident that coastal storms are the driving mechanisms of dune change along this shoreline. Despite observing the linear retreat of the dune, there are spatial variations in the dimension of elevation change. The areas of deposition within the overall area of erosion are related to the existence of blowouts through the foredunes (Fig. 4.5). The openings between the beach and backdune serve as transport corridors, facilitating the movement of sediment to areas landward of the foredune (Fig. 4.5).

4.5 Conclusion

This chapter outlined the use of an established, statistically based method to estimate sediment volume variations and geomorphic response patterns in the coastal dune site of Chandrabhaga Beach in Puri district. A geostatistical approach based on Kriging was used to model statistically significant changes within discrete, linked geomorphic units (beach, berm, foredune, and backdune) at the sub-landscape scale, which captures and incorporates inherent spatial structure in measured elevation datasets as controlled by distinctly different process-response dynamics. The method yielded low cross-validation error, a robust interpolation method accounting for DEM creation, statistically significant volumetric changes, and accurate surface change (erosion/deposition) maps. To generate additional survey points (X, Y, and Z) logistically defined over the beach-dune system for the use of directed variograms for DEM creation, the best potential sample routine was considered. Seasonal shifts in beach-dune volumetric states (monsoon to pre-monsoon and post-monsoon to monsoon) were well recorded by this study. During the pre-monsoon, monsoon, and post-monsoon seasons, the erosion on Chandrabhaga Beach (CB) increased unidirectionally. The loss of sand from the dune exceeded the overall losses suffered by the beach and berm during the post-monsoon due to the unexpected advent of the Phailin storm along the Odisha Coast. The current work visualizes coastal dunes in a far more comprehensible way than prior studies on Indian coastlines, which used simple two-dimensional profiles. Three-dimensional representations can help us better understand the complicated coastal features. This study demonstrates the stochastic three-dimensional spatial variability of the meso-scale sediment budget rather than a two-dimensional deterministic process defined by systematic observations along predetermined transects. The chapter concluded by emphasizing that the topographic data presented here merely indicates the inter-annual state of volumetric change. Long-term elevation change data with seasonal intervals is required for a better understanding of the effects of dominant morphodynamic variables on the beach-dune system in the study area as well as other parts of the sandy coastlines.

References

Aagaard, T., Davidson-Arnott, R., Greenwood, B., & Nielsen, J. (2004). Sediment supply from shoreface to dunes: Linking sediment transport measurements and long-term morphological evolution. *Geomorphology, 60*(1–2), 205–224. https://doi.org/10.1016/j.geomorph.2003.08.002

Andrews, B., Gares, P. A., & Colby, J. D. (2002). Techniques for GIS modeling of coastal dunes. *Geomorphology, 48*(1–3), 289–308. https://doi.org/10.1016/S0169-555X(02)00186-1

Anthony, E. J., Vanhee, S., & Ruz, M.-H. (2006). *Short-term beach–dune sand budgets on the north sea coast of France: Sand supply from shoreface to dunes, and the role of wind and fetch.*

Arens, S. M., Jungerius, P. D., & van der Meulen, F. (2001). *Coastal dunes* (pp. 229–272). Managing the Physical Environment.

Carter, R. W. G. (1988). Chapter eight – Coastal dunes. In R. W. G. Carter (Ed.), *Coastal environments* (pp. 301–333). Academic.

Carter, R. W. G., & Woodroffe, C. D. (1994). *Coastal evolution: Late quaternary shoreline Morphodynamics*. Cambridge University Press.

Cowell, P. J., Roy, P. S., & Jones, R. A. (1995). Simulation of large-scale coastal change using a morphological behaviour model. *Marine Geology, 126*(1–4), 45–61. https://doi.org/10.1016/0025-3227(95)00065-7

Das Majumdar, D. (2016). *Formation of coastal sand dunes their geomorphology ecology degradations and management A study at Orissa Coast India (Vidyasagar University)*. Retrieved from http://hdl.handle.net/10603/302982

Davidson-Arnott, R. G. D., & Law, M. N. (1996). Measurement and prediction of long-term sediment supply to coastal foredunes. *Journal of Coastal Research, 12*(3), 654–663.

Erikson, L. H., Larson, M., & Hanson, H. (2007). Laboratory investigation of beach scarp and dune recession due to notching and subsequent failure. *Marine Geology, 245*(1–4), 1–19. https://doi.org/10.1016/j.margeo.2007.04.006

Gares, P. A. (1992). Topographic changes associated with coastal dune blowouts at Island Beach State Park, New Jersey. *Earth Surface Processes & Landforms, 17*(6), 589–604.

Gares, P. A., Wang, Y., & White, S. A. (2006). Using LIDAR to monitor a beach nourishment project at Wrightsville Beach, North Carolina, USA. *Journal of Coastal Research*, 1206–1219.

Hesp, P. (2002). Foredunes and blowouts: Initiation, geomorphology and dynamics. *Geomorphology, 48*(1–3), 245–268. https://doi.org/10.1016/S0169-555X(02)00184-8

Hesp, P. A., & Walker, I. J. (2013). 11.17 Coastal Dunes. In J. F. Shroder (Ed.), *Treatise on geomorphology* (pp. 328–355). Academic Press.

Houser, C., & Ellis, J. (2013). 10.10 beach and dune interaction. In J. F. Shroder (Ed.), *Treatise on geomorphology* (pp. 267–288). Academic Press.

Mitasova, H., Overton, M., & Harmon, R. S. (2005). Geospatial analysis of a coastal sand dune field evolution: Jockey's Ridge, North Carolina. *Geomorphology, 72*(1–4), 204–221. https://doi.org/10.1016/j.geomorph.2005.06.001

Nordstrom, K. F., & Jackson, N. L. (1993). The role of wind direction in eolian transport on a narrow sandy beach. *Earth Surface Processes & Landforms, 18*(8), 675–685.

Orford, J., Carter, R., & Forbes, D. (1991). Gravel barrier migration and sea level rise: Some observations from Story Head, Nova Scotia, Canada. *Journal of Coastal Research, 7*, 477–489.

Orford, J. D., Murdy, J. M., & Wintle, A. G. (2003). Prograded Holocene beach ridges with superimposed dunes in north-east Ireland: Mechanisms and timescales of fine and coarse beach sediment decoupling and deposition. *Marine Geology, 194*(1–2), 47–64. https://doi.org/10.1016/S0025-3227(02)00698-9

Paul, A. K. (2002). *Coastal geomorphology and environment* (pp. 1–342). ACB Publication.

Paul, A. K., Paul, A., & Majumdar, D. D. (2023). Coastal sand dunes along the Western and eastern shores of India. In *Sand dunes of the northern hemisphere* (pp. 350–382). CRC Press.

Psuty, N. P. (1988). Sediment budget and dune/beach interaction. *Journal of Coastal Research, 3*(3 SPEC. ISSUE), 1–4.

Psuty, N. P. (1992). *Spatial variation in coastal foredune development* (pp. 3–13). Geomorphology, Ecology and Management for Conservation.

Psuty, N. P. (2004). *The coastal foredune: A morphological basis for regional coastal dune development* (pp. 11–27). Ecology and Conservation.

Ramesh, R., Purvaja, R., & Senthil Vel, A. (2011). *National Assessment of shoreline change* (p. 57). NCSCM, Ministry of Environment and Forests, Government of India.

Reitz, M. D., Jerolmack, D. J., Ewing, R. C., & Martin, R. L. (2010). Barchan-parabolic dune pattern transition from vegetation stability threshold. *Geophysical Research Letters, 37*(19). https://doi.org/10.1029/2010GL044957

Ruz, M.-H., & Meur-Ferec, C. (2004). Influence of high water levels on aeolian sand transport: Upper beach/dune evolution on a macrotidal coast, Wissant Bay, northern France. *Geomorphology, 60*(1–2), 73–87. https://doi.org/10.1016/j.geomorph.2003.07.011

Sherman, D. J., & Bauer, B. O. (1993). Dynamics of beach-dune systems. *Progress in Physical Geography, 17*(4), 413–447.

Short, A. D., & Hesp, P. A. (1982). Wave, beach and dune interactions in southeastern Australia. *Marine Geology, 48*(3–4), 259–284. https://doi.org/10.1016/0025-3227(82)90100-1

Thornton, E. B., MacMahan, J., & Sallenger, A. H., Jr. (2007). Rip currents, mega-cusps, and eroding dunes. *Marine Geology, 240*(1–4), 151–167. https://doi.org/10.1016/j.margeo.2007.02.018

Van Gent, M. R. A., van Thiel de Vries, J. S. M., Coeveld, E. M., de Vroeg, J. H., & van de Graaff, J. (2008). Large-scale dune erosion tests to study the influence of wave periods. *Coastal Engineering, 55*(12), 1041–1051. https://doi.org/10.1016/j.coastaleng.2008.04.003

Walker, I. J., & Hesp, P. A. (2013). 11.7 Fundamentals of aeolian sediment transport: Airflow over dunes. In J. F. Shroder (Ed.), *Treatise on geomorphology* (pp. 109–133). Academic Press.

Walker, I. J., Eamer, J. B. R., & Darke, I. B. (2013). Assessing significant geomorphic changes and effectiveness of dynamic restoration in a coastal dune ecosystem. Geomorphology, 199(0), 192–204. https://doi.org/10.1016/j.geomorph.2013.04.023.

Wheaton, J. M., Brasington, J., Darby, S. E., & Sear, D. A. (2010). Accounting for uncertainty in DEMs from repeat topographic surveys: Improved sediment budgets. *Earth Surface Processes and Landforms, 35*(2), 136–156. https://doi.org/10.1002/esp.1886

Woolard, J. W., & Colby, J. D. (2002). Spatial characterization, resolution, and volumetric change of coastal dunes using airborne LIDAR: Cape Hatteras, North Carolina. *Geomorphology, 48*(1–3), 269–287. https://doi.org/10.1016/S0169-555X(02)00185-X

Chapter 5
Evidence of an Ancient Coastline in the Southwest Sundarban, India, with Investigation of Geoarchaeological Remains

Suman Saren and Ashis Kumar Paul

5.1 Introduction

The Sundarban is the world's single largest mangrove forest, and it is a deltaic depositional ground produced by the Ganga River and Brahmaputra River jointly extending over the countries of Bangladesh and India (Paul, 2002, 2011). The bushy and muddy part of the present Sundarban is not conducive enough for human habitations, but sufficient evidence of ancient human settlement has been recovered from the entire region. The British rulers would have liked us to believe that people began to settle in the area only after they began to colonise the region and divided it into lots and plots in the early nineteenth century (Dutta, 2002). In other words, according to them, human habitation in the Sundarban does not have much antiquity. This research brings together a set of geoarchaeological studies in the diverse coastal zones of the study area, encompassing a deep history of human occupation from the terminal Pleistocene through the Holocene. The recent discovery of ancient artefacts deep in the heart of the Sundarban in West Bengal indicates that the region had human habitation as early as the third century BC and once again refutes the claim by colonial historians that it was the British who made the Sundarban habitable. From as early as the third century BCE to as late as the eleventh century AD, around 500 antiquities were found in an exploration carried out by the Directorate of Archaeology and Museums, Government of West Bengal, in Govardhanpur and its adjacent Uttar Surendraganj, near the mouth of the Ganga in the interiors of the Sundarban in South 24 Paraganas District, from as early as the third century BCE

S. Saren
Department of Geography, Swami Niswambalananda Girls' College,
Uttarpara, Hooghly, West Bengal, India

A. K. Paul (✉)
Department of Geography, Vidyasagar University, Midnapore, West Bengal, India
e-mail: akpauleastcoast@gmail.com

A. K. Paul, A. Paul (eds.), *Crisis on the Coast and Hinterland*,
https://doi.org/10.1007/978-3-031-42231-7_5

(Chattopadhyay, 2015; Bandopadhyay, 2002, Bhattyacharya & Mukherjee, 1977). These are evidence of continuous habitation in the region from the ancient period until the early mediaeval period. Semiprecious stones were found in this region that were not available in this part of the country. Thousands of stone beads and many potteries from early historic periods were found in the southwestern Sundarban region. Such beads were also obtained from the Tilipi excavation site (22°13′ 40 N, 88°33′ 55 E), which has been assigned to the Mourya period (Islam, 2008). Tilipi is located on the bank of the Piyali River, and the site was excavated by the Directorate of Archaeology and Museum, West Bengal (2005, 2006, and 2007). Pottery with blackware and redware has been reported from the area, with basket impressions and ornamental designs. Terracotta plaques depicting female figures in Sunga style, toy carts depicting elephants, divine figures, and a moulded figurine of an elephant were recovered (Sengupta et al., 2015). Walls of brick tiles, countless potteries, cast copper coins, and a seal engraved with pre-Bengali scripts were also discovered in the antiquities field of Mandirtala, bearing witness to a huge chapter of Bengal's silent history between the pre-Mourya period and the eleventh and twelfth centuries A.D. (Khanra, 2007). Punch-marked and cast copper coins are also important findings in the region. Bones and ivory objects, terracotta seals, beads of semi-precious stones, copper and iron objects, terracotta figurines, and plaques comprise a rich material culture in this phase (Roychowdhury, 2011).

The early historic period gave rise to several large settlements, mostly in the southern part of West Bengal (Lower Bengal), especially in the coastal areas. The term "Early Historic period" in Bengal can be assigned to the time period approximately between the middle of the first millennium BC and 500 AD (Sengupta et al., 2015). Urbanization in Bengal probably started in the Mauryan period (Chakrabarti et al., 1994; Chakraborty, 2001, Chattopadhyay et al., 2005, IAR, 1954). It is known after the discovery of such archaeological evidence that an advanced civilization was developed in this region, like in other provinces of Bengal (Dutta, 2002). Collected archaeological evidence indicates that Mourya and pre-Mourya cultures were developed in the entire Sundarban region, including Sagar Island (Chowdhury, 2009). From the discovery of early historic fossils, pottery, etc., in the creek and forest dominated modern Sundarban region, it is known that there is a footprint of a foreseeable past civilization in the geological formation of this country (Atul S, 2008 , Mandal, 2010). The history of human habitations in the part of the southwest Sundarban dates back to the early-historic period and, somewhere, the proto-historic period as well.

5.2 Study Area

The study area is located in the southwestern part of the district of South 24 Paraganas in West Bengal, India (Fig. 5.1). Actually, the southwest part of the Indian Sundarban consists of four distinct administrative blocks, viz., Sagar, Namkhana, Patharpratima, and Kakdwip. This part of the Sundarban is surrounded by the Hooghly River in the west, the Matla River in the east, and the Bay of Bengal in the

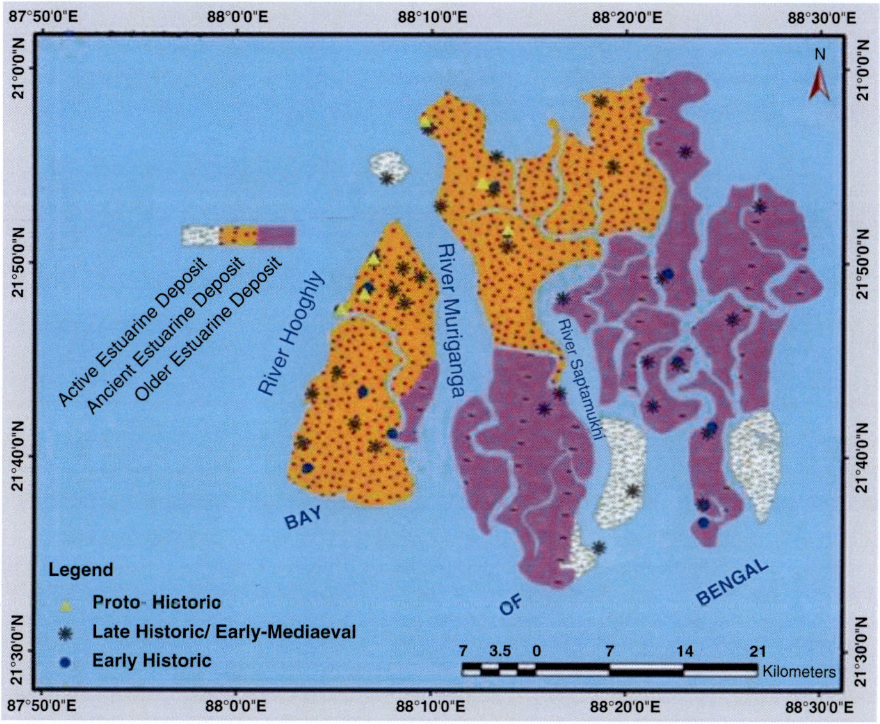

Fig. 5.1 The archaeological evidences of the part of south western Sundarban

south. The Sundarban region is a low-lying delta plain with an average elevation of 3–4 m in height above sea level. Therefore, tidal water often covers it with the rise and fall of water levels twice daily (Paul, 2002; Chattopadhyay et al., 2005).

5.3 Materials and Method

The present study depends on an extensive literature review for the archaeological identification of places in the study area. The information was used from different institutes of the IAR (Indian Archaeology: A Review), published by the Department of Archaeology, Government of India. In the pre-field stage of the research, a collection of historical documents in the form of charts, maps, primary literatures, and books were considered for primary mapping and measurements. An exhaustive literature survey on the geomorphology, palaeontology, and cultural history of the study area, the museum survey from different stations using the significant archaeological tools available from different localities, as well as databases were constructed for the present research. Fieldwork was conducted during the winters of 2012, 2013, and 2014 to record the present geographical situation of the archaeological sites and to track them with the GPS survey. In addition to the primary field survey, collecting archaeological information and taking photographs of archaeological evidences from different museums located in the study area were also very helpful for the study.

5.4 Results and Discussion

5.4.1 Archaeological Places of Southwestern Sundarban Coastal Plain

Part of the southwest Sundarban is significant from the perspective of archaeology because enough archaeological evidence has been discovered in different parts of the region (Table 5.1). The traces of ancient settlements were uncovered after the land reclamation in the British period of this swampy and forested tract of the Sundarban. Sri Kalidas Dutta deserves credit for discovering archaeological evidence for the first time in almost the entire Sundarban region. The colonial theory of the British that it was they who made the Sundarban habitable was first challenged in the 1920s by Kalidas Datta, an amateur archaeologist from Joynagar-Majilpur in South 24 Paragana district. Datta, a wealthy landowner in the area, began scouring the Sundarban for artefacts dating from the second century BCE to the ninth century CE. Datta was adamant about establishing a Bengali identity under the yoke of colonial rule (Basak, 2014). Datta's work threw new light on the region and refuted the colonial claims (Chattopadhyay, 2015) (Fig. 5.1). Other institutional and local researchers also focus on the archaeological evidence of these regions at different times (Chakraborty, 1998; Hunter 1998; Maity 2002; Middya 2002; Banerjee 2005; Haldar 2007; Islam 2008; Chowdhury, 2009; and Mandal, 2010). After the review of all the archaeological evidence that has been discovered from this region, it indicates the temporal expansion of this evidence is from the Mourya period to the Pala–Sena period (Middya, 2002; Khanra, 2007; Chowdhury, 2009; Mandal, 2010). Mandirtala is a significant archaeological site in the part of south western Sundarban located on the east bank of the Hooghly River. The discovery of Neolithic Celts in this region provides important evidence of human habitation during that time period (Middya, 2002; Bandopadhyay, 2002; Naskar, 2002; Chakraborty, 1998, Chattopadhyay, R.K. 2018). Apart from these, lots of small terracotta pots, thousands of stone beads, non-Aryan-Matrika or Mother Goddess sculptures, portraits of Hindu goddesses, ancient CORRIEs, silver and gold coins, and animals (deer) were found in the present study area (Fig. 5.2).

The handmade archaic mother goddess existed between the second and first centuries BC, as archaeological evidence demonstrates the antiquity of Mandirtala (Dutta, 2013). Available archaeological evidence of this place depicts the true story of the forgotten history of human civilizations from the Chalcolithic to Mediaeval periods. Except for Mandirtala, almost the same archaeological evidence was found

Table 5.1 The archaeological evidences of the part of South West Sundarban

Location	GPS records	Archaeological evidences	Archaeological periods	References
Mandirtala	21°48′22.015″N, 88°6′39.922″E	A Neolithic Shelt. Non-Aryan "Matrika" sculpture, sculpture of goddess Vishnu and Sourya, thousands of beads, potteries, terracotta vessels, corrie, Ganesha sculpture terracotta figure of first and second AD, seal engraved with pre-Bengali word, copper coin, gold coin	Proto-historic to early mediaeval	Naskar (2002), Middya (2002), Mandal (2010), Chakraborty (1998), Haldar (2007), and Chowdhury (2009)
Sapkhali	21°50′10.82″N, 88°7′4.491″E	Pre-historic Mahakala portrait, punch coin, stone weapon	Proto-historic to early mediaeval	Middya (2002)
Kochuberia	21°49′46.251″N, 88°8′28.728″E	Terracotta portraits, stylish pottery, remains of building most of them from Pala and Sena period	Late historic or early mediaeval	Chowdhury (2009)
Fuldubi	21°47′46.917″N, 88°5′19.196″E	So many beads, sculpture of terracotta Ganesha, sculpture of goddess Sourya (5 to sixth AD), sculpture of Mahishamardini (4 to sixth AD)	Proto-historic to early historic	Middya (2002) and Mandal (2010)
Bamankhali	21°48′26.707″N, 88°8′3.82″E	Brocken sculpture of goddess Vishnu (10 to 11th AD), sculpture Vishnu (8 to ninth AD), and Ganesha (eighth AD)	Late historic or early mediaeval	Mandal (2010)
Harinbari	21°44′23.345″N, 88°5′5.156″E	Terracotta potteries, different stylish potteries, most of them from Pala and Sena period	Late historic or early mediaeval	Chowdhury (2009)

(continued)

in Sapkhali, Kochuberia, Rudranagar, Mansadwip, Buraburirtat, and Gobordhanpur (Table 5.1). These are significant archaeological sites in the part of the southwest Sundarban, along the coast line of the Bay of Bengal (Fig. 5.2). Stone and bone weapons, ancient-Matrika sculpture, gold coins of the Gupta period, sculptures of

Table 5.1 (continued)

Location	GPS records	Archaeological evidences	Archaeological periods	References
Manasadwip	21°41′13.813″N, 88°7′53.629″E	Terracotta brick (size-12”x7x1 ¾”), terracotta potteries, remains of building most of them from Pala and Sena period	Late historic or early mediaeval	Maity (2002) and Chowdhury (2009)
Rudranagar	21°43′13.148″N, 88°6′39.922″E	Sculpture of terracotta “Yakshmini” numbers of pottery pot	Early historic	Maity (2002) and Chowdhury (2009)
Sagar	21°39′28.518″N, 88°3′23.37″E	Broken part of burn brick, sculpture of goddess Vishnu, pottery pots, bronch sculpture	Early historic	Maity (2002) and Chowdhury (2009)
Ghoramara	21°54′23.53″N, 88°7′32.57″E	Terracotta potteries, different stylish potteries, most of them from Pala and Sena period	Late historic or early mediaeval	Chowdhury (2009)
Radhakrishnapur	21°43′30.697″N, 88°3′40.92″E	Terracotta portraits, stylish pottery, remains of building, most of them from Pala and Sena period	Late historic or early mediaeval	Chowdhury (2009)
Kirtankhali	21°41′55.931″N, 88°5′50.784″E	Terracotta lady sculpture	Late historic or early mediaeval	Maity (2002) and Chowdhury (2009)
Kalimasthan	21°35′32.007″N, 88°18′41.221″E	So many baskets pottery	Late historic or early mediaeval	–
Lothian	21°38′15.931″N, 88°20′20.194″E	So many baskets pottery, remains of ancient building	Late historic or early mediaeval	Barman (2007) and Field Survey: Maity (2002)
Iswaripur	21°43′12.849″N, 88°16′25.134″E	Sculpture of terracotta “Yakshmini” numbers of pottery pot	Late historic or early mediaeval	Maity (2002) and Chowdhury (2009)
Dwarikanagar	21°42′32.642″N, 88°15′51.112″E	Different stone sculpture	Late historic or early mediaeval	Barman (2002) and Maity (2002)

(continued)

Yakshas and Yakshini, and a portrait of the Hindu goddess on a terracotta plaque were found in the aforementioned places (Maity, 2002; Barman, 2002, 2007; Sahu, 2007; Chowdhury, 2009, Saren et al., 2016). Rakshaskhali is another notable

Table 5.1 (continued)

Location	GPS records	Archaeological evidences	Archaeological periods	References
Pakurtala	21°53′37.615″N, 88°13′10.281″E	Ancient sculpture of goddess Bishnu (7th to 89th AD), Ganesha sculpture (ninth to tenth AD), ancient coins, basket pottery, cast copper coin	Proto-historic to early mediaeval	Mandal (2010), Haldar (2007), and Middya (2002)
Pukurberia	21°50′53.692″N, 88°13′53.582″E	Terracotta sculpture, portrait of "Banabibi"	Proto-historic to early mediaeval	Chowdhury (2009)
Lat no-6	21°57′4.84″N, 88°9′46.15″E	Terracotta "Matrika" sculpture, sculpture with Roman and Greeks style	Proto-historic to early mediaeval	Chowdhury (2009)

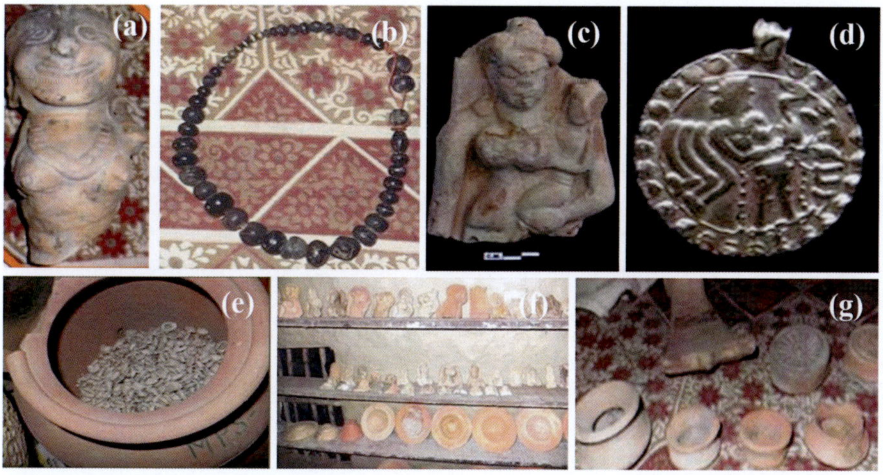

Fig. 5.2 (**a**) Handmade Archaic Mother-Goddess, (**b**) stone beads of the mother goddess, (**c**) description of the mother goddess, (**d**) sculpture of Yakshas|| and Yakshini||. (**e**) Ancient CORRIE of Hindu goddess, (**f**) gold coins of Gupta and Kushana periods, beautiful terracotta dolls, and (**g**) pots of mother goddess. (From Bharatmata Sevasadan Museum, Mandirtala, South 24 Paragana, an amateur archaeologist)

archaeological site, from which a copper plate from 1196 A.D. of the king Domman Paul was found. Except for the mentioned places, other important sites are Brajaballavpur, Banashyamnagar, Pakurtala, Lat No. 6, etc. From the available archaeological evidence from the entire South West Sundarban, it can be understood that the surface of this region was suitable enough for human habitation a few 1000 years ago. In particular, the discovery of stone and bone weapons extends the history of human settlement in this region up to the Neolithic period. From these relics, there is no doubt about the antiquity and past human settlement of the Sundarban region (Fig. 5.2).

5.4.2 Geographical Distribution of Archaeological Sites of Sundarban Coastal Plain in Relation to Local Environment

Fluvial regimes in deltaic settings are known for frequently changing erosional and depositional locations caused by changing hydrological and sedimentation patterns. These archaeological records preserved in alluvial settings are dramatically affected by temporal changes in fluvial processes; moreover, cultural adaptation also depends on the nature of fluvial terrain (Rajaguru et al., 2011). Humans are elements of their physical environment, and the settlement pattern of the people is governed by the morphogenic evolution of the landscape from the early times. The location of any archaeological site is deeply rooted in the local geography of any region because the history of a country is inseparably connected with its geography (Bhattyacharya & Mukherjee, 1977).

In the absence of inscriptions and epigraphic records of the history of the Sundarban between the fourth century A.D. and the twelfth century A.D., researchers are mostly dependent on archaeological findings. According to many early scholars, the temporal expansions of the available archaeological evidence of the South West Sundarban are the Neolithic-Chalcolithic to the Pal-Sena periods (Table 5.1). However, the amount of Neolithic evidence is considerably lower, which is not enough to prove the existence of settlement in that period. The archaeological evidence of the proto-historic to early mediaeval period was mainly found in the northern segment of the South Western Sundarban (Fig. 5.2). The majority of the archaeological evidence found in the rest areas, on the other hand, dates from the early historic period to the late mediaeval period, and most of the ancient evidence is located in the northern section (Fig. 5.2). Following a comparative analysis of both, it is possible to investigate the inherent relationship between historical or archaeological elements and regional geographical factors. Three distinct morphogenic surfaces can be identified after the study of the geological map of the southwestern Sundarban, viz., ancient, older, and active estuarine deposits (Fig. 5.1). The ancient estuarine deposit surface is constituted by an interdistributary supratidal flat with a soil profile underlain by silty clay interbanded with sand and silty horizons, and the absolute age determined by C14 technology is 2900 ± 20 YBP (Ray, 1997; Chattopadhyay et al., 2005). It is very interesting that the majority of the earliest archaeological sites in the southwestern Sundarban, such as Mandirtala, Sapkhali, and Pakurtala, are located on the surface of ancient estuarine deposits (Fig. 5.2). Ancient evidences like Neolithic Celt, non-Aryan Matrika (Mother Goddess) sculptures, CORRIE, and thousands of stone beads were found in these parts. The lithological configuration of old estuarine deposits is an intertidal, distributary, and supratidal flat underlain by a greyish black clay blanket of sandy layers of different tidal epochs (Chattopadhyay et al., 2005). This is an upper Holocene surface, and the archaeological evidence recovered from it dates from the early historic to the early mediaeval periods (Ray, 1997). Notable archaeological sites in this part are Gobordhanpur, Buraburirtat, north Surendraganj, and Rakshaskhali (Table 5.1). From the aforementioned discussion, it is clear that most of the ancient evidence was found on the earliest geological surface of the southwestern Sundarban. This means that there is a harmonic relationship between the geological oldness and the archaeological antiquity of this region. The southern part's average elevation is only 3 m, and the archaeological evidence is relatively recent. It is seen that there is a deep relationship between the most elevated surface and the most ancient archaeological sites.

Since ancient times, the evolution and extension of natural landscapes have governed the development and diversity of human civilizations. An adverse environment acts as a barrier to the development processes of human civilizations. The unearthed archaeological evidence of the South Western Sundarban continues to carry the memory of the pre-Christian era to the Pal-Sena era (Chowdhury, 2009; Atul S., 2008; Khanra, 2007). Available archaeological evidence from the entire Sundarban can be divided into three major historical phases, namely: the earliest is "proto-historic," and the other two are "early historic" and "early mediaeval." Various proto-historic and early historic evidences have been discovered in places such as Mandirtala, Sapkhali, Gobordhanpur, and Pakurtala, and among others, part of the South West Sundarban is encircled by the Bay of Bengal in the south and the Hooghly River in the east. The availability of ancient "corrie," copper coins, and thousands of stone beads from early historic periods from places like Mandirtala, Sapkhali provides strong evidence in favour of a culturally and economically prosperous trading civilization. Archaeological evidences like gold coins of the Gupta and Kushana periods, beautiful terracotta dolls, and sculptures of "Yaksha" and "Yakshini" were found in Gobordhanpur, Buraburirtat, and Surandraganj near the Bay of Bengal coast, a glorious city of early historic times. For the past 25 years, they (Biswajit and Bimal Sahu, two fishermen of the Gobardhanpur) have been collecting and preserving these artefacts, that is, potsherds, terracotta figurines, and other items they find on their fishing trips or during their expeditions to the deep forest around them. The coins, in their own time, faithfully serve the purpose, for which they were made, but beyond that, they retain their value and importance (Fig. 5.2). They provide an almost unparalleled series of historical documents (Gupta, 2013). Substantial archaeological evidence was found below the high tide line, including the remains of an ancient temple on the bank of the Hooghly River at Mandirtala.

Countless archaeological remains were found in Gobordhanpur and Buraburirtat during low tide along the Bay of Bengal coast. The availability of such archaeological evidence in a mangrove dominated estuarine environment is enough to cause surprise. Apart from the discovery of archaeological evidence, what various literatures say about this region is very significant; for example, the author of "Padmapurana" writes, "once upon a time the kings of Chandra dynasty reign over the Sagar Island." The Gangasagar region had been mentioned as Rasatala in the Ramayana (Haldar, 2007). It is understood from the above-mentioned facts that the southern part of the southwest Sundarban was extended a few kilometres towards the sea before a 1000 years ago. Sarkar and Sen (2014) discovered a clay sample at a depth of 0.90 m on these surfaces from Gangasagar at (C14) 3076.52 YBP. When the low tide line dropped below 6–7 m, then the archaeological evidence was explored in the Mandirtala region, and the maximum temporal extension of this evidence is the proto-historic period, which means 3000–4000 years earlier. It is clearly understood from previous information that there is a positive relationship between old geology and the antiquities of human settlement in the southwestern Sundarban Region. Early records of human habitation were found in the northern and northwestern parts of the study area along the ancient estuarine geological surface. The archaeological evidence of the proto-historic and early-historic periods was found primarily in the northern segment of the study area; on the other hand, the majority of the archaeological evidence of the rest areas dates from the early historic period to the late mediaeval period.

5.5 Conclusion

Overall, the findings of the present research prove that the natural landscape and cultural history of the Sundarban deltaic surface have changed over the geological period. Over a geological timescale and archaeological time period, the ancient people occupied the early landscapes of the changing environment to maintain their traditional livelihood as well as shift and adjust with the modified occupational ground from the northwest to the south and southwest. The southwest Sundarban delta plain surface was separated into many island units by larger tidal rivers. However, with the gradual accretion of fluvio-tidal sediments, the islands were agglomerated, and many tidal rivers were modified into narrow channels. The people of the proto-historic and early historic periods made their habitations in the northern part of the southwest Sundarban region about 4000 years ago, and they gradually shifted towards the southern part of the study area with the advancement and maturity of the deltaic surface area.

References

Atul, S. (2008). *"Bangalir Nritatwik Porichay" (in Bengali) an entho-cultural history of Bengal* (pp. 75–77). Sahityalok.

Bandopadhyay, A. (2002). *Ganga port (In Bengali), memorandum of archaeo-history convension, Baruipur, south 24 Paragana*.

Banerjee, M. (2005). *Archaeology of Sundarban (In Bengali)*. In Sreekhanda sundarban.

Barman, S. K. (2002). Archaeological places of Kakdwip sub division (in Bengali). In K. Mandal (Ed.), *Memorandum, South Twenty Four paragana archaeo history convention*. South Twenty Four Paragana.

Basak, B. (2014). Interpreting historical archaeology of coastal Bengal: Possibilities and limitations. In *The complex heritage of early India, essays in memory of RS Sharma* (pp. 153–178).

Bhattyacharya, A., & Mukherjee, B. N. (1977). *Historical geography of ancient and early Medival Bengal, Sanskriti Patra*. Vol. IV, June, 2004.

Chakrabarti, D. K., Goswami, N., & Chattopadhyay, R. K. (1994). Archaeology of coastal West Bengal: Twenty-four Parganas and Midnapur districts. *South Asian Studies, 10*(1), 135–160.

Chakraborty, D. K. (1998). *The issues in East Indian archaeology* (p. 151). Munshiram Manoharlal.

Chakraborty DK (2001). *Archaeological geography of the ganga plain: The lower and the Middle Ganga*. Permanent Black/Distributed by Orient Longman. ISBN- 8178240165 9788178240169, https://www.worldcat.org/title/archaeologicalgeography-of-the-ganga-plain-the-lower-and-the-middle-ganga/oclc/49750346

Chattopadhyay, R. K. (2018). *The Archaeology of Coastal Bengal*, Oxford Scholarship Publication, ISBN-13:9780193481682.

Chattopadhyaya, S. S. (2015). *Settlement of history*. FRONTLINE. http://www.frontline.in/artsandculture/heritage/settlementofhistory/article5486821.ece

Chattopadhyaya, B. D., Sengupta, G., & Chakraborty, S. (2005). *An annotated archaeological atlas of West Bengal, Vol. I, prehistory and protohistory*. Centre for Archaeological studies and Training.

Chowdhury, K. (2009). *24 Parganas, Uttar-Dakshin* (In Bengali), A cultural History of 24 Parganas, Subhas Chandra Dey's Publishing House (pp. 24–28).

Dutta K (2002). Dakshin-Paschim Sundarbaner Pratnasampad (in Bengali). In *Archaeological Resource of southwest Sundarban. "Smaranika, Dakshin 24 Pargana Pratna-Itihas Sammelan"* p-220 (Memorandum, South Twenty Four Paragana Archaeo- history Convention) Mandal, K. (Ed), Baruipur, South Twenty Four Parganas.

Dutta, A. (2013). *The Cultural Significance of Early Historic Terracotta Art of West Bengal: An Ethnoarchaeological Approach*. A unpublished thesis submitted to Deccan College Post-Graduate and Research Institute. http://hdl.handle.net/10603/27148

Gupta, P. L. (2013). *Coins*. National Book Trust (1st Ed., p. 301) ISBN:81-237-1887-0. https://indianculture.gov.in/ebooks/coins

Haldar N (2007). *Archaeology of Kakdwip sub-division* (in Bengali), Memorandum, south 24 Paragana Archaeo-History Convention, Board of Editors, Pratna Itihas Charcha Samiti, 2007.

Hunter, W. W. (1998). *A statistical account of Bengal, Vol. I, Part-II: Sundarbans*. Government of West Bengal (Original work published 1875).

IAR. (1954). *Indian archaeology: A review*. Annual Publication of the Archaeological Survey of India, New Delhi. 55: 1973–1974.

Khanra A (2007). Searching of Gemstone in a archaeo – island (in Bengali). In *Memorandum, south 24 Paragana Archaeo-History Convention*, Board of Editors, Pratna Itihas Charcha Samiti.

Maity, J. (2002). Archaeological evidences of Sagar Island, (in Bengali). In K. Mandal (Ed.), *Memorandum, south 24 Paragana Archaeo-. History convention*. South Twenty Four Paragana.

Mandal, K. (2010). *Some portrait – Sculpture of Sundarban, (in Bengali)*. Nabachalantika.

Middya, D. (2002). The south 24 Paraganas, from the light of archaeology, (in Bengali). In K. Mandal (Ed.), *Memorandum, south 24 Paragana Archaeo history convention*. South Twenty Four Paragana.

Naskar, S. K. (2002). Archaeological resource of south 24 Paragana. In K. Mandal (Ed.), *Memorandum, south 24 Paragana Archaeo-history convention*. South Twenty Four 1Paragana.

Paul, A. K. (2002). *Coastal geomorphology and environment* (pp. 1–342). ACB Publication.

Paul, A. K. (2011). Man-environment interaction in the single largest mangrove along the shoreline of the bay of Bengal: A case study of Sundarban. In Sharma & Kale (Ed.), *Geomorphology of India* (pp. 263–286). Prayag pustak Bhawan.

Rajaguru, S. N., Deotare, B. C., Gangopadhyay, K., Sain, M. K., & Panja, S. (2011). Potential Geoarchaeological sites for luminescence dating in the ganga Bhagirathi – Hughli delta, West Bengal. *India. Geochronometria, 38*(3), 282–291. https://doi.org/10.2478/s13386-011-0041-6

Ray, A. K. (Ed.). (1997). *Gangasagar quadrangle, West Bengal*. Publish under Director General, Geological Survey of India.

Roychowdhury, A. (2011). *Prehistoric India: Stone age to iron age, (in Bengali)*. Mitram.

Saad-Ul, I. (2008). "Tilipi thake Tilogram" (In Bengali). In *A journey from Tilipi to Tilogrammam and its beyond: An archaeo-historical quest for glory* (pp. 125–132). Subhas Chandra Dey's Publishing House.

Sahu, B. K. (2007). "Atit Savyata O Sundarban" (in Bengali), "Smaranika, Dakshin 24 Pargana Pratna-Itihas Sammelan" (memorandum, South Twenty Four Paragana Archaeo- history convention). In K. Mandal (Ed.), *Baruipur, South Twenty Four Parganas. Memorandum, south 24 Parganas Archaeo-history conventions*. Board of Editors, Pratna Itihas Charcha Samiti.

Saren, S., Paul, A. K., & Chatterjee, S. (2016). Identification of archaeological remains around the "Rarha" region and littoral tract of Midnapur, West Bengal, India. *International Journal of Research in Social Sciences, 6*(3), 463–482.

Sarkar, A., & Sen, P. K. (2014). Occurrence of Holocene mammalian remains and Mulhouse shells from Indian Sundarban with environment of deposition. *Journal of Pharmacy & Bioallied Sciences, 9*, 6.

Sengupta, G., Chakraborty, S., & Roy Chowdhury, S. (2015). *Eloquent earth: Early terracotta's in the state archaeological museum* (pp. 10–45). West Bengal.

Chapter 6
Beach Stage and Dune Stage Modelling Approach in the Geomorphological Evolution of Beach Dune Landscapes, Mandarmoni Coast, India

Abhinanda Bal and Ashis Kumar Paul

6.1 Introduction

Beach morphodynamics is the interaction and mutual co-adjustment of intertidal beach-surf zone morphology, the hydrodynamic variables including the wind, beach gradient and sediment (Short, 1979). Beach stage dynamism is a model put forward by A. D. Short in 1979 on selected coastal stretches of Australia. Based on his model and also multifaceted attributes, a beach stage dynamic model has been put forward in this work. However, the formation and dynamism of sand dunes as a sedimentary depositional environment also behave differently in different parts of the land by process variables. Such continuous modification of the sedimentary depositional environment over the time creates or builds up the specialized land surface supported by tropical floral structure and community pattern on the basis of available moisture condition and wind regime. The loss of vegetation from this natural built-up surface increases the dynamism of surface stability of sand dunes. In the present study, the dune features and their dynamism are identified to build up a dune stage morphodynamics model. The beach profile is shaped by various factors acting upon the coast, mainly the energy of the wave, swash and backwash intensity and balance, sediment texture and compaction, wave generated currents, and wind speed. According to Bryant (1982), beaches may have seasonal shifts from reflective to dissipative in response to storm and swell conditions.

In case of our present study, this is a sandy beach with low gradient and mesotidal effect and also with low to moderate energy of waves (Paul A.K. 2002). The angle of wave approach is mainly from south-west. It has been proved by Santra

A. Bal
K.D. College of Commerce and General Studies, Midnapore, West Bengal, India

A. K. Paul (✉)
Department of Geography, Vidyasagar University, Midnapore, West Bengal, India
e-mail: akpauleastcoast@gmail.com

A. K. Paul, A. Paul (eds.), *Crisis on the Coast and Hinterland*,
https://doi.org/10.1007/978-3-031-42231-7_6

Mitra (2013) that there is a constant rate of shoreline shift in this area. Hence proved that this could be an erosional beach, but the eastern portion of the selected area is showing accretional features especially at the Soula River mouth. According to Pethick (1984), the driving force behind almost every coastal process is due to waves. Wave size gets varied by the fetch (the distance over which the wind blows), wind velocity, and wind direction (Carter 1995). The power of a given wave is approximately proportional to its amplitude squared and to its period (Mueller et al. 2007). According to Paul, A.K. (2002, 2022), a barrier bar setup with back water areas, wide sandy tract, multiple beach ridges with high dune platforms fringed by extensive beach plain surface was existing in the previous centuries (1800,1900 A.D.) in the region from Digha Estuary to Pichhaboni Estuary of the Kanthi Coastal Plain.

Some authors like Santra Mitra (2013), Sahoo and Bhaskaran (2016), and others have tried to estimate the shoreline shift in a temporal scale at this dynamic coastline but only considering the physical aspect and also in a 2-dimensional view point. But dune has never been a factor for consideration in these rhythmic dynamisms of the foreshore and backshore region of the coast. In case of Mandarmani beach, the dune is a highlighting base for protection of the back dune wetland areas and the flourishment of coastal development projects. This study is an attempt to categorise sediment transportation budgets with micro-structural features under consideration and to showcase the relationship between fluvio-tidal and marine processes.

6.2 Study Area

The area selected for this study is the part of the extensive shoreline of Bay of Bengal along the West Bengal coast (157 km including islands), popularly known as Mandarmani Beach, a fast developing sea side resort village. The coastal stretch extends from Jaldah estuary in the west to Pichhaboni inlet or Soula estuary in the east (Fig. 6.1). Out of 11 villages, former 8 villages fall under Ramnagar II Block and last 3 villages come within Ramnagar I Block of the Contai Subdivision of the District of Purba Medinipur, West Bengal. Sedimentation from the Hugli River Estuary is also a prominent influence on its mechanism. The local alluvium coast has evolved during and after the Late Holocene stage. There are mainly two types of deposition. Mandarmani is related to the recent Digha formation (1000–2900 years). South-west monsoon and north-east monsoon blows blow alternatively throughout the year in the region. However, the wind velocity and direction of the coastal wind system are also affected by the influence of land breeze and sea breeze. These cyclonic storms clubbing up with high tides may render in the overtopping of natural protective embankments or coastal dunes, thereby spoiling the standing crops and rendering soil salinity.

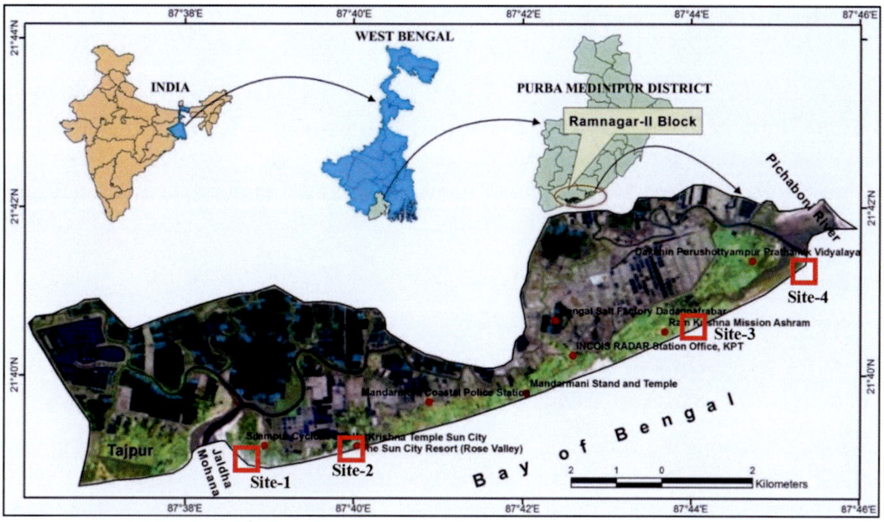

Fig. 6.1 Location map of the study area

6.3 Methodology

For this micro-level survey of beach and dune morphodynamics, few parameters are selected: Pre-existing morphology, Gradient of the beach and the dunes, Sediment texture, Beach-surf zone hydrodynamics, Dune ecological community and Beach faunal community. This spatial survey is then again compared temporally in a three-year time span. This varying survey has been carried out for 4 years in order to prove the research model. Tidal data and meteorological data are also collected and implemented in this work. Moreover, large-scale episodic events that create a major change in the coastline of Mandarmani are also taken into consideration and analysed during the field work.

After this in the on-field survey, many analyses like temporal analysis of the morphology of the region, identification of various micro-terrain features, preparation of DEM and applying various morphometric techniques on it have been done, which can ultimately divide the whole coastal area of Mandarmani into many zones of erosion prone area, depositional area and equilibrium area of erosion and accretion. An ideal model of beach dynamism is recreated and a beach-dune dynamic model is generated based on the series of analysis. The modern shoreline systems are modified and processed with the effects of alternate wet and dry seasons and events of storms. Sediment transportation paths along the shore fringes are guided by effective monsoon currents (Northeast and Southwest monsoon drifts) in the region. Therefore, survey profiles are conducted in a repetitive manner in monsoon and in non-monsoon months. Understanding the dynamic process is immediately needed to estimate and to manage the coastal resources of the sedimentary depositional environment.

6.4 Results and Discussion

For this research purpose, we have used the Shuttle Radar Topography Mission (SRTM) Digital Elevation Model (DEM) and various field-based data to generate a DEM for the study area and apply different morphometric techniques. Moreover, we have collected field-based hydrodynamic and aerodynamic data and analysed various statistical methods for the final composition of the beach and dune stage models.

6.4.1 Estimation of Sediment Transportation Budget

From the DEM of the study area (Fig. 6.2a), it can be assured that the broad physiographic regions of the study area are the beach, the dunes and the back dune wetlands. The beach lying at the southernmost part of the region has an elevation of 2 m on an average and a slope of 1° towards the sea. The topographical highs, that is, the barrier dunes, have a height ranging from 6 to 15 m. The seaward gradient of the dunes is much steeper than the landward slope. The back dune wetlands with the tidal inlets are forming the lowest physiographic expression with some depressions having depth less than the mean sea level. These tidal depressions are the main source of moisture, which again helps in the formation of dune on the shoreline on one hand and also restricts the extension of dune towards the back dune area. This is the reason for the low dunal rise in this particular tropical climatic condition with limited and local extent (Bird, 2008). The 50 cm contour interval map is depicting another vivid layout of the morphogenetic zoning of the region ranging from less than sea level up to 15 m elevation. This map (Fig. 6.2b) shows the extended dune belt garlanding the beach area and protecting the back-dune fringe area. Moreover, there is a concentration of contour in the northern most frontiers but with variations. South-western concentric contours highlight the extension of the topographic highs of the Tajpur area adjusting and being dissected by the Jaldha Tidal Mouth and its associated creeks. The remaining concentrations are the deepest parts of the shallow wetland depressions both in the north-western and north-eastern parts except the northern topographic high in the eastern part of the map. This comes with the

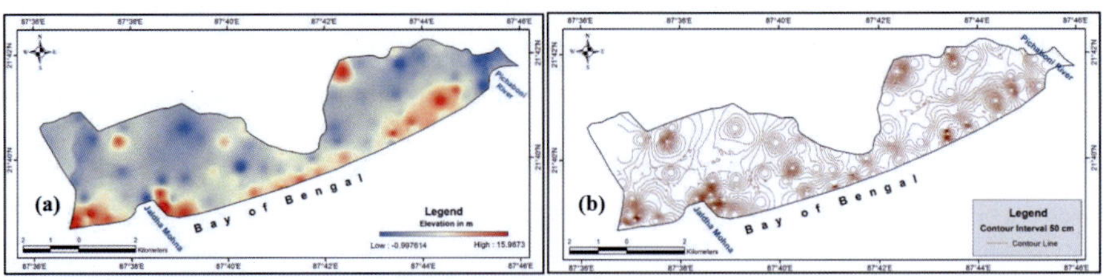

Fig. 6.2 (**a**) High resolution interpolated Digital Elevation Model (DEM) and (**b**) the 50 cm contour map has been generated by using DEM

history of the formation of the whole Contai or Kanthi coastal dune, because this is argued to be the part of the extended Chenier belts of Kanthi Coastal Dune. The selected study area is a more or less flat surface (Fig. 6.2a, b) with low to very low slope not more than 2°, comprising of the tidal flats, back dune marshy lands and estuary mouths of the tidal creeks and river inlets. The maximum slope ranges from 5° to 15°, which concentrates mainly around the dunal slopes and inclinations of the deeper wetlands to the backshore areas. This down slope direction is the representation of the maximum rate of change in value from each cell to its neighbouring cell. History can only justify this type of aspect ratio, which is the extended part of Chenier Dunes of Contai coastal belt. The remaining part of the area has another highlight of dunal shift or extension dune chains near the eastern part of Jaldha Mouth.

The total stretch of the land selected for the study has diversified characteristics in respect of selection of sample sites. For the purpose of beach and dune study, mainly there are two types of influences, one is the morphodynamic impact and another one is the anthropogenic variations. Sample Site 1, that is, the Jaldha at the Jaldha Mohona, and Site 4, that is, the Soula Mouth near the Pichaboni River Mouth, are selected on the ground of morphological influence, hydrodynamism, aerodynamism, floral diversity, faunal presence, etc. On the other hand, Site 2, that is, near Rose Valley or Sun City Resort, is suited to fit the rapidly rising influence of Tourism Industry and gradual change of morphology, ground water, environment and ecological scenario and influence of it. Last but not the least Site 3, that is, Khoti or boat anchoring ground for fishermen community of Dakshin Purushottampur village, reveals the character of the livelihood present at this vulnerable coastal scene and its impact on the physical setup of the beach and dune fringe area. The selection of these sample sites is really valuable because the research work is not only studying the physical setup but also the relationship between the existing community and the governmental regulation regarding the conflict between the expansion of territory for the native villagers or the outside tourism sector influencers. The comparative survey at each sample stations is carried out between 2 years 2015 and 2018. This comparative survey is mainly done with the Total Station (TS) for estimation of the changes in morphology and micro-geomorphic features, rate of change and identification of the stage of the beach and dune. The total station survey helped to estimate the calculation for sediment transportation volume and moreover evaluation for the rate of erosion per annum. So, these results can give us a clear picture of the physical setup of beach and dune fringe area and its variations due to morphodynamism or anthropogenic activities.

The two Total Station (TS) DEMs of sample Site 1 from 2015 and 2018 have shown that a drastic change has happened over three years of time spans (Fig. 6.3). There are morphological changes with dune crestal peak concentration during 2015, whereas dune flattening is observed on 2018, 2015 shows pockets of dune blowout, but 2018 has shaped as dune cliffing, the ridge and runnel topography is prominent for 2015 and micro-cusp feature overtook the lower beach face area during 2018. These features are exposed at Site 1 and Site 2 where underlying muddy layers are also exposed at few pockets of beach. If we talk about the erosion aspect, then it is definite that there is a huge reduction of elevation of nearly 3 m concentrating

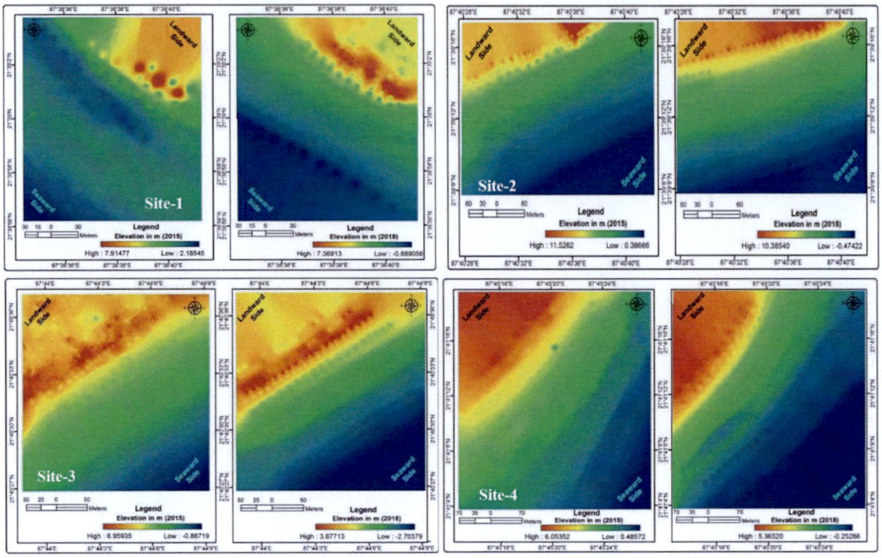

Fig. 6.3 Digital Elevation Model for the selective four sample sites (2015 and 2018) by using total station survey for studying the micro-features of the beach and dune stage model in the study area

contour mainly over the beach area. The dune has not drastically changed its elevation because this area is dominant with hydrodynamic character and has very little influence by human activities. But the erosion rate is high because of the tidal influences at the Jaldha Mouth, high wave energy and eastward moving longshore currents. The zone of erosion is nearly 90% relative to the accretion zone in sample number of Site 1, and following Jeanks' Natural Break Method, Volume Zonation has been done. These concentration zones are estimated by trend method through Interpolation techniques. The map (Fig. 6.4) produced hereby shows that among five zones, which are delineated according to topographic unit, four zones put up negative values, that is, Erosion, and one zone has positive value, which means concentration of accretion. In this site, total volume of sand transported (positive + negative) is 64142.51 cu m within 3 years' time span (2015–2018). However, the net accretion is of much lower value, that is, 2409.19 cu m (3.75%) in comparison with net erosion 61733.32 cu m (Table 6.1).

Thus, rate of erosion per year has been estimated to be 20,578 cu m, which is a huge volume of sand being transported from the beach and dune fringe area due to dune flattening process. This means the peaks of the dune crests are lowered and the dune furrows are filled up by the sediments from the adjacent dune peaks. Otherwise, this whole area is an erosive site with very few pockets of accretion. By comparing the contrast between 2 years (2015 & 2018) for Sample Site 2, it is evident that there is a reduction of 1 m elevation (Fig. 6.3). If we consider the deeper details about morphological aspect, then it is very much clear that there is a spreading of dune sediment parallel to the shore. But that does not mean the accretion rate has

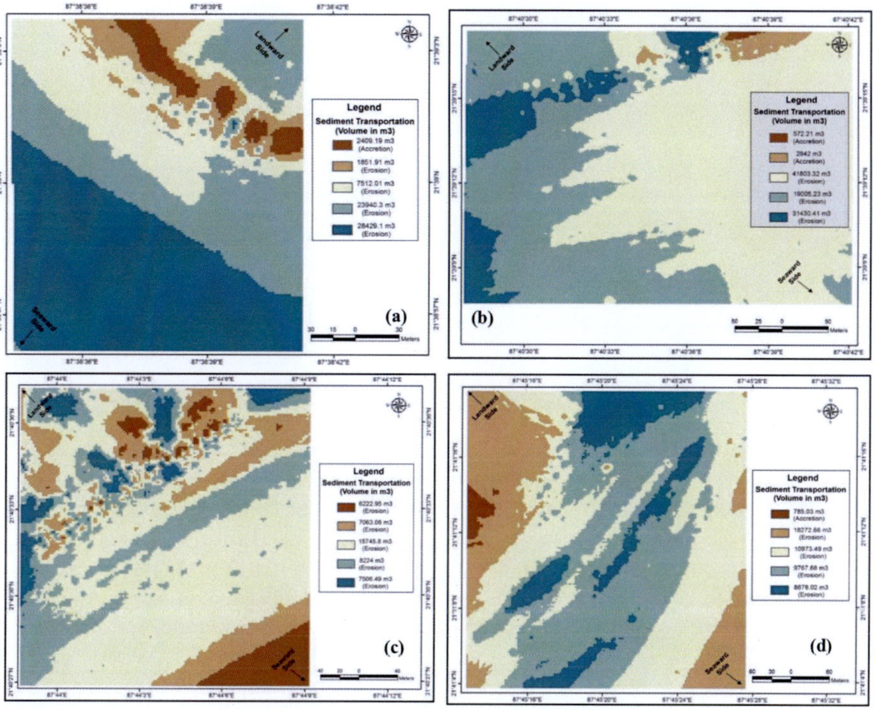

Fig. 6.4 Sediment volume zonation maps of the Ramnagar I Block of the Contai Subdivision of the District of Purba Medinipur (a. Site-1, b-Site-2, c-Site-3 & Site-4)

Table 6.1 Estimation of change in volume of sand at Site 1 from 2015 to 2018

Sl. No	Topographical zone	Elevation in m	Area in Sq.m.	Volume of sand in cubic meter	Erosion/ accretion	Net erosion cubic m	Rate of erosion per year cu m
1	Dune crest	2.2572	1916.31	2409.19	Accretion	61733.32	20577.77333
2	Dune slope	1.1603	1596.06	1851.91	Erosion		
3	Upper beach face	0.7569	9924.71	7512.01	Erosion		
4	Middle beach face	1.6592	14428.8	23940.3	Erosion		
5	Lower beach face	1.5393	18468.8	28429.1	Erosion		
Total		7.3729	46334.68	64142.51			

increased, instead it means that the sediment has been washed out by the highest high tide (HHT) or the storm surge and re-deposited at the dune slopes and furrows. The presence of runnel at the fore dune base was prominent during 2015, but the runnel has spread its limit and has lowered the elevation of the upper beach face area.

Table 6.2 Estimation of change in volume of sand at Site 2 from 2015 to 2018

Sl. No	Topographical zone	Elevation in m	Area in Sq.m	Volume of sand in cubic meter	Erosion/ accretion	Net erosion cubic m	Rate of erosion per year cu m
1	Dune crest	3.2487	176.1354	572.21	Accretion	92238.97	30746.32333
2	Dune slope	1.8877	1505.5383	2842	Accretion		
3	Upper beach face	0.8195	51010.7689	41803.32	Erosion		
4	Middle beach face	0.6768	28081.0152	19005.23	Erosion		
5	Lower beach face	4.873	6449.9106	31430.41	Erosion		
Total		11.5057	87223.3684	95653.18			

This is established that Sample Site 2 is erosion prone. The net erosion volume in 3 years is 92238.97 cu m, which is of much greater intensity compared to the previous site. Here the accretion volume is 3414.21 cu m (Table 6.2), which is only 3.5% of the total volume of sediment transported (95653.18 cu m) in the region. The accretion rate is lower than Site 1, which is an alarming situation because of the fact that this is a Tourism dominated site. From late 2009, the development of tourism sector had gradually accelerated and never retarded, so the beach and dune of that particular site must be quite supportive for this kind of infrastructural establishment. But as the time passes, the encroachment of construction and development sector overtakes the natural scenery of the Mandarmani Beach and especially the beach and dune of Silampur and Sona Muhi village units. Another alarming situation can be noticed from Fig. 6.4 is that the western part of the Site has been excessively eroded. This is an alarming concern to think about because the topographic rise as dunes are totally washed off, which can aggravate to coastal flood at each and every HHT and at any kind of storm surge. The worst affected stakeholders are the villagers who are wholly and solely backing up the resorts and hotels responsible for the economic growth of the tourism sector. Thus, if the rate of erosion continues to remain nearly 30,746 cu m per year, then all the dwellers and tourism area will be inundated regularly, that is, at every spring tide and in the event of high waves.

The Sample Site 3 (Fig. 6.3) was selected to show the relationship of the intensity of utilization of resources by local inhabitants and quest for stability of the beach and dune against them. There is extreme reduction of volume of sediments from the beach, dune and back dune area; nearly 3.5 m elevated surface was cut off within 3 years over the dune and beach, which is showing a reduction of elevation for about 2 m (Fig. 6.4) in the region. The inhabitants of Dakshin Purushottampur utilize these kinds of back dune slope, dune crest, fore dune slope, beach face for various purpose of their livelihood like boat manufacturing, dry fish processing, livestock rearing, marine fish extracting, etc. This man induced activities continuously turning the erosive coast of Mandarmani to much more erosive. The total volume of sediment transported in 3 years is 44762.32 cu m (Table 6.3). Yet we

Table 6.3 Estimation of change in volume of sand at Site 3 from 2015 to 2018

Sl. No	Topographical zone	Elevation in m	Area in Sq.m	Volume of sand in cubic meter	Erosion/ accretion	Net erosion cubic m	Rate of erosion per year cu m
1	Dune crest	1.0639	5849.19	6222.95	Erosion	44762.32	14920.77333
2	Dune slope	0.4581	15418.2	7063.08	Erosion		
3	Upper Beach face	0.3207	49098.2	15745.8	Erosion		
4	Middle Beach face	0.3635	20899.6	8224	Erosion		
5	Lower beach face	1.4865	5049.78	7506.49	Erosion		
Total		3.6927	96314.97	44762.32			

Table 6.4 Estimation of change in volume of sand at Site 4 from 2015 to 2018

Sl. No	Topographical zone	Elevation in m	Area in sq.m	Volume of sand in cubic meter	Erosion/ accretion	Net erosion cu m	Rate of erosion per year cu m
1	Dune crest	0.7958	986.4635	785.03	Accretion	47891.8537	15963.95123
2	Dune slope	1.0973	16652.3834	18272.66	Erosion		
3	Upper beach face	0.3786	28984.3998	10973.4937	Erosion		
4	Middle beach face	0.3366	29018.669	9767.68	Erosion		
5	Lower beach face	0.9677	9174.3544	8878.02	Erosion		
Total		3.576	84816.2701	48676.8837			

came to realize that the rate of erosion reaches near about 14,920 cu m, which is the lowest among all the four selected sites. So, the question remains, why the rate of erosion is still lower than the previous two sites. Long shore current flowing eastward is the main reason behind this retarded rate of erosion. The sediments eroded from the coasts of western portion of the study area, that is, the coast near Jaldha Estuary, are drifted towards the east and so the eastern part is much more accreted than the west. Moreover, the Pichaboni River is also providing some sediment to the eastern portion, which feed them with sand and help to reduce the erosion rate.

The elevation (Fig. 6.3) has reduced by 1 m in the case of Sample Site 4 (Fig. 6.4). The most important point is the location of this site, which is adjacent to the Soula Mouth. As eastern part of the coast is much more depositional in nature, the estimation is also showing a smaller zone of accretion (785.03 cu m) and also less intensive erosion 47891.8537 cu m, and this results to a rate of erosion of 15,963 cu m (Table 6.4). The rate of Site 3 is still showing lesser compared to this region at the mouth of Pichaboni, which is influenced by large hydrodynamic effect. Though the whole coastline is an erosive coast, still the accretion rate is also balancing out with it. The reason is also clear that the two mouza Silampur and Sonamuhi are moulded at its top by the tourism sector since 2009. Another output is that the eastern part of the coastal stretch has a trend of accretion relative to the others due to the least presence of human activities and play positive role to protect the nature not destroy it, and at the same time the long shore current is present from the west to the east. Thus, the natural marine processes of the region also force the sediment to drag from Jaldha to Pichaboni.

The exchange of volume of sand as per estimation at Site 2, that is, at Sonamuhi mouza near Rose Valley or Sun City resort, is showing 95,653.18 cu m of sediment (Table 6.5) and minimum volume at Site 3, Dakshin Purushottampur at Khoti with 44,762 cu m of sediment. This leads to the prediction that the rate of transportation of sediment for beach-dune interface is 78,40,563.21 tonnes per year. The rate of erosion of sediment per cubic metre per year is highest at Site 2 (30,746.32 cu m), and the rate of accretion is also at its top at this site (1138.07 cu m). Though there is a trend of healing with nourishment of sand by accretion process, the erosion rate is so high that it is not able to compensate the preceding process. The accretion pockets are seen only as an area of sand refilling in the dune furrows, which have been a prominent feature in the region 3 years ago (Fig. 6.5a). The beach or fore shore face has been so utilized with tourist activities like walking, biking, water sports (water bike, speed boat, paragliding, etc.), movement of jeeps and cars that the stability of the beach is also hampered in the Site 2.

6.4.2 Estimation of Wave Hydrodynamics

Three consecutive observations (10 min each comprising 30 min) of wave hydrodynamic parameters in different tide levels (Rising tide, High tide and Falling tide) conducted at each sample site confer a significant result on response to beach morphology and sediment distribution with spatio-temporal variation (Bird, 2008). The length of Swash (L_s) and Backwash (L_b) waves was also measured along with the sediment carried on-shore and off-shore with them, respectively, was weighed (Table 6.6). This would give a proper insight of the strength of the wave and current at each and every sample site. From this, it can be deduced that whether the waves are constructive or destructive. As this is a shallow environment so the wave-dependent variable changes according to the depth of the wave.

There is a steady relationship between wave depth and wave amplitude, wave breaking height, wave height and wave relative height (Fig. 6.6a), which gradually and slowly diminishes as the wave depth increases. This negative relation is much stronger for Site 3 and 4 because the eastern part is much more accretion prone. This is evident that there will be a negative relationship between wave depth and wave length, but it is much flatter in case of Site 3 (Fig. 6.6b). This may be due to the near-shore bathymetry, which is quite deeper in this case, that's why this site is used by the local fishermen for the boat landing station. But the exception is for Site 2 of course due to anthropogenic influences. The Swash & Backwash length data set is

Table 6.5 Sediment transportation budget

Site No.	Name of the places	Total change of volume of sand	Accretional sand volume cu m	Erosional sand volume cubic m	Rate of erosion per year cubic m	Rate of accretion per year cubic m	Area in sq. m	Transportation rate of sediment per year cubic m	Transportation rate of sediment in tonnes per year
1	Jaldha Mohona (Silampur)	64142.52	2409.19	61733.32	20577.77	803.06	46334.68	21380.84	50906.7619
2	Rose Valley (Sonamuhi)	95653.18	3414.21	92238.97	30746.32	1138.07	87223.36	31884.39333	75915.22222
3	Khoti (Dakshin Purushottampur)	44762.32		44762.32	14920.77		96314.97	14920.77333	35525.65079
4	Soula Mohona (Dakshin Purushottampur)	48676.88	785.03	47891.85	15963.95	261.67	84816.27	16225.62667	38632.44444
Total					82208.81	2202.8	314689.28	84411.63333	200980.0794
Total erosional/accretional rate in cubic meter per year for the total dune coverage area					2120551.44	56820.56	8,117,315	2177372.61	5184221.5
Total/erosional/accretional rate in cubic meter per year for the total beach dune complex					3207100.79	85934.85	12,276,546	3293036.55	7840563.21
Total/erosional/accretional rate in cubic meter per year for the total beach shore face					1086549.35	29114.29	4,159,231	1115663.94	2656342.71

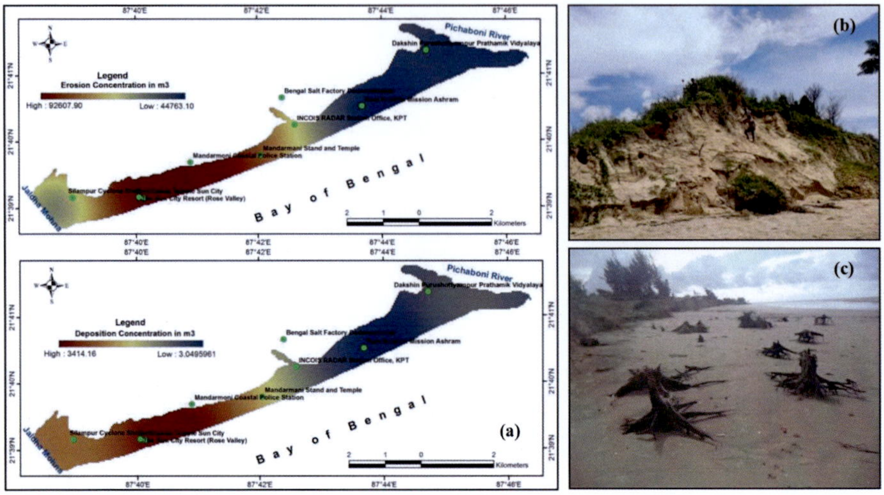

Fig. 6.5 (**a**) Erosion and accretion concentration zones along the beach-dune interface; (**b**) dune cliff, and (**c**) destruction of flora in the present study area

Table 6.6 Sediment weights of swash and backwash current

Site No	Place	Weight of swash sediment (gm)	Weight of backwash sediment (gm)
1	Jaldha (Site 1)	672	622
2	Rose Valley (Site 2)	1240	1484
3	Khoti (Site 3)	545	578
4	Soula (Site 4)	722	185.48

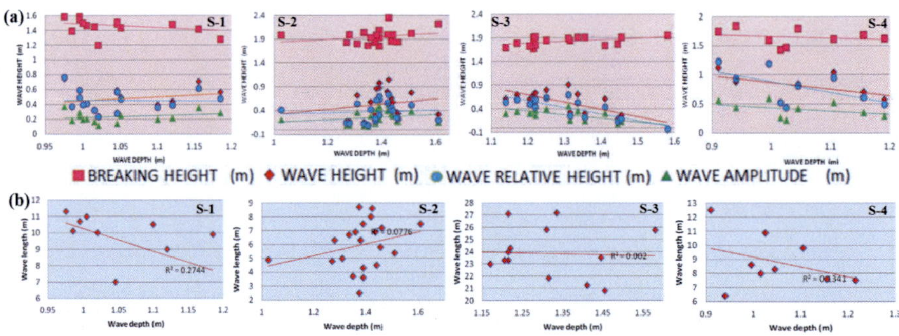

Fig. 6.6 (**a**) Relationship between various hydrodynamic variables and (**b**) relationship between wave length and wave depth (Key: S-1-Site-1, Site-2, Site-3 and Site-4)

showing the fact that the Swash lengths are longer than the Backwash length (Fig. 6.6), which means there is scope for accretion for this coastal stretch. But if we look into the mass of the sediments transferred by this circulation, it is seen that for Site 1 and 3, the sediment brought in is compensated with the sediment washed out. In case of Soula, the swash is much stronger than the backwash, but for Site 2 backwash is quite stronger making the beach unstable inside out.

6.4.3 *Estimation of Aerodynamics*

The wind direction and velocity play an important role in sediment transportation so the wind has to be measured with the help of anemometer and prismatic compass. The four sample sites show a variation of wind direction mainly concentrating as Eastern wind with the variation between east, south-east and south-south east. Keeping parity with the 3 consecutive observation of wave hydrodynamics, wind speed was estimated in 3 observations with 10 mins interval of each observation. Wind velocity ranges from 5 to 7 m/sec to 11-13 m/sec in the region as the coastal region having open sea ahead with varieties of unstable weather conditions so the wind speed fluctuates very often. Site 2 and 3 having higher velocity of wind speed and Site 1 and 2 are towards the lower range. This can directly affect the mechanics of the wave and current.

The striking feature is that the breaker height is much higher than the average height in case of Site 2 and 3, which is the tourism and fishing dominated area (Fig. 6.7a). This is an alarming feature showing the rage of the breaker waves, which can smash against any odds in front of them. Due to this character of the vertical cliff generated at the sample site shows the stronger backwash effect and gradually penetrating sea water landward. The wave velocity is mainly guided by the wind speed so depending on the fetch length as the wind speed increases the wave velocity increases and this is a very direct and positive relation between the two (Figs. 6.7b and 6.8). But the graph depicting the retarded wave velocity is showing either a phase of wind gap or a phase of counter current of wind from the land. The value of

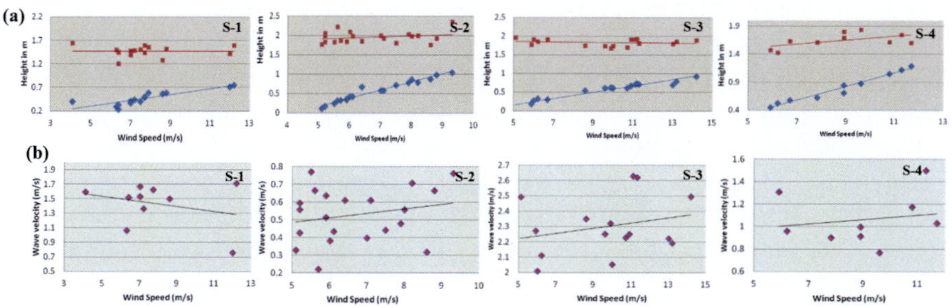

Fig. 6.7 (**a**) Relationship between wave height and wave breaking height with wind speed and (**b**) Relationship between wind speed and wave velocity (Key: S-1-Site-1, Site-2, Site-3 and Site-4)

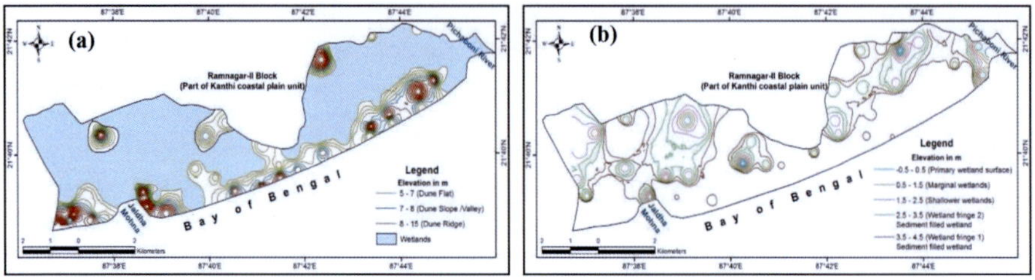

Fig. 6.8 (**a**) Micro-geomorphic features of dune and (**b**) micro-geomorphic features of wetland

Pearson's correlation coefficient of the relation is 0.44954319, and the critical value of 't' at 60 degree of freedom is 1.67 (the hypothetical value). But the computed value of 't' is 3.930602 at 60 degrees of freedom. As the computed value of 't' at 60 freedoms is more than the hypothetical value, the null hypothesis is rejected at favour of alternative hypothesis. The relation between wind speed and wave height is significant and it is concluded that the wind is the only dominant control factor of wave parameters.

6.4.4 *Micro-geomorphic Features*

Micro-geomorphology is evidently a crucial indicator for identification of the stage of a beach, and the rhythmic nature of the beach and changing dune features depend on the change of season (Fig. 6.9b–d). Beach mapping is the only method through which the micro-features can be identified from the rhythmic features located on the beach. This mapping process is done through crude method by shooting photographic snap shots in a grid method. So an area at the sample Site 4 is taken into consideration because this area received impact from 3 elements: the morphogenetic factor or the influence of the wind, wave, tides, currents, etc.; impact of local inhabitants; and also due to pressure from tourism sector. Likewise, 10 grid (1 × 1 m area) were plotted in east-west direction, that is, parallel to the shoreline, and in this way, it was proceeded from land towards the sea in the field. The time taken was during neap-tide in order to get lowest low tide and full exposure of the beach. In this manner, a strip of land was photographed covering in total 1040 sq. m on the surface of the beach. These photographs made into a strip were later cropped and merged, and a micro-feature identification map (Fig. 6.9a) of the beach area was formed through this grid method.

From beach mapping, 10 zones were identified on the beach as well as during analysis. The adjacent dune region has an elevation of not more 2 m from the mean sea level. The dune shape shifts from sharp cliff (Fig. 6.5b) with concave slope facing the sea during monsoon and post-monsoon season to a convex mound of sand during pre-monsoon season. The zones were identifying from the base of the dune along the HTL, and these 10 zones were again divided into 3 beach faces as upper

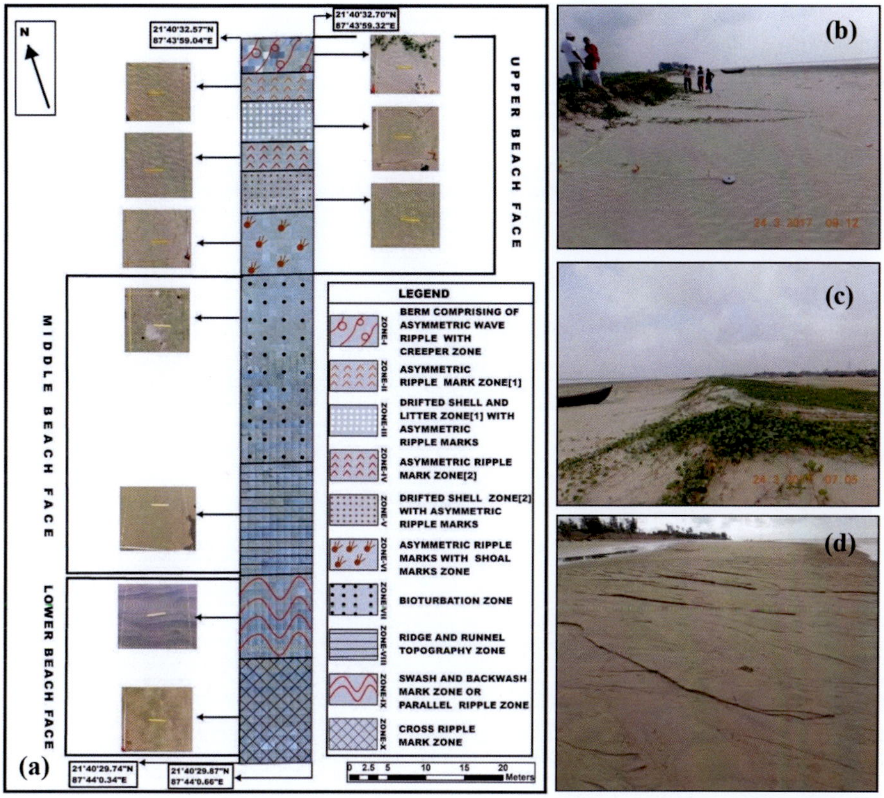

Fig. 6.9 (**a**) Beach mapping through grid method; (**b**, **c**) rhythmic nature of beach and dune; and (**d**) Rip channels

beach face, middle beach face and lower beach face. The upper beach face has characteristics like berm surface covered with creepers, asymmetric ripple marks, drifted shells, litter zone and shoal marks. Middle beach face comprises mainly with ridge and runnel topography (higher slope angle 6°–8°) and bioturbation zone, and the lower beach face (beach slope of 5°) has alternate bands of swash and backwash marks (parallel ripples) or cross ripple marks meeting the near-shore bathymetry. There is a seasonal shift of shore parallel bands of feature showing mainly in the lower beach face. During the heavy downpour from south-west monsoon, strong and defined rip channels (Fig. 6.9d) generate here accelerating erosion process across the shoreline. But during the month of January to May, alternate micro-cusp and bay features were evolved in a rhythmic process. So the beach is showing a shift of features from dissipative to reflective beach type. Though the dunes are acting as a topographic rise in this region, with the accelerated influence by the tourism flourishment, few breach points and dune blowouts (Fig. 6.8a) are noticed through which saline water intrusions are happening. It has been traced out with its morphological behaviour that this region has been developed in a back spit environment so the sediment filling processes are still going on. This process is also one reason for the filling up of the back dune wetland surface (Fig. 6.8b) and making the shallow depressions inhospitable for pisciculture other than the anthropogenic influences.

6.4.5 *Estimation of Beach-Dune Morphodynamism*

All the above analysis and estimation leads to the final level of morphodynamism, which includes all the factors, causes behind the factor, environmental influence, anthropogenic impacts. Now in Table 6.7, each sample site was graded and assimilated to deduce the final result of each of its condition and conclude the stage at which the beach-dune morphology is acting at present time. All these analysis leads

Table 6.7 Morphodynamic indices of the sample sites

Sl. No.	Parameters	Site 1	Site 2	Site 3	Site 4	Average
1	Month	August	August	August	August	
2	Inter-tidal width/m	250	180	175	200	201.25
3	R.L difference between HWL and LWL	4	3.5	4.2	4.5	4.05
4	Beach face angle (β) (degree)	5	2.5	3	2.8	3.325
5	Swash line bearing (degree)	23	34	36	30	30.75
7	Mean water depth/m	1.05	1.39	1.29	1.06	1.1975
8	Mean wave height(h) in m	0.47	0.55	0.53	0.79	0.585
9	Significant wave height (H1/3) in m	0.67	0.97	0.82	1.12	0.895
10	Mean wave period/s	6.5	10	12.6	14.9	11
11	Mean wave approach angle (α) in degree	292	247	290	242	259.6667
12	Velocity of longshore current/(m.S-1)	1.039	0.673	0.417	0.704	0.70825
13	Wave form velocity © in m/sec	1.03	0.98	1.25	1.36	1.155
14	Wave length (L) in m	9.45		23.87	8.88	12.5
15	Wave steepness/(H.L − 1)	0.5	0.43	0.2	0.95	0.52
16	Max. & min. Wave height/m	0.75 & 0.29	1.05 & 0.25	0.92 & 0.27	1.19 & 0.45	0.97 & 0.35
17	Surf scaling factor/ϵ	39	59.99	46.87	52.65	49.6275
18	Breakers type	Spilling	Spilling	Spilling	Spilling	Spilling
19	Rate of sediment transport/ (m.S-1)	21380.84	31884.39	14,920	16225.62	21102.71
20	Wave energy (E) in/ (J. M.S-2)	1228	1153	927	1407	1178.75

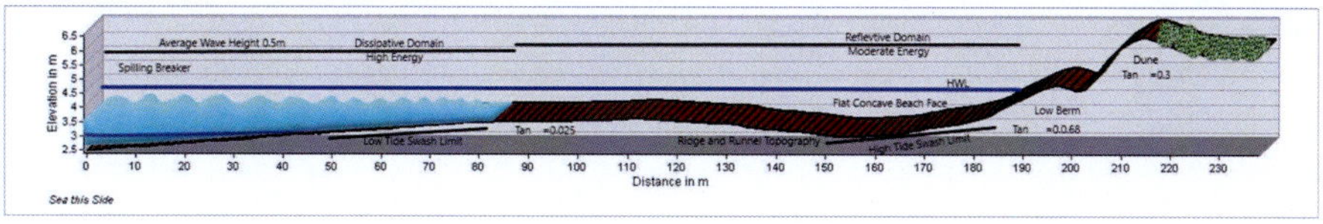

Fig. 6.10 Schematic framework of the Beach-Dune Stage Model

the research to the outcome, which have the final say about the physical setup of the region. Beach stage model and dune dynamism estimates have an immense effect on the livelihood options and expansions, which has to be dealt in an inter-dependent manner. In the beach or dune that has high risk from the stability point of view, there must be some preservation of buffer zone to cope up with the crisis along the shoreline.

All the analysis proves the fact that this coast is an admixture of dissipative and reflective beach (Fig. 6.10). The beach and dune adjust among themselves for their progression and also adjust with the high wave energy at the lower beach face and moderate wave energy at the upper beach face. The beach stage model can never be separated from the dune dynamism in case of Mandarmani coast because beach forms the dune and dune protects the beach and the back dune region. So, all the attributes acting upon the landform is very much interlinked with each other. In between two tidal channels, that is, Jaldha tidal inlet and Pichaboni River, lies the Mandarmani tidal system. Under this system, there are four subsystems as Beach, Dune, Barrier and Wetland. Further smaller subsystems are seen within wetland subsystem as Channels, Mangroves, and Tidal Flats. These subsystems interact with each other and within themselves creating a morphodynamic setup of the whole tidal system.

This interaction between system and subsystems made by all the hydrodynamic and aerodynamic attributes lead to the landscape evolution of the area. Moreover, the system-subsystem setup is again influenced by external forces or the entropy factors, that is, the storm, sea level rise, fluvial influence (from the Hugli Estuarine system or Subarnarekha Fluvial system), geology and tectonics, and also anthropogenic pressures. All these characteristics lead to the adjustments of the equilibrium within the Mandarmani Tidal or Coastal System. Each entropy force influences each subsystem individually and collectively, which distorts the equilibrium condition. The recovery phase taken for adjustment of this equilibrium setup if took a longer period calling a positive feedback system may take years or decades for balancing this imbalance situation. But if there occurs a negative feedback system among the entropy, then the recovery phase will be smaller and equilibrium can be easily reinstated (Masselink et al. 2011). From the Beach Stage Model (Fig. 6.10), it is evident that the Mandarmani beach is in a dissipative stage, but with seasonal fluctuation, the character of the beach shifts from dissipative to reflective nature. Dune stability is interlinked with the beach status and the progradation and advancement of dune solely depends upon the moisture condition and the floral community. However, 7840563.21 tonnes of sediment is being transported per year through the long shore current from west to east along the shoreline. The erosion and accretion concentration zone of the beach dune complex are both concentrated near the tourism sector. This is not a state of balance but an immense state of imbalance because this is the signature of dune flattening (3–8 m dune height) and deposition only in the dune furrow regions. The sediment texture is clearly showing that the sands are medium to fine-grained and the fine-grained percentage gradually increases from west to east which also proves that longshore transport is more prominent from west to east with maximum effect on the coast. But keeping parity with the cell circulation system, the long shore movement of sediment also transports from east to west during north-east Monsoonal wind. The two morphogenetic regions of dune and back dune wetlands are facing two different crisis situations with 4 breaching points and dune blow outs and sediment filling process in the shallow depressions. The hydrodynamics and aerodynamics are directly related but with little changes particularly in the tourism sector. The *Ipomea* sp., *Launia* sp. and *Sesuvium* sp. are the dominant floral species which try to stabilise the unconsolidated dune structure and arrest fluctuation of sediments. The bioturbation method by beach fauna is continuously replenishing the beach sediment and helping the stabilisation process. The *Casuarina* sp. plantation leads to dune cliffing, dune erosion in the shore front side and dune advancement in the lee side of the dune. Sometimes it generates the formation of dune blowouts and promotes overwash deposits in the landward side, and the relation between wind speed and wave height is showing moderately positive relation in the section of the shoreline. The final result can be explained as first outcome is considering mainly the excessive exploitation of coastal resources by the tourism industry through sand mining, constructing concrete buildings and roads within few metres of High Tide Line (HTL), clearing the natural forest cover, restricting the natural movement of sand dunes seasonally, extracting ground water and lowering it immensely resulting on the salt water intrusion. The second output shows that the eastern part of the coast is more protected and mainly a zone of accretion, which can be explained along with the other results.

6.5 Conclusion

Understanding the dynamisms of shoreline processes of the local and external systems indicate there should be a well-kept wide and open space immediately behind the shoreline and two active tidal streams to interplay the marine, coastal, tidal, aeolian and biological processes for their natural adjustment with dynamic systems. Such dynamisms of coastal processes with climate variables and sea level variations will produce damages on the land and resources (erosion, inundation, salt water encroachment, etc.) that can influence directly over the livelihood patterns and dynamic adjustments of the local people. Integrated Coastal Zone Management (ICZM) is the only possible strategy to be used for a sustainable economy and environment for improved coastal development projects. The above study will supply a significant finding with database to the coastal managers and administrators for the appropriate management of the beach dune systems to support the emerging tourism recreation sites of the region.

References

Bryant, E. (1982). Behavior of grain size characteristics on reflective and dissipative foreshores, Broken Bay, Australia. *Journal of Sedimentary Research, 52*(2), 431–450.

Bird, E. C. F. (2008). *Coastal geomorphology: An introduction*. Wiley. 2nd illustrated annotated edition.

Carter, R. W. G. (1995). *Coastal environments: An introduction to the physical, ecological, and cultural systems of coastlines*. Elsevier.

Masselink, G., Hughes, M. G., & Knight, J. (2011). *Introduction to coastal processes and geomorphology*. Hodder Education.

Mueller, M. A., Polinder, H., & Baker, N. (2007, May). Current and novel electrical generator technology for wave energy converters. In *2007 IEEE international electric machines & drives conference* (Vol. 2, pp. 1401–1406). IEEE.

Pethick, J. S. (1984). *An introduction to coastal geomorphology*. Department of Geography, University of Hull.

Paul, A. K. (2002). *Coastal geomorphology and environment: Sundarban coastal plain, Kanthi coastal plain, Subarnarekha Delta Plain*. ACB publications.

Paul, A. K. (2022, August). Dynamic behaviour of the estuaries in response to the phenomenon of global warming in the coastal ecosystems of West Bengal and Odisha, India. In *Transforming coastal zone for sustainable food and income security: Proceedings of the international symposium of ISCAR on coastal agriculture, March 16–19, 2021* (pp. 907–931). Springer.

Short, A. D. (1979). Three-dimensional beach-stage model. *The Journal of Geology, 87*(5), 553–571.

Santra Mitra, S. S., Santra, A., & Mitra, D. (2013). Change detection analysis of the shoreline using toposheet and satellite image: A case study of the coastal stretch of Mandarmani-Shankarpur, West Bengal, India. *International Journal of Geomatics and Geosciences, 3*(3), 425.

Sahoo, B., & Bhaskaran, P. K. (2016). Assessment on historical cyclone tracks in the Bay of Bengal, east coast of India. *International Journal of Climatology, 36*(1), 95–109.

Chapter 7
Spit Dynamics and Hydrogeomorphic Processes Along the Chilika Lagoon Estuary of Odisha, India

Damodar Panda, Rashmi Rani Anand, and M. Devi

7.1 Introduction

Spit-platforms have not drawn much attention, despite their expected importance for understanding sediment transport mechanisms leading to siltation of tidal channels. The sediment enters the spit platform as sandbars oriented perpendicular to the shore, migrating alongshore because of wave activity, until the sediment reaches an ebb-dominated tidal channel (Paul, 2014, 2022; Paul et al., 2014). In and near this channel, the sediment is transported downdrift by the ebb-tidal currents until the sand enters a shallow dissecting ebb channel, where the sand is transported obliquely offshore and deposited on a small ebb-tidal delta. Dominated by the combined influence of fluvial and marine processes as well as human activities, estuarine regions are facing serious challenges. Spits are found growing along the river mouths and lagoons on the coast. Spit growth is responsible for the shifting of the river and lagoon mouths, resulting in a rapid change in estuary morphology and the sedimentation process. The opening and closure of the inlets depend on the discharge of water and sediments from the lagoon to the sea as well as the input of sediments from the sea to the channel, depending on the dominance of the flood or ebb flow. The barrier spits were capped by low dunes and were separated by migrating inlets connecting the ocean with the backing estuaries and rivers discharged into the

D. Panda (✉)
Department of Geography, Utkal University, Bhubaneswar, Odisha, India
e-mail: damodar_65@rediffmail.com

R. R. Anand
Saheed Bhagat Singh College, New Delhi, Delhi, India

Department of Physics, KiiT University, Bhubaneswar, Odisha, India

M. Devi
Department of Physics, KiiT University, Bhubaneswar, Odisha, India

A. K. Paul, A. Paul (eds.), *Crisis on the Coast and Hinterland*,
https://doi.org/10.1007/978-3-031-42231-7_7

estuaries at times, decreasing salinities and depositing sediments (Mohanty et al., 2008; Engstrom, 1974).

Strong storm waves combined with the high sea level caused substantial changes in the coastal geomorphology of depositional shores. The most exceptional changes occurred in the areas that were well exposed to the storm winds and wave activity, which resulted in the elongation of a spit (Kokot et al., 2005). The spits and inlets vary relatively according to the influence of the physical processes, such as river discharge, wave power, and longshore sediment transport. During higher energy conditions, the sediment transport is more affected by wave-induced currents, leading to a high transport rate in the shallow part of the spit platform and less transport in the horizontally restricted channel. A dominant southward longshore transport system is responsible for the formation of three spits. Another smaller spit grew northward. Currently, there is no evidence of any northward longshore transport.

7.2 Study Area: Chilika Lagoon

The Chilika Lake is enclosed by the Mahanadi delta in the north, the Eastern Ghat mountain range in the west, the Rushikulya plain in the south, and the Bay of Bengal in the east. The lagoon is drained by 52 rivers in its catchment. It experiences a tropical monsoon climate. A large number of islands developed in the lagoon due to sediments deposited by the rivers and tidal currents. It covers parts of the Puri, Khurda, and Ganjam districts of Odisha. The water spread area of the lagoon varies from 1165 to 900 km^2 during monsoon season and summer season, respectively (Paul et al., 2014; Paul, 2022; Vivek et al., 2019). The lagoon is connected with the Bay of Bengal by a narrow inlet. The lagoon is well connected by roads and railways from different directions. On account of its rich biodiversity and ecological significance, Chilika was designated as the 1st “Ramsar Site” of India in 1981 (Fig. 7.1).

7.3 Database and Methodology

The geomorphological evolution of the Chilika lagoon has been studied on high-resolution Landsat images, Google Earth images (1984–2022), and the Chilika Development Authority (CDA) Report. Wetland Atlas, SAC, Ahmedabad, and the Central Water Commission report on the water discharge and sediment transport of the Mahanadi River. Shoreline changes represent different years (1984–2022) of the Chilika spit by overlaying images of multiple years. Quantification of the erosion and accretion rates is done by digitization using Q GIS. Geomorphologic features like sandbars, spits, barrier islands, islands, and channels are denoted on the images and plotted as a geomorphologic map for each station in the period of the study. A Geographic Information System (GIS) is designed to work with data referenced by

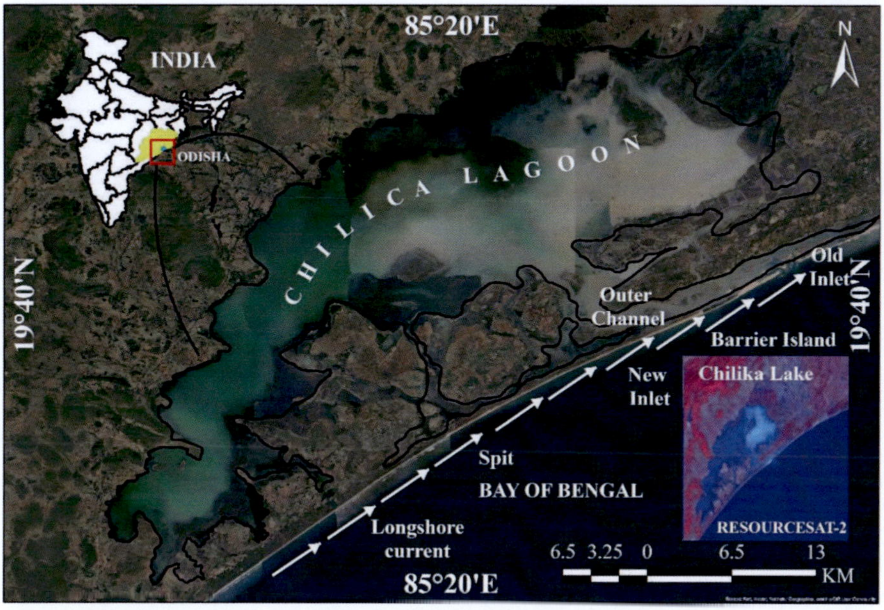

Fig. 7.1 The location of lake Chilika in Odisha coast showing with the geomorphological features and longshore current direction along the shoreline of the Bay of Bengal

spatial or geographical coordinates. Remote sensing satellite images have been effectively used for monitoring shoreline changes at different locations. A geo-database is created for the extracted spit and inlet locations.

7.4 Results and Discussion

7.4.1 Geomorphological Evolution of the Chilika Lagoon

The lagoon was formed during the Upper Pleistocene and Upper Holocene periods due to the regression of the sea (Ahmad et al., 2022; Paul et al., 2014). The whole northern region above the lagoon was under the sea in the geological past. Chilika is a tidal lagoon, developed by a barrier or sand spit that developed by the accretion of coastal sediments following the stabilization of sea levels some 3000–4000 years ago. Most lagoons were formed due to the worldwide rise of sea levels over the last 6000–8000 years. There was a break in the rise in sea levels about 7000 years ago, when a sandy beach might have formed near the coast in the southern sector. As the sea rose further, this sand beach grew gradually. A recent finding of the fossil from the south-western edge of the spit has been dated back to about 3500–4000 years ago. Some rocks in the southern sector are marked by a band of white formed by the remains of coral (which are exclusively marine) at 8 meters above the current sea level, and these rocks are of marine origin. The coastline has moved considerably eastward with the growth of the spit and barrier bars in the northeast direction. The longshore sand supplied from the south helps to extend the barrier spit and form a small lagoon. The opening end of the inlet shifted towards the north, and the spit very often joined the northern coast by closing the lagoon. Due to the impact of the landfall of cyclonic storms, the spit has been broken into segments by storm waves to modify the barrier islands and create new inlets. The sediments delivered to the lake through the Daya and Bhargavi distributaries of the Mahanadi River continued after the coastal barrier was formed and the lake was developed over time, leaving between 900 and 1200 sq. km of water surface area at present (Fig. 7.1).

7.4.2 Spit Dynamics

The spit, associated with Chilika lagoon, is about 49 km long, extending from south to north along the direction of the longshore drift. In between the coast and the spit, a long, narrow, and shallow outer channel is formed through which fresh water and saline water exchange takes place. The outer channel gradually became shallow and narrow due to sediment deposits and a reduced flow of water in and out of the lagoon. The inner channel shows a large number of islands. It has been observed that the maximum width of the outer channel is 2.18 km and its length is 28.38 km; the depth varies from 1 to 4.5 m. The sediment deposits along the spit are maximum during the monsoon season, from June to September, and lowest in December and January. The average annual sediment deposited by the longshore drift is about 975,375 m^3/month (Sahoo et al., 2018). The growth of the spit northward joined the northern coast of the lagoon by closing the inlet in the past. To maintain the salinity of the lagoon for the fish population, the local fishermen, who depend on the lagoon for their livelihood, very often cut open an artificial inlet. Thus, the spit was divided into two parts: the south spit and the north spit. The south spit is higher, with a maximum height of 18 m, than the north spit, with a maximum height of 6 m. The south spit is wider, with a maximum width of 0.97 km, than the north spit, with a maximum width of 0.34 km distance. Both the south and north spits became narrower towards the inlet. Due to the growth of the southern spit towards the north, the inlet has become narrow and shallow, restricting the inflow of tidal water into the lagoon. Large waves associated with cyclonic storms coupled with a rise in the sea level result in erosion of the spit in the northern sector, and waves sometimes cross the spit at some locations. The spit is widening, and the position of the mouth is constantly shifting, generally towards the north. The Chilika lagoon is becoming gradually shallower due to being silted up by the rivers flowing into it. The gain or loss of spits and the closure or opening of inlets is significantly controlled by the high wave power, longshore drifts, and river discharge. Currently under the impact of severe cyclonic storms, the spits are highly eroded, resulting in a sudden widening of the inlet. Foreshore and backshore erosion play important roles in the landward

Table 7.1 Change of the length of the spit in the Chilika lagoon

Year	Length of south spit in km.	Length of north spit in km.	Total length of spit in km.	Change of length of south spit in km	Change of length of north spit in km	Change of length of spit in km
1984	39.91	6.65	46.56	–	–	–
1990	40.99	4.36	45.35	+1.08	−2.31	+1.21
1995	43.59	3.02	46.61	+2.6	−1.34	+1.26
2000	45.34	0	45.34	+1.75	−3.02	−1.26
2001	27.24	0	27.24	−18.1	0	−18.1
2003	27.45	17.97	45.42	+0.21	+17.97	+18.18
2010	23.06	15.83	38.89	−4.38	−2.14	−6.53
2013	29.11	12.89	42	+6.05	−2.94	+3.11
2017	31.55	13.97	45.52	+2.44	+1.08	+3.52
2022	33.11	8.89	42	+1.56	−5.09	−3.52
Average	34.14	8.36	42.89	+1.97	+2.22	−2.13

migration of spits in response to rising water levels or increasing wave heights, normally during the passage of high-intensity cyclonic storms. In 1984, the length of the south spit was 39.91 km and the north spit was 6.65 km. From 1984 to 2000, the south spit increased from 39.91 to 45.34 km, and the north spit decreased from 6.65 km in 1984 to completely disappear in 2000 (Table 7.1).

To maintain the salinity of the lagoon, the Chilika Development Authority (CDA) and the government of Odisha opened a new inlet by dragging Sipakuda near the main channel of the lagoon. The spit was divided into two parts: the southern part, a spit of 27.24 km, and the northern part, a barrier island of 17.26 km. From 2000 to 2002, one spit in the south and an elongated barrier island in the north formed. The old inlet was located on the north coast; hence, no north spit was found from 2000 to 2002. The old inlet closed in 2003, and the barrier that joined the north shore of the lagoon became the north spit. In 2003, the south spit length was 27.45 km and the north spit was 17.97 km. Since 1984, the length of the north spit has been at its maximum of about 17.97 km. However, the south spit decreased from 27.45 km to 23.06 km, and the north spit decreased from 17.97 to 15.83 km in 2010 due to erosion of the spit by high energy waves associated with the 2008 solar eclipse. From 2010 to 2022, south spit increased and north spit decreased. The south spit increased from 23.06 km in 2010 to 33.11 km in 2022, and the north spit decreased from 15.83 km in 2010 to 8.89 km in 2022. The Very Severe Cyclonic Storm Phailin crossed the Orissa coast near Gopalpur on October 12, 2013, and caused severe damage to the coastal structures, shoreline erosion, and the natural opening of the new inlet in the Chilika lagoon. Due to constant erosion and the impact of cyclonic storm Phailin, the Phailin inlet developed, thus expanding the inlet to 3.28 km and the south spit to 12.89 km (Table 7.1, Figs. 7.2 and 7.3a).

In the last 38 years, the average length of the south spit is 34.14 km and the north spit is 8.36 km. The total length of the spit was 46.56 km in 1984 and decreased to

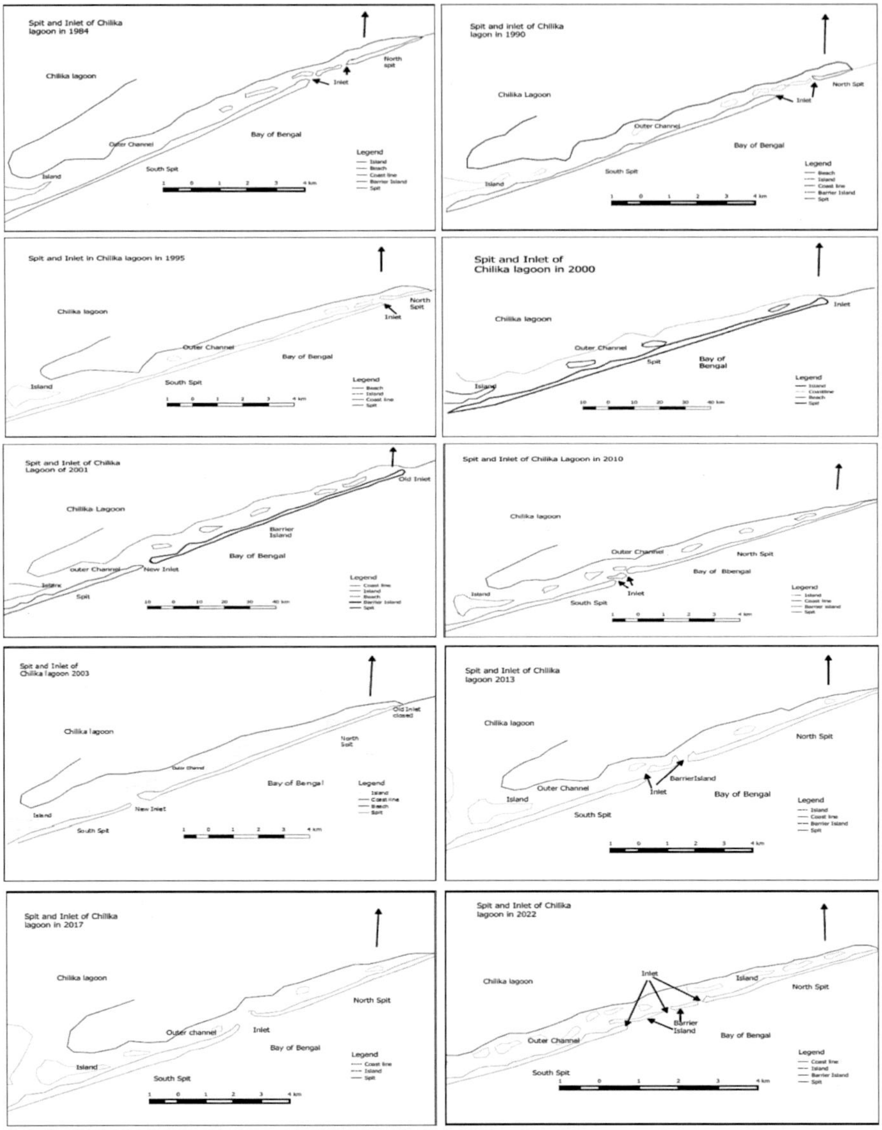

Fig. 7.2 Spatio-temporal changes of the spit and inlet in the Chilika lagoon from 1984 to 2022

27.24 km. in 2001. It increased to a maximum of 45.42 km. in 2003. The total length of the spit in 2022 is 42 kilometers. During this period, the average length of the spit is 42.89 km. The average annual growth of the south spit is +1.97 km. The north spit is +2.22 km long, while the total spit is −2.13 km long. (Table 7.1, Figs. 7.2 and 7.3a).

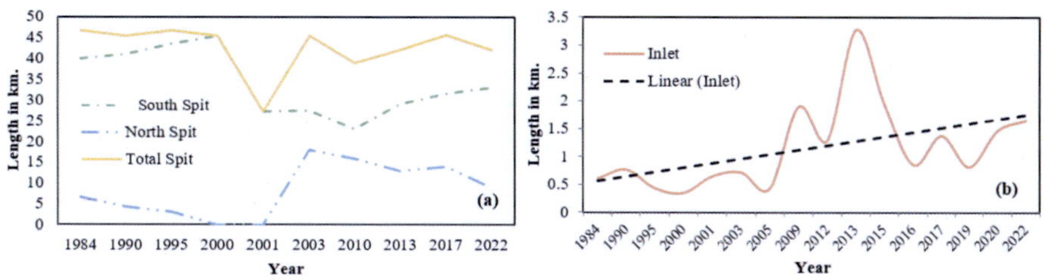

Fig. 7.3 (**a**) Change of the length of the spit in the Chilika lagoon (1984–2022) and (**b**) change of the length of the inlet in the Chilika lagoon (1984–2022)

7.4.3 Changing Inlet of the Lagoon

The inlet that allows the exchange of sea water between the sea and the smaller water bodies such as bays and estuaries is a complex hydraulic system that is in dynamic equilibrium with tidal range, inlet geometry, wave energy, and a variety of ever-changing sand shoals and bars. The location, width, and depth of the inlet depend on various factors. The new inlet becomes the main channel for the exchange of the tidal flow between the sea and the bay or estuary. The segmented end of the spit is then eroded by waves and driven shoreward, which transfers an enormous volume of sand from the updrift side to the downdrift side of the inlet. If the elongated transport of the sediment along a barrier island is great, a spit can form and grow into the downdrift inlet, causing it to constrict and change its location. Spit elongation redirects the flood and ebb tidal currents against the opposite side of the inlet, which erodes. Normally, the cycle of spit elongation, breaching, and sand bypassing takes decades to complete. The barrier breaching is more likely to originate from the bay or lagoon than the sea side. The cutting of an inlet and the resultant segmentation of the barrier result from a combination of factors. In most cases, the breached inlet is short-lived because the longshore drift of sand quickly closes the passage as sand drifts across the inlet mouth. Some breached inlets, however, may deepen and widen to become well-established features of the coastal zone. The coast of the Chilika lagoon is concave due to coastal erosion. The coastal erosion is higher on the north, resulting in a changing inlet and spit. There is a clear distinction between the inlet before 2000 and after 2000. Before 1984, in 1971, due to the impact of a super cyclone, three inlets were developed in the lagoon. The inlets were located at the extreme end of the inner channel at Arakhakuda village as a result of the migration of inlets due to the increasing length of the southern sand spit. The barrier island was formed by the breaching of the north spit due to storm action, forming two inlets separated by the barrier island. Two inlets separated by a barrier island continued until 1990. In 1991, one inlet closed as the barrier island joined the northern spit due to sand deposited by the longshore current. Due to the increasing length of the southern spit, the inlet length decreased from 0.60 to 0.34 km. In 1996, the north spit was breached by tidal waves, forming a new inlet and a barrier island.

In 1997, the new inlet closed as the barrier island again joined with the north spit. The inlet further migrated, and the south spit extended more to the north. In 1999, the north spit vanished due to erosion, and the south spit extended further north, close to the north shore of the lagoon, reducing the length of the inlet. The 1999 super cyclone that devastated coastal Odisha did not develop a new inlet, and the north spit was completely eroded. The flow of tidal water into the lagoon drastically reduced, affecting the lagoon ecosystem due to the narrow and shallow inlet at the extreme end of the outer channel. To protect the estuarine character of the lagoon, the Chilika Development Authority (CDA) opened a new inlet by dredging the southern spit by 0.34 km. On September 23, 2000, at Sipakuda, near the inner inlet of the lagoon, after the new inlet opened, the flow of lagoon water in the old inlet channel reduced, leading to the closure of the old inlet in 2003 due to the extension of the barrier island to the north shore of the lagoon (Table 7.2, Fig. 7.3b).

Initially the inlet was narrow but subsequently expanded by the natural process due to the increasing frequency of tropical cyclones originating over the Bay of Bengal and making landfall in the Odisha coast. In 2008, a second inlet opened by breaching the north spit at a distance of 0.64 km north of the old inlet at Gabakunda due to high energy waves associated with a solar eclipse condition. A barrier island is of 0.64 km^2. The length of the new inlet was 0.81 km, and the old inlet was 0.38 km. Due to erosion of the barrier island, the two inlets joined to form one inlet with a length of 1.37 km. The inlet at Dhabali opened naturally north of the Gabakunda inlet in September 2012. As the barrier island eroded and became smaller, two inlets joined to form one wide inlet in 2012. In 2013, due to the impact of severe cyclonic storm Phailin, a new inlet opened at a distance of 1 km from the

Table 7.2 Change of length of the inlet in the Chilika lagoon

Year	Length in Km	Change of length in km.	Trend
1984	0.60	–	–
1990	0.76	+0.16	Increase
1995	0.43	−0.33	Decrease
2000	0.34	−0.09	Decrease
2001	0.64	+0.3	Increase
2003	0.71	+0.07	Increase
2005	0.42	−0.29	Decrease
2009	1.90	+1.48	Increase
2012	1.29	−0.61	Decrease
2013	3.28	+1.99	Increase
2015	1.94	−1.34	Decreased
2016	0.85	−1.09	Decreased
2017	1.37	+0.52	Increased
2019	0.81	−0.56	Decreased
2020	1.46	+0.56	Increased
2022	1.65	+0.19	Increased
Average	1.52	+0.96	Increase

old inlet by breaching the north spit. A barrier island of 1.86 km developed. The length of the new inlet was 1 km. The total length of the inlet was 3.28 km. The new inlet expands more than the old inlet. Due to the erosion of the barrier island, the north inlet became wider, and due to the growth of the southern spit, the south inlet became narrower in 2014. Further growth of the southern spit and erosion of the barrier island's north and south inlets merged to form one wide inlet of 1.94 km length. In 2016, the inlet became narrower than 0.85 km and both spits bent towards the inner channel. In 2017, the width of the inlet increased to 1.37 km. The inlet expanded in 2018 due to the impact of Cyclone Titli. In 2019, the south spit grew and the north spit breached and formed two more inlets due to the impact of super cyclone Fani, which developed two long barrier islands of 2.60 km and 1.54 km. Three inlets developed at 0.24 km, 0.29 km, and 0.13 km, respectively, and both ends of the barrier island bend towards the inlet channel. In 2020, one north inlet closed and two inlets with a long barrier island developed. The length of the barrier island was 2.72 km, and two inlets are 0.31 and 0.35 km length. The inlet of the lagoon is constantly changing its location at different rates. From 1984 to 2002, the inlet shifted towards the north by about 5.06 km. On an average, annually the inlet shifted 0.31 km. After 2000, the new inlet shifted about 7.72 km; from 2000 to 2022, on average, the inlet shifted 0.35 km. The southern spit is accreted, and the barrier islands and the northern spit are eroded. The southern spit is becoming wider and higher, while the northern spit is becoming narrower and lower. The southern spit head is bending towards the south, whereas the northern spit head is bending towards the north due to the impact of storm waves. The barrier island thus formed subsequently joins with the southern spit (Table 7.2).

7.4.4 Annual Flow of Water and Sediment in the Mahanadi River

Due to climate change coupled with the construction of water reservoirs in the upper catchment, the flow of water both on and off season has reduced, leading to heavy siltation of the lagoon and river beds. The distributaries of the Mahanadi river draining into the lagoon are silted up, leading to reduced discharge of water to the lagoon; hence, the flow of the water from the lagoon to the sea has been reduced, leading to a shallow outlet and intensifying the spit and barrier island formation. The outlet is more controlled by the lake discharge than the sea waves. The water flow in the Mahanadi River has been reduced from 70,586 MCM in 1973–27,963 MCM in 2018 in the last 45 years, which is about 60% less. The sediment load of the river is also drastically reduced, from 24,307,629 MT in 1973 to 8,768,664 MT in 2018, which is about 64% less (Das 2021; Das et al. 2022). Accordingly, the lagoon receives less fresh water and sediment annually. The inlet gradually reduced and closed. Tidal action dominates the northern area of the lagoon, where tidal plains and tidal channels are evident in the island sector. Silt and clay are the prevailing sediments deposited in the low-energy settings of the lagoon (Paul 2014). The

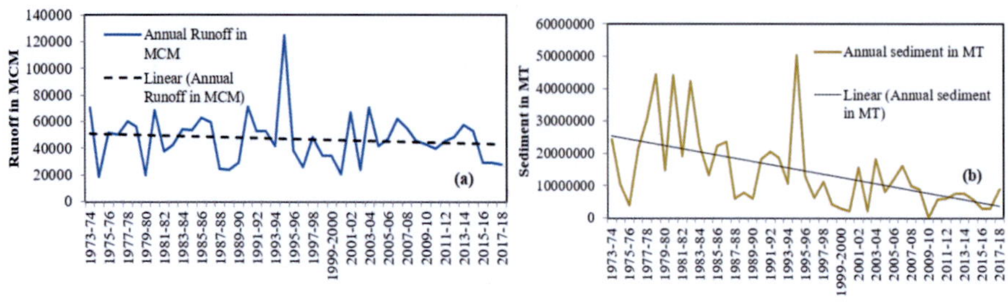

Fig. 7.4 (**a**) Annual runoff in the Mahanadi River (1973–1918) and (**b**) annual variation of sediment in the Mahanadi River (1973–2018)

morphological changes depend not only on the littoral drift but also on the tide dynamics and strong currents in the lagoon (Fig. 7.4a, b).

7.5 Conclusion

The lagoon receives fresh water and sediment mainly from the distributaries of the Mahanadi River. Due to the reduced flow of water and sediment through the distributaries of the Mahanadi River, the spit extends to the north and closes the inlet with the sediments deposited by the longshore current and tidal waves. Due to an inadequate discharge of water from the lagoon to the sea, the inlet gradually silted up by the littoral drift, which reduced the inlet and increased the spit. The spit extends from south to north and joins the north coast by completely separating the lagoon from the Bay of Bengal. Cyclonic storms associated with strong wave energy beached part of the spit, thus opening the wider inlet. With increasing spit, the inlet migrates towards the north. The old inlet closed as the artificial inlet opened. The survival of the lagoon depends on both the natural process and anthropogenic factors in the development of the inlet. In the absence of a strong cyclonic storm and tidal surge, the growth of the spit continues uninterrupted to the north and closes the inlet. To maintain the salinity of the lagoon, opening of the inlet artificially is the only alternative. The flow of a huge volume of sea water into the lagoon upsets the salinity, causes ecological imbalance, and floods human settlements. The artificial inlet was opened after the natural inlet was closed to maintain the brackish water quality of the lagoon. The river mouths need to be dredged to ensure an increased flow of flood water into the lagoon during monsoon season. Increasing the flow of lagoon water to the sea will remove the spit and keep the inlet wide and deep.

References

Ahmad Shah, R., Khan, I., Rahman, A., Kumar, S., Achyuthan, H., Shukla, A. D., et al. (2022). Holocene climate events and associated land use changes in the eastern coast of India: Inferences from the Chilika lagoon. *The Holocene, 32*(10), 1081–1090.

Das, S. (2021). Dynamics of streamflow and sediment load in Peninsular Indian rivers (1965–2015). *Science of the Total Environment, 799*, 149372. https://doi.org/10.1016/j.scitotenv.2021.149372

Das, S., Kandekar, A. M., & Sangode, S. J. (2022). Natural and anthropogenic effects on spatio-temporal variation in sediment load and yield in the Godavari basin, India. *Science of the Total Environment, 845*, 157213. https://doi.org/10.1016/j.scitotenv.2022.157213

Engstrom, W. N. (1974). Beach foreshore sedimentology and morphology in the Apostle Islands of northern Wisconsin. *Journal of Sedimentary Research, 44*(1), 190–206.

Kokot, R. R., Monti, A. A., & Codignotto, J. O. (2005). Morphology and short-term changes of the Caleta Valdés Barrier Spit, Argentina. *Journal of Coastal Research, 21*(5), 1021–1030.

Mohanty, P. K., Panda, U. S., Pal, S. R., & Mishra, P. (2008). Monitoring and management of environmental changes along the Orissa coast. *Journal of Coastal Research, 24*(10024), 13–27.

Paul, A. K. (2014). Morphology of estuaries and tidal inlets: An emphasis on Hugli, Subarnarekha and Chilika systems along the Bay of Bengal shoreline. In *Proceedings of the national conference on modern trends in coastal and estuarine studies*. Tilak Maharashtra Vidyapeeth.

Paul, A. K. (2022, August). Dynamic behaviour of the estuaries in response to the phenomenon of global warming in the coastal ecosystems of West Bengal and Odisha, India. In *Transforming coastal zone for sustainable food and income security: Proceedings of the international symposium of ISCAR on coastal agriculture* (pp. 907–931). Springer.

Paul, A. K., Islam, S. M., & Jana, S. (2014). An assessment of physiographic habitats, geomorphology and evolution of Chilika lagoon (Odisha, India) using geospatial technology. In C. Finkl & C. Makowski (Eds.), *Remote sensing and modeling. Coastal research library* (Vol. 9). Springer. https://doi.org/10.1007/978-3-319-06326-3_6

Sahoo, R. K., Mohanty, P. K., Pradhan, S., Pradhan, U. K., & Samal, R. N. (2018). *Bed sediment characteristics and transport processes along the inlet channel of Chilika Lagoon (India), 47*(2), 301–307.

Vivek, G., Goswami, S., Samal, R. N., & Choudhury, S. B. (2019). Monitoring of Chilika Lake mouth dynamics and quantifying rate of shoreline change using 30 m multi-temporal Landsat data. *Data in Brief, 22*, 595–600.

Chapter 8
Barrier Spit Morphology and Beach Ridge Formation in the Subarnarekha Delta: A Review of the Protective Functions of the Low-Lying Coast

Ashis Kumar Paul, Sudip Dasgupta, Anurupa Paul, and Joydeb Sardar

8.1 Introduction

The present-day Subarnarekha delta represents successive beach ridges and wide swales on the right bank and left bank of the river system in parallel to the Bay of Bengal shoreline. The landward beach ridges chenier extend up to Khejuri on the bank site of the Hugli estuary, and to the west, the chenier sand ridges extend up to Chandipore on the bank margin of the Burahbolong River. There are nine sets of chenier-beach ridge sands on the right bank and four sets with other subsets of beach ridge cheniers extending towards the left bank of the Subarnarekha River. The wider swales between successive beach ridges provided the low-lying flat surfaces for acting as linear tidal basins during the late Holocene evolution of the Ganga delta towards the east (Paul, 2002; Chakrabarti, 1991, Dalabehera et al., 2020). Marine transgression and regression phases with temporal shifts of the shorelines from early Holocene to mid-Holocene, mid-Holocene to late Holocene, and late

---

A. K. Paul (✉)
Department of Geography, Vidyasagar University, Midnapore, West Bengal, India
e-mail: akpauleastcoast@gmail.com

S. Dasgupta
Department of Geography, Asutosh College, Kolkata, West Bengal, India

A. Paul
Department of Remote Sensing & GIS, Vidyasagar University, Midnapore, West Bengal, India
e-mail: anurupapaul2017@gmail.com

J. Sardar
Department of Remote Sensing & GIS, Vidyasagar University, Midnapore, West Bengal, India

Centre for Environmental Studies, Vidyasagar University, Midnapore, West Bengal, India
e-mail: joydebsardar18@gmail.com

A. K. Paul, A. Paul (eds.), *Crisis on the Coast and Hinterland*,
https://doi.org/10.1007/978-3-031-42231-7_8

Holocene to recent Sub-recent stages played a major role in the modifications of the top set of the Subarnarekha delta in the form of beach ridge cheniers and swales (Paul, 2002; Jana & Paul, 2020; Jana et al., 2014). In the later stage, the swales were modified by Subarnarekha flood deposits and the occurrences of annual floods in the region. The underlying delta plain of Subarnarekha below the marine depositional surface evolved with the extension of fluvial distributary flood basins over wide areas along the Bay of Bengal shoreline. Various types of drainage features, like the exposures of shoreface tidal river mouths from west to east, paleo-channels, abandoned river courses, older natural levees beheaded streams, and channel meanders at the river courses, justify the idea of an underlying pre-Holocene delta plain below the chenier delta in the region (Paul, 1997, 2002; Niyogi, 1968; Maiti, 2013; Kamila et al., 2021).

Chenier means the deposition of linear sand ridges over the muddy tracts in a coarsening upward sequence of the sediments (Tamura, 2012). Beach ridges, on the other hand, represent high-energy wave deposits with the nucleus of beach berm formation and modification by washover processes along the active shorelines of different phases. A barrier bar emerges when a sand spit extends outward from the main land with active longshore current deposits and separates the low-lying backshore wetland from the impacts of an open marine environment. As the linear sand bar acts as a barrier against marine influences to protect the backwater environment, the depositional feature is known as a barrier bar. A barrier spit or bar may be converted into a barrier island if beaches occur across the linear deposit by occasional landfalls of cyclonic storms and detouch from the mainland into a separate unit of landmass.

Barrier spit morphology and beach ridge formation are demonstrated in the present work, considering their sequences of development in the right bank of the

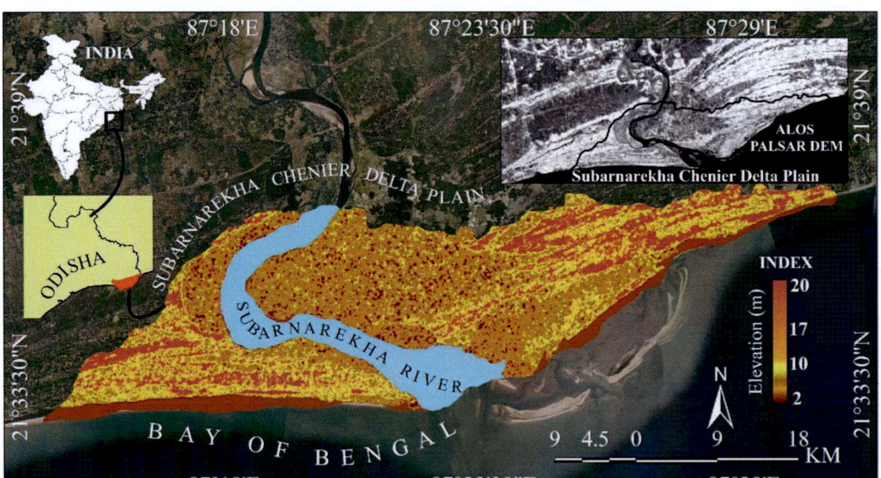

Fig. 8.1 The study areas of beach ridge chenier deposits in Subarnarekha delta plain landscapes represented with ALOS PALSAR DEM and HH polarisation band image

Subarnarekha delta (Fig. 8.1). Presently, the cross-sectional forms of the four sets of chenier sand ridge formations are well represented in the region as a linear offshore bar, barrier sand spit and overlying sand dune, paleo beach ridge feature, and the landward extension of the a paleo beach ridge capped with sand dunes. They are parallel to the present-day shoreline of the Bay of Bengal and separated by linear tidal basins, swale flats dominated by tidal basins with salt marshes, and swale basins with mangrove wetlands and tidal creeks. Related morphology, processes involved, and spatio-temporal changes over a timescale are explained to highlight the possible mechanism of their formation during the presence of active shorelines.

8.2 Materials and Method

Application of the beach ridge chenier deposits in paleo environmental reconstruction needs local-level studies with field-based topographic surveys, sediment analysis, and assessments of long fetch waves and event frequencies of tidal inundations into the back shore swales. The younger beach ridge chenier deposits of the south eastern bank margin environment of the Subarnarekha estuary along the Bay of Bengal shoreline demonstrate the mechanism of their formation in the present-day conditions. Survey of India's toposheets (1930–1931 and 1973), aerial photographs (1950), Google Earth Images, Landsat 7, and ALOS PALSAR DEM (2008) are utilised to delineate the boundaries of the linear deposits and other geomorphic features associated with them and to prepare a contour plan of the region. Field surveys during the pre-storm and post-storm phases, as well as during the tidal emersion and tidal submergence periods to identify the sedimentary depositional sequences from different depths using the hand piston auger, provided information regarding the mechanism of the beach ridge formation. Existing radiocarbon dates are utilised to reconstruct the paleo shoreline of the delta plain.

8.3 Results and Discussion

The section discusses present-day shoreline processes, beach ridge formation, chenier deposits, modification of the wide spaced linear swale between successive beach ridges, paleo beach ridges and shoreline shifting characters, and climate change reworking of the beach ridge chenier deposits in the Subarnarekha delta.

8.3.1 *Multiple Shoreline Processes*

The Subarnarekha deltaic shorelines are influenced by long swell waves with steeper and moderately steeper gradients due to their long fetch distance location towards south southwest and south east directions across the Bay of Bengal.

The configurations of Chandipore Bay with occupational south-west monsoon winds stress and impacts of Coriolis force over the hydraulic surface helped to generate a strong long-shore current parallel to the shoreline from southwest to northeast directions in the region to transport sediments and to accumulate them after crossing the Subarnarekha estuary mouth in shore parallel direction. Thus, sediment delivery into the system is guided by events such as river floods, cyclonic storms, and associated surge waves across the shallow shelf, and littoral drift currents provide materials and various energy levels for accumulation of bars. The alternate clay deposits and sand sheet deposits over the island flats, estuary fringes, shorelines, and swale basins are influenced by tidal currents, inundation events, and tidal surge waves, as well as the occurrences of tidal bores into the estuary. The plumes of suspended loads thrown into the shallow offshores from the mouth of the Subarnarekha River during the mouth of October take time to resettle within February by acting longshore currents and tidal currents in the coastal zones (Fig. 8.2). Signatures of recurved deposits at the river mouth and sea face are actually guided by the dynamic diversion of tidal currents. Active wind waves of steeper gradient occur along the shoreline during the southwest monsoon season and play a significant role in beach deposits.

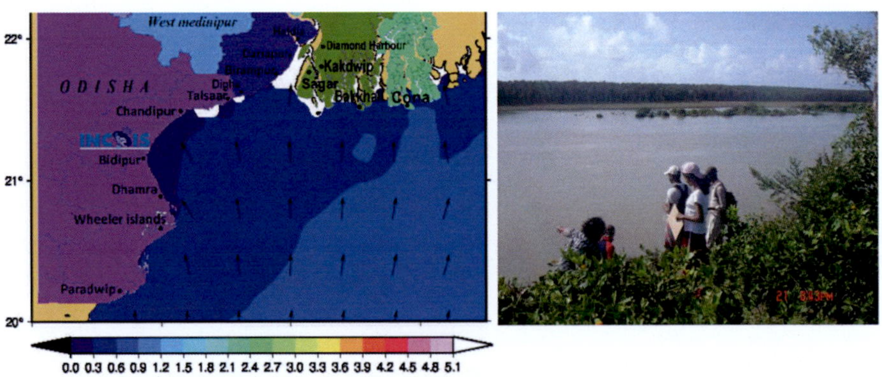

Fig. 8.2 Swell wave height (m) and direction (degree) in the shallow marine shelf in the northern part of Bay of Bengal approaching towards the shoreline (INCOIS-2023), and the linear tidal basin in the backwater lagoonal setting behind the barrier and spit morphology of Talsari in high tide phase, Odisha, India

8.3.2 Beach Ridge Formation

Beach drift materials are supplied by swash deposits with steeper gradient waves in an angular direction after breaking the incident wave energy, which supplies medium-sized sand particles with fragments of sea shells into the shoreline beaches. Particularly during the monsoon months, at the storm breaks, and in the events of approaching swell waves, longshore current-derived sands are transported across the beaches to form beach berm platforms in the upper and middle parts of the beach face along the shorelines. Gradual accumulation of medium- to coarse-sized sands over wide areas will build up a relatively higher platform on the landward side than the seaward dipping beach face in the region. Such a built-up platform can modify the beach gradient on the sea face by producing a typical rhythmic feature with beach cups, horns, rip channels, and swash marks.

The higher water levels caused by storm surge and tidal surge waves can drift the materials further landward through ridge push mechanisms during the washover process, but they are unable to shift or push them as they are backed by foredune ridges parallel to the shoreline. Thus, dune fronts are marked by cliffs and sometimes breached by washover encroachments at the lower, elevated fore dunes (Fig. 8.3). However, paleo beach ridges of the delta plain surface had restored the wider beach ridge platforms before the accumulation of chenier sand ridges over them during the presence of active shorelines in that region (Paul, 1996a). Beach ridge sediments are compacted and oxidised with rainwater processes in the Holocene warm and humid environment. Dune sands over the beach ridges bear the signatures of cross-stratification, with finer and sorted sediments of different sets.

Beach ridges develop over the shoreline beaches, barrier bars, back spits, and also over the offshore bars due to the impacts of steeper gradient waves (Paul,

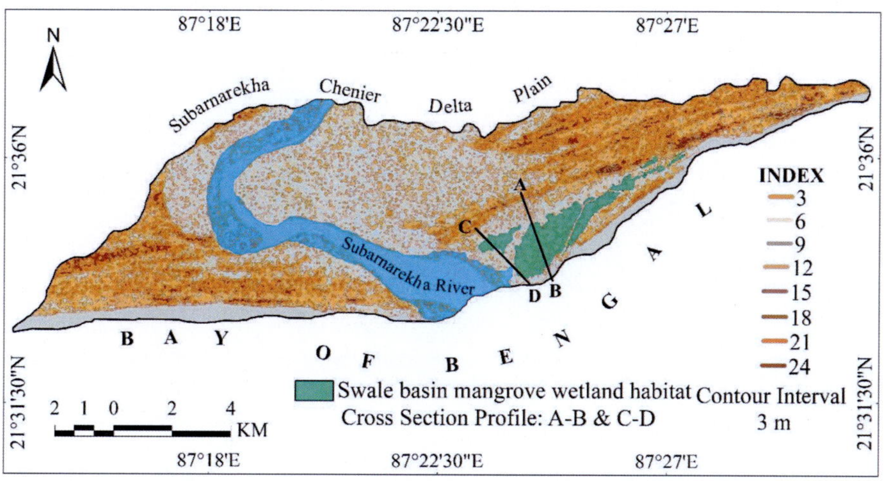

Fig. 8.3 The contour plan of the beach ridge chenier delta prepared from the ALOS PALSAR DEM to identify the morphological configurations

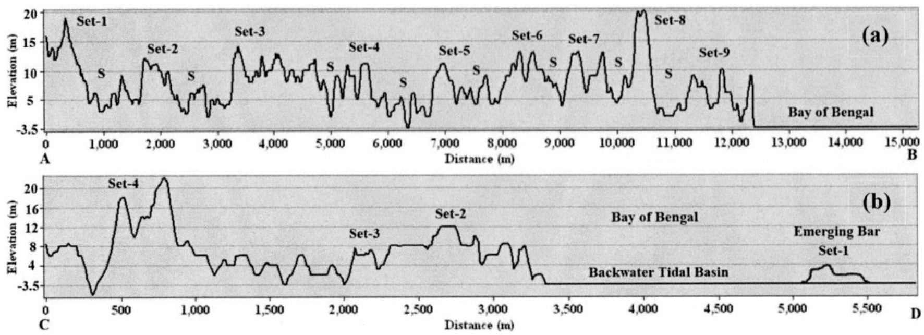

Fig. 8.4 (**a**) The cross-sectional forms of the nine set of beach ridge chenier deposits interspaced with series of swale flats and (**b**) the younger beach ridge chenier deposits backed by a paleo beach ridge chenier deposit surface separated by swale basins (based on ALOS PALSAR DEM-12.5 m spatial resolution, Key: S-Swale)

1996b, 2002). Subarnarekha estuary front shallow offshore bank accumulated liner bars parallel to the present-day shorelines, in which beach ridges have developed with breaker waves over wide areas in the form of turtleback surfaces. They are wave reworked, broken, and recurved at one end and pushing inland with overwash process during the storm attacks (2009, 2013, 2016, 2019, 2020, and 2021) in the region. Modifications are also going on in the linear deposits by longshore currents and steeper gradient swell waves. There are three sets of paleo beach ridges behind the offshore bars, and among them, the third set represents much higher deposits than the first and second sets in the right bank of the estuary. The paleo beach ridge surface is ranging from 1.65 to 2.50 m in height from the local surface and terraced in a few places at Talsari (Fig. 8.4a, b). The outcrop section of the paleo beach ridge formation bears compact sands with shell lines and bioturbated beds of the upper sequence. A large dune row with a 12 m to 17 m elevated surface of unconsolidated sands is overlying the paleo beach ridge platform in the region (Figs. 8.3 and 8.4a, b).

8.3.3 Chenier Sand Ridge Deposits

The relatively finer sands of wind-blown deposits in the forms of sandy hills, dune slopes, dune flats, and dune valleys are overlying the paleo beach ridges parallel to the present-day shoreline behind the linear tidal basin of swale topography. The entire paleo beach ridge chenier deposits are accumulated over the muddy banks of the paleo delta plain surface in coarsening upward sequence. The linear sand dune

Fig. 8.5 Younger beach ridge with berm platform and the paleo beach ridge capped chenier sand ridge deposit by aeolian process in Talsari, Odisha, India

ridges are covered by vegetation and hold the near-surface dune aquifers by rainwater filtering process. Stratigraphic exposure sites of the dune ridge at Kirtania demonstrated six to seven sets of cross beds and an underlain layer of paleosols (Fig. 8.5). They are indicative of two different distinct formations of older beach ridge platforms, and after a gap in sea level, still stand the overlying younger chenier dune ridge deposits. The dune sands are finer in size, highly sorted in character, and bioturbated in the upper parts.

8.3.4 Linear Swales with Tidal Basins

Paleo beach ridges and chenier deposits are separated by wide and narrow swales in the chenier delta. Present-day active swales are studied in the south-eastern part of the Subarnarekha estuary to reconstruct the paleo environmental characteristics of older swale flats in the region (Fig. 8.6). The swale-forming processes continued over time through different stages of development.

The four sets of swales of different ages demonstrated their evolutionary steps in the region. The 'youngest swale' is forming as a backwater setting immediately behind the bar-built beach ridges (2007–2023) on the sea face. Now the linear basin is modified with the filling of silty sands and finer sands over the basement clay (Fig. 8.7). The tidal flat is drained by three to four major tidal creeks, and basin margin emerged flats are colonised by smaller patches of mangroves. Similarly, the swale basin is inundated by tides twice daily, keeping it in contact with the estuary section and open marine environment through inlets and embayments. The next younger 'linear tidal basin' is emerging in between the two sets of beach ridge chenier deposits. Such a swale basin is filled up with tidal sediments and by washover sand deposits. The relatively higher flats of a swale basin are colonised by saltmarsh vegetation, drained by tidal creeks of entrenched valleys built by compact mud banks, and inundated by spring tide events.

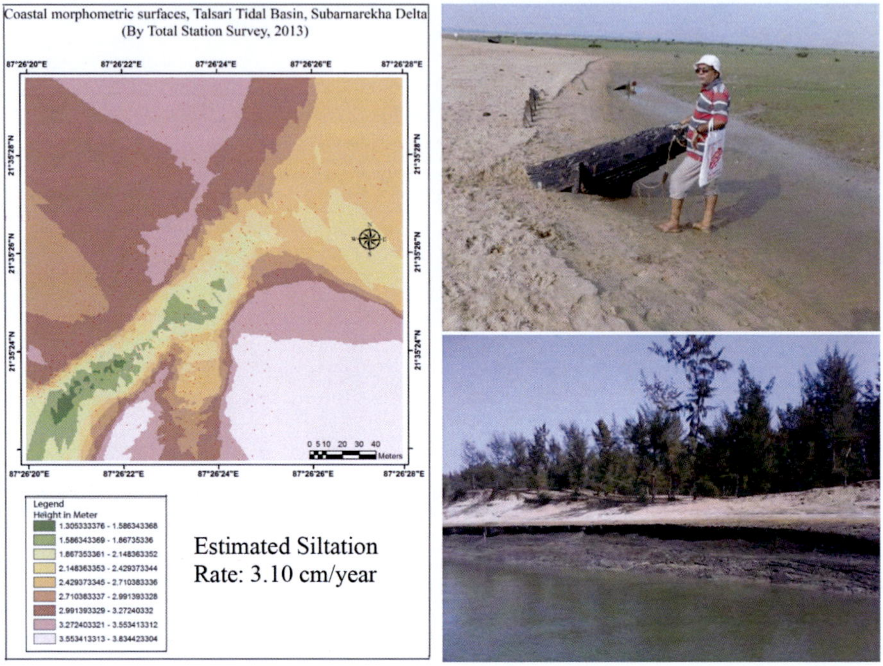

Fig. 8.6 Morphogenetic surfaces of the linear tidal basin represented by total station survey in Bichitrapur and Subarnapur areas on the right bank of Subarnarekha estuary. Advancing sand fan lobes over the marshy tract of Kirtania Island; and beach ridge chenier deposit over the paleo mudbank surface in the left bank of Subarnarekha estuary

However, the "third set of swale basins" (1950–2023) between the second, third, and fourth sets of beach ridge chenier deposits represent a linear tidal basin of backwater settings with mangrove forests, tidal mudflats, tidal creeks, and saltmarsh tracts. The basin is usually inundated by tides through the larger tidal creeks and inlets connected with the estuary mouth of Subarnarekha and the Bay of Bengal. Topographically, the basin fringe areas close to the successive beach ridges occupy higher flats (3.5 m above the low tide level), and silted channel beds expose at low tide along the middle and eastern parts of the tidal basin except the entrenched channel valley sections of the estuary fringe flat to the west. Alternate tidal, marine, and fluvial influences are well preserved in the sediment depositional facies along the basin fringe surface. Alternate sand-clay beds in off-lap-on-lap sequences highlight the impacts of tidal and marine overwash deposits. Sequences of compact mud

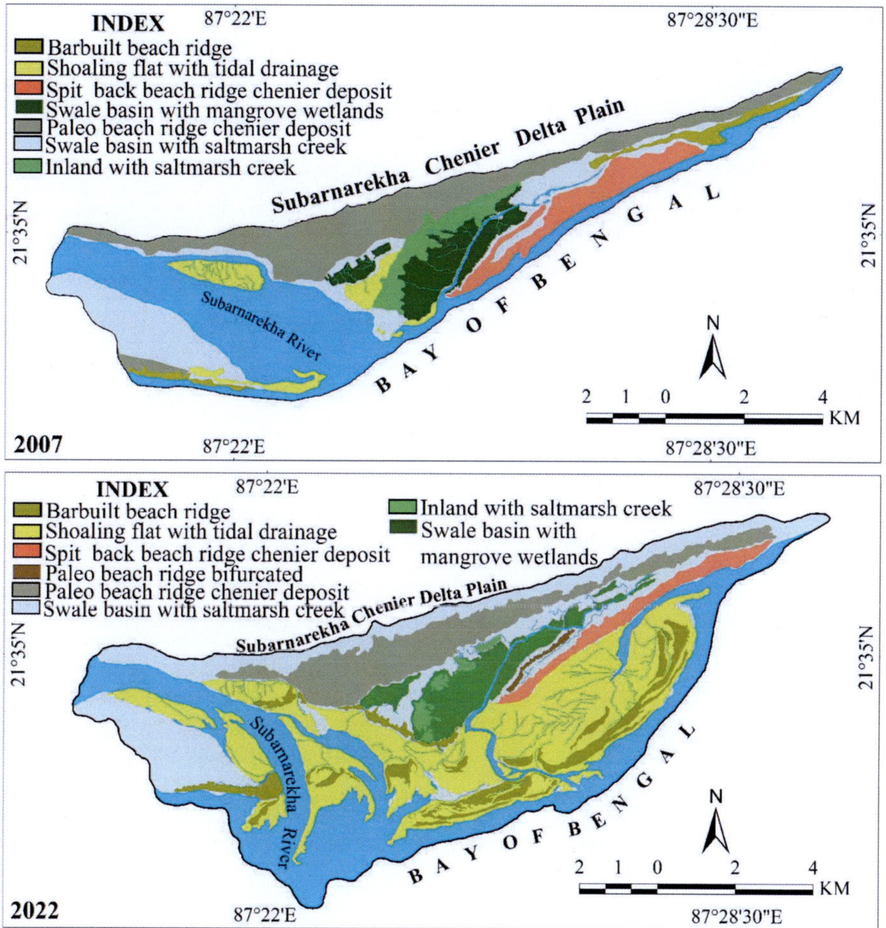

Fig. 8.7 Types of beach ridge chenier deposits in younger and older formations at the south eastern corner of Subarnarekha estuary mouth with their spatio-temporal diversity (2007–2022). Longshore current directions, tidal currents, and storm signatures played importance roles in the offshore bar and barrier bar formations at the initial stage of beach ridge deposits along the Bay of Bengal shoreline in the deltaic coast

layers with fluvial, tidal, and estuarine brackish water deposits are found in the exposure banks of deeper entrenched channel cut valleys in the swale basin (Tables 8.1, 8.2 and 8.3, Figs. 8.8 and 8.9).

The fourth set of swale basins represented an older sequence of sand, silt, and clay deposits that emerged flat and were usually influenced by Subarnarekha River flood alluviums behind the paleo beach ridge chenier sands of the landward side. Thus, the paleo environment of the mature swale basin is reconstructed as evolutionary stages through embayed shores, backwater settings, a sediment-filled tidal

Table 8.1 Stratigraphical sections of Kirtania Island in younger deposition surface and Talsari back water salt marsh terrace of matured mud bank surface of linear swale basin

Layers	Thickness (cm)	Soil type	Characteristics
Younger mud bank surface of Kirtania Island platform			
S2A	−20	Liquid mud	Swampy clay (slightly oxidised)
S2B	−15	Moderately compact	Dark clay with rusting effect
S2C	−10	Compact mud	Dark grey to black in colour
S2D	−55	Compact clay with mica flakes	Moderately dark to grey clay
S2E	−35	Thinly laminated clay	Alternate dark and grey colour
S2F	−40	Multiple sets of laminated clay with mica flakes	Dark black colour
Matured mud bank surface of the Talsari back water salt marsh terrace			
S1	−13	Mud	Dark grey in colour
S2	−17	Sand	Deep brown in colour
S3	−30	Sand	Whitish in colour
S4	−32	Sand	Yellowish in colour
S5	−8	Sandy mud	[a] Oxidised sandy mud with rusting colour in 1 m depth
			[a] In 92 cm from top surface decomposed organic layer of 8 cm thickness
S6	−2	Sand	Whitish grey in colour
			Oxidised
S7	−28	Muddy sand	
S8	−25	Muddy sand to sandy mud	Moderately oxidised and brownish in colour
S9	−45	Sticky clay	Black clay layer

[a] This a marker point between transgressive and regressive sequence of the marsh depositional basin

Table 8.2 Sediment textures of Kirtania Island over the younger depositional surface at different depths

	Parameter	(Depth wise sediments samples in Kirtania Island)					
	Soil texture	Clay loam	Clay loam	Clay loam	Clay loam	Clay loam	Clay loam
Sl. No	Depth (cm)	0–20	20–35	35–45	45–100	100–135	135–175
a	Sand (%)	27.5	25.4	29.3	28.7	31.9	27
b	Silt (%)	38.3	37.9	37.2	39.9	37.7	38.5
c	Clay (%)	34.2	36.7	33.5	31.4	30.4	34.5

basin with mangrove wetlands, an emerged flat with flood plain alluviums, and a wide-spaced mature swale basin with terrestrial wetland vegetation.

Table 8.3 Sediment textures of mature depositional surface at Talsari backwater salt marsh terrace in different depths

	Parameter	(Depth wise sediment samples in Bichitrapur)								
Sl. No	Depth (cm)	0–13	13–30	30–60	60–92	92–100	100–102	102–130	130–155	155–200
a	Sand (%)	44.1	86.2	87.1	85.9	86.7	45.9	25.3	47.3	28.2
b	Silt (%)	37.2	9.5	9.2	10	10.5	48	43	46.8	41.2
c	Clay (%)	18.7	4.3	3.7	4.1	2.8	6.1	31.7	5.9	30.6

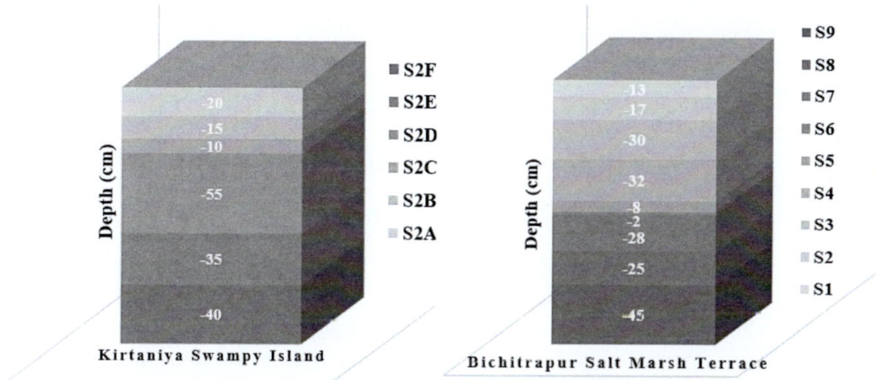

Fig. 8.8 Sediment sample cores of 1.75–2.0 m depths from Kirtania island and Talsari backwater saltmarsh terrace of the linear swale basin between successive beach ridges in Subarnarekha delta represent the alternate high energy and low energy depositional sequences

Fig. 8.9 Sediment sampling from different depths (1.75–2.50 m) in the salt marsh terrace of linear swale basin using Hand Piston Auger

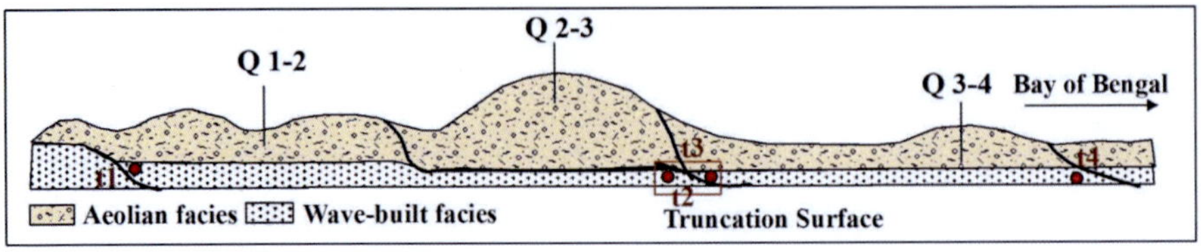

Fig. 8.10 Schematic representation of the sedimentary depositional environments with aeolian facies over the wave-built facies in paleo beach ridge chenier deposits in Subarnarekha delta with regressive phases of the sea level and the truncation by erosive phase at t2-t3. Maximum aeolian deposits took place in Q2-3 than the inland portion of Q1-2 and the sea facing barrier spit aeolian deposits at Q3-4 along the Bay of Bengal shoreline

8.3.5 Paleo Beach Ridges with Shifting Shoreline Characters

The study indicates that all the beach ridge chenier deposits were initiated on the sea face as linear offshore bars parallel to the shoreline. Similar depositional features are produced in the shallow offshore to the south east of the Subarnarekha estuary mouth after arresting the sediments derived from the major river flood of 2008 and the major event of the Aila storm surge of 2009 in the northern Bay of Bengal. From the temporal image data (ALOS PALSAR and Google Earth images), it is observed that steeper gradient wave breakers and surge waves accumulated sand over the bars parallel to the shoreline. Increasing event frequency of overwash has pushed the linear bar built beach ridge surface towards the land margins parallel to the present-day shoreline in the region. The possibility of a ridge collision may arise if the frontal beach ridge pushes further landward by the energy events in the near future. In such a condition, the swale basin will be narrowed or narrowly spaced between two successive beach ridge chenier deposits, and as a result, shoreline shifting will take place with ridge collision. In the late Holocene stage, similar incidents took place between the second and third sets of beach ridge chenier deposits (Fig. 8.10). Based on the above idea, along with the present-day shoreline, three more additional Paleo shorelines are demarcated in parallel to the linear beach ridge chenier deposits of the past (Fig. 8.10). However, the series of paleo beach ridge chenier deposits indicate multiple shorelines shifting phases in the delta plain with the past fluctuation of Holocene Sea levels.

As a result of sea level rise, fall, and standing periods of unequal timescale, the paleo shoreline shifting demonstrated irregular spacing between successive beach ridge chenier deposits and their bifurcation from the major linear deposits. Towards east of the Subarnarekha River, the growth of beach ridge chenier deposits is restricted, possibly due to the insufficient sediment supply from the fluvial discharges between 3000 and 5500 YBP, the influences of the Hugli delta lobes, and the relative upliftments of the western delta plain of the Subarnarekha.

8.3.6 Climate Change Reworking

Currently, the morphological signatures emerging on the sea face demonstrate the nature of climate change reworking of the beach ridge chenier deposits in the coastal setting. They are identified as (i) beach ridge truncation, (ii) overwash event frequencies, (iii) breach points across the spit-back beach ridge chenier deposits, (iv) beach berm narrowing and cliffing the chenier sand deposits, and (v) landward ridge pushing of bar-built beach ridge deposits, etc., phenomena after the emerging storminess of the sea.

Overstepping of the paleo beach ridge cheniers on the sea face by cut and fill processes and aeolian transport of sediments over the beach ridges are indicative of transgressive sequence in the coastal setting at present. The older beach ridge chenier (fourth set from the sea towards land) deposits extending from Kirtania to New Digha represents such a type of over stepped deposit when the shoreline environment was much more dynamic. As the anthropogenic controlled discharge and sediment load into the estuary reduced the sediment supply along the coastal setting, the longshore currents, storm surges, and tidal surges will rework the younger beach ridge chenier deposits by cut and fill deposits and roll over processes with washover encroachments. By planting vegetation, attempts have been made to trap the aeolian

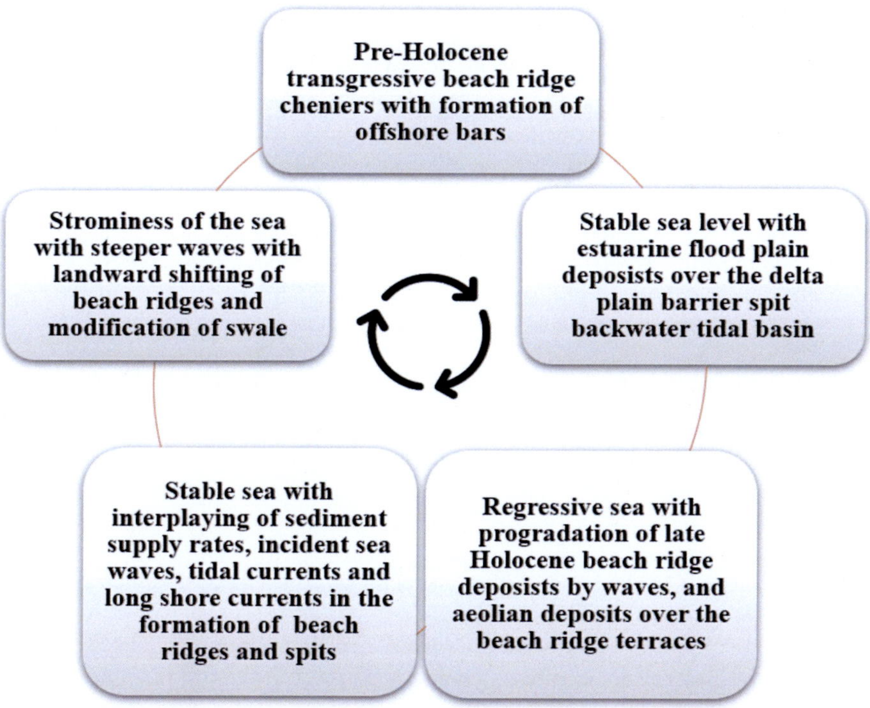

Fig. 8.11 Paleo depositional environments reconstructed on the basis of physiographic characters, sedimentary characters, and process variables of the beach ridge chenier deposits under sea level fluctuations

sediments over the beach ridge chenier deposits, but the incidents of high storm surge levels are pushing the linear deposits by active overwash process into the swales. Emerging shoaling bars and beach ridge formations are fragmented and dispersed in the estuary fringes by such reworking of spits and older beach ridges in the setting. The mangrove wetlands and salt marsh terraces along the swale basins are susceptible to increasing rates of sediment deposition and channel decaying processes through reworking of the landscapes.

8.3.7 Reconstructing the Depositional Environment

Considering the present-day formations and existing morphologies of the beach ridge chenier deposits, a conceptual model is formulated to reconstruct the paleo environment in the coastal settings of the Subarnarekha Delta. The coastal settings of the beach ridge chenier delta evolved through transgressive and regressive stages of the sea at different levels with sediment supply, marine and coastal processes, fluvio-tidal events, aeolian deposits, coastal plain gradients, and storminess of the sea (Fig. 8.11).

The northern most transgressive ridge with paleo beach ridge chenier sand deposits took place during the early Holocene period, when the shoreline was parallel to the linear depositional features for a long time. The next eight sets of beach ridges were developed in the regressive stages of the sea up to the present-day shoreline in the coastal setting during the late Holocene to recent-sub recent period. Beach ridges emerged as offshore bars on the active shoreline of the past, and the rate of sediment supply and level of the sea with surge waves played a significant role in their initial mechanism of formation. Successive aeolian sand deposits and overwash processes have made them wider and higher in elevation along the shorelines. Swales were modified with flood plain deposits by distributary channels of Subarnarekha during events of floods. The narrow spaced and wide-spaced swales were affected by shaping and reshaping with the influences of marine waves and tidal backwater lagoonal settings.

8.4 Conclusion

Subarnarekha Chenier Delta plain sedimentary depositional landforms consist of multiple sets of beach ridge chenier deposits on the right side and left side of the estuary with widely spaced and narrowly spaced swales. The paleo environmental reconstruction of the depositional features is carried out by the study. Morphogenetically, the features were evolved with the combination of sea level fluctuations, sediment supply rates, storminess of the sea, fluvio tidal, and coastal marine processes in the coastal settings. The low-height beach ridges were possibly developed as offshore bars parallel to the shorelines of the past. In many cases, the beach ridges are terraced and capped by aeolian dune ridge sediments and elevated up to 20 m in height above the local sea level. Such wind-blown deposits dominated by finer sands over the beach ridges represent chenier formation during the stable sea levels. Swales originated in backwater settings with the accretion of tidal deposits and the development of mangrove wetland habitats. Later, they were filled by flood plain deposits and emerged as valley flats between successive beach ridges.

Following the morphology, pattern of deposits, overstepping and downstepping characters, interspacing of the swales, and curvatures of the beach ridge cheniers, the sea level positions and sediment supply rates are reconstructed. The reworking of the beach ridge cheniers represented various geomorphic signatures. The schematic models formulated for the beach ridge formations and chenier deposits represent five stages of formation in coastal settings.

References

Chakrabarti, P. (1991). Morphostratigraphy of coastal quaternaries of West Bengal and Subarnarekha delta, Orissa. *Indian journal of earth sciences, 18*(3–4), 219–225.

Dalabehera, L., Hazra, A., Pattanayak, S., & Pal, T. (2020). The paleo beach ridges of Digha coastal tract, West Bengal, India: Observation and implication for sea regression from 500 YBP to 200 YBP. *Journal of the Geological Society of India, 95*(2), 131–144.

Jana, S., & Paul, A. K. (2020). Chronological evolution of the channel functional units in association with palaeo-hydrogeomorphological environment in the ancient delta fan of Subarnarekha basin, India. *Environmental Earth Sciences, 9*, 331. https://doi.org/10.1007/s12665-020-09093-1

Jana, S., Paul, A. K., & Islam, S. M. (2014). Morphodynamics of barrier spits and tidal inlets of Subarnarekha Delta: A study at Talsari-Subarnapur spit, Odisha, India. *Indian Journal of Geography and Environment, 13*, 23–32F.

Kamila, A., Paul, A. K., & Bandyopadhyay, J. (2021). Exploration of chronological development of coastal landscape: A review on geological and geomorphological history of Subarnarekha Chenier delta region, West Bengal, India. *Regional Studies in Marine Science, 44*, 101726.

Maiti, S. (2013). Interpretation of coastal morphodynamics of Subarnarekha estuary using integrated cartographic and field techniques. *Current Science India, 104*(12), 1709–1714.

Niyogi, D. (1968). Morphology and evolution of the Subarnarekha Delta, India. *TridsskriftSaertrkyafGeografisk, 67*, 230–241.

Paul, A. K. (1996a). Chenier beach ridge and chenier sand ridge formations around Subarnarekha estuary. *National Geographer, XXXi*(1&2) Allahabad, 143–153.

Paul, A. K. (1996b). Identification of coastal hazards in West Bengal and parts of Orissa. *Indian Journal of Geomorphology, I*(1) New Delhi, Academy & Law serials, 1–27.

Paul, A. K. (1997). Coastal erosion in West Bengal. *MAEER MIT Pune Journal, IV*(15k &16k) (special issue on coastal environmental management), 66–84.

Paul, A. K. (2002). *Coastal geomorphology and environment* (pp. 1–342). ACB Publication.

Tamura, T. (2012). Beach ridges and prograded beach deposits as palaeoenvironment records. *Earth-Science Reviews, 114*(3–4), 279–297.

Part II
Assessment Through Environmental Approaches

Chapter 9
The Degradation of Coastal Habitats in Andhra Pradesh and Tamil Nadu: An Environmental Approach

Ashis Kumar Paul, Anurupa Paul, Joydeb Sardar, and Bubay Bid

9.1 Introduction

The geographical habitats of Andhra Pradesh coast and Tamil Nadu coast are supported by four major deltas (Godavari, Krishna Cauvery delta and Penna R. delta), bay fringed shores of Kakinada, Machilipatnam (Masulipatnam), Chirala, Palk Bay, Mandapam Bay and other embayed shores, rocky shores of hill ridges, river systems, various lakes and backwaters, and beach dune fringed shorelines. They harbour mangrove ecosystems, coral reef ecosystems, sea grass ecosystems, beach dune ecosystems, and other wetland ecosystems along the coastal stretches of the region. The purpose of this chapter is to investigate the types of coastal habitat degradation and their causes in the regional settings of the long coast. For this study, the entire coast is categorised into ten different types of geographical sections, from Kanyakumari to Ichchapuram. The geospatial techniques helped to identify the specific coastal habitats in the region. The nature of habitat degradations in the coastal stretch was revealed by geospatial changes in the coastal habitat. The assemblage of plants and animals together with the abiotic environment of the tropical coast represented productive ecosystems in the transition between land and sea. They are categorised as (i) rock intertidal shore platform ecosystems, (ii) mangrove ecosystems of the deltaic shores, (iii) coral reef ecosystems at the shore fringe shelf banks, (iv)

A. K. Paul
Department of Geography, Vidyasagar University, Midnapore, West Bengal, India

A. Paul (✉) · B. Bid
Department of Remote Sensing and GIS, Vidyasagar University, Midnapore, West Bengal, India
e-mail: anurupashis@gmail.com

J. Sardar
Centre for Environmental Studies, Vidyasagar University, Midnapore, West Bengal, India

A. K. Paul, A. Paul (eds.), *Crisis on the Coast and Hinterland*,
https://doi.org/10.1007/978-3-031-42231-7_9

beach dune ecosystems along the sandy shores, (v) sea grass ecosystems along the bay fringed shores, (vi) riverine ecosystems along the coastal drainage channels, (vii) other wetland ecosystems of the coastal zones, and (viii) marine aquatic ecosystems of the shallow seas. The natural drainage flows into the coast, along with favourable sea conditions that harbour such diverse ecosystems, but human factors and climate change-induced physical factors are causing the degradation in quality of coastal habitats. Earlier studies (Das, 2021, Das et al., 2022, Mukhopadhyay, 2007, Ramkumar et al., 2016, Paul, 2005; Paul et al., 2018, 2023b; Jagtap, 1996; Jagtap et al., 2003, Thangaradjou & Kannan, 2005; Nageswara et al., 2015; Manikandan et al., 2011; Murty et al., 2022; Sengupta & Ghosal, 2017) demonstrated the reduction in sediment discharges and stream flows of the east flowing river systems of the peninsular and extra peninsular India. So far as, 17 factors (11 human factors and 6 physical factors) are considered to assess the habitat quality index in the present study. The study highlighted that beach dune ecosystems, mangrove ecosystems, riverine ecosystems, and other wetland habitats of the Coromandel coast (Andhra Pradesh and Tamil Nadu coasts) are degraded in comparison to the other habitats.

9.2 Study Area

The present study areas are extended in Andhra Pradesh and Tamil Nadu coast along the shorelines of 1030 km-long sections in the Bay of Bengal fringe areas. The south-eastern coast of the subcontinent represents the most diverse coastal habitats in India. Human encroachment on coastal processes and resource exploitation have increased significantly in the region. The study area is divided into 17 sections to identify the 8 types of coastal habitats and also for assessing their quality (Fig. 9.1). There are five major shipping ports and seven major cities located on the shore fringe coasts of Andhra Pradesh and Tamil Nadu. The southeast coast is also drained by a number of major (Godavari, Krishna, Cauvery, and Penner) and minor river systems (Vamshadhara, Nagavali, Gosthani, Sarada, Vaigai, Gundar, Vaipar, Thamirabarani, Manimutthar, Vellar, Vellaru, Gadilam, Palar, Penar, Varagali, Swarnamukhi, and Kondurpalem, etc.).

9.3 Materials and Methods

The current study investigates coastal habitat degradation on temporal and spatial scales using Sentinel-2A and Landsat 8 (multispectral data). Environmental data of fresh water flows and sediment discharges by the river systems, cyclone landfalls into the coastal zones, sea level rise rates, shoreline erosion-accretion rates, salt water encroachments into the coastal aquifers, 2004 tsunami impacts, and other human factors are considered from different reports and published papers for the

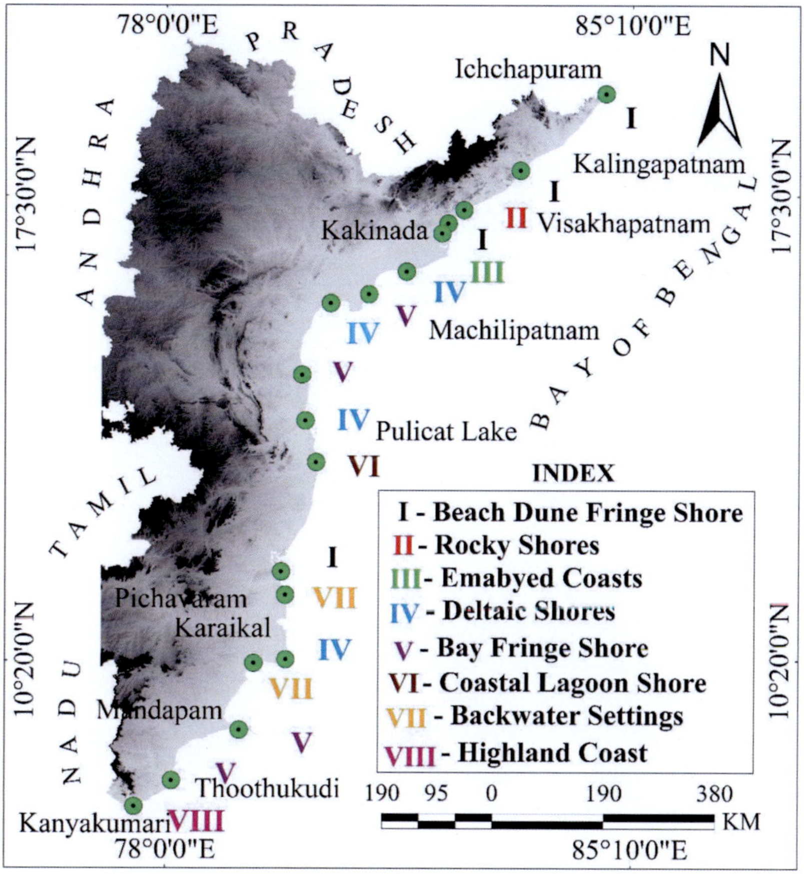

Fig. 9.1 The diversity in coastal habitats along the Coromandel Coast (Andhra Pradesh, Puducherry and Tamil Nadu, India)

study. Multicriteria decision analysis (MCDA) is conducted using the weighted sum method for assessing the habitat quality index for the coastal sections of Andhra Pradesh and Tamil Nadu. Considering the MCDA for each section, a final map is prepared to represent the habitat diversity and quality along the coasts. Finally, the validation is done through field studies in site-specific areas and through the Google Earth image of the coastal belts.

9.4 Results and Discussion

The following sections will demonstrate the coastal habitat types and their locations, the temporal changes of the coastal habitats and degradation trends, the identification of multiple factors causing the degradation of coastal habitats, the estimation method of the habitat quality index, and an attempt to conserve the degraded habitats of the sensitive coastal environment (Ghorai et al., 2016, 2017a, b). The southeast coast of India will suffer from climate change impacts and resource use conflicts due to the alarming rate of population density in the coastal districts of Andhra Pradesh and Tamil Nadu.

9.4.1 Coastal Habitat Diversity in Andhra Pradesh and Tamil Nadu

The south-east coast is extended from Kanyakumari in Tamil Nadu to Ichchapuram in Andhra Pradesh along the Bay of Bengal across 27 coastal districts. A significant number of major and minor rivers are crossing the coastal districts from west to east and debouching into the Bay of Bengal. A few of them have built up wide and extensive deltas at their mouths with sedimentary depositional landforms. Such deltaic and marine alluvium plains harbour freshwater wetlands and coastal wetlands in the region. The fertile alluviums are also cultivated for growing crops, inhabited by population, and provided the space for urbanisation along the coastal zones. The inland coasts, the shorelines, and the shallow offshores supported eight types of coastal habitats, which are sensitive and fragile in consideration of their environmental significance under fluvial, tidal, and marine processes (Paul et al., 2023a). Southern Tamil Nadu and southern Andhra Pradesh harbour multiple habitats in the coastal sections (Table 9.1, Fig. 9.2).

Table 9.1 The location of habitat types and their characters in the Coromandel coast along the south-eastern shoreline of India

Habitat types	Geographical location	Characters
River mouth ecosystems along the drainage channels	Coastal drainage systems	Freshwater ecosystems
Other wetland types	Inland floodplains	Wetland ecosystem
Rocky intertidal shores	Shore platforms of the rocky coasts	Intertidal organisms
Beach dune habitat of the sandy shores	Sandy shorelines	Dune plants and benthic fauna
Tidal flat wetland system	River delta, estuaries, backwaters, lagoons	Mangrove ecosystem
Sea grass patches	Subtidal shores	Seagrass colony
Coral reefs	Shore fringe tidal banks	Coral reef ecosystem
Marine habitats	Open marine offshores	Aquatic ecosystem

Fig. 9.2 A few habitat types in Coromandel coast of India: (**a**, **b**) Rocky intertidal shores in Manapad and Vishakhapatnam; (**c**, **d**) Beach dune habitats at Cape Comoran and Red hills of Vishakhapatnam; and (**e**, **f**) Sea grass patches along the Palk Bay fringe shorelines of Rameswaram

9.4.2 Temporal Changes in Coastal Habitats and Trends of Degradation

Coastal habitats are changing temporarily in different sections due to the impacts of multiple human factors (11 factors) and physical factors (6 factors). Land use and land cover changes, beach placer mining, cropping intensity, reductions in sediment and freshwater discharges by the river systems, and coastal urbanisation rates are major human factors causing habitat degradation in the coastal belts of Andhra Pradesh and Tamil Nadu (Fig. 9.4). The temporal changes of coastal habitats are recorded through satellite remote sensing studies. Mangrove ecosystems of the tidal flat habitats are affected due to the conversion and spread of fish farm plots and salt processing ponds along the coastal belts. The distribution of such coastal wetland habitats was recorded and documented in the coastal districts through the geospatial techniques in 2011 by MOEF, Government of India for Andhra Pradesh and Tamil Nadu (NWIA, 2011). The study reveals that total 28.22% coastal wetlands are represented in the coastal districts of Andhra Pradesh, and total 13.56% coastal wetlands are distributed in the coastal districts of Tamil Nadu in India (Fig. 9.3). However, most of the wetlands are distributed in 4 coastal districts of Andhra Pradesh (east Godavari, west Godavari, Krishna, and Nellore) and only 3 coastal districts of Tamil Nadu (Kanchipuram, Pudukkottai, and Ramanathapuram) as per the previous estimation by MOEF.

All the river systems at their lower courses in Andhra Pradesh (Vamshadhara, Nagavali, Gosthani, Krishna, and Penneru) and in Tamil Nadu (Cauvery, Palar, Vaigai, Pennaiyar R., and Tammrabarani R.) demonstrated depositional features by extensive channel bars and less discharges of water and sediments into the coastal zones. Human activities in the catchment areas of the rivers and climate change impacts are mainly responsible for such reductions of fresh water and sediment discharges into the coastal zones. During the past five decades, a significant decrease in stream flow has been found in the Krishna and Cauvery rivers, and there has been

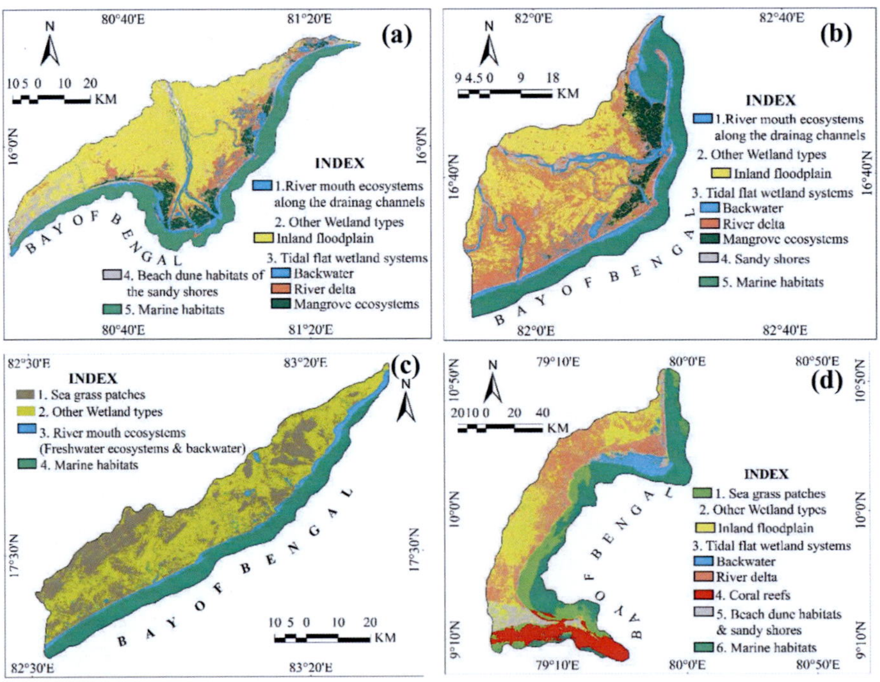

Fig. 9.3 Spatial diversity of coastal habitats in (**a**) Krishna delta, (**b**) Godavari delta, (**c**) Vishakhapatnam shoreline, and (**d**) Gulf of Mannar

a variation in precipitation in the peninsular river basins (Das et al., 2022). Hypersalinity in the tidal flats of the deltaic mangroves ecosystems resulted from shortage of fresh water and sediment accretion levels. The Palk Bay and Gulf of Manner sections of Tamil Nadu coast are noted for extensive patches of sea grasses and shelf fringe coral reefs. They were affected by erosion with dynamic marine processes, illegal coral mining activities, and various shore fringe activities by humans with placer mining, urbanisation processes, tourism recreation pressures, marine fishing, and port activities during the previous decades. The sandy shores of beaches, bars, and sand dunes are severely affected by placer mining activities on the Andhra Pradesh and Tamil Nadu coasts. Land use and land cover changes and tourism-related recreational activities along the seashore cause the degradation of sand dunes by diminishing the habitat areas with dune floral structures. Marine waves and currents have accelerated shoreline erosion along alluvial coasts in recent decades (Fig. 9.4).

Fig. 9.4 Coral reefs and mangroves: (**a**, **b**) Living corals on the bank of shallow seas in the Palk Bay and Gulf of Mannar (*Acropora Valida and Goniopora Stutchburyi*); (**c**) Coral mining by the poor people in Rameswaram Island; and (**d**) Recreation and tourism activities in Pichavaram mangroves

9.4.3 Selection of Multicriterion Factors for Assessing Habitat Degradation

This part of the study reveals the identification of reasons for habitat degradation with the help of temporal change detection and repeated field observations during the previous decades in the coastal sections under study. So far as, multiple human factors (11 numbers) and various types of physical factors (6 numbers) have been selected as major reasons for the degradation of coastal habitats in the two eastern coastal states of Peninsular India. During the field survey, these reasons were considered in the questionnaire schedule to ask the selected number of locals to justify the factors playing a significant role in habitat degradation. However, it is validated that coastal urbanisation, land use and land cover changes, marine fishing, tourism and recreation activities, and the reduction of stream flows and sediment transport played the most significant role as human factors compared to others. Among the physical factors affecting cyclone landfalls, shoreline erosion accretion rates and saltwater encroachment into the coastal aquifers played the most significant roles in the region (CGWB, 2014). On the basis of habitat types and their diversity, the entire coastal stretch is classified into 17 coastal sections from Srikakulam to Kanyakumari for assessment of habitat degradation (Table 9.2).

9.4.4 Assessment of the Habitat Quality Index

For assessing the status of coastal habitats, the total 17 factors are applied for the each section of the coastal districts with 10 point assigned values by estimating the roles played by them in a regional framework of the coastal system. Thus, after giving the assigned values for 17 factors to the 17 coastal sections, a total value for each section will be estimated to compare the regional variation in stressors along the coastal system. Similarly, the separate rate values for human factors and physical factors out of the total assigned values of the coastal section will come out of the assessment to identify the dominance of group factors over others for each section in the coastal framework. The final ranking is given after the summation of total scores for each section of the coastal system in the study. However, the coastal habitats are ranked as very high, high, moderate, and low in terms of stressors by the integrated scores estimated by the study. The two sections of very high stressed habitats are located in northern Andhra Pradesh and southern Tamil Nadu coasts of the Peninsular India. Similarly, high-stress habitats are also located in four sections along the northern Andhra Pradesh and southern Tamil Nadu coasts. The low- and moderately stressed sections of coastal habitats are represented as better-quality habitat in the regional framework of coastal systems (Fig. 9.5).

Another observation is made by the study when the dominance of factors is estimated separately for human and physical factors. In three cases, the physical factors are dominating over the human factors in the degradation of coastal habitats in the coastal sections. Among them, two sections are located in Andhra Pradesh and only

Table 9.2 Assessment of habitat quality by the weighted sum method with 11 human factors and 6 physical factors applied in 17 coastal sections of the Coromandel Coast

Sl No	Human Factors	S-1	S-2	S-3	S-4	S-5	S-6	S-7	S-8	S-9	S-10	S-11	S-12	S-13	S-14	S-15	S-16	S-17
1	Coastal Urbanization processes	0.7	1.2	0.1	1	0.1	0.1	0.1	0.4	0.3	0.4	1.3	0.9	0.5	0.1	0.9	0.9	1
2	Land Use and Land Cover changes in the coast	0.1	0.1	0.1	0.1	1.1	1	1.1	0.65	0.65	0.5	0.7	0.2	1.1	0.1	0.5	1	1
3	Coastal tourism recreation activities	0.5	1.7	0.5	1	0.1	0.1	0.1	0.1	0.1	0.5	1.2	1	0.1	0.1	0.5	1	1.4
4	Shipping ports and harbours	1.2	1.2	0.2	1	0.2	0.2	0.2	0.2	0.2	0.7	1.5	0.2	0.2	0.2	0.5	1.5	0.6
5	Marine fishing ventures into the sea	1.5	0.2	0.2	1.4	1.5	1.5	0.2	0.2	0.2	0.2	0.1	0.2	0.1	0.1	1	1	0.4
6	Fish farming and salt processing in the coast	1.3	0.1	0.2	1.2	0.6	1.2	0.5	0.4	0.4	1.2	0.2	0.5	0.5	0.5	0.5	0.5	0.2
7	Beach placer mining in the shore	1.5	1.2	0.2	0.2	0.2	0.2	0.2	0.6	0.6	0.2	0.5	0.5	0.2	0.2	0.5	1.5	1.5
8	Coral mining activities in the near shores	0	0	0	0	0	0	0	0	0	0	0	0	0	0	3.2	3.8	3
9	Thermal power station and Atomic power station in the coastal site	0	1.5	1.5	0	0	0	0	0	0	1.5	0.5	0	0	0	0.5	0.5	4
10	Intensive cropping in the coastal zones	0.1	0	0.5	1.5	1.9	1.5	0.5	0.1	0.5	0	0	1.5	1.9	0	0	0	0
11	Reduction in sediment discharges and stream flows into the coastal zones	1	0	0	1.5	2	0	1.5	0	0.5	0	0.5	0.7	1.5	0	0.8	0	0
	Physical Factors	S-1	S-2	S-3	S-4	S-5	S-6	S-7	S-8	S-9	S-10	S-11	S-12	S-13	S-14	S-15	S-16	S-17
12	cyclone landfalls and impacts	0.5	0.5	1.5	0.5	1.75	1.75	0.4	0.5	0.5	0.3	0.2	0.4	0.3	0.3	0.2	0.2	0.2
13	Sea level rise rates	2	1.5	0.9	0.9	0.9	0.9	0.3	0.3	0.3	0.3	0.3	0.3	0.3	0.3	0.3	0.1	0.1
14	Shoreline erosion and accretion rates	1	1	0.9	1	0.7	0.4	1	0.3	0.4	0.2	0.5	0.5	0.5	0.5	0.2	0.2	0.7
15	Sediment deposition in the river beds	1	1	0.5	0.5	2	0.2	0.8	0.5	0.5	0.2	0.4	0.4	0.4	0.3	0.2	0.6	0.5
16	Salt water encroachments into the aquifers	0.2	0.2	0.2	0.2	1	1	0.7	0.7	0.7	0.7	0.4	0.4	1	1	0.8	0.4	0.4
17	Tsunami waves impacts (2004)	0.4	0.4	0.4	0.4	0.4	0.4	0.4	0.4	0.4	0.4	0.5	0.5	1.3	1.3	0.5	0.7	0.7
	Integrated scores of habitat quality index	13	11.8	7.9	12.4	14.5	10.5	8	5.35	6.25	7.3	8.8	8.2	9.9	5	11.1	14	15.7
	Habitat degradation ranking	VH	H	M	VH	VH	H	M	L	L	M	M	M	H	L	H	VH	VH

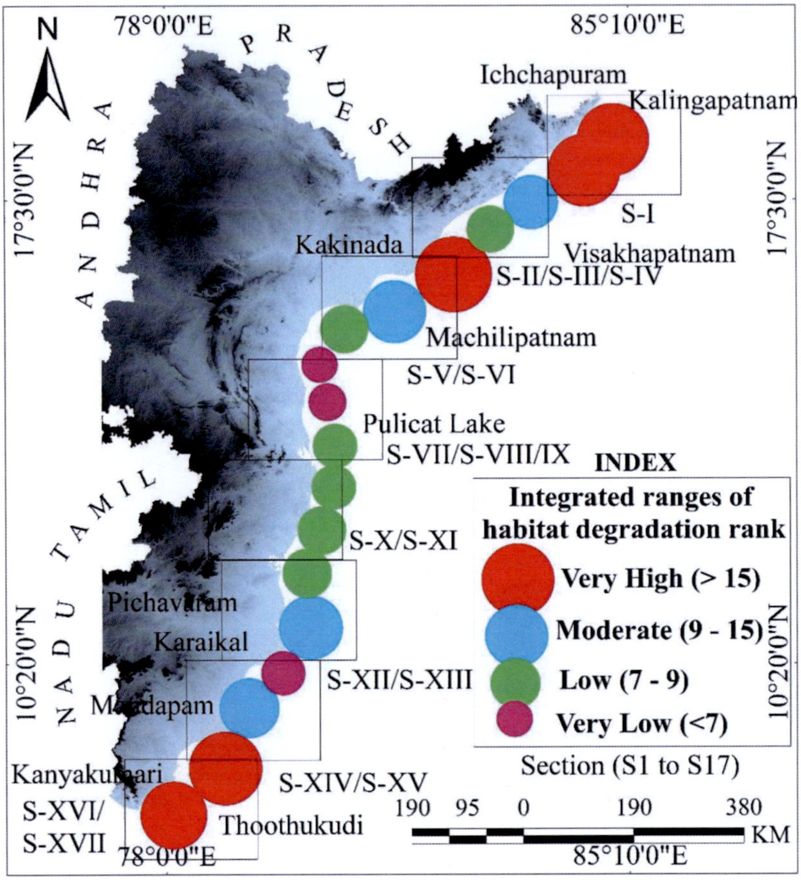

Fig. 9.5 The rate of habitat degradations assessed by integrated scores of habitat quality in seventeen coastal sections of the study area

one section in Tamil Nadu coasts under study. However, in fourteen sections of coastal habitats, human factors are playing the most significant role in the degradation of the habitat quality. Types of multiple habitats are distributed all along the coastal sections, except in five places where a single habitat is extended. The current database will help to find out the quality of coastal habitats for appropriate management practises in marine and coastal environments.

9.4.5 Management Approaches for the Habitat Conservation in the Coastal Areas

The Gulf of Mannar and Palk Bay coasts of Tamil Nadu were declared as Marine National Parks in 1988 for the location of two significant habitats as the coral reef ecosystem and the seagrass ecosystem, in the regional settings of the coasts.

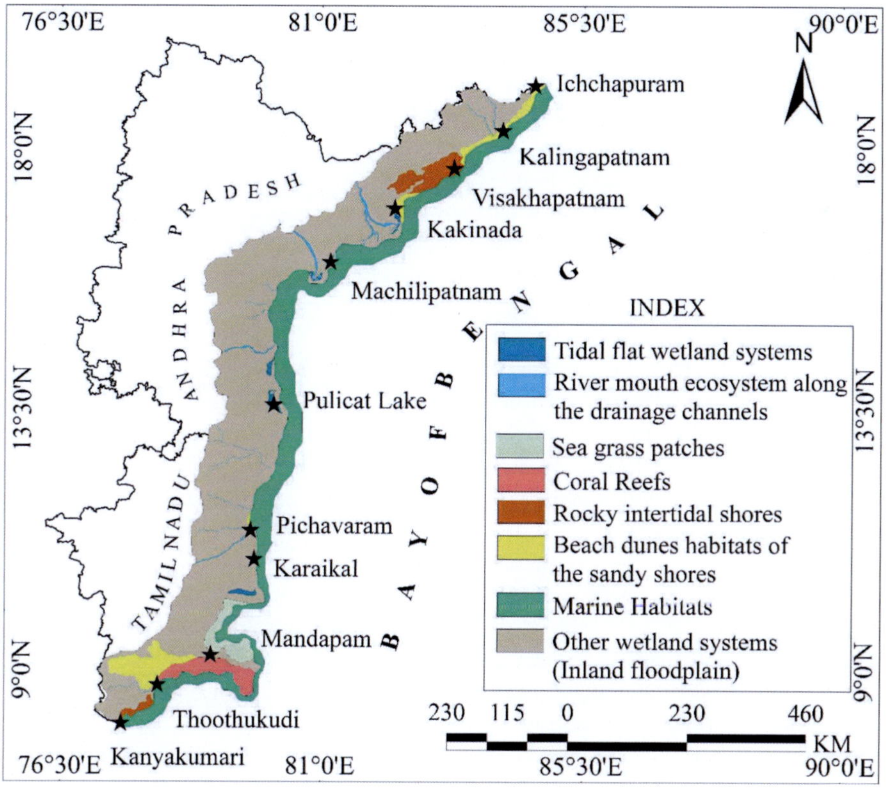

Fig. 9.6 Proposed coastal habitat management and conservation sites to meet the demand for the long-term goal in environmental crises

Significant environmental regulations are in place to limit tourist visits to the areas in order to conserve the region's sensitive coastal habitats. During the previous decades, overexploitation of marine resources resulted in the vast destruction of marine flora and fauna in the coastal marine sections. The coral reef areas that once thrived have been exploited for industrial uses, and only a few sections remain today, particularly from Tuticorin to the Krusadai Islands of the Tamil Nadu coast. Earlier studies (Patterson et al., 2004; Edward et al., 2008; Mohanraj et al., 2010) also demonstrated that the Tuticorin section of the Gulf of Manner coast is the most environmentally stressed area, mainly due to various human activities such as destructive trawl fishing, coral mining, and pollution from industries along the coastal belt. The mangrove forests were replaced along the estuary fringe coasts and backwater fringes by the extension of aquaculture ponds and salt processing ponds along the coastal sections of Andhra Pradesh and parts of Tamil Nadu. Since 2006, ecological engineering methods have been used in the Godavari, Krishna, Cauvery delta, and Pichavaram coast to increase mangrove cover areas and to restore habitat from hypersalinity in coastal sections (Fig. 9.6).

The Ministry of Water Resources (2014) has prepared a report on the status of ground water quality in coastal aquifers, which reveals that the alluvial coasts of Andhra Pradesh and Tamil Nadu are significantly affected by the encroachment of salt waters into the aquifers. Kakinada in Andhra Pradesh and Rameswaram Island in Tamil Nadu suffer mostly due to the degradation of groundwater quality. Human factors in terms of overextraction of ground waters due to rapid urbanisation in the coastal parts, sea level rise impacts, and salt water inundations due to the impacts of storm surges and tsunami waves in 2004 are primarily responsible for the degradation of ground water quality in coastal aquifers. However, such a groundwater quality status report will serve as a database for future management of water resources in coastal zones.

A national wetland atlas on the district level was also prepared by MOEF, Government of India, in 2011 with the help of ISRO, Ahmedabad, for understanding the current status of various wetlands. The coastal wetlands of Andhra Pradesh and Tamil Nadu are well represented at the district level in the status report. As per the report, the coastal wetlands include saltpans (inland), lagoons, creeks, sand bars, intertidal mudflats, saltmarshes (coastal), mangroves, coral reefs, saltpans (coastal), aquaculture ponds, etc., as components in the region. The report also demonstrated that a total of 28.22% of areas in Andhra Pradesh and 13.56% of areas in Tamil Nadu occupy coastal wetlands of different types. However, the present study includes more other types, for example, riverine wetlands, sea grasses, shallow marine water spaces, islands, barriers, and coastal sand dunes, including with the coastal wetlands types considered in the MOEF report. Human overexploitation of such natural resources, particularly through changing LULC activities along the coasts, untreated waste dumping through urban drainage outlets, coastal aquaculture ponds, tourism-recreational pressures, expanding port cities, agricultural runoff, and other industrial centres, created stress factors in coastal habitats. For management and conservation of coastal habitats, such reports and research activities are needed to meet the demands of the sustainable goal in the sensitive coastal environment of the region.

9.5 Conclusion

The present study indicated that the degradation of habitat quality in the coastal belts or coastal sections indicated a significant environmental crisis. The majority of the coastal habitats are highly stressed due to human factors and physical factors at present in Andhra Pradesh and Tamil Nadu. The coral reefs and sea grasses attenuated wave energies in the past to protect the coasts from a high-energy marine environment and harbour a wide variety of assemblages of marine animals in the coastal sections of the Gulf of Mannar and Palk Bay. Now they are restricted only to a few islands and coastal sections after overexploitation of these natural resources by the local people. Mangrove ecosystems of deltaic, estuarine, and lagoonal habitats, on the other hand, were acting as a very good buffer against wind waves and storm surges in coastal resiliencies while also supporting a rich biodiversity in the swampy forests of Peninsular India's south eastern mega deltas. The ecosystem services are affected due to the reduction of freshwater and sediment discharges into the deltaic coasts, as well as land use and land cover conservation efforts in the wetlands, and also by emerging threats from sea level rise impacts and growing hypersalinity patches in the tidal flats of tropical coastal environments. Coastal sand dunes and sandy beaches are highly exploited for placer mining activities in many coastal sections of Andhra Pradesh and Tamil Nadu as they are the sources of rare earth minerals. The habitat supported a specific community of plants and animals on the seashores, but they are seriously affected by overexploitation of natural resources, particularly in the Vishakhapatnam, Vimlipatnam, Ramanathapuram, and Srikakulam districts of the region.

Coastal wetlands habitats are highly stressed in the northern Andhra Pradesh and southern Tamil Nadu coasts. Immediate attempts are needed to restore the habitats of mangrove ecosystems, coral reefs, sea grasses, and coastal sand dunes to save the people and their livelihoods from the threat of coastal risk associated with cyclones, tsunamis, and sea level rise impacts on the low-lying coasts of India.

References

CGWB. (2014). *Report on status of ground water quality in coastal aquifers of India*. Ministry of Water Resources Government of India. Assessed 2014.

Das, S. (2021). Dynamics of streamflow and sediment load in Peninsular Indian rivers (1965–2015). *Science of the Total Environment, 799*, 149372. https://doi.org/10.1016/j.scitotenv.2021.149372

Das, S., Kandekar, A. M., & Sangode, S. J. (2022). Natural and anthropogenic effects on spatio-temporal variation in sediment load and yield in the Godavari basin, India. *Science of the Total Environment, 845*, 157213. https://doi.org/10.1016/j.scitotenv.2022.157213

Edward, J. P., Mathews, G., Patterson, J., Ramkumar, R., Wilhelmsson, D., Tamelander, J., & Linden, O. (2008). Status of coral reefs of the Gulf of Mannar, southeastern India. *Coastal Oceans Research and Development in the Indian Ocean, 45*.

Ghorai, D., Mahapatra, M., & Paul, A. K. (2016). Application of remote sensing and GIS techniques for decadal change detection of mangroves along Tamil Nadu Coast, India. *Journal of Remote Sensing & GIS, 7*(1), 42–53. https://doi.org/10.37591/.V7I1.531

Ghorai, D., Devulapalli, S., & Paul, A. K. (2017a). Cyclone vulnerability assessment of Tamil Nadu Coast, India using remote sensing and GIS techniques. *Journal of Remote Sensing Technology, 5*(1), 32–43.

Ghorai, D., Roy, S., & Paul, A. K. (2017b). *Land use/land cover assessment of Tamil Nadu Coast*. India using Remote Sensing and GIS Techniques.

Jagtap, T. G. (1996). *Some quantitative aspects of structural components of seagrass meadows from the southeast coast of India*.

Jagtap, T. G., Komarpant, D. S., & Rodrigues, R. S. (2003). Status of a seagrass ecosystem: An ecologically sensitive wetland habitat from India. *Wetlands, 23*(1), 161–170.

Manikandan, S., Ganesapandian, S., & Parthiban, K. (2011). Distribution and zonation of seagrasses in the Palk Bay, southeastern India. *Journal of Fisheries and Aquatic Science, 6*(2), 178.

Mohanraj, J., Johnson, J. A., Ranjan, R., Johnson, L., Pandi, U., & Shunmugaraj, T. (2010). Coral reef associated gastropods in Tuticorin coast of gulf of Mannar biosphere reserve, India. *Indian Journal of Science and Technology, 3*(2), 204–206.

Mukhopadhyay, S. K. (2007). The Hooghly estuarine system, NE coast of Bay of Bengal, India. In *Paper presented at the workshop on Indian estuaries*. CSRI-National Institute of Oceanography.

Murty, M. R., Reddy, K. M., & Swamy, K. V. (2022, December). Flood response to geomorphic setup and coastal land use patterns: A case study in Krishna River Delta, Andhra Pradesh, India. In *5th World Congress on disaster management* (pp. 107–119). Routledge.

Nageswara Rao, P. V., Appa Rao, S., & Subba Rao, N. (2015). Suitability of groundwater quality for drinking, irrigation and industrial purposes in the Western Delta Region of the River Godavari, Andhra Pradesh. *Journal of the Geological Society of India, 86*, 181–190.

National Wetland Atlas. (2011). *As a part of the project on National Wetland Inventory and Assessment (NWIA), Ministry of Environment and Forests*. Government of India. Assessed 2011.

Patterson JKE, Patterson J, Venketesh V, Mathews G, Challam C and Wilnelmisson D (2004). *Afield guide to stony coral (scleractinia) of Tuticorin in Gulf of Mannar*. Southeast coast of India. SDMRI, Special Publ., No- 4 (pp. 192).

Paul, A. K. (2005). *Tsunami: An assessment of the disaster over the nations in the Indian Ocean. Kolkata, India* (pp. 1–125). ACB Publications.

Paul, A. K., Kamila, A., & Ray, R. (2018). Natural threats and impacts to mangroves within the coastal fringing forests of India. In *Threats to mangrove forests* (Coastal Education and Research Foundation, JCR) (Vol. 25, pp. 105–140). Springer Nature. https://doi.org/10.1007/978-3-319-73016-5

Paul, A., Paul, A. K., & Sardar, J. (2023a). Carbonate landforms of Rameswaram Island, Tamil Nadu coast, India: A review. *Arabian Journal of Geosciences, 16*(3), 176. https://doi.org/10.1007/s12517-023-11222-6

Paul, A. K., Paul, A., & Majumdar, D. D. (2023b). Coastal sand dunes along the western and eastern shores of India. In *Sand dunes of the northern hemisphere* (pp. 350–382). CRC Press.

Ramkumar, M., Menier, D., Mathew, M., & Santosh, M. (2016). Geological, geophysical, and inherited tectonic imprints on the climate and contrasting coastal geomorphology of the Indian peninsula. *Gondwana Research, 36*, 65–93. https://doi.org/10.1016/j.gr.2016.04.008

Sengupta, D., & Ghosal, S. (2017). Environmental implications of Mining of Beach Placers for heavy minerals. *Fish & Ocean Opj, 2*(3), 12.

Thangaradjou, T., & Kannan, L. (2005). Marine sediment texture and distribution of seagrasses in the Gulf of Mannar biosphere reserve. *Seaweed Research and Utilization, 27*, 145–154.

Chapter 10
Mangrove Sensitivity to the Ephemeral Rise of Sea Waters in the Western Sundarban

Debasmrity Mukherjee and Ashis Kumar Paul

10.1 Introduction

The western Sundarban region is the most densely populated part of the mangrove region in India, and it is extended from the estuary of the Hooghly River to the eastern boundary of the Indian Sundarban, the Ichhamati Raimangal estuary complex. Significant parts of the mangroves in this region are cleared through historical reclamations to make space for human habitations and enterprises (Paul, 2002; Paul et al. 2017; Gopal & Chauhan, 2006). However, the local population there depends considerably on the resources obtained from the existing mangroves for economic sustenance (Paul et al., 2018; Chaudhuri & Choudhury, 1994). The ongoing climate change activities with sea level rise impacts affect the Sundarban mangroves, along with the rest of the world (Choudhury et al., 2019; Ghosh & Roy, 2022), very significantly. The low-lying deltaic tracts of the Sundarban region are liable to flooding due to the expected impacts of sea level rise threats in the near future. Many researchers have investigated the impact of sea-level rise threats on the mangroves in the Sundarban (Chowdhury et al., 2016; Payo et al., 2016; Mukhopadhyay et al., 2018). However, most of these studies are about the long-term sea-level rise and its impacts on the mangrove ecology. Short-term, temporary rises in the sea level also have a significant impact on the mangrove characteristics in the Sundarban region.

Being situated in the macro-tidal estuary of the regional setting, the mangroves in the Sundarban are subject to the effects of tides twice daily, and a vast portion undergoes tidal inundation. The spring tides, due to their higher levels and landward-advancing wave energy, have particularly high impacts on the low-lying areas above

D. Mukherjee (✉)
Department of Geography, Asutosh College, Kolkata, India
e-mail: debasmrity4email@gmail.com

A. K. Paul
Department of Geography, Vidyasagar University, Midnapore, West Bengal, India

A. K. Paul, A. Paul (eds.), *Crisis on the Coast and Hinterland*,
https://doi.org/10.1007/978-3-031-42231-7_10

the mean sea level. Climate change is also associated with an increase in cyclone frequency and severity in the Bay of Bengal (Paul, 2022; Balaguru et al., 2014; Reddy et al., 2021), and the cyclones often destroy vast tracts of mangroves in the Sundarban region. Cyclones in particular tend to cause over wash deposition on the coast, which has its own impacts on the local mangroves. On the other hand, the local population plays its part in the dynamics of the influence of these temporary rises in the sea level on the mangrove characteristics by constructing barriers and embankments along the creek boundaries (Paul, 2002; Ghosh & Mistri, 2020). The local geomorphology determines how tide water intrusion affects inland mangroves via saltpan formation or encroaches tidewater through the tidal creeks into the inner parts of the islands. The present work reveals the effect of such a transient rise in water level on the mangroves of the western Sundarban region, which is being considered in the study.

10.2 Study Area

The study is being conducted to assess the effects of water level and concentration at various levels of the island surfaces occupied by mangrove zones in the south-western Sundarban's Patibunia, Sushuni, and Henry Islands along the Muriganga estuary and Saptamukhi estuary complex (Fig. 10.1). The surface elevation of the islands ranges from 1.7 to 4.26 m above the mean sea level and fringed with estuary channels and Bay of Bengal shorelines in the lower Ganga delta.

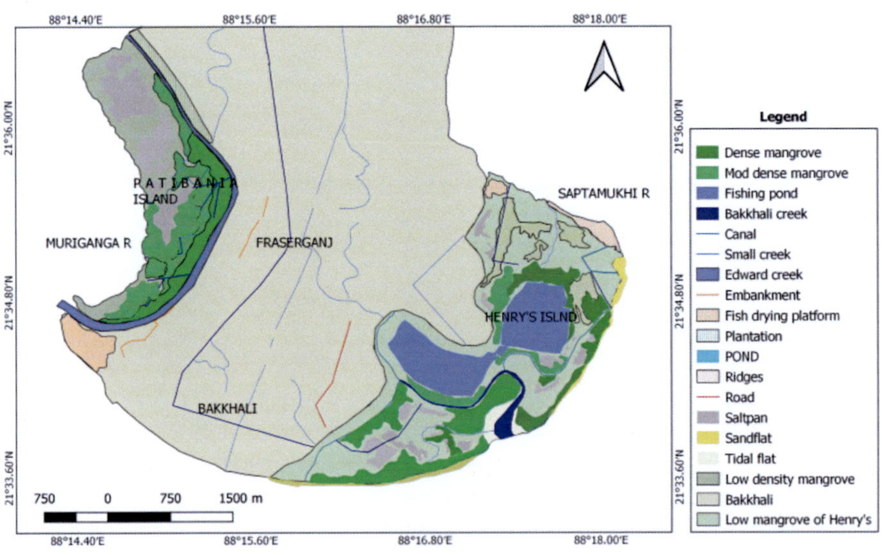

Fig. 10.1 Mangrove areas of the western Sundarban under stress

10.3 Methods of the Study

Cartosat 2 A DEM, Landsat 8 OLI, SOI toposheets, Google Earth images, and seasonal field survey methods are applied to carry out the present work in the region. The high-resolution data is used to prepare contour plans of the low-lying islands with a 1 m interval for the assessment of inundation frequency and sensitivity of mangroves for the islands considered in the study. The contour plan is further used to classify the surface elevation zones of the tidal flats on the islands. Regular fieldwork techniques with sediment sampling, mangrove status, photographic documents, and degradation characteristics of the mangroves are carried out in different seasons to validate the satellite-based information for the present study. Finally, the study identified areas with degradation zones and healthy mangroves. To investigate the tidal range in this region, daily tidal data was collected from the Kolkata Port Trust and recorded at Sagar station in the year 2018.

10.4 Results and Discussion

10.4.1 Geomorphology of the Mangrove Shores

The Sundarban mangroves grew on the Ganga-Brahmaputra estuary's tropical shores (Sahana & Sajjad, 2019) with available sediment and freshwater supplies. The deposition of sediments has created a complex web of creeks and channels, which in turn offered a region that was sheltered or semi-sheltered from the direct tidal waves, making this region ideal for the growth of mangroves. The geomorphic features of this tidal estuarine region played a vital role in determining the effects of short-term rises in the sea level due to storms and tides and the resulting inflow of saline water and inundation frequencies on the mangrove zones of the low-lying islands (Fig. 10.2). The major geomorphic features include tidal flats, beach ridges, inner island tracts, and depressions separated by inland tidal creeks. The shore-fringed sand dunes were diminishing in the previous decades due to the impacts of devastated cyclones (Paul et.al., 2023a), and as a result of the removal of the dune barrier, the mangrove wetlands were frequently inundated by seawater at the astronomic tides across the coastal zones of the Bay of Bengal (Paul et al., 2023b). These parts undergo frequent inundation with brackish water from the diurnal tides. The tidal flats have a very gentle slope and are divided into three parts based on geomorphic characteristics: the lower tidal flats, the middle tidal flats, and the upper tidal flats. The lower tidal flats are the lowest part of the tidal flat, next to and below the average water level. During low tides, this part gets exposed completely, but during high tides, it gets completely inundated. The tidal energy is highest in this part, and the accompanying erosion is also the highest. Due to these reasons, the lower tidal flats are generally unsuitable for the growth of large, woody mangrove trees, and mangrove seeds and seedlings often get washed away in the tides from these areas (Fig. 10.2).

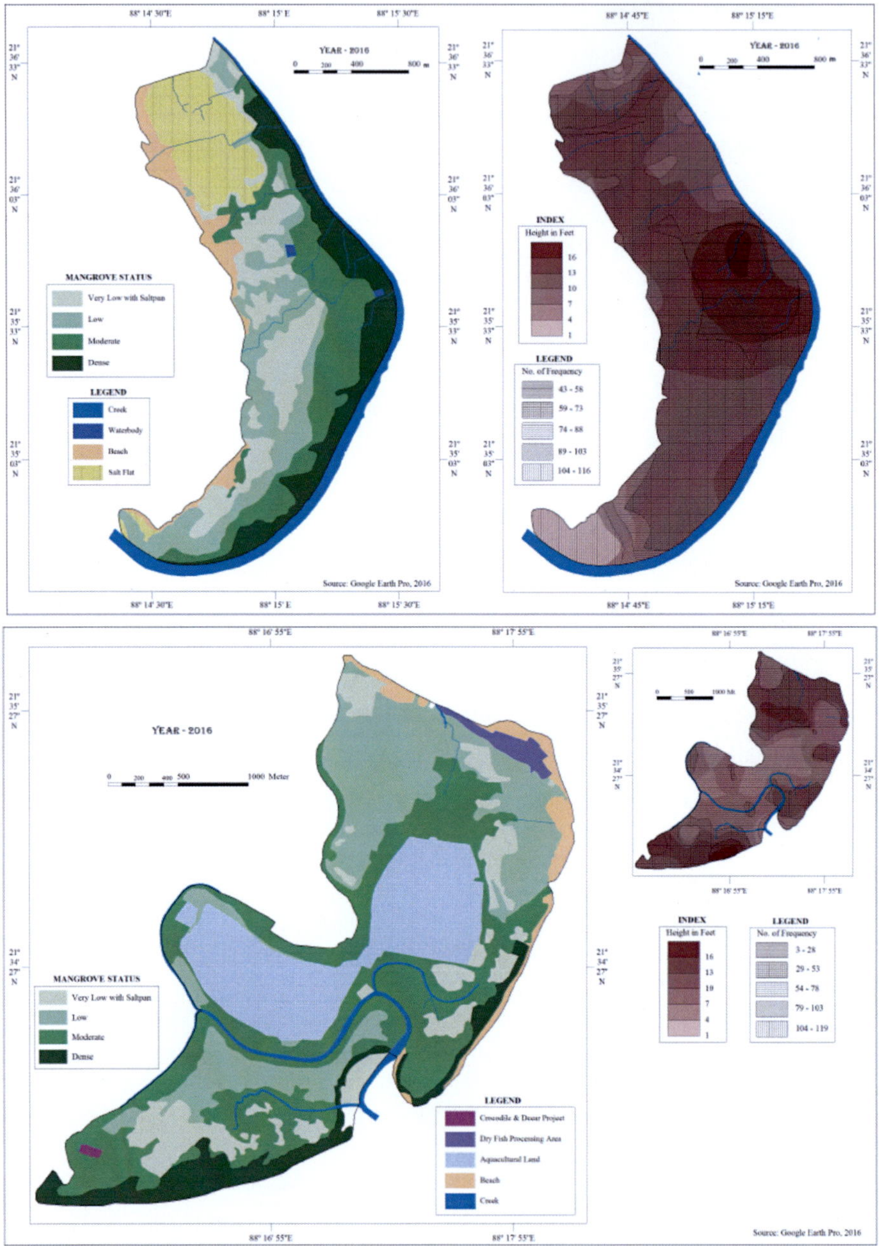

Fig. 10.2 The inundations of Patibunia Island and Henry's islands delineated using the CARTOSAT 2 DEM and contour patterns. Mangroves of the lower parts are liable to degradations and tidal inundations

The middle tidal flats start at the upper boundary of the lower tidal flats. During the daily high tides, these areas typically get only partially inundated. The tidal energy and the accompanying erosion are lower in these areas compared to the lower tidal flats. The mangrove tree lines typically start at the upper ends of the middle tidal flats, with the lower areas covered in herbaceous mangrove plants. The upper tidal flats are rarely inundated during the typical daily tides, but are inundated during the spring tides and storm surges. The very low tidal energy results in low erosion and sediment build up. This area is usually dominated by the larger mangrove plants, as they receive an adequate supply of tidal water but less disturbance from tidal waves. On the upper boundary of the upper tidal flat, there is often a beach ridge, which is a relatively elevated platform created from the interplay of sedimentation and tidal erosion. Beach ridges are not inundated except during the highest spring tides. However, storm surges can overflow them and also topple them to create over wash deposition. Behind the beach ridges, the inland areas often have a lower elevation. The beach ridges serve to protect these enterprises from the forces of storm surges and spring tides. Due to this, artificial dams and barriers are often constructed along the beach ridges in the Sundarban (Dhara & Paul, 2016). Sometimes, larger mangrove trees are planted along the beach ridges.

This is done to create a stronger barrier against the forces of storm surges and cyclones. Such mangrove plantations along the beach ridge can be seen on Henry's Island, which is located at the southern end of the western Sundarban at the mouth of the Saptamukhi river (Paul, 2002; Mukherjee & Paul, 2020). Often, regions with low elevation form depressions that are liable to be inundated. But when these are inundated, which happens during spring tides and storm surges, the saline tidal water gets trapped in the depressions due to a lack of drainage avenues. Over time, this saline water gets evaporated, leading to build up of salt-encrusted ground surfaces and the subsequent formation of saltpans (Paul et al., 2017). The salinity there is frequently too high for even halophytic mangroves to tolerate, resulting in local mangrove degradation. Inland regions with relatively high elevations can be suitable for supporting mangroves if they have adequate water supplies through inland creeks. They are generally too narrow and shallow to carry much of the force of tidal waves and witness negligible erosional activities. However, the tidal water that flows through them is laden with sediments, which are frequently deposited on the channel beds, eventually filling up the already shallow channels.

10.4.2 Climatic Phenomena and Tides Behind the Ephemeral Rise of Sea Level

The Bay of Bengal is among the regions witnessing the highest number of tropical cyclones in the world. The cyclones of the northern Bay of Bengal usually occur in two periods, before and after the monsoon season. The massive wind energy of the cyclones causes devastating waves and torrential rainfall, causing extensive damage to human settlements and mangrove ecosystems alike. Thousands of mangrove trees are uprooted during cyclones, despite their extensive root network in the deltaic islands. Damage to larger woody mangrove trees, on the other hand, is more severe because they take longer to grow. The herbaceous mangroves, on the other hand, are relatively fast growing. After a devastating event, the faster-growing herbaceous mangrove plants may occupy the niche previously occupied by the larger mangrove trees and thus hinder their reestablishment. In this way, frequent cyclones cause changes in the composition and character of the mangrove ecosystem, tilting the vegetation density downward.

The storm surges associated with the cyclones also cause extensive coastal flooding. The destruction of the natural and artificial ridges and embankments leads to further saline water intrusion, which worsens the salinity build-up in the inner parts of the islands. The western Sundarban tidal range is frequently greater than 4 m (Fig. 10.3). The Hugli-Muriganga estuary, which forms the western boundary of the Sundarban region, only has a source of upstream fresh water. Saptamukhi, Matla, Bidyadhari, and other streams in the Indian Sundarban to its east are entirely influenced by tides. The Muriganga River brings higher levels of sediment due to its upstream connectivity. Tidal action can be seen not only in the areas adjacent to these streams and creeks but also on the seashores in the western Sundarban (Chatterjee et al., 2013). The erosional activities of the tidal waves often negatively affect the mangrove ecosystem in the low-lying tracts. It can be seen that each month witnesses two spring tides with higher wave energy and thus a higher potential for damage. It can also be seen that the tidal ranges are relatively higher in the monsoon season. This is because of the enormous extra water available through the monsoon rains and related upstream discharge.

Rainfall is one of the critical sources of freshwater in this region. It works towards balancing the level of salinity in the soil caused by the saline tidal water encroachments. Rainfall occurs primarily during the monsoon season, but there is also heavy rain during the pre- and post-monsoon seasons. To investigate the trends and characteristics of rainfall, daily rainfall data were collected from the Diamond Harbour Station of the Indian Meteorological Department over a 20-year period from 2000 to 2019. The yearly rainfall is plotted for the years 2000–2019 in the western Sundarban, which indicates that the yearly rainfall has a negative trend (Mukerjee & Paul, 2021).

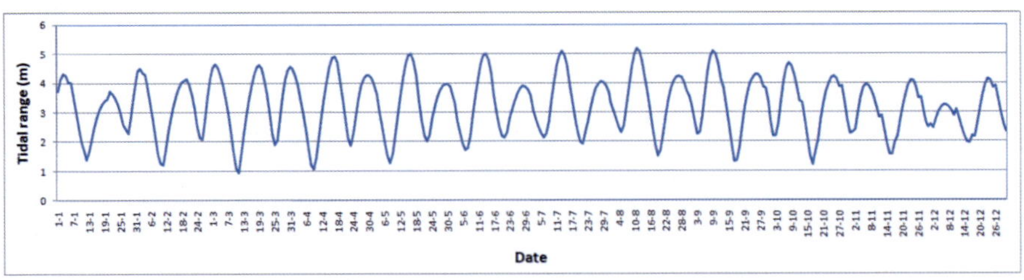

Fig. 10.3 Tidal range of Sagar station, based on tidal data in the year 2018

10.4.3 Anthropogenic Influence

Mangrove plantations are undertaken principally to withstand the wind forces of storms and cyclones. Apart from increasing the quantity of mangroves in an area, mangrove plantations also improve the quality of the mangrove ecology by adding and aiding the growth of larger mangrove trees, which might be relatively difficult otherwise. The role of adequate water supply in the health and growth of mangroves is paramount. Natural water supply in the inland regions of the islands is frequently hampered by creek and stream sedimentation. Due to a lack of water, which is often associated with a buildup of salt in the soil, the mangroves in that area have declined (Fig. 10.4a).

Human settlements have already substantially decreased the mangrove ecosystem. On the other hand, the existing patches of mangrove forests are generally declared reserve forests to protect the environment. Patibania Island, a small island lying along the western boundary of the western Sundarban region, is such a protected forest that human activity is prohibited there. However, several artificial streams have been built there to bring tidal water inland on that island, so that mangroves flourish well there (Paul, 2002; Paul & Paul, 2022). The drainage characteristics of two islands are investigated in the western Sundarban region (Henry's Island and Patibania Island). A significant portion of the inland creeks and streams on Patibania Island are man-made, connecting to Edward's Creek to bring tidal water inland. Henry's Island also has extensive artificial streams connected to the

Fig. 10.4 (**a**) Salt extraction processes in western Sundarban; (**b**) Saltpans in the Henry's Island; (**c**) overwash deposits in the Henry's Island in 2020; and (**d**) Overwash deposits at the beach ridge of the Patibania Island

natural creeks. Saltpans are one of the most unfavourable areas for mangrove growth in the Sundarbans. However, due to the very high concentration of salt in the soil of these regions, it is economically viable to extract salt here, which has opened avenues for economic enterprises and employment for the local population (Fig. 10.4a, b).

10.4.4 Saltpan Formation

The formation of saltpans is one of the principal ways through which the temporary rises in sea levels harm the mangrove ecosystem. Due to being surrounded by regions with higher elevation, the depression areas have no direct connection to tidal streams. When the time of the high tide passes, or the storm is over, the saline water from the spring tides and storm surges gets trapped in these depressions. This stagnant saline water gets evaporated in the hot sub-tropical environment over time, leaving the salt on the ground. Repetition of this process leads to the gradual accumulation of salt on the soil surface, eventually leading to the formation of saltpans (Fig 10.4b).

Mangroves are halophytic by nature, but the excessive levels of salinity found in the saltpans are too high for most mangroves to thrive. Saltpans are thus visible as abruptly barren surfaces in an otherwise densely vegetated region. Henry's Island is situated along the southern seashore in western Sundarban, and Patibania Island is situated along the western boundary of western Sundarban. The amount of salinity is not uniform throughout a saltpan. Typically, salinity increases from the boundary of the saltpan towards its centre. Mangrove characteristics are influenced accordingly in different parts of a saltpan. Water supply to a saltpan reduces its salinity level, and that water may come through rainfall and floods. Typically, the monsoon rainfall washes the saltpans and decreases their salinity levels. The characteristics of mangroves in a saltpan region thus change seasonally with dryness and wetness. Thus, the saltpans tend to slowly expand over time and harm the adjacent mangroves (Paul et al., 2017). Such saltpan expansion was prominently observed following Cyclone Aila in 2009. The flooding of saltpans by storm surges and cyclones also changes their shapes. The prolonged inundation caused by floods associated with cyclones and subsequent evaporation aids in the emergence and expansion of new saltpans. Due to their barrenness, saltpan regions receive little organic matter in the form of mangrove detritus. The very high salinity also interferes with the decomposition of any fresh organic matter received. In this way, the organic nutrients in a saltpan region get affected over time. On the other hand, most of the salt content in a saltpan is concentrated within the top 30–50 cm from the soil surface because of the evaporative mechanism of saltpan formation (Mukherjee & Paul, 2020). So, it is relatively easy to artificially remove most of the salt by scraping off the top surface of a saltpan, which then may be again amenable to vegetation growth if a suitable water supply is arranged.

10.4.5 Overwash Deposition

The high energy of the waves, coupled with the loose cohesion of the soil of the sand dunes and the sea beaches, washes away into the low-lying tracts behind the beach barriers. This process of waves washing off the top surface of sand dunes and other sediment deposits and then depositing the mass behind the area of the dunes is called overwash deposition (Paul et al., 2017; Donnelly et al., 2006). Overwashing occurs rarely and is associated with storm surges that are strong enough. But when they occur, they can substantially change the topography of the beach. This process of erosion of the sand dunes and sediment deposits decreases the elevation of those topographic forms, which increases the likelihood of further overwash (Figlus et al., 2011). This process, when repeated, can effectively flatten the sand dune system by lowering its elevation. The overwash deposition occurs inland, where the wave energy is depleted. This type of overwash process produces fan-shaped deposits known as overwash fans. On the other hand, if the wave crest is sufficiently high to overflow the entire sand dune system, a curtain of overwash deposition occurs, called an "overwash terrace" (Paul, 2022). In the coastal areas along the beaches in the western Sundarban, overwash deposits are quite common. Overwash deposition is expected to increase as sea levels rise due to ongoing climate change (Paul, 2002; Paul et al., 2017, 2018). Overwash deposits destroy the existing mangroves in the area of their formation by burying them underneath. They also alter the nutrient availability and suitability of reestablishing mangroves in the areas (Fig 10.4c).

10.4.6 Effects on Mangrove Characteristics

As a result, as saltpans form, the established, larger mangroves decline and eventually die away. The niche vacated by them is filled up by smaller mangrove shrubs. Saltpans tend to expand in size over time due to the continual build-up of salinity. This causes proportionate harm to the pre-existing mangrove ecology in that region and decreases the vegetation level. The mangrove shrubs yield almost no economic resource comparable in value to wood, honey, and other resources obtained from the woody mangrove trees. So, the expansion of saltpans in a region also diminishes the economic role that mangrove ecology plays, and harms the livelihood of the local human communities. The bigger mangrove plants may still survive in the fringe areas of a saltpan due to lower salinity. In those areas, the vegetation density is also higher in comparison. As one moves towards the centre of the saltpan, with increasing salinity, the locally extant mangroves change to a smaller size. Sometimes the same mangrove species may achieve a larger size in the fringe areas and become progressively stunted towards the centre of the saltpan.

The monsoon season brings a lot of freshwaters in the form of rainfall, upstream discharge, and occasional floods. This decreases the salinity level in the saltpan somewhat, and the vegetation density increases there in the monsoon season.

However, the relatively short duration of a season is insufficient for the re-establishment of larger mangrove species that could have existed at lower salinity levels. The decrease in salinity is thus only accompanied by the rejuvenation of the existing plant life, as well as the proliferation of mangrove grasses and small shrubs. Sometimes this proliferation of vegetation may happen even in the centre of the saltpan. After the monsoon, the freshwater level decreases substantially and the salinity level starts to increase, which causes the decline of the new growth and some of the vegetation to die off. In the summer, intense heat causes further depletion of moisture in the saltpan areas, which further harms the surviving mangroves.

Overwash deposits bury the existing vegetation behind sand dunes and beach ridges beneath a thick layer of sand and sediments. This destroys much of the buried mangroves and severely degrades the area's ecology. However, sometimes lower amounts of overwash deposits may help in the growth of herbaceous mangroves in salt marshes (Paul et al., 2017). Overwash forces a migration of mangroves towards inland areas. After an event of overwash deposition, the pioneer mangrove species establish themselves on top of the deposits, which have buried the existing mangrove plants. This, along with the larger mangrove trees poking out from under the deposits, transforms the composition and characteristics of the mangrove ecosystem of the area. On Patibania Island and along the beach in Bakkhali near Henry's Island, the bigger mangrove trees are partially buried and protrude out of the thick layers of sand and sediments of the deposits. However, the herbaceous mangroves are completely buried, and those will die off.

10.5 Conclusion

Ephemeral rises in the sea level have an overall detrimental influence on the current state of the mangrove ecology in the Western Sundarban. Over time, hypersalinity in those depressions degrades the mangrove health and biodiversity, even leading to the emergence of barren saltpans devoid of vegetation. On the other hand, the large storm surges associated with the numerous cyclones in the Bay of Bengal frequently topple the sand dunes and other deposition along the beaches, burying the existing mangroves situated there. This results in the devastation of the existing mangrove ecology.

Tides are also a source of saline water, which is required by the mangroves. It is observed that the accumulation of salt on the topographic depressions occurs only because the trapped saline water has no way out, leading to its evaporation and the deposition of its salt content in the soil. This phenomenon can be mitigated by digging man-made canals connecting those topographic depressions and the main creeks, rivers, or sea. In the presence of such connecting canals, tidal water from each high tide would reach the depressions, and even though the tidal water is saline, it would flush away excess salt from the hypersaline patches. Mangroves are also observed to thrive on the banks of these canals due to the regular availability of water. The hypersaline patches develop due to an infrequent influx of water, and this development can be stopped or reversed by providing frequent supplies of water to those depressions. Commercial extraction of salt from the saline patches is another possible way to reduce their development. This would also provide economic livelihoods to some of the local population. However, this approach would not likely have an effect on the restoration of mangroves on those saline patches, unlike the previous one. The deleterious effects of overwash deposition on the existing mangroves cannot be mitigated. However, overwash also spreads nutrients from the swept depositions, and this nutrient availability can have a positive impact on the reestablishment of mangroves in the overwash areas. New mangroves will naturally establish themselves over the overwash deposits that buried the existing mangroves.

References

Balaguru, K., Taraphdar, S., Leung, L. R., & Foltz, G. R. (2014). Increase in the intensity of post-monsoon Bay of Bengal tropical cyclones. *Geophysical Research Letters, 41*(10), 3594–3601.

Chatterjee, M., Shankar, D., Sen, G. K., Sanyal, P., Sundar, D., Michael, G. S., et al. (2013). Tidal variations in the Sundarbans estuarine system, India. *Journal of Earth System Science, 122*(4), 899–933.

Chaudhuri, A. B., & Choudhury, A. (1994). *Mangroves of the Sundarbans. Volume One: India.* International Union for Conservation of Nature and Natural Resources (IUCN).

Choudhury, M., Fazli, P., Pramanick, P., Gobato, R., Zaman, S., & Mitra, A. (2019). Sensitivity of the Indian Sundarban mangrove ecosystem to local level climate change. *Science and Education, 5*(3), 24–28.

Chowdhury, A., Sanyal, P., & Maiti, S. K. (2016). Dynamics of mangrove diversity influenced by climate change and consequent accelerated sea level rise at Indian Sundarbans. *International Journal of Global Warming, 9*(4), 486–506.

Dhara, S., & Paul, A. K. (2016). Embankment breaching and its impact on local community in Indian Sundarban: A case study of some blocks of South West Sundarban. *IJISET, 3*(2), 23–32.

Donnelly, C., Kraus, N., & Larson, M. (2006). State of knowledge on measurement and modeling of coastal overwash. *Journal of Coastal Research, 22*(4), 965–991.

Figlus, J., Kobayashi, N., Gralher, C., & Iranzo, V. (2011). Wave overtopping and overwash of dunes. *Journal of Waterway, Port, Coastal, and Ocean Engineering, 137*(1), 26–33.

Ghosh, S., & Mistri, B. (2020). Geo-historical appraisal of embankment breaching and its management on active tidal land of Sundarban: A case study in Gosaba Island, South 24 Parganas, West Bengal. *Space and Culture, India, 7*(4), 166–180.

Ghosh, S., & Roy, S. (2022). Climate change, ecological stress and livelihood choices in Indian Sundarban. In *Climate change and community resilience* (pp. 399–413). Springer.

Gopal, B., & Chauhan, M. (2006). Biodiversity and its conservation in the Sundarban mangrove ecosystem. *Aquatic Sciences, 68*, 338–354.

Mukherjee, D., & Paul, A. K. (2020). Biophysical and chemical properties of soil in the mangrove habitat of the hypersaline tracts: An assessment in the Henry Island of southwestern Sundarban, India. *Journal of Indian Geomorphology, 8*, 90–104.

Mukherjee, D., & Paul, A. K. (2021). Mangrove sensitivities to climate change and its impacts in the Sundarbans: A case study in the Patibania Island of south western Sundarbans, India. In *Modern cartography series* (Vol. 10, pp. 353–385). Academic Press.

Mukhopadhyay, A., Payo, A., Chanda, A., Ghosh, T., Chowdhury, S. M., & Hazra, S. (2018). Dynamics of the Sundarbans mangroves in Bangladesh under climate change. In *Ecosystem services for well-being in deltas* (pp. 489–503). Palgrave Macmillan.

Paul, A. K. (2002). *Coastal geomorphology and environment* (pp. 1–342). ACB Publication.

Paul, A. K. (2022, August). Dynamic behaviour of the estuaries in response to the phenomenon of global warming in the coastal ecosystems of West Bengal and Odisha, India. In *Transforming coastal zone for sustainable food and income security: Proceedings of the international symposium of ISCAR on coastal agriculture* (pp. 907–931). Springer International Publishing.

Paul, A. K., & Paul, A. (2022). Adjustment of the coastal communities in response to climate variability and sea level rise in the Sundarban, West Bengal, India. In Climate Change (Ed.), *Disaster and adaptations: Contextualising human responses to ecological Change* (pp. 201–217). Springer International Publishing.

Paul, A. K., Ray, R., Kamila, A., & Jana, S. (2017). Mangrove degradation in the Sundarbans. In *Coastal wetlands: Alteration and remediation* (pp. 357–392). Springer. https://doi.org/10.1007/978-3-319-56179-0

Paul, A. K., Kamila, A., & Ray, R. (2018). *Natural threats and impacts to mangroves within the coastal fringing forests of India* (pp. 105–140). Hazards, Vulnerability, and Management.

Paul, A. K., Paul, A., & Majumdar, D. D. (2023a). Coastal sand dunes along the Western and eastern shores of India. In *Sand dunes of the northern hemisphere* (pp. 350–382). CRC Press.

Paul, A. K., Paul, A., & Sardar, J. (2023b). Susceptibility of the climate resilient landforms of the coastal ract of odisha and West Bengal, India. *Journal of the Geological Society of India, 99*(6), 827–839.

Payo, A., Mukhopadhyay, A., Hazra, S., Ghosh, T., Ghosh, S., Brown, S., Nicholls, R. J., Bricheno, L., Wolf, J., Kay, S., Lázár, A. N., & Haque, A. (2016). Projected changes in area of the Sundarban mangrove forest in Bangladesh due to SLR by 2100. *Climatic Change, 139*(2), 279–291.

Reddy, P. J., Sriram, D., Gunthe, S. S., & Balaji, C. (2021). Impact of climate change on intense Bay of Bengal tropical cyclones of the post-monsoon season: A pseudo global warming approach. *Climate Dynamics, 56*(9), 2855–2879.

Sahana, M., & Sajjad, H. (2019). Assessing influence of erosion and accretion on landscape diversity in Sundarban biosphere reserve, Lower Ganga Basin: A geospatial approach. In *Quaternary geomorphology in India* (pp. 191–203). Springer.

Chapter 11
Land Degradation and Its Management Approaches in the Middle and Lower Courses of the Subarnarekha River Basin, India

Ratan Kumar Samanta and Ashis Kumar Paul

11.1 Introduction

The middle part and lower part of the Subarnarekha basin are extended over the states of Odisha, West Bengal, and parts of Jharkhand in eastern India. The river Subarnarekha flows across the states boundaries and empties into the Bay of Bengal in Odisha. Geomorphologically, the region is grouped into parts of lateritic upland tracts, the extensive floodplain region of the Subarnarekha River basin, and the deltaic part of sandy chenier tracts. The river system comprises nine different watershed areas to support the tributary drainage systems within the basin area. The major types of land degradation are identified as soil erosion, landscape dissections, open bare surfaces with stoney, rocky, and lateritic hardpan surfaces, saline tracts of the low-lying delta plain surface, and coastal sandy soil surfaces in the middle and lower catchment areas of the Subarnarekha basin. There is a strong relationship between morphogenetic regions and land degradation types in the river basin. The lateritic upland tracts were densely forested in 1973 (SOI Toposheet), but the encroachment of agricultural plots and settlements into the forest area by occupying forest land through forest cutting has resulted in soil erosion and the formation of hardpan lateritic tracts in the western part of the river basin. However, the riverine flood plains with alluvial soil support agricultural activities, but during catastrophic floods, soil erosion takes place due to the impact of floods and surface runoff characters in the region. The lowermost part of the deltaic beach ridge and Chenier plain of the coastal fringe area are affected by soil erosion due to marine and aeolian processes (Paul, 2002).

R. K. Samanta
Swarnamoyee Jogendranath Mahavidyalaya, Amdabad, West Bengal, India

A. K. Paul (✉)
Department of Geography, Vidyasagar University, Midnapore, West Bengal, India
e-mail: akpauleastcoast@gmail.com

A. K. Paul, A. Paul (eds.), *Crisis on the Coast and Hinterland*,
https://doi.org/10.1007/978-3-031-42231-7_11

The land degradation processes have been identified by the study, which shows that the slope of the land, surface runoff trends, forest cutting and other human activities, composition of the soil, flood events, and wind-blown activities are the major factors that control the rate of spatial degradation process in the basin. Hence, riling, gullying, and ravening processes of the fluvial activities during the wet monsoon phase produce a significant loss of soil from the top surface of the morphogenetic regions (Paul et al., 2022). Salinization and alkalization are also found in the coastal areas and flood plains of the lower part of the Subarnarekha basin. Using remote sensing and GIS tools, the land degradations under different morphogenetic regions are estimated and monitored in the study area. For the lateritic upland tracts, contour bunding and check dams with suitable cropping patterns may be recommended. Agroforestry, pasture land with cultivation of fodder crops, and crop rotation techniques may be followed for the reduction of land degradation rates in floodplain areas. Finally, the coastal areas need the restoration of habitats (e.g., mangroves, salt marshes, sand dunes, beach ridges, and tidal flats) to protect the high rate of soil loss by planting mangroves, afforestation, and promoting agroforestry in the region. ICZM policies should be adopted in the region to improve the capacity development of the coastal tracts. The objectives of the study are: (i) to identify the different land degradation processes in different geomorphic environments, (ii) identify the degraded lands or watery wastelands, and (iii) pursue the management strategies to combat the degradation of land for its sustainable uses.

11.2 Study Area

The middle and lower catchments of the Subarnarekha River Basin extend from 21°30′N, 22°23′N latitude and from 86°42′E to 87°30′E longitude (Fig.11.1). This part of the river, including its tributaries, runs through the extreme south-western part of Paschim Medinipur district of West Bengal and the easternmost part of Mayurbhanj and Baleswar districts of Odisha. The study area includes the administrative blocks of Gopiballavpur-I and II, Sankrail, Nayagram, Keshiary, and Dantan-I of Paschim Medinipur (West Bengal), Moroda, Betnoti, and Rasgobindapur blocks of Mayurbhanj district (Odisha), and Jaleswar, Basta, Bhograi, and Baliapal blocks of Baleswar district, Odisha. The entire study area covers an area of 3745 square km. Topographically, the area consists of lateritic uplands, valley cuts, and river valleys of the Rahr Plain in the north-west and the Coastal Plain in the south. In the northwestern part of the study area, topography is moderately undulating, and in the middle part, it is gently undulating, while the southern part is characterized by the flat alluvial plain of the Subarnarekha delta, marked by a number of parallel palaeobeach ridges and sand dunes along the coast line of the Bay of Bengal (Fig. 11.1).

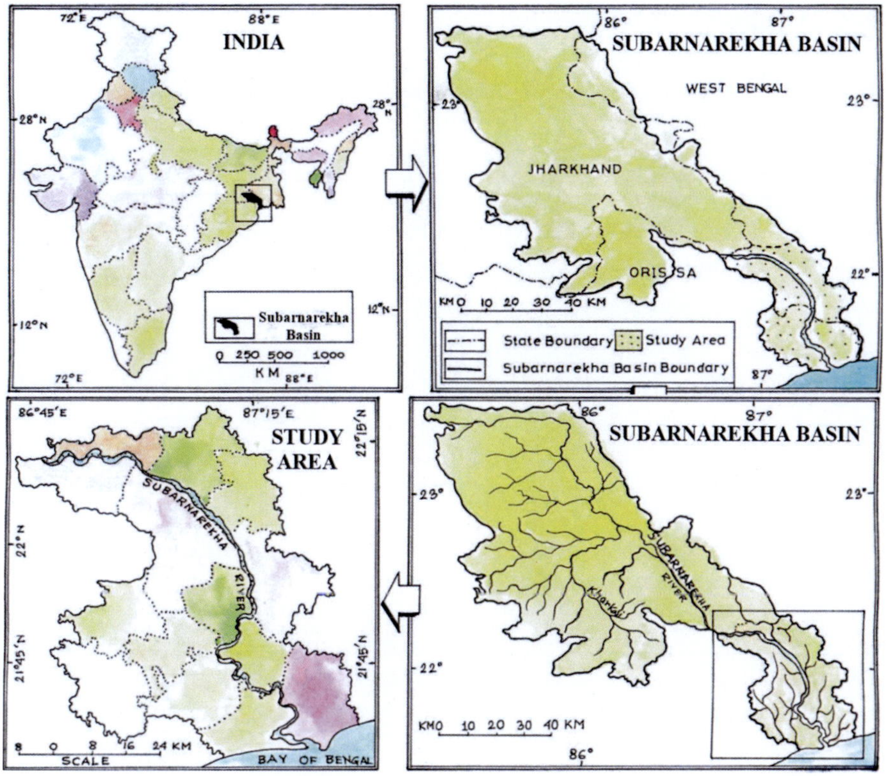

Fig. 11.1 The location Subarnarekha River basin with its middle and lower catchment areas in Odisha and West Bengal, India

11.3 Methodology

Land degradation processes have been studied on the basis of field surveys, Landsat satellite image analysis (2008), study of SOI toposheets (1973), use of a GIS platform, etc. A field survey was conducted across the Singari Khal (Singalila Canal), located at Pindragariya village in Gopiballavpur-I block. The survey reveals that the upland margins (68–70 m in height) with loose soils are highly susceptible to water erosion. The deforested tracts of upland with loose soil are susceptible to sheet, rill, and gully erosion. The highly eroded tracts have been converted into bad land. The lowlands (66–67 m in height), along the ditches, are the depositional surface. But during the flash flood, water erosion at a higher rate is also observed in this zone.

11.4 Results and Discussion

11.4.1 Identification of Land Degradation Types (Regional Diversity of Land Degradation)

Land degradation remains an important global issue in the twenty-first century because of its adverse impact on agronomic productivity, the environment, and its effect on food security and the quality of life (Eswaran et al., 2019; Johnson & Lewis, 2007; Bajocco et al., 2012; Bossio et al., 2010; Blaikie & Brookfield, 2015; Gisladottir & Stocking, 2005). In the diverse landforms of the middle and lower catchments of the Subarnarekha basin, different types of land degradation have been identified. The different types of land degradation processes and water erosion are dominant in the bare laterite uplands and upland margins. Duricrust surfaces, the product of seasonal rainfall and the alternate wetting and drying of laterite upland, are found in the laterite upland. Upland surfaces, upland margins, and valley fill surfaces are more susceptible to soil acidification. Rills and gullies are formed due to fluvial erosion. Land degradation is a process that can occur in any physiographic environment. So far as the bare duricrust surface, laterite mining pits, rills, gullies, and ravines with dissected surface and soil erosion affected areas are identified as major land degradation types in the region. But its type, intensity, and magnitude vary from place to place in the regional settings of the present study area (Table 11.1, Fig. 11.2).

11.4.1.1 Land Degradation in the Upland Surface with Dissected Valleys

The laterite duricrust surfaces are found in the upland tracts of the interfluves in the middle catchment area. Here laterization is the main process, except in the narrow tracts of river channels. The most affected areas lie in the upland areas of Gopiballavpur I and II, Nayagram, Moroda, Betnoti, and Rasgobindapur blocks. In

Table 11.1 Spatial diversity of land erosion rates in the different morphogenetic surfaces under multiple watershed areas of the Subarnarekha River basin (middle and lower catchment)

Morphogenetic surface	Name of the watershed	Rate of soil erosion (ton/ha/yr)
Deltaic low land sandy tract	Khaljori watershed	22
	Subarnarekha watershed	39.5
	Hanskura watershed	26.8
Lateritic upland dissected tract	Burahbolong watershed	15.8
	Bayesmukha watershed	14.6
	Jamira watershed	14.2
Riverine flood plain	Dulung water shed	5.8
	Kaliaghai watershed	5.2
	Kushumri watershed	6

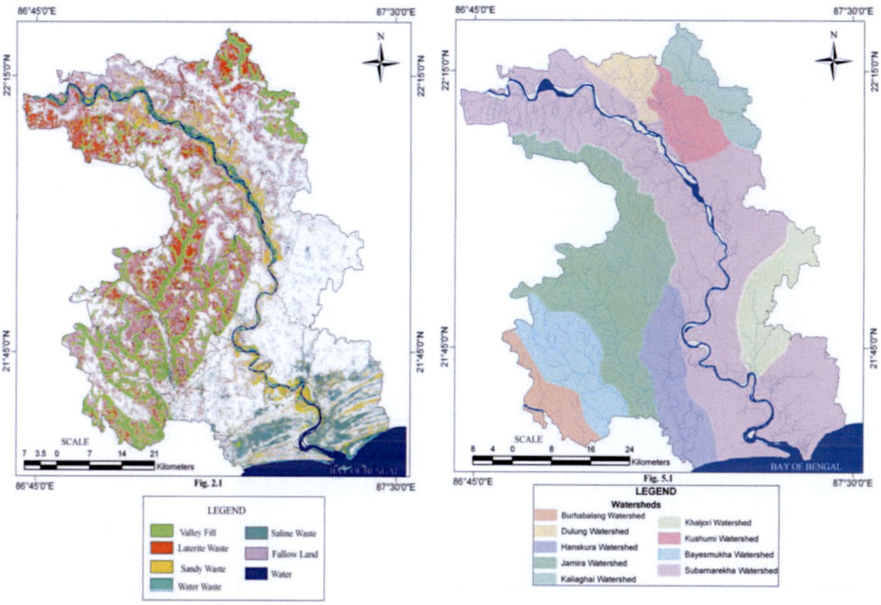

Fig. 11.2 Various types of land degradation characters under nine watershed regions of middle and lower catchment area of Subarnarekha River basin (using Landsat Image and SOI toposheets)

this area, other types of land degradation include rills, gullies, and sheet erosion. It was noticed that the gradient of the land surface was higher (1°–3°) in compared to other physiographic regions of the study area. Deforestation, weathering, and erosion processes over a long period accelerated the rate of degradation. Upland margins are the most vulnerable areas in this respect (Table 11.1).

11.4.1.2 Land Degradation in the Flood Plains Dominated by Subarnarekha and Other Streams

Land conversion for agricultural uses is the main degradation type in this region. During the field survey, it was observed that most of the farmers use fertilizers, pesticides, and insecticides unscientifically. Besides, they also use less organic manure, leading to the degradation of agricultural land. This picture is acute, especially in the forest agricultural land interface where most of the farmers belonging to Schedule Caste and Schedule Tribe communities have little knowledge about the fertility status of the soil. Besides, more than 50% of the families fall into the BPL category. Probably, illiteracy and poverty are the main causes of ignorance. The soil analysis report and maps prepared from these data reflect that if such a process continues; in the near future, the agricultural land in this area will be gradually degraded. Besides, soil erosion of lower magnitude (less than 5 ton/ha/year) is observed in the areas with a 1–2-degree average slope (Table 11.1).

11.4.1.3 Land Degradation in the Deltaic Flat with Coastal Features

The low-lying land along the seashore and Subarnarekha River side in Bhograi and Baliapal blocks gets inundated by seawater during floods. As a result, lowlands remain uncultivated due to the high salinity of the soil. The aeolian process plays an important role in this area of land conversion. Aeolian erosion and accretion are going on in the coastal belt of the study area. As a result, land in this tract remains in a degrading state. Bare sand dunes are more susceptible to water and wind erosion. The rate of soil erosion in this area is 20–40 tons per acre per year (Table 11.1). Consequently, a number of sandy wastes have formed in this belt. Sandy wastes are also found along the river channels in the form of deposits of sand brought down by the rivers flowing from the uplands.

11.4.2 Process of Development of Land Degradation Types

The processes of land degradation are as follows: (i) physical degradation (comprising crusting, compactation, erosion by water and wind, etc.) and (ii) chemical degradation (comprising acidification, salinization, alkalization, leaching, and faulty agricultural practices, etc.). Some lands or landscape units are affected by more than one process, such as water and wind erosion, salinization, and crusting or compacting the surfaces. However, different land degradation processes are discussed herewith.

11.4.2.1 Water Erosion

Over-cutting of forests, over-grazing, removal of protective plant cover by tillage operations, etc., are contributing to soil erosion through the water erosion process in the study area. The types of water erosion may be categorized as: (i) rain drop splash soil erosion is noticed all over the study area. (ii) Sheet erosion is dominant along the valley fill sites of the upland and floodplain regions of the study area. (iii) The upland tract of the study area is more susceptible to riparian erosion compared to the floodplain and coastal delta's flat surfaces. (iv) The network of gully channels and erosion is well reflected in the true-color Landsat OLI image of Chandabila and Betnoti-Moroda. Very small to medium gullies are found in the study area, and most of them are located in the laterite upland of Gopiballavpur-I and II, Sankrail, Keshiary, Nayagram, Moroda, Betnoti, Rasgobindapur, and Jaleswar block. (v) Stream channel erosion is the scouring of material from the water channel and the cutting of banks by flowing or running water. The rate of erosion accelerates during the rainy season, especially at the time of floods. Bank erosion along the stream channel is common in the Nayagram, Dantan, Jaleswar, and Gopiballavpur-I and II blocks.

11.4.2.2 Flood in the Deltaic Flats with Coastal Features

Flood is an attribute of the physical environment and acts as both an eroding and depositing factor. It promotes the washing away of surface soils by its great force, leading to soil degradation (Paul, 2002). Silting on the river bed, construction of roads and a railway network against the natural gradient, a large number of sprung-up fish farms with protective bunds, heavy monsoon rains, cyclones, tidal waves, and the sudden release of dam and barrage water from the upstream section are the factors of flood in the low-lying coastal tract and rolling flood plain of the study area. In the flood plain area, flooding is caused by heavy rainfall, the sudden release of water from the dams and barrages upstream, and the heavy monsoon. The villages on both sides of the river Subarnarekha are flood-prone. The parts of Gopiballavpur-I and II, part of Nayagram, part of Dantan-I, and part of Keshiary are affected by the flood havoc in the previous decades. The laterite upland is devoid of flooding due to its higher altitude and comparatively steep slope.

11.4.2.3 Wind Erosion

Wind erosion takes place in the deltaic flat of Bhograi and the Baliapal block of the study area. The most serious damage caused by wind erosion is the change in soil texture. Since the finer soil particles are subject to movement by the wind, wind erosion gradually removes silt, clay, and organic matter from the top soil, leaving the coarse soil material. Soil particles are carried by the wind in three ways: salination, suspension, and surface creep. However, some factors influence the wind erosion; these are: (i) characteristics of wind: wind speed, direction, structure, temperature, humidity burden carried, etc., (ii) characteristics of surface: roughness, vegetation cover, obstruction, and temperature, (iii) topography: flat, undulating, broken, and (iv) nature and properties of soil: Texture, structure, organic matter, and moisture content. In the coastal area, wind erosion takes place on bare sand dunes. Casuarina trees and different mangrove trees are planted along the coastal tract of the study area. But recently, due to illegal forest cutting, the sand dunes are susceptible to wind erosion (Paul, 2002).

11.4.2.4 Salinization Problems

Salt-affected soils are unproductive unless excess salts are removed or reduced. From the field study and laboratory analysis of soil samples collected from different parts of the study area, it was found that salt-affected soils are mainly concentrated in the coastal area, where soils suffer from inadequate drainage and seawater inundation. Marshy lands in the Subarnarekha estuary are highly affected by salinity. Soils are said to be saline if they contain an excess of soluble salts.

11.4.2.5 Soil Acidity and Land Degradation

Soil samples collected from agricultural fields in the blocks of the study area were tested to get information regarding the soil acidity and alkalinity. On the basis of pH values obtained from soil sample analysis, it is found that in the laterite upland with valley cuts, soil pH ranges between 4.6 and 6.5, indicating acidity in nature, whereas in the floodplain and delta, the pH ranges between 6.5 and 7.2, indicating neutrality in nature. It must be mentioned that the marshy area in the deltaic flat has a pH higher than 7.5. Moderately and strongly acid soil is noticed in the laterite uplands of Gopiballavpur-I, Nayagram, Sankrail, Rasgobindapur, Moroda, Betnoti, and Jaleswar. In other blocks, the soils are neutral in character. Acid soils are not suitable for further cultivation to grow agricultural crops.

11.4.2.6 Overgrazing and Land Degradation

The livestock population of the study area was 10,81,122 in 2002–2003 and rose to 12,47,495 in 2005–2006. An increasing trend is also found in every block of the study area. So, there is an increase in livestock population with the passage of time. The fodder requirement is met from the nearby forests in the form of grazing and fodder cut for stall feeding. Overgrazing and over-extraction of green fodder both lead to forest and land degradation through a loss of vegetation and physical deterioration in the form of compaction, reduced infiltration, and an increase in soil erodibility.

11.5 Management Strategies

Land management cannot solve the problem of land degradation. By integrating sustainable management of land and watersheds, the problem of land degradation can be mitigated. The entire study area has been divided into eight watersheds. These are: (i) Subarnarekha, (ii) Jamira, (iii) Dulung, (iv) Bayes Mukha, (v) Kusumi, (vi) Burabalang, (vii) Haskura, and (viii) Khaljori. Geospatial studies and field surveys reveal that all the watersheds suffer from different types of degradation. Sustainable watershed management has emerged as a new paradigm for planning, development, and management of land, water, and biomass resources with a focus on social and environmental aspects and a participatory approach. We should keep in mind that it is not merely an anti-erosional and anti-runoff approach; it is a comprehensive, integrated approach. The approach is preventive, progressive, corrective, and curative. Watershed management involves the judicious use of natural resources with the active participation of institutions and organizations in harmony with the ecosystem. According to Mani (2005), watershed development (WSD) is one such approach that is not only eco-friendly but also a hydrologically correct approach for efficient use of soil and water for more agricultural production.

11.5.1 Land Management

The different measures of land management are grouped as follows: (i) structural measures, which include interventions like contour bunds, compartmental bunds, earthen dams, graded bunds, contour terrace walls, contour trenches, bench terracing, stream bank stabilization, check dams, etc. (ii) Vegetative measures include vegetative cover, plant cover, mulching, vegetative hedges, grassland management, agroforestry, etc. Watersheds may contain natural ecosystems like grasslands, wetlands, mangroves, marshes, and water bodies. All these ecosystems have a specific role in nature. (iii) Production measures include interventions aimed at increasing the productivity of land like mixed cropping, strip cropping, cover cropping, crop rotations, the cultivation of shrubs and herbs, contour cultivation, conservation tillage, land leveling, the use of improved varieties of seeds, horticulture, etc. (iv) Gully plugging, runoff collection, land slide control, etc., may be considered as protective measures. Adoption of all the interventions discussed above should be done strictly in accordance with the characteristics of the land taken for management.

11.5.2 Water Management

The broad interventions for water resource management are: (i) rainwater harvesting, (ii) groundwater recharge, (iii) maintenance of water balance, (iv) preventing water pollution, and (v) economic use of water. Some simple and cost-effective rainwater harvesting structures have been suggested in the study area, including: (i) percolation pits or tanks, (ii) recharge trenches or rain pits, (iii) recharge wells, (iv) farm ponds, (v) "V" ditches, and (vii) bench terracing. Economic use of water and the avoidance of affluence in the use of water at individual and community levels may be the major concerns for water management in the years to come. Different agencies have been operating in the study area with the aim of improving rural socio-economic standards and conserving the natural resources, especially land and water resources.

11.6 Conclusion

The intensive study based on fieldwork, remote sensing data, and other relevant information reveals that the study area faces a variety of land degradation and water resource management problems. Developmental activities without much attention to environmental issues have resulted in great pressure on natural resources. However, the following suggestions have been made for the sustainable development of land and water resources in the study area. There is no well-defined land use policy in the whole study area. This lacuna is highly responsible for the current stage of land degradation and wasteland formation. The land use management policy to be set involves working with government organizations (Ministry of Environment and Forests, Ministry of Water Resource, Ministry of Rural Development, Ministry of Information, etc.), NGOs, and other stakeholders to develop a coordinated approach for sustainable land use management in the context of related cross-cutting issues. The dispute over land ownership and tenure is an age long problem in the region. Proper steps should be taken to increase mass awareness of land use planning and land management, land fertility, and farming systems, including their impacts and sustainability. Afforestation, reforestation, social forestry, and agroforestry should be given special attention as they are major tools for soil conservation, ecological balance, and the conservation of biodiversity. Embankments, along with the sluice gates in the coastal tract, should be higher and stronger to protect the reclaimed agricultural lands behind them. Pesticides, insecticides, and chemical fertilizers are to be used with utmost caution to avoid the possibility of land and water pollution, and more attention should be given to the integrated nutrient and biological pest management strategies. Farmers should use organic manures, green manures, different composts, organic chemicals, eco-friendly pesticides and insecticides, and biofertilizers in addition to chemical fertilizers to protect agricultural land in the particular environment as a whole from degradation. Municipal waste, industrial waste, and domestic waste must all be treated before disposal to prevent environmental degradation.

Rainwater harvesting and groundwater recharge through the formation of storage tanks and the construction of check dams across the drainage channels are essential throughout the study area to reduce the rate of soil erosion and increase the soil moisture retention capacity in the locality. In this regard, the government has adopted a few management plans for the restoration and conservation of forests and for the storage of rainwater in the artificial ponds. Population should be controlled to reduce the pressure on natural resources, especially land, water, and forest resources. Indigenous land and water management practices should be encouraged in the region to reduce soil erosion and land degradation activities. Simultaneously, local knowledge systems and environmental values of landscapes should be given due honor.

References

Bajocco, S., De Angelis, A., Perini, L., Ferrara, A., & Salvati, L. (2012). The impact of land use/land cover changes on land degradation dynamics: A Mediterranean case study. *Environmental Management, 49*, 980–989.

Blaikie, P., & Brookfield, H. (2015). *Land degradation and society*. Routledge.

Bossio, D., Geheb, K., & Critchley, W. (2010). Managing water by managing land: Addressing land degradation to improve water productivity and rural livelihoods. *Agricultural Water Management, 97*(4), 536–542.

Eswaran, H., Lal, R., & Reich, P. F. (2019). Land degradation: An overview. *Response to land degradation*, 20–35.

Gisladottir, G., & Stocking, M. (2005). Land degradation control and its global environmental benefits. *Land Degradation & Development, 16*(2), 99–112.

Johnson, D. L., & Lewis, L. A. (2007). *Land degradation: Creation and destruction*. Rowman & Littlefield.

Mani, N. D. (2005). *Watershed management – Principles, parameters and programs* (p. 110002). Dominant Publishers and Distributors.

Paul, A. K. (2002). *Coastal geomorphology and environment* (pp. 1–342). ACB Publication.

Paul, A., Mallik, I., Sardar, J., & Bandyopadhyay, J. (2022). Soil loss risk assessment of lateritic badland surface of Garhbeta Block-I, West Bengal, India using an integrated approach of Revised Universal Soil Loss Equation (RUSLE) algorithm and geospatial techniques. *Safety in Extreme Environments, 4*(2), 149–170.

Chapter 12
Issues and Management Strategies of the Riparian Corridor of the Dulung River Basin, India

Lila Mahato

12.1 Introduction

In any watershed, the floodplains and the riparian buffers play a significant role in providing environmental benefits. The riparian zone along rivers is a specific biome composed of three main compartments: the watercourse, the terrestrial environment, and the riparian vegetation. It balances the ecosystem's functions, which depend on many interacting factors. The issues related to riparian corridor are water quality, flood prevention, wildlife habitat, economics, and various other ecological, physical, biological, and chemical processes (Agar et al., 2016; Amper et al., 2019). Riparian corridors reduce erosion, filter sediments, and pollutants out of overland runoff, block solar radiation to moderate water temperature, provide habitat, and store water (Atkinson & Lake, 2020; Boyce, 1995). This riparian area, like a green belt, has an impact on the groundwater table, soil health, soil moisture content, etc. Functionally, it is an area with three-dimensional ecotones consisting of both terrestrial and aquatic ecosystems. It encompasses the groundwater situation, the floodplain, the near-slopes area, the lateral terrestrial ecosystem, and the longitudinal watercourse at a variable width (Erős & Lowe, 2019). This is the area where fluvial geomorphology and stream ecology meet and play a significant role in providing a new dimension to the river continuum concept (Naiman et al., 1993; Forman, 1995; Frazier, 2019; Paudel & Yuan, 2012; Sanderson & Harris, 2000; Turner & Gardner, 1991).

The problems of river bank erosion, occasional flooding, and shrinkage of vegetative cover due to human interference are common in almost all tropical Indian river corridors. The corridor includes all perennial streams with defined channel flow and encompasses adjacent terrestrial areas influenced by the river and its land,

L. Mahato (✉)
Post Graduate Department of Geography, Krishnagar Govt. College, Krishnagar, Nadia, West Bengal, India
e-mail: lilamahato2071@gmail.com

A. K. Paul, A. Paul (eds.), *Crisis on the Coast and Hinterland*,
https://doi.org/10.1007/978-3-031-42231-7_12

soil, slopes, and vegetation features. Management strategies call for maintaining and improving ecological functions to improve the sustainability and productivity of the area. The river Dulung, a fifth-order stream, portrays a unique perspective in its catchment. The area is surrounded by several hills in the northwest corner, for example, Kanaisore (370 m), Choukisole (320 m), and some other residual hills. Some large-scale gullies associated with the middle parts of the river add diverse dimensions to the rolling plains. Some small streamlets, after crossing the 170 m contour elevation of the catchment area, reach the plains and form a distinct flow. The area at the confluence is bounded by a 30-meter contour line or less. It ultimately joins the Subarnarekha River near Rohini in the Sankrail block in the Jhargram district of West Bengal. The riparian area in the Dulung watershed includes streams, ponds, dams, jhils, large water bodies, groundwater resources, and irrigation canals traversing the watershed. The corridor along the river has been divided into three reaches – upper, middle, and lower – based on the morphometric assessment of the DEM of the basin.

12.2 Rationale of the Study

The Dulung River Basin, a part of the Subarnarekha River System, being a hydrological unit, provides a congenial environment for agricultural pursuits and other economic activities. The riparian belt is at stake in some places due to recent anthropogenic activities (Fig. 12.1). Both aquatic and riparian systems are negatively influenced, and biodiversity is hampered, but the waterbodies with terrestrial land surfaces must be assessed for this. The study falls into two basic categories: (1) Functional assessments of riparian function exist at present and (2) identifying issues and stating management strategies from an environmental standpoint. The present work focuses on the systemic study of the uniqueness of the area based on morphometric parameters and analyses how far it can act as a buffer for the entire catchment to restore its ecosystem services. Strategies have been formulated to tackle different issues relating to the environment. The work also highlights different accomplishments that are going on for the welfare of the stakeholders.

12.3 Materials and Methods

Thematically, the present research work is an integrative one and is linked with various branches of geography. Landsat 8 OLI multispectral data also used to classify the LULC activities by human encroachments into the river basin. The adoption of statistical techniques is essential for analysing different parameters of land and water resources. The Basin boundary map is overlaid on the administrative one and forty-six (46) Gram Panchayats and one municipal area that includes parts of East Singbhum district in Jharkhand and parts of Jhargram district in West Bengal. The

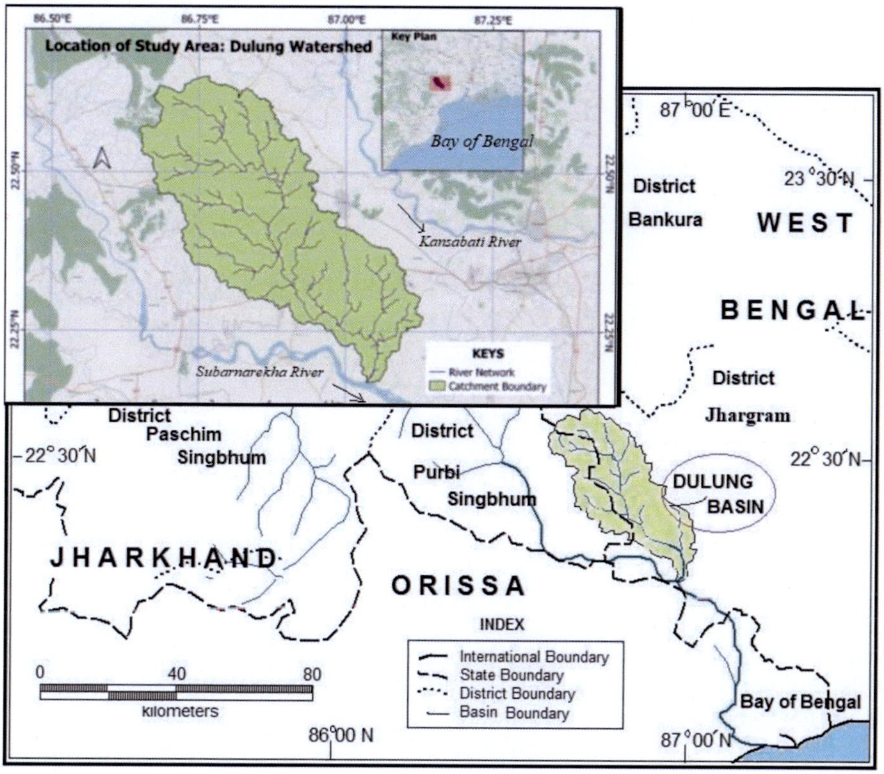

Fig. 12.1 Dulung river basin under the catchment area of Subarnarekha River

river basin is identified and demarcated both topographically and cartographically. The required database is constructed by acquiring data from both primary (field survey) and secondary sources (Survey of India toposheets, reports, journals, books, gazetteers, maps, satellite images, etc.). The database is managed (storage, editing, manipulation, and analysis) entirely with statistical software using a PC and has been mapped and modelled primarily with the help of GIS software. General cartographic principles have been followed for representing the output of the analysis.

12.4 Results and Discussion

12.4.1 Geomorphological Perspective

Dulung Watershed, with all its physiographic personalities, reveals a distinctive landscape. Kupan nala, Deb nala, and Palpala khal are the main right bank tributaries, while Simana and Champa khal join from the left and enhance its total

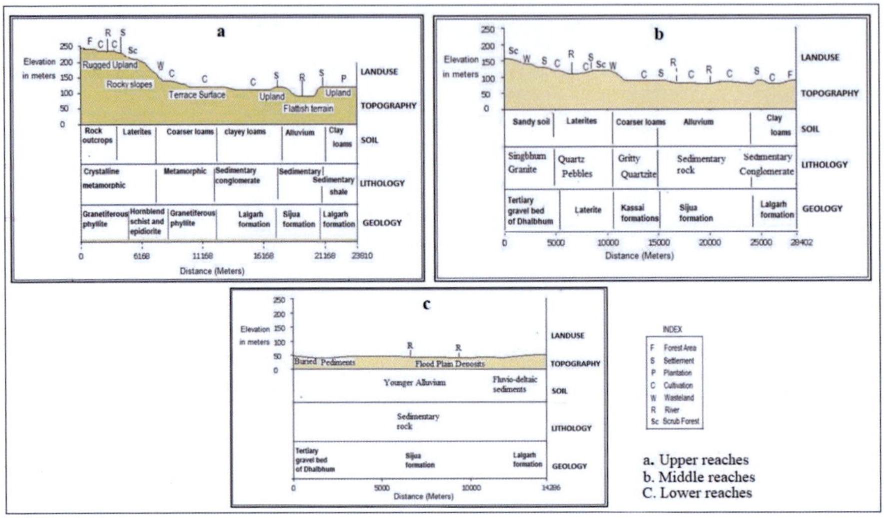

Fig. 12.2 Characteristics of land use in relation to geology, soil, and topography

discharge. The river bears several orders with respective stream numbers (Graziano et al., 2022). Geographically, the basin is divided into four units (Fig. 12.2), which are as follows: (a) upland surfaces with hard rock exposure, (b) lateritic uplands with valley cuts, (c) valley fill surfaces, and (d) floodplain alluviums of lower catchment areas. According to the hydro-system approach, it can be divided into three separate zones.

The *headwaters or production zone* supply water, sediment, organic matter, and solutes come down the slopes to stream channels. The area with more than 100 m of elevation represents the production zone, which is the result of the process-response relationship system of the region. The hill-slope processes determine the runoff and sediment yields. The *transfer zones* highlight all the materials that flow downstream with water as dissolved loads. The area lies between 40 and 100 m. elevation and collects all the water to channel into a higher order stream. These zones act as a 'black box' and control the primary input of the hydro system. The *storage zone* is where transported materials are deposited and retained for a long time. Sediments and organic matter are in constant interaction with the floodplain aquifer, which makes this zone suited for all kinds of agricultural pursuits. The agricultural landscape was more fertile in the basin near the confluence with alluvial fills than in the other parts. These three zones interact with their immediate environments and constitute an 'eco-complex' (Groom et al., 2011; Lawler, 2004; Jiménez-Carmona et al., 2020). Thus, the corridor comprising the channels, adjacent floodplains with aquatic and terrestrial habitats, and underlying aquifer exerts a great impact on its biological features and the people living within it.

12.4.2 The River Corridor Complex

Dulung watershed (approx. 1270 km^2) possesses important agricultural soils, floodplain wetlands, fisheries, and wildlife habitats. The corridor includes lands adjacent to rivers of all ordered streams except the first, which has a visible flow only during rains. The entire corridor complex has been divided into three zones, keeping in parity with the hydro system zones, like the upper, middle, and lower reaches. The headwater zone is defined by close-spaced contours of much higher slopes, rugged upland, and hummocky surfaces and corresponds to the upper reaches with more than 100 m elevation. Geological formations, for example, laterite formations, are often merged with Sijua and Kasai formations near the river valley. Denudation terraces with rocky outcrops are the main characteristics of the landscape. Lateritic uplands with valley cuts bestow the area with distinct erosional features produced by undercut and bank erosion. Coarser loams and sandy soil can be found far from the main channel, while alluvial soil can be found along the river.

The middle reaches are usually associated with comparatively low relief and valley-fill surfaces. The area is characterised by third- and fourth-order streams with less slope. Deb Nala and Palpala Khal cross from the west and join Dulung. Other major tributary streams flowing through narrow valleys have deposited a considerable amount of denuded material, and these can be delineated as shallow valley fill surfaces. Deep to moderately buried sediment and laterite capping also characterise some areas. Lower reaches are the floodplain alluviums of lower catchment areas and possess a good prospect of groundwater availability. Older alluvium includes medium- to coarse-grained sands with a brownish tint, poorly sorted ferruginous nodules, and calcareous concretion aggregates. The river valley contains finer to medium sand to clay, which belongs to the newer alluvium. The relief is much less, ranging from 64 to 30 m, and the slope varies from 6° to 2° in the area. Near the confluence, the corridor span is enriched by cultivated fields and settlements in the hinterland areas of Subarnarekha middle catchment area. All the three cross sections (Fig. 12.2) drawn over the upper (Fig. 12.2a), middle (Fig. 12.2b), and lower reaches (Fig. 12.2c) of the basin clearly indicate that topographic landscape is the resultant effect of soil, lithology, and geology, which ultimately controls the land use and land cover situations.

12.4.3 Landscape Ecology Along the River Corridor

Landscape ecology emphasizes the pattern and interaction among ecosystems of varied land use and land cover. The interactions affect ecological processes, especially the unique effects of spatial heterogeneity over the entire Anthropocene (Ryszkowski, 2002). The landscape feature along the two-kilometre buffer zone (Fig. 12.3a) of the river Dulung influences the plant, animal, and human populations. The study of 2014 and 2022 Landsat 8 data classification reveals that

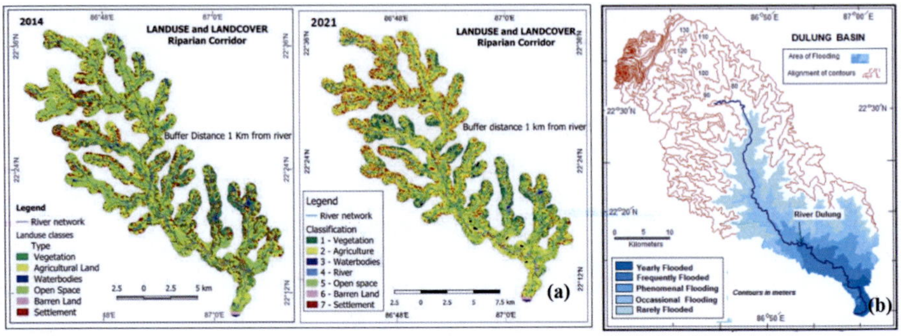

Fig. 12.3 (**a**) Land use and landcover in the riparian corridor (2014–2021); and (**b**) flood situation analysis with characteristics and zonation along the river corridor

Table 12.1 Patch analysis in landscape studies (2014 and 2021)

Classes	Land cover (Area in km^2)	Land cover in %	Edge length in km	Number of patches	Greatest patch area	Mean patch area (m^2)
Vegetation	126.20	20.46	5141.88	7,014	7,306,200	17992.94
Agricultural land	313.99	50.90	9642	9,944	11,135,700	22222.96
Waterbodies	10.85	1.76	640.02	1,953	451,800	5553.46
Open space	78.50	12.73	5229.42	5,561	8,761,500	35695.41
Waste land	2.33	0.38	115.2	428	963,000	5454.67
Settlement	84.93	13.77	8045.34	10,146	2,723,400	11005.32
Total (2014)	616.80	100.00	28813.86			
Vegetation	105.96	17.18	5300.4	9,073	5,076,900	16087.31
Agricultural land	373.58	60.57	10564.32	2,324	359,034,300	177531.58
Waterbodies	5.51	0.89	339.78	1,309	429,300	4210.54
Open space	23.39	3.79	697.56	2,659	551,700	3474.76
Waste land	9.36	1.52	197.82	685	685,800	5438.10
Settlement	99.00	16.05	6057.18	5,432	1,435,500	12303.63
Total (2021)	616.80	100.00				

landscape processes affect evolutionary procedures, especially in human-modified landscapes, and create various consequences in the surrounding environment. Landscape metrics (Table 12.1) measure the landscape composition and describe each land cover class quantitatively, as well as the spatial pattern of the ecological processes over time and space. The open space with scattered scrubs in 2014 has been converted and cleared for agricultural practises in the later period.

Several interesting features emerge from the analysis of landscape metrics along the 2 km buffer corridor. The number of patches in each category class changes in 2021 compared to 2014. Such an increase in the number of vegetated areas suggests that the patches are being broken up and that other economic activities are infiltrating the land class (Table 12.1). Simultaneously, the total edge length increases, proving the same fact. Waterbodies show an increase in patches, but edge length and a total percentage decrease show a decrease in water surface area in the 2 km corridor complex. This patch analysis clearly identifies the stress area in terms of the ecological needs of the riparian corridor.

12.4.4 The Issues of the River Corridor

12.4.4.1 Flooding Events

The river corridor and floodplains are synonymous in some places, especially for the people of the middle and lower sections of the area. The Sankrail block is the worst affected area, where villages suffered a lot, houses got damaged, and cultivated plots were converted to sand deposits, rendering the area unsuitable for agriculture. The flood zonation (Fig. 12.3b) shows the inundation scenario of the floodplain area near the confluence zone. In years of excessive rainfall coupled with depressions, cyclones cause a great problem, as occurred in 2007 and 2008 (a significant flood event). In most cases, inundation occurs below 70 m of elevation and on a 3-degree slope. A flood situation in an area with an elevation of less than 30 metres causes significant damage to agricultural fields (Table 12.2). The villages of Rohini, Kharbandhi of Sankrail, and Belebera panchayats of Gopiballavpur II blocks are the worst sufferers. The area above 100 m elevation is rarely flooded, so it is less disturbed.

12.4.4.2 Bank Failure and Riverbank Erosion

The gushing and whirling action of river water, particularly during the monsoon season, causes bank erosion in cliff bank areas. The channels in the upper and middle reaches have constricted their course on rugged laterite and intrusive rocks. In the lower reaches, it flows through a sinuous course, sometimes meanders, creating point bars and slip-off slopes. The thalweg section and river cliff frequently cause bank erosion, gulping agricultural land, and culturable wasteland. Although no catastrophic stream bank erosion occurs, several incidences of bank failure change the

Table 12.2 Flood nomenclature in the lower reaches

Elevational zones (m)	Area (km^2)	Nomenclature
<30	19.36	Yearly flooded
30–40	55.97	Frequently flooded
40–50	53.04	Phenomenal flooding
50–60	110.9	Occasional flooding
60–70	140.4	Rarely flooded

bank lines. It is considered as toe erosion, planer failure, and the sliding down or toppling of the materials that are responsible for the removal of bank material by failure processes.

12.4.4.3 Shrinkage of Vegetation

Bushy and scrub vegetation and thorny plants like babla, phani mansa are observed in areas with a low soil moisture regime. Climbers form dense vegetative growth in the Chakulia, Baharagora, and Jhargram areas, which are areas of pure stands of the species Sal trees, kendu, palash, and other epiphytes. Neem, khejur, imli, peepal, sajinaa, bargad, and kathal are seen. The vegetation along streambanks and in floodplains diminishes the velocity of flood waters, lessens the erosive power, and helps to accumulate nutrient-laden sediment. The temporal LULC proves the shrinkages of vegetations in lower section and middle section of the river basin.

12.4.4.4 Soil Degradation

Lateritic soil cover area occurs in the panchayats of Binpur-II, Jhragram, the western portion of Jambani Block in West Bengal, and Chakulia and Baharagora in Jharkhand. The soil morphology profiles across the basin show that sandy loam is the predominant soil, followed by sandy soil in the western part of the basin. Next to sandy loam, loam soil is observed along the valley fill area. Clay loam dominates the areas along the main course of the Dulung River, where paddy is the predominant crop in upland and lowland situations, though other crops grown with irrigation facilities are maize, oilseeds, pulses, and vegetables. The agricultural productivity of the area is rather low because traditional methods of cultivation and local varieties of crops are being mostly used by small and marginal farmers. Sediment deposits and fertilizer-washed water are responsible for deteriorating the water quality of the region, ultimately leading to problems for the biodiversity of the aquatic ecosystem.

12.4.5 Management Strategies

Comprehensive approaches involving stakeholders from across the fluvial landscape are required along the corridor, including the entire basin, for the restoration of river corridor resources. Restoration includes efforts to re-establish, the former healthy state so that all processes and functions can be operated effectively. Disturbance regimes in the buffer zone have been sorted out. The expansion of vegetative patches along the river is a priority area that requires management attention.

Hydrologic disturbance may occur due to check dams, landscape alteration, natural processes like global climate change, accelerated erosion, channel incision, and other impacts. Management strategies must include re-creating the composition, structure, and functions of an environmentally friendly landscape for the stakeholders. Positive environmental changes include: (a) the creation of green coverage to ensure the viability of native riparian vegetation with the improvement of orchards, (b) the prevention of bank failure to ensure river bank stability, (c) the expansion of the edge effect by enhancing landscape patches to maintain balance in both aquatic habitat and adjacent terrestrial ecosystems, and (d) the provision of a continuous corridor for the movement of faunal communities to ensure the food chain. The watershed, being a hydrological unit of study, is a healthy network of intermittent and ephemeral streams that is helpful for agricultural purposes and for supporting plant communities. Sometimes the replacement can make beneficial changes to the structural composition of landscape patches. The introduction of some exotic alternative species of flora on the floodplain and stream banks may check the erosion. Reclamation of highly degraded land or habitat is another way to achieve normalcy in ecosystem services. The process of mitigation assists in alleviating some of the detrimental effects that arise from human actions in riparian areas. Levee management along the river is beneficial for protecting human-induced developments and maintaining hydrologic connectivity of near-channel floodplains and side channels. In forested systems, large and small tree plants influence the groundwater recharge system, and thus, the water quality of the stream will improve in the near future.

12.5 Conclusion

Riparian corridor management is multifaceted in nature and necessitates coordination of research, planning, and implementation efforts. The Dulung River corridor is plagued by micro-level environmental issues that are mostly zone-specific and, in some cases, site-specific. The upper reaches need attention related to enhancing the soil-moisture situation. The construction of gully plugs and contour bunds may be an effective tool for planting small shrubs and grasses. The middle part of the corridor requires valley-side treatment along its less sinuous course in the channel networks. Agricultural pursuits sometimes have a negative impact on the aquatic habitat, and this must be checked in a scientific way. The lower parts of the corridor, that is, the near-confluence zone, are the area of occasional flood incidents, causing a great loss to agricultural production. The sand deposits during flood submergence change the land's character and pose a threat to its cultivation for subsequent production. All of this necessitates a strategic approach to restoring and maintaining a healthy riparian corridor in which all stakeholders can enjoy and reap the greatest benefit from their available natural resources in the riverine landscape.

References

Agar, P., Ortega, M., Pablo, D., & Tomás, C. (2016). A procedure of landscape services assessment based on mosaics of patches and boundaries. *Journal of environmental management., 180*, 214–227. https://doi.org/10.1016/j.jenvman.2016.05.020

Amper, R., Puno, G., & Puno, R. (2019). Rapid assessment of the riparian zone habitat of river. *Global Journal of Environmental Science and Management, 5*(2), 175–190. https://doi.org/10.22034/gjesm.2019.02.04

Atkinson, S. F., & Lake, M. C. (2020). Prioritizing riparian corridors for ecosystem restoration in urbanizing watersheds. *PeerJ, 8*, e8174.

Boyce, S. G. (1995). *Landscape forestry*. Wiley.

Erős, T., & Lowe, W. H. (2019). The landscape ecology of rivers: From patch-based to spatial network analyses. *Current Landscape Ecology Reports, 4*, 103–112. https://doi.org/10.1007/s40823-019-00044-6

Forman, R. T. (1995). *Land mosaics: The ecology of landscapes and regions*. Cambridge University Press.

Frazier, A. (2019). Landscape metrics. In J. P. Wilson (Ed.), *The geographic information science & technology body of knowledge* (2nd Quarter 2019 Edition). https://doi.org/10.22224/gistbok/2019.2.3

Graziano, M. P., Deguire, A. K., & Surasinghe, T. D. (2022). Riparian buffers as a critical landscape feature: Insights for riverscape conservation and policy renovations. *Diversity, 14*(3), 172. https://doi.org/10.3390/d14030172

Groom, J. D., Dent, L., Madsen, L. J., & Fleuret, J. (2011). Response of western Oregon (USA) stream temperatures to contemporary forest management. *Forest Ecology and Management, 262*(8), 1618–1629. https://doi.org/10.1016/j.foreco.2011.07.012

Jiménez-Carmona, E., Herrera-Rangel, J., Renjifo, L. M., & Armbrecht, I. (2020). Restoration of riparian forest corridors: Eight years monitoring the diversity of soil ants in an Andean rural landscape. *Insect Conservation and Diversity, 13*(4), 384–392.

Lawler, D. (2004). Bank Erosion. In A. Goudie (Ed.), *Encyclopaedia of geomorphology* (pp. 48–52). Routledge.

Naiman, R. J., Decamps, H., & Pollock, M. (1993). The role of riparian corridors in maintaining regional biodiversity. *Ecological Applications, 3*, 209–212. https://doi.org/10.2307/1941822

Paudel, S., & Yuan, F. (2012). Assessing landscape changes and dynamics using patch analysis and GIS modeling. *International Journal of Applied Earth Observation and Geoinformation, 16*, 66–76. https://doi.org/10.1016/j.jag.2011.12.003

Ryszkowski L, ed. (2002). *Landscape Ecology in Agroecosystems Management*. Florida, USA: CRC Press, Boca Raton.

Sanderson, J., & Harris, L. D. (2000). *Landscape ecology: A top-down approach*. Lewis Publishers.

Turner, M. G., & Gardner, R. H. (Eds.). (1991). *Quantitative methods in landscape ecology*. Springer-Verlag.

Chapter 13
The Importance of Shoreline Beaches for Coastal Tourism Potential and Their Diversities on the Odisha Coast

Soumita Guha, Ashis Kumar Paul, and Joydeb Sardar

13.1 Introduction

There are a number of sea beaches (>18) along the shorelines of the Bay of Bengal in Odisha. The sloping beach faces of Gopalpur on sea, Puri, Toshali Sands, Chandrabhaga, Paradip, and Dagara on the sandy coast are backed by wide beach berms and sand dunes. They are wave-dominated and reshaped by swell waves with seasonal longshore currents along the Bay of Bengal shorelines in a seaward setting of elongated fetch distance. The wide beach berms provide ample space for the visitors and others to regulate the system of recreational activities in a commercial and aesthetic mode. Sun, sea, and sand are the major ingredients of the shore-attached beaches, and they are used for sun bathing, sea bathing, beach walking, sea viewing, animal riding, sand biking, sailing, and nature watching for a period of 16 hours per day. Similarly, beach stalls, beach festivals, activities of sand arts, religious gatherings with holy deep into the sea, and sea winds or strong breezes supported additional attractions for the visitors. Rating of the beaches depends on the nature of their physical, biological, and human factors in the respective coastal sections of the regional settings. The earlier researchers carried out their work on the coastal tourism aspects and potential environments of the Odisha coast and other coastal belts (Paul et al., 2014, 2016, 2017, 2019; Paul & Guha, 2018; Shaw & Williams, 1992; Sahoo, 2014; Das, 2013; Baud-Bovy & Lawson, 1998).

S. Guha
Sishutirtha Sukanta Vidyaniketan, Kolkata, West Bengal, India

A. K. Paul (✉)
Department of Geography, Vidyasagar University, Midnapore, West Bengal, India
e-mail: akpauleastcoast@gmail.com

J. Sardar
Centre for Environmental Studies, Vidyasagar University, Midnapore, West Bengal, India
e-mail: joydebsardar18@gmail.com

A. K. Paul, A. Paul (eds.), *Crisis on the Coast and Hinterland*,
https://doi.org/10.1007/978-3-031-42231-7_13

However, the detached beaches on the sand spits and barrier bars provide ideal sites for recreational values and the significance of environmental values for habitats of marine turtles, red crabs, and shore fringe forests on the Odisha coast. Some important sites of this class represented the open marine shores of Udaipur, Talsari, Bichitrapur, Choumukh, Chandipore, Gahirmata, Marine Drive, Pearl Beaches, New Mouth Spit, Rushikulya Spit, and Gopalpur Port fringe beaches of the region (Fig. 13.1). They are relatively remote and attractive for water sports, nature tourism, and ecotourism of the destinations. The present work focuses on the tourism significance of the beaches of different categories, the heritage characters of the coastal region, and the tourism earnings from the shoreline beaches of the state of Odisha (Fig. 13.1).

13.2 Materials and Method

13.2.1 Tourist Visits in the Coastal Destinations

The number of tourist arrivals from domestic and foreign sources is tabulated on the basis of statistical bulletin data supplied by the Department of Tourism, Government of Odisha (2008, 2009, 2010 and 2013, 2014, and 2015) for the calculation of tourism inflow of money in the coastal destinations of the state (Fig. 13.3a, b). It is an important task to calculate the earnings from tourists visiting coastal destinations in

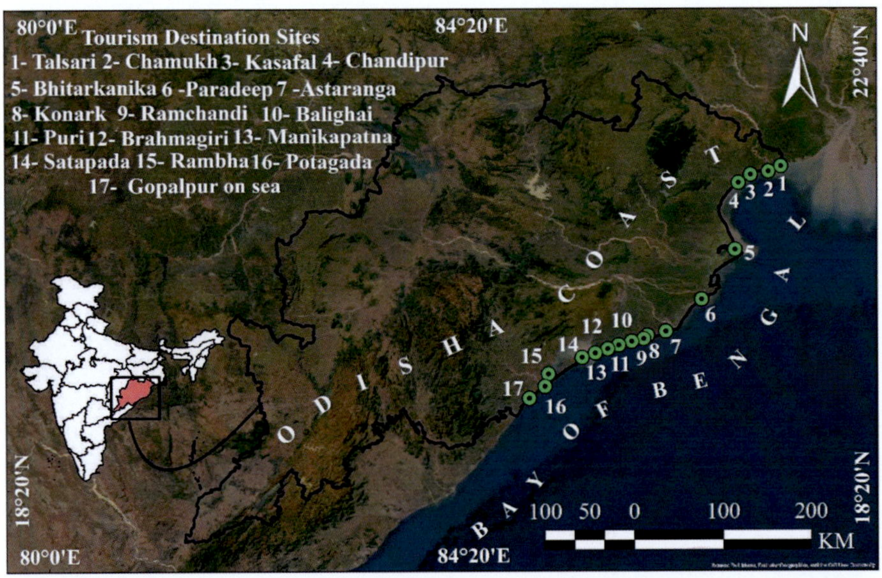

Fig. 13.1 Coastal beach tourism destination sites in Odisha state along the Bay of Bengal Shoreline, India

order to realize the economic benefit of the tourism process. The average inflow of money through tourist expenditure has been calculated on the basis of a tourist profile survey conducted by the tourism department of Odisha. The survey was conducted by the department in 2004–2005, 2008–2009, and in 2013–2014.

A tourist profile survey was conducted by the Department of Tourism, Government of Odisha, during the years 2004–2005. At that time, the average duration of stay was found to be 12.7 days in the case of foreign tourists and 3.9 days in the case of domestic tourists. Similarly, the average per capita expenditure was found to be Rs. 1944 for a foreign tourist and Rs. 1275 for a domestic tourist. In the years 2008–09, another tourist profile survey was conducted by the Department of Tourism to find out the changes, if any, that occurred in the previous survey. It was found that the average duration of stay is 14.2 days for foreign tourists and 3.7 days for domestic tourists. Similarly, the average per capita expenditure was found to be Rs. 2255 in the case of a foreign tourist and Rs. 1357 in the case of a domestic tourist. The average per capita expenditure for the foreign tourists is estimated at Rs. 4167 and for the domestic tourists at Rs. 2763 in the year 2013–2014, as per the information recorded by the Tourism Department, Govt. of Odisha. The destination-wise calculation is done for the years of 2008, 2009, 2010, 2013, 2014, and 2015. However, the method of calculation to estimate the inflow of money (tourist expenditure, or TE) follows in the Eq. (13.1).

$$\begin{aligned}\text{TE} &= \text{Number of Tourists visiting} \times \text{Average duration of stay} \\ &\quad \times \text{Average per capita expenditure}\end{aligned} \tag{13.1}$$

13.3 Results and Discussion

13.3.1 Beach Tourism Significances

Beaches on the seafront or any waterfront location act as magnets of attraction for tourism and recreation activities. The beaches of Odisha are well connected with the urban centers by railway routes, surface routes, and sea routes. Both domestic and foreign tourists spent their vacations on the sea beaches of Odisha for 3 to 4 nights and 12 to 14 nights per destination. Beaches are categorized into three classes as per the available amenities, emerald beauty of nature, and tourist attractions (Table 13.1).

The category-one beaches provide more amenities, more attractions, and easier communication with other destinations. Puri, Konark, Chandipore, and Gopalpur on sea jointly attracted over 80% visitors among the total tourist inflow in Odisha (2008–2015) during the previous decades. There is a strong conflict between environment-friendly tourism and mass tourism activities along the coastal sections. Thus, a selected group of visitors and educated tourists visit the eco-tourism destinations where sufficient amenities are not available, but the natural beauty and wild life habitats attract them in such remote areas of the coast. As a result of tourism

Table 13.1 Classification of beach tourism destination sites in Odisha

Beaches	Locational characters	Available tourism recreational amenities	Beauty of nature
Shore attached beaches (category-I)	Shore fringe sandy alluvium coast	Large and diverse amenities available	Dune ridge, beach berm, beach face, and longshore trough
Barrier Beaches (category-II)	Barrier fringe alluvium coast	Moderate amenities available	Low dune, beach berm surface, and beach face
Other category of beaches in Odisha (category-III)	Low land bay side beaches	Moderate to few amenities available	Beach ridge, beach face, and tidal flat

earnings, the beaches of Odisha hold the key to the tourism development of the region. Lake Chilika, Bhitorkanica, Hukitola Bay, and Bichitrapur areas along the Bay of Bengal shorelines represent sensitive coastal habitats. Thus, tourism activities and infrastructural development are restricted to promoting the eco-tourism destinations of the coastal areas.

13.3.2 Odisha Coastal Heritage and Tourism

These types of heritage sites have become cultural tourism attractions in recent years in the coastal region. They are grouped as (i) Built-up Heritage Attractions: various archaeological sites along the fringe of Chilika Lagoon, ancient sea ports of the earlier silk route, Konark Sun Temple or the ancient black pagoda, historic buildings of Puri-Gopalpur, etc. (ii) Natural Heritage Attractions: Bhitarkonika National Park, Bichitrapur Mangrove Coast, Chandrabhaga (abandoned stream), Gahirmata Spit, Chilika Spits, Rishikulya Spit, Balukhand Spit, GhantashilaPahar with caves, other residual hills with pedimented surfaces along the south western Chilika, wide sandy shorelines (Chandipore, Dogara, Gopalpur, Puri, etc.), (iii) Cultural Heritage Attractions: International Sand Art Competition; various handicrafts of wood, stone, and silver (Utkalika); marine shells, coconut cells, bamboo, Odisha handlooms and textile goods; Rathajatra festivals; Odishi dance; Gotipura dance, etc. (iv) Industrial Heritage Attractions: Beach placer mining of Gopalpur on Sea, shell and lime manufacturing processes of Balasore shores, and Ganjam coast, handicrafts villages, and various handloom product centers (Raghurajpur). (v) Religious sites and attractions: Puri Temple, Puri Town, Balukhand Temple, Kalijai Temple, Chandaneswar Temple, Bhusundeswar Temple, Bhabakundaleswar Temple, Manikpatna Mosque, Pilgrimage Route of Odisha, etc. (vi) Military Heritage Attractions: Chandipore missile testing centers, rocket launching stations, etc. (vii) Literary or artistic heritage attractions: Nandan Kanan, Udaigiri-Khandagiri, Konark Temple, and the temple sites of Chandaswar, Bhusundeswar, Balikhand, etc.

Heritage is defined as a view from the present, either backward to the past or forward to the future. Essentially, heritage is the contemporary use of the past, including both its interpretation and representation; for example, Bhabakundaleswari Temple was recently excavated from a buried sand dune of an ancient spit back surface along the shores of Chilika Lagoon, in which an ancient seaport was located very adjacent to the historical temple site. The Archaeological Survey of India is now taking care of the temple to interpret its history and represent the past art and culture of the site. Traditionally, festivals were first and foremost religious celebrations involving ritualistic activities. For example, in Puri, the Rathajatra festival and the other 12 festivals of the Puri temple every year afford an opportunity to worship the Lord Jagannath by means of reviving the local culture or tradition, and they also offer the Odisha people the chance to celebrate their cultural identity. Hence, several festivals during the periods of Puri and Konark attract a higher concentration of visitors from the country and abroad that are already established tourist destinations along the coastal zones of Odisha. There are five different aspects of a heritage site as identified by Tunbridge and Ashworth (Tunbridge & Ashworth, 1996), and they are listed as follows: (i). A synonym for any relict physical survival of the past, (ii) individual and collective memories in terms of non-physical aspects of the past as seen from the present, (iii) all accumulated cultural and artistic productivity, (iv) the natural environment, (v) a major commercial activity (e.g., the heritage industry).

The heritage-based tourism of the relict lake margins of Chilika and estuary banks of Rishikulya rivers may have been brought into focus for the existence of history and culture in the region through exploration of ancient silk routes, location of the ancient sea ports in and around the shore fringe areas, ancient temple sites in the nearby coastal hinterlands, activity of marine trades, and artistic silk production of the region, as well as the heritage landscapes and seascapes of the region between parts of Chilika lake shores, Bay of Bengal nearshores, and Rishikulya river mouth section (Odisha State). Archaeologically, several places in the Chilika Lake region reveal ancient habitational remains dating from the Neolithic-Chalcolithic period to the third millennium BC with datable records (Tripathi & Vora, 2005). The number of shipping ports and their existence with maritime activities in the historical past in and around the Chilika Lake region proved the glorious maritime heritage of this part of Odisha state. Modern landscapes and seascapes of the region provide enough evidence for the identification of geomorphological changes, sea level fluctuations, several relict features, and erosion-accretion dynamics of the region that played a major role in the decline of navigational traditions around Chilika Lake and the Rishikulya River Mouth.

13.3.3 Festivals Organized by Odisha Tourism Department

There are many other products and commodifications of tourism-based activities in the coastal destinations for attracting the demand of tourists. They are included as: (i) Odisha State Emporium: Odisha State Co-operative Handicrafts Corporation

Limited; (ii) revitalizing the tourism pride through the spread of cultural tourism in Odisha, with special reference to the typology of cultural tourism attractions; (iii) Konark-Chandrabhaga heritage tourism; (iv) beach tourism and temple tourism; (v) Chilika lake-based nature tourism and eco-tourism; (vi) natural heritage landscape tourism and mangrove eco-tourism; (vii) the heritage-based tourism of relict lake margins around Chilika and the estuary banks of the Rishikulya River (Gourangapatna, Potagrah, Palurport, Manikpatna, Bhabakundaleswari Temple Site, and other ancient sea ports around Chilika Lake); and (viii) the foundation of sand art on the Odisha coast and international sand art competitions. Festivals and carnivals in Odisha also attracted a number of visitors to coastal tourism destinations.

They are grouped as follows: (i) The Konark Festivals (Dance and Music): Konark, Chandrabhaga; (ii) The International Sand Art Festival: Puri, Gopalpur, Chandrabhaga; (iii) Gotipura Dance Festival: Chandipur, Puri; (iv) International Odissi Dance Festival: Puri, Chandrabhaga; (v) Megha Utsav and World Tourism Day celebration: puri; (vi) Beach festivals at Puri, Chandrabhaga, and Gopalpur on Sea beaches; (vii) Rathajatra Festival: Puri; (viii) Snan Jatra Festival: Puri; (ix) Ulta Ratha Jatra Festival (Gundichabari): Puri; (x) Nabakolebar Festivals (month of May): Puri; and (xi) The Colorful Holi Festivals (March): Puri Temple and all other temples along the coastal belt.

13.3.4 Ancient Maritime Heritage as Tourism Significance

The growing access to the coast has led to the development of more specialized resorts in the undiscovered destination sites of heritage significance present in Odisha's coastal districts. Such specialized tourism demand is emerging among the visitors to explore the history, archaeology, heritage landscapes, sites of ancient trade and commerce with far of countries, the past glory of culture, and maritime tradition of the coastal communities. The modern-day tourism industry is guided by the government and private entrepreneurs with the emergence of commercially organized tourism package tours to promote newer forms of recreation by running parallel to the mass tourism phenomena of coastal resorts. The relict lake margin landscapes and seascapes were fully utilized by ancient people on the Odisha coast for spreading Indian culture to other countries. On the basis of available literature and archeological records, the cultural sites of the ancient people of coastal Odisha can be classified as (i) Chilika Lake margins, (ii) Rishikulya estuary bank, (iii) Daya and Prachi valleys, and (iv) Hills and caves of coastal districts (Tripathi & Vora, 2005; Patra & Patra, 1993; Paul, 1985). The buried fort of Potagrah, a 350-year-old historic structure, is currently being restored and renovated with the remains of old buildings for a heritage tourism destination under the World Bank funded ICZM Project of Odisha. The other sites are Bhaba Kundaleswar Temple of Manikpatna, Balihara Chandi Temple near Puri, Hariharadeva Temple, Nairi ancient port, Bateswara Temple, Kantiagada (Ganjam), Jagannath Temple at Puri, Pentha, and the Jamboo colonial building of Kendrapara. Potagarh was built in 1768 as a pentagon shaped fort that is surrounded by a wide and deep moat. Currently, the number of tourist attractions is increased in the heritage destination of the coast by good access road and by the implementation of ICZM in Odisha coast.

13.3.5 Coastal Evolution and Decline of Cultural Heritage Sites

Various evidences of geomorphic features of the relict landscapes and seascapes provide opportunities to reconstruct the past (Fig. 13.2). The coastal landscape evolution involving surrounding environmental conditions is primarily responsible for the decline of glorious heritage sites in the coastal districts. From studying the geomorphology of relict landscapes, the following features can be identified with the delineation of their boundaries. They are grouped as ancient shoreline along the shelf break slope, hill planation surfaces, basal weathering surfaces, pedimentation of the hill slope, wash deposit surfaces, scarps, lake fringed terraces, extensive promontories, and embayments. The other features include abandoned channels and valleys, spit growth and lake deposits, shoreline advancement towards the sea, dune mobility with transgressive sands, and wash over fan deposits. Further, they can be sequentially arranged in a chronological order with geomorphic signatures and other available dated records of archeological tools explored from various landform units (Paul et al., 2014).

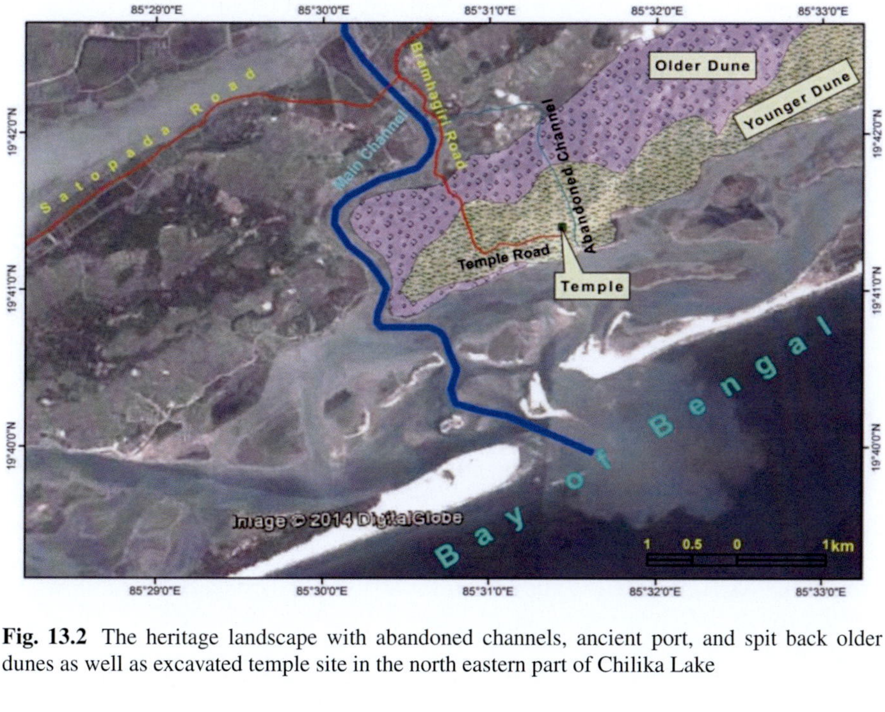

Fig. 13.2 The heritage landscape with abandoned channels, ancient port, and spit back older dunes as well as excavated temple site in the north eastern part of Chilika Lake

13.3.6 Sand Art Culture in Sea Beaches

The creation of sculptures with materials is inherent among people since the primitive period. Senses of art and skill among the ancient people were reflected in sculpturing on the materials of stone, soil, wood, cave rock, etc., but the invention of sand art is a great achievement of the people in the history of sculpture in India. Attractive sea beaches with berm-surface sand storage can provide the sources of sand materials for the application of dormant inspiration to create and culture the sand art on the sea beaches visited by tourists. Such sand art creatures can provide mental peace, pleasure, and amusement on seashores. The hidden ideas or pictures in the mind about the people, environment, livelihood, animals, and wildlife are created through the sculpturing of sand materials on the sea beaches of Odessa's coastal districts. Art lovers usually participate in such inherent art forms as beach sand sculptures on the Puri coast, Gopalpur coast, Chandrabhaga beach, and Kendrapara shoreline in Odisha. The intrinsic values of such sand art culture may be exhibited on the sea beaches, in the festivals, in the fields of competitions, through field documentation by photography and videography, by circulating information through printing and publications, social networks, and other electronic media for generating interest among visitors, local people, and researchers. The sand art research and development center (A.S.A.C.) had come up in 1987 at Puri through the initiative taken by the master of sand sculpture, Mr. Manmohan Mahapatra of Odessa State. A new dimension in the field of art has developed at Puri Sea Beach and is named "Sand Art," after winning the "Best Sculpture" award in the World Competition in 1994. The tradition of Utkaliya Art is now well preserved through the creation of sand sculptures on the seashores. The open art exhibition at Puri had paved the way for him to develop a new composition of sand art. The International Sand Art Competition is now spreading the art and culture of Odisha among the participants, who are coming from abroad and different states of the country. The communication network system of the coastal belt provides easy access, frequent movement of tourists, selection options of tourism destinations by the tourists, and opportunities for the development of transit routes between departing tourists and returning tourists of the destination sites and source areas.

13.3.7 Inflow of Money Through Tourist Spending (Expenditure)

On the basis of the tourist profile survey conducted in 2008–2009 and in 2013–2014, the estimated inflow of money through tourist spending (expenditure) during 2008–2010 is given in the table (Table 13.2). The data on the inflow of money through tourist spending expenditure in Odisha reveals that the inflow of money increased from Rs. 3270.27 crores to Rs. 12,356.03 crores from the years 2008 to 2015. Though the inflow of money from foreign tourists has increased from Rs. 108.54 crores to Rs. 306.97 crores, the inflow of money from domestic tourists has increased manifold during the same period. The destination competitiveness as well as the international competitiveness is also needed to emerge as high-quality tourism infrastructure-dominated destinations for foreign tourists in coastal Odisha (Fig. 13.3a, b). The inflow of domestic and foreign tourists is increasing rapidly in the state due to improvements in the attractions, stability, safety, and infrastructure of tourism and leisure activities. The 8 years of growth in the inflow of tourists (2008–2015) indicate the above facts about tourism-related issues. The Department of Tourism of the Government of Odisha played a significant role in making the state an ideal destination for tourists in the eastern part of India.

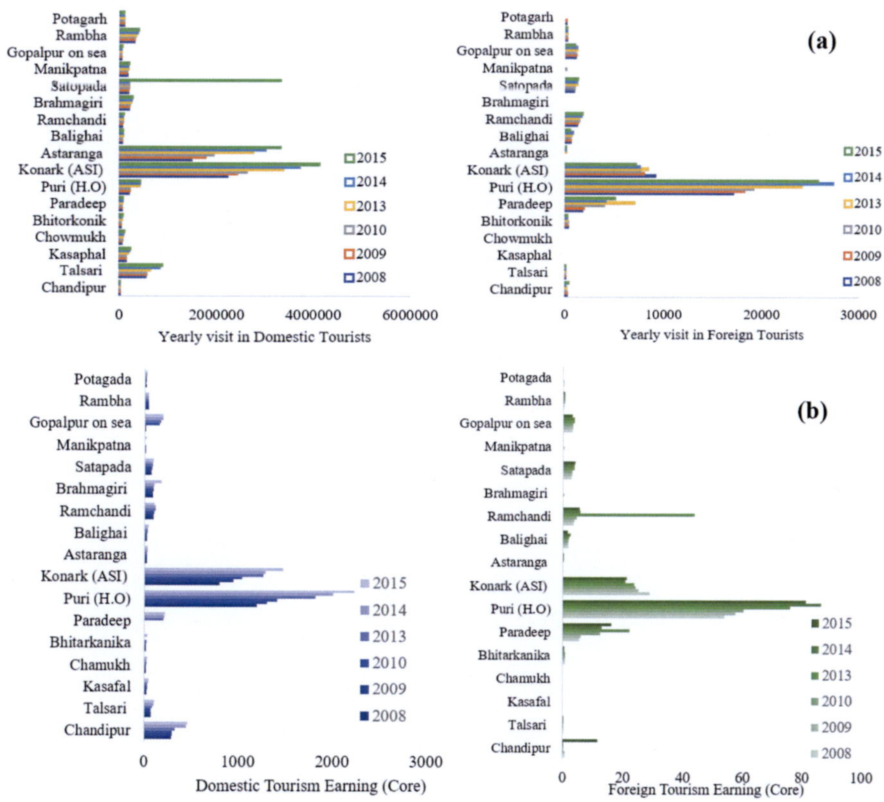

Fig. 13.3 (**a**) Number of domestic and foreign tourists arriving in the beach tourism destination sites in Odisha during 2008–2015 and (**b**) tourism earnings calculated on the basis of tourists' arrival and staying in the destination sites of the Odisha coast (2008–2015)

Table 13.2 Tourism earnings calculated for the destination sites of coastal fringe beaches of Odisha (2008–2015)

Estimation of tourism earnings from domestic (D) and foreign (F) tourists in coastal destinations of Odisha State (2008–2015)												
	2008		2009		2010		2013		2014		2015	
Location	D	F	D	F	D	F	D	F	D	F	D	F
Chandipur	289.86	0.7	298.56	0.74	299.27	0.35	331.93	0.36	452.14	0.3	467.6	11.52
Talsari	70.73	0.03	72.14	0.04	73.62	0.09	87.43	0.11	105.17	0.09	114.87	0.06
Kasafal	26.16	0	26.68	0	32.09	0	35.99	0	42.07	0	49.03	0
Chamukh	17.96	0	18.32	0	18.79	0	24.14	0	27.32	0	33.13	0
Bhitarkanika	20.05	0.9	23.49	0.8	23.21	0.9	24.53	0.7	3.33	0.6	33.55	0.6
Paradeep	0.9	5.53	1.02	5.88	2.12	12.29	210.2	22.09	218.26	12.77	225.27	15.92
Puri (H.O)	1200.9	53.74	1309.8	57.33	1419.9	60.18	1822.9	75.78	2011.2	86.19	2235.06	81.24
Konark (ASI)	798.5	28.62	945.67	24.98	1043.7	23.95	1267.1	23.63	1292.2	20.73	1476.3	21.22
Astaranga	24.39	0	25.57	0	26.08	0	30.24	0.006	32.92	0.02	33.99	0.006
Balighai	25.75	1.67	27.02	1.76	28.86	1.68	31.91	1.97	35.49	2.44	41.36	1.56
Ramchandi	98.08	3.55	101.7	3.84	104.5	4.41	118.9	43.56	125.04	5.79	104.01	5.41
Brahmagiri	88.04	0	92.72	0	94.57	0.009	100.59	0	102.88	0	180.08	0
Satapada	77.02	2.71	74.94	2.93	81.87	2.9	87.2	3.69	89.66	3.63	100.13	3.93
Manikpatna	9.49	0	9.68	0	9.88	0.002	1.61	0	1.81	0	20.89	0
Gopalpur on sea	15.29	3.11	15.75	3.33	162.23	3.14	176.38	3.73	196.69	3.62	203.06	3.02
Rambha	39.98	0.4	41.18	0.4	42.01	0.5	45.19	0.5	46.03	0.5	46.52	0.6
Potagada	21.2	0.002	14.96	0.003	22.17	0	23.49	0	25.92	0	29.22	0

13.3.8 Tourism Earnings from the Coastal Destinations of Odisha

Over time, tourism and leisure activities in the coastal belt have modified the shoreline environment by occupying the natural coastal habitats in the places of destinations. The revenue earnings from the sensitive coastal destinations should be utilized partially for the environmental management and conservation of sensitive shoreline destinations in Odisha state. Sustainable tourism infrastructures for the eco-tourism sites are needed immediately for the coastal belt of mangrove dominated shorelines, Chilika Lake and back waters, islands, and coastal sand dunes covered by vegetations. Therefore, the share of revenue earnings is estimated from the tourist spending expenditures for every coastal destination in the state.

13.3.9 Estimated Results of Tourism Earnings from the Coastal Destination

On the basis of tourist arrivals and the tourist profile survey method, the tourism earnings from the coastal destinations are calculated in the following ways to identify the share of income of the coastal belt destinations in the tourism industry of the state (Table 13.2). The above-mentioned estimated earnings from the coastal destinations of Odisha account for more than 80% of the state's earnings from tourism. Among them, the earnings from Puri and Konark destinations for the year 2010 represented 50% of the total earnings registered from the total destinations of the state. However, in 2015, the share of earnings from Puri and Konark destinations was more than 60% of the total earnings (State of Odisha). The remaining destinations of Coastal Odisha share more than 20% of the total earnings from the inflow of money into the state (Fig. 13.3b). Therefore, the above study proves the significance of coastal tourism as a major attraction.

13.4 Conclusion

The beaches of Odisha's coast are very attractive to inbound tourists from outside of the state. Available amenities, easy access, cultural heritage, historical monuments, temples, wildlife habitats, the natural beauty of the landscape, and low-cost budget tourism attracted tourists to the coastal area of the state. The category-one beaches of Puri, Chandrabhaga, Toshali Sands, Gopalpur on Sea, and Chandipore provide ideal beach berm platforms for playing recreational activities and good amenities with easy access. Remaining beaches are ecologically sensitive for their natural habitats and attract a few visitors in which environment-friendly activities are followed. Other categories of beaches are also attractive to tourists as they are well connected with inland destinations for travel and offer options of low budget cost tourism activities, but their access is not as easy as the category of beaches in Odisha. The study reveals that beaches earned 80% of the total earnings from Odisha's tourism activities. Among all the beaches and other coastal destinations, a few sites like Puri, Konark, Chandrabhaga, Chandipore, and Gopalpur on sea jointly earned the most money in the previous decades. The expenses incurred by the tourists in the beach tourism activities include lodging, fooding, transport, traveling, and sea bathing, and other recreational services provided along the coast. The domestic tourists spent Rs. 2763 per head per stay in the tourism destinations, and foreign tourists, on the other hand, spent Rs. 4167 per head per stay in the coastal destinations during the years 2014–2015. Following the per capita expenditure, it is estimated that Rs. 2235.06 crores were earned for the Puri beach destination during 2015, particularly by the domestic tourists (417,659 visitors) arrived at Puri. However, the foreign tourists spent a total of 81.25 crores during their stay in the destination, as per the per capita expenditure provided by the tourism department of Odisha. An attempt is needed to restore the habitats of sand dunes, barrier beaches, and dune forestry, as well as the environmental zoning approach to be introduced in the coastal destinations to achieve sustainable coastal tourism at the sites of mass tourism activities.

References

Baud-Bovy, M., & Lawson, F. (1998). *Tourism and recreation: Handbook of planning and design*. Butterworth-Heinemann Ltd.

Das, S. K. (2013). Growth and prospects of Odisha tourism: An empirical study. *Odisha Review, 125*, 134.

Paul, A. K. (1985). The development of accumulation forms and erosion in coastal tracts of Balasore, Medinipur and South 24 parganas districts of Orissa and W.B, India. *Indian Journal of Landscape Systems and Ecological Studies*, 19–32.

Paul, A. K., Islam, S. M., & Jana, S. (2014). *An assessment of physiographic habitats, geomorphology and evolution of Chilika Lagoon (Odisha, India) using geospatial technology* (pp. 135–160). Advances in Coastal and Marine Resources.

Paul, A. K., Guha, S., & Kamila, A. (2016). The regional geomorphology and characteristic heritage coastlines with adjacent seascapes of Southwest Chilika Lake and Rushikulya River Mouth, Odisha, India. *INDIAN CARTOGRAPHER*, 295.

Paul, A. K., Soumita, G., & Amrit, K. (2017). Drivers of coastal tourism in Odisha state: A case study of Puri-Konark sites along the Bay of Bengal coast. *Journal of Coastal Sciences, 4*(1), 6–19.

Paul A. K., & Guha, S. (2018). Geo-tourism prospects with geomorpho site assessment in an around Chilika Lake, Odisha, *Eastern Geographers*, 11–19.

Paul, A. K., Soumita, G., & Amrit, K. (2019). Physical carrying capacity of the beach tourism and recreational activities in Odisha coast. *Journal of Coastal Sciences*, 1–9.

Patra, S. K., & Patra, B. (1993). Archaeology and the maritime history of Ancient Orissa. *Orissa Historical Research Journal, 47*(2), 107–117.

Shaw, G., & Williams, A. (1992). Tourism, development and the environment: The eternal triangle. In *Tourism, development and the environment: the eternal triangle*, pp. 47–59.

Sahoo, D. (2014). A Case Study on 'Beach-Tourism potential of Odisha. *Journal of Kashmir for Tourism and Catering Technology, 1*(2).

Tripathi, S., & Vora, K. H. (2005). Maritime heritage in and around Chilika Lake, Orissa. *Current Science, 88*(7), 1175–1181.

Tunbridge, J. E., & Ashworth, G. J. (1996). Dissonant heritage. *The Management of the Past as a Resource in Conflict, 40*, 547–560.

Chapter 14
Climate Variability and Agricultural Modifications in Purulia and Bankura Districts of West Bengal

Asutosh Goswami and Ashis Kumar Paul

14.1 Introduction

There are many variables that will shape worldwide food security as well as the farming business system in the forthcoming years. Weather and climate systems are the most crucial among them. As a result, agriculture is likely to be affected by climate change, which may pose a threat to established farming practises but also present opportunities for advancement. Climate change is viewed as a genuine danger to life, which unfavourably influences different frameworks on the planet, starting from primary activities to the psycho-social behaviour of human beings (Sivakumar et al., 2005; Mall et al., 2006; Lobell & Gourdji, 2012; Manyeruke et al., 2013; Salvo et al., 2013; Ladan, 2014; Mahato, 2014; Babar et al., 2015; Cherian & Khanna, 2018; Liu & Basso, 2020). Total population is supposed to increase roughly 10 billion by 2050 (Arora, 2019), which would enhance strain on agrarian terrains to fulfil the developing needs for food previously impacted by the environmental changes. Precipitation designs in many areas of the globe have been moved because of environmental change and fluctuation. Horticulture assumes a significant role in the general monetary and social prosperity in India. In the tropical nations, especially, the ranchers depend vigorously on normal precipitation for the creation of harvests (Aninagyei & Appiah, 2014). Indeed, even the slightest deviation from typical atmospheric conditions genuinely hinders the proficiency of food production. The variation methodology should be reinforced to reduce the effect of environment changeability. Current rural practises are viewed as unreasonable, on the

A. Goswami (✉)
Department of Earth Sciences and Remote Sensing, JIS University,
Agarpara, Kolkata, West Bengal, India
e-mail: goswamiasutosh@gmail.com

A. K. Paul
Department of Geography, Vidyasagar University, Midnapore, West Bengal, India

A. K. Paul, A. Paul (eds.), *Crisis on the Coast and Hinterland*,
https://doi.org/10.1007/978-3-031-42231-7_14

grounds that they exploit important assets to degrade the ecological quality (Pareek et al., 2020). In a true sense, the farming action of an area is essentially constrained by the climatic state of that area, apart from the issue of irrigational water requirements, which is also a significant parameter for the growth of a great variety of crops. Water quality is also a significant issue for Indian farming. An exceptionally high pace of land debasement is brought about by environmental change, leading to desertification and supplementing soil deficiencies. Studies in regards to changes in events and the conveyance of precipitation are exceptionally essential for the appropriate administration of water assets and horticultural perspectives (Chakraborty et al., 2013). A huge portion of the world's total land area is comprised of dry terrains, where more than one billion individuals reside (Soro et al., 2016). Purulia and Bankura are the western-most districts of West Bengal, where precipitation is thought to be lacking to accomplish the ideal degree of rural efficiency. In any case, it could be brought up that the normal yearly precipitation of these two districts is in excess of 1200 mm, which is similar with different locale of the state (Goswami, 2019). The dryness of the region cannot be overlooked, where after precipitation there is high likelihood of surface run-off because of undulating nature of landscape of the area. Based on the proportion of precipitation to potential evapotranspiration (PET), it is noticed that in excess of 600 mm of precipitation after the satisfaction of the need of PET moves as overland flow during rainstorm coming about into dampness pushed soil (Bera et al., 2017; Goswami, 2019). The region is not viewed as reasonable for the development of long span assortment of rice all through the year due to presence of a quantities of restricting variables viz. dry spell inclination, limit of climate and environment, dry top soil, high surface run-off, and so on. The pre-rainy season is set apart by troublesome proportion of precipitation to PET. The precipitation in the pre-rainy months is likewise set apart by higher changeability. The intensity wave peculiarities during pre-storm additionally make a few dangers for the farming exercises in the locale. Thus, a broad examination is exceptionally important to present some modern cash crops and green items in the review region with the assistance of existing geo-climatic circumstances.

The study area gets in excess of 1300 mm of precipitation on a normal for the time span from 1976 to 2017. Much of this amount is concentrated in the four monsoon months from June to September. Because of non-appearance of any dependable wellspring of precipitation during pre-rainstorm, winter, and post-rainy seasons, the precipitation amount is small. The events of tempests and nor westers' in some cases make precipitation over the region during pre-rainy season. Be that as it may, a significant part of this water moves as surface run-off exploiting undulating territory of the area. In addition, PET is considerably higher than the precipitation amount during non-rainy seasons. However, the geo-climatic states of the locale have the possibility to help an extensive variety of customary and modern cash crops. Mishra (2008) proposed some transformation procedures in horticulture against environmental change for the 'western tract' of West Bengal and recommended taking on water preservation measures at all levels to build the water system capability of the plots. The current chapter is novel in that sense, as it has assessed the possibility of the current climatic state of the area and its ability to help a few customary and contemporary green items and cash crops apart from the computation of water budget based on the complex relation between precipitation and PET. The present study is aimed at identifying the impact of climate variability on the agricultural prospects of two districts (Purulia and Bankura) in West Bengal based on both primary and secondary observations. The present chapter emphasizes the trend of climatic parameters, mainly temperature and rainfall, and the extremeness of climate through the computation of SPI and NDVI, and it also deals with the impact of these climatic extremes, including their trends, on the agricultural practices in the district. The study is also unique as it calculates the irrigation water requirement of some selected crops on the basis of the relation between photosynthetically active radiation (PAR) and stress degree day index (SDDI).

14.2 Materials and Methods

14.2.1 Study Area

These two districts are marked by varying climatic conditions and hydro-geological properties. The low-lying alluvial plains to the east and north-east are similar to thc predominant rice lands of the southern tip of West Bengal (Fig. 14.1a). The land gradually rises to the west, becoming undulating with rocky hillocks scattered throughout. There are forests covering a lot of the districts. In the western part of the region, the soil is poor and the beds are hard lateritic, with scrub forests and sal woods. Seasonal cultivation is evident on long, broken ridges with irregular patches of more recent alluvium. The eye always rests in the eastern region on vast expanses of rice fields that are green in the rain but dry and parched in the summer. The climate is much drier in the upland areas to the west than in the eastern or southern tracts, especially. The agro-climatic conditions are ideal for horticultural and plantation crops. Though the region is marked by a dry climate and a lot of wasteland, it has the potential for plantations and horticultural activities. Both traditional and non-traditional plantations, such as mango, guava, cashew nut, jackfruit, banana, papaya, and others, can be grown on a large scale.

14.2.2 Trend Analysis of Rainfall and Temperature Using Mann-Kendall Test

As a matter of some importance, test for the pattern in yearly series is made to get a general perspective on the potential changes in information processes. To decide whether the patterns found are critical, the Mann-Kendall pattern test has been utilized. In the Mann-Kendall trend statistics, S variance is given as (Eq. 14.1);

$$S = \sum_{n-1}^{i=1} \sum_{n}^{j=i+1} sgn\left(Xj - Xi\right) \tag{14.1}$$

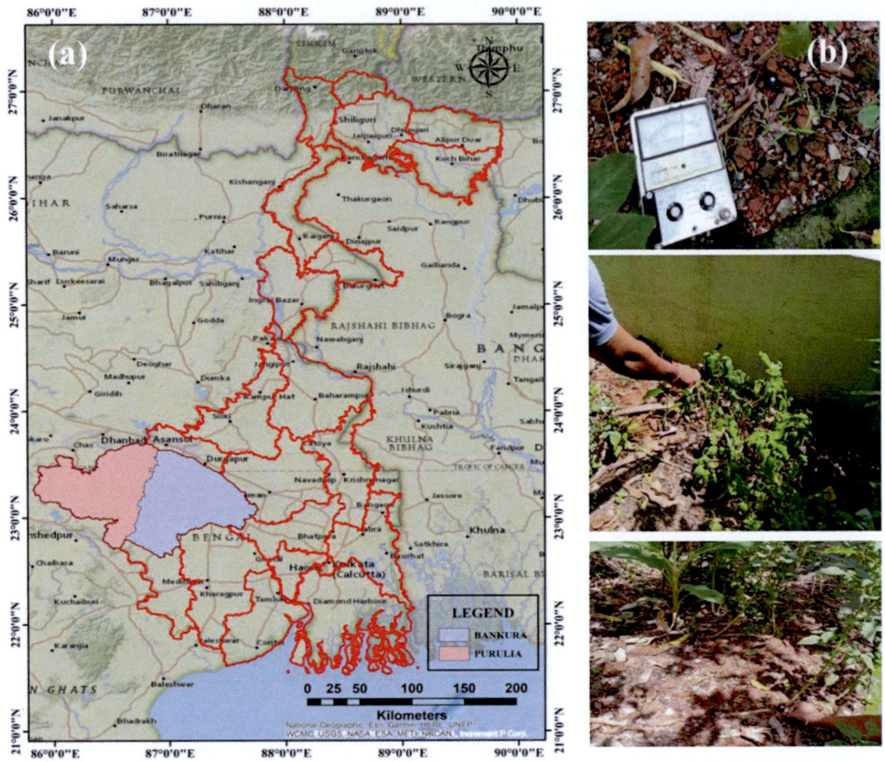

Fig. 14.1 (**a**) Location map of the study area and (**b**) measurement of top, bottom, and intercepted incident PAR (from left)

It is without appropriation and not impacted via occasional changes (Eq. 14.2). The applicability of the trend test rely heavily on the time series estimation that has been ranked from $i = 1, 2\ldots\ n - 1$ and xj; furthermore, ranking has been done from $j = i + 1, 2,\ldots, n$. The present equation employs individual data point xi as a point of reference, and it is compared with the remaining point observation xj (Eq. 14.2) so that,

$$\mathrm{Sgn}\left(Xj - Xi\right) = +1, > \left(Xj - Xi\right).0, = \left(Xj - Xi\right). -1, < \left(Xj - Xi\right) \tag{14.2}$$

The size of pattern was anticipated by the Sen's assessor. For direct pattern, the slant was generally assessed by figuring the least squares gauge utilizing direct relapse (Eq. 14.3).

$$\frac{n\left(n-1\right)\left(2n+5\right) - \Sigma \frac{m}{i=1} ti\left(i\right)\left(i-1\right)\left(2i+5\right)}{18} \tag{14.3}$$

If $S > 0$: then

$$Zc = \frac{S-1}{\sqrt{\text{Var}(S)}}$$ If $S < 0$: then $$Zc = \frac{S+1}{\sqrt{\text{Var}(S)}}.$$

14.2.3 Standardized Precipitation Index (SPI)

The standardized precipitation index (SPI) is a somewhat new dry season record based just on precipitation. Since the SPI is standardized, wetter and drier environments can be addressed similarly, and wet periods can likewise be observed utilizing the SPI. Following formula is used to detect the SPI (Eq. 14.4):

$$\text{SPI} = \left(Xi - \overline{X}\right) / \text{SD} \tag{14.4}$$

where Xi = precipitation, $\overline{X}$ = mean value of precipitation and SD = Standard Deviation However, the computation of SPI has a few intricacies. For this situation, long haul precipitation record is fitted with likelihood circulation to change it into a typical dispersion to get the mean SPI for the area along with the ideal time frame zero. For the said reason, following equation is utilized (Eq. 14.5);

$$\text{SPI} = (a - b) / \text{SD} \tag{14.5}$$

where a = individual gamma circulation, b = mean worth, SD = Standard Deviation; In the current research, this precipitation-based list has been utilized to demonstrate the events of dry spells on fluctuates timescales (one month, 90 days, a half year, and a year).

For the said reason, precipitation information has been gathered from the site (http://archive.indiawaterportal.org/metdata) and state agricultural department, Government of West Bengal. DrinC programming has been utilized for the calculation of SPI on 3 months timescales.

14.2.4 NDVI

NDVI is determined as the proportion of the red (RED) and near infrared (NIR) groups of a sensor framework. Following formula is used to determine the NDVI (Eq. 14.6). The target of this chapter is to identify agricultural drought in the two districts with rural terrains where biomass concentration is habitually low. The districts in the review region with the most noteworthy biomass thickness mass are the forested and the flooded horticulture regions.

$$\text{NDVI} = (\text{NIR} - R)/(\text{NIR} + R) \tag{14.6}$$

14.2.5 Water Budget

In the present study, an attempt has been made to identify the water budget of the said districts on the basis of the difference between precipitation (P) and potential/possible evapotranspiration (PET). In the first stage, it is very essential to ascertain the yearly value of the intensity list (I) in view of the month-to-month heat file (*j*) also, adding all the yearly heat indices data (Eq. 14.7 and 14.8).

$$j = (T_A / 5)^{1.51} \tag{14.7}$$

$$I = j1 + j2 + \ldots + j12 \tag{14.8}$$

T_A is the monthly average value of temperature. In the second stage with $a = 67.5 \times 10\text{-}8I\ 3\text{–}77.1 \times 10\text{-}6I\ 2 + 0.0179I + 0.492$, unadjusted PE' value (mm) is computed utilizing the accompanying condition (Eq. 14.9).

$$\text{PET} = 16(10.T_A / I)^a \tag{14.9}$$

In the third stage, assuming the sunlight span information is known, the accompanying condition can be utilized to compute the changed PET (Eq. 14.10).

$$\text{PET} = PE'(d/12)(N_d/30) \tag{14.10}$$

where N_d is the quantity of days in a month and d is sunlight duration (in hour).

14.2.6 Stress Degree Day Index (SDDI), Photosynthetically Active Radiation (PAR), and Irrigation Water Requirement (IWR)

Covering temperature was estimated at 10.00 hours with the assistance of infrared tele-thermometer (AG-42Telatemp infra-red thermometer, Australia). Stress degree day index (SDDI) has been determined with the accompanying equation (Eq. 14.11):

$$\text{SDDI} = T_C - T_A \tag{14.11}$$

where T_C and T_A represent canopy and air temperatures, respectively. Three extensively produced crops in particular tomatoes, green beans, and small vegetables have been chosen from the said districts of West Bengal to recognize the variety in radiation circulation among the three yields in the pre-monsoon and monsoon seasons, 2019 to 2022 (Fig.14.1b). The relation between PAR and SDDI has been utilized for the detection of irrigation water requirement. For the calculation of irrigation water requirement (IWR), CROPWAT 8.0 model has been utilized, which is a choice help program in view of a few numerical conditions. This model was created and customized by FAO for the computation of reference evapotranspiration (ET0), irrigation water requirement (IWR), and water system utilization employs the information of soil, environment, and yield. Assessment of harvest water necessity (crop water requirement (CWR)) is another more capability of this model. Crop water necessity is acquired from crop evapotranspiration (ETC), and it is determined by the accompanying equation (Eq. 14.12)

$$\mathrm{KTC} = \mathrm{KC} \times \mathrm{ET0} \tag{14.12}$$

where ETC = Crop evapotranspiration; KC = Crop co-productive and ET0 = Reference evapotranspiration. IWR can be designated as the distinction between successful precipitation (mm) and harvest water prerequisite (mm).

14.2.7 *Cobb Douglas Production Function*

For concentrating on the information yield relationship of developing various harvests, Cobb-Douglas creation capability has been fitted by utilizing the given underneath (Eq. 14.13).

$$\log Y = \log a + b1 \log XL + b2 \log Xc \tag{14.13}$$

where Y = yield, XL = temperature, Xc = rainfall, $b1$ and $b2$ = elasticity coefficients, a = constant term. In this theory, it is investigated the capability application in development plan crashing and project risk examination connected with span of development projects.

14.3 Results and Discussion

14.3.1 *Trend Analysis of Rainfall and Temperature*

Obviously, all seven agro-meteorological stations show declining patterns of precipitation for the long periods of February and March; however, the processed R2 values shift over time from one month to another and from one station to the next.

Table 14.1 Non-parametric trend statistics of rainfall for the Purulia and Bankura districts of West Bengal

Stations	Statistics and significance	Months											
		Jan	Feb	Mar	Apr	May	Jun	Jul	Aug	Sep	Oct	Nov	Dec
Hatwara	Trend statistics	−0.077	−0.222*	−0.074	−0.137	0.213*	−0.120	0.006	−0.022	0.001	0.044	−0.187	−0.123
	Significance	0.484	0.042	0.493	0.204	0.047	0.264	0.957	0.837	0.991	0.680	0.101	0.285
	R^2	0.0033	0.06	0.026	0.029	0.1446	0.022	0.028	1E-05	0.0028	0.013	0.033	0.026
Santuri	Trend statistics	0.068	−0.207	−0.001	0.083	0.215*	−0.071	0.173	0.006	0.171	0.169	−0.179	−0.037
	Significance	0.541	0.057	0.991	0.441	0.045	0.509	0.106	0.957	0.111	0.116	0.110	0.748
	R^2	0.002	0.0181	0.0239	0.0223	0.0471	0.0346	0.0279	0.01	0.0112	0.0084	0.0095	0.021
Jhalda	Trend statistics	−0.078	−0.220*	−0.085	−0.048	0.072	−0.069	0.148	0.084	0.131	0.015	−0.058	−0.210
	Significance	0.490	0.045	0.433	0.656	0.502	0.523	0.169	0.435	0.221	0.888	0.608	0.071
	R^2	0.0131	0.0114	0.0395	0.0273	0.0843	0.0164	0.0375	0.0592	0.0005	0.0684	0.019	0.0251
Barabazar	Trend statistics	−0.077	−0.053	−0.105	−0.108	0.171	−0.034	0.069	0.062	0.027	−0.017	−0.172	−0.173
	Significance	0.501	0.633	0.338	0.322	0.111	0.753	0.523	0.566	0.803	0.871	0.132	0.142
	R^2	7E-05	0.0203	0.0202	0.019	0.0354	0.0285	0.0195	0.0105	0.0007	0.008	0.0189	0.0239
Joypur	Trend statistics	−0.271*	−0.205	−0.180	−0.26*	0.057	−0.187	0.016	−0.166	0.146	0.051	−0.267*	−0.137
	Significance	0.017	0.064	0.099	0.019	0.595	0.081	0.879	0.121	0.178	0.633	0.020	0.254
	R^2	0.023	0.117	0.0078	0.0097	0.77	0.0057	0.666	0.0014	0.015	0.069	0.0032	0.022
Taldangra	Trend statistics	−0.086	−0.092	−0.028	−0.172	0.163	0.082	0.110	0.175	0.066	0.176	−0.104	−0.13
	Significance	0.452	0.411	0.794	0.113	0.129	0.442	0.303	0.102	0.537	0.102	0.359	0.264
	R^2	0.0022	0.0524	0.018	0.0062	0.034	0.0006	0.035	0.0133	0.0167	0.0049	0.0001	0.0838
Barjora	Trend statistics	−0.092	−0.118	−0.097	−0.093	0.137	−0.15	0.073	0.002	0.022	0.051	−0.135	−0.15
	Significance	0.415	0.285	0.390	0.390	0.201	0.169	0.495	0.983	0.837	0.633	0.228	0.203
	R^2	0.402	0.055	0.1006	0.0615	0.0034	0.0542	0.0145	0.0278	0.0045	0.0033	0.0301	0.0188

** Significant at 1% level, *Significant at 5% level

The figured R2 values for the long stretch of February are viewed as 0.06, 0.117, 0.0524, 0.0181, 0.055, 0.0114, and 0.0203 for the stations Hatwara, Santuri, Jhalda, Barabazar, Joypur, Taldangra, and Barjora separately (Table 14.1). For the three stations, specifically Hatwara, Joypur, and Taldangra, the pattern of precipitation in the long stretch of January is viewed as expanding with the registered R2 upsides of 0.0033, 0.023, and 0.0022, respectively. In spite of the fact that the majority of the stations show declining patterns of precipitation in the pre-rainy season of March and April, the precipitation pattern in the period of May is viewed as expanding, with the computed R2 values of 0.14, 0.77, 0.034, 0.0471, 0.0034, 0.0843, and 0.0354 for the stations Hatwara, Joypur, Taldangra, Santuri, Barjora, Jhalda, and Barabazar, separately. Chosen stations show rising patterns of precipitation in the month of July. The figured R2 values for the month of July fluctuate from 0.666 for the station Joypur to 0.0145 for the station Barjora (Table 14.1). The depressions and deep depressions subsequent to those originating from the Bay of Bengal move towards the western part of the state (West Bengal) and stay fixed over Chhota Nagpur level to cause heavy precipitation over the eastern edge of the plateau. A greater part of the stations shows declining patterns of precipitation in the rainy month of June (however, this is not critical). The station Para shows a declining pattern of precipitation in the post-rainy month of November, which is critical at a 5% level (Table 14.1).

The studied stations show expanding patterns of most extreme temperatures from 1979 to 2014 with the registered R2 upsides of 0.0196, 0.0155, 0.0182, and 0.0222 for the stations Barabazar, Joypur, Taldangra, and Barjora, respectively. The stations show expanding patterns of maximum temperatures for the period of March and April with the differing rates. For the stations Barabazar, Joypur, Taldangra, and Barjora, the most extreme temperatures in the period of March increment at the paces of 0.0315 °C/year, 0.0335 °C/year, 0.0316 °C/year, and 0.0282 °C/year, respectively (Table 14.2). Another pre-rainy month, May, shows declining patterns of the most extreme temperatures for every station in the stretch of time from 1979 to 2014 (Table 14.2).

Table 14.2 Slope of maximum temperature during pre-monsoon season

Stations	March	April	May
Barabazar	0.031545689	0.026258687	−0.011936937
	Slight increase	Slight increase	Slight decrease
Joypur	0.033509653	0.025402831	−0.020682111
	Slight increase	Slight increase	Slight decrease
Taldangra	0.031619048	0.027351351	−0.014646075
	Slight increase	Slight increase	Slight decrease
Barjora	0.028213642	0.030480051	−0.012302445
	Slight increase	Slight increase	Slight decrease

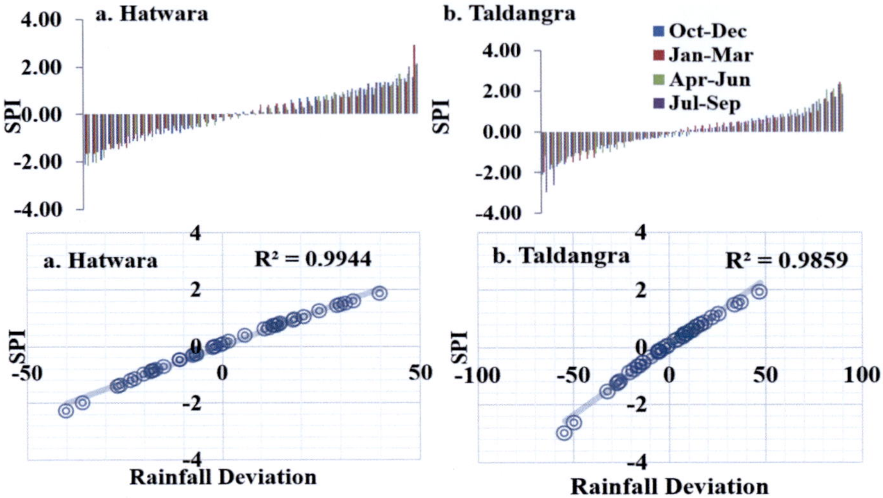

Fig. 14.2 Correlation between rainfall deviation and SPI

14.3.2 SPI

The station Hatwara has encountered a less number of dry spell occasions during the winter, pre-rainy, and post-rainy seasons (Fig. 14.2). The absolute quantities of dry spell occasions are viewed as 9, 13, and 24 during post-monsoon, pre-monsoon, and monsoon, separately. The station has revealed very wet circumstances in the years 1979, 1987, 1991, 1998, 2007, 2013, 2015, 2016, and 2017 with SPI values of 2.0 or more. The station Taldangra likewise encounters occasional varieties in dry and wet circumstances (Fig. 14.2). For this station, the winter, pre-monsoon, and post-monsoon seasons have encountered fewer dry spell occasions when contrasted with the monsoon season.

The complete quantities of dry spell occasions are identified as 25, 21, and 6 during the monsoon, pre-rainy, and post-rainy seasons, respectively. Also, extremely high negative deviations of precipitation address SPI values varying between −1.50 and −1.99. Positive precipitation deviations are related to the positive SPI values demonstrating the absence of a dry spell. Extremely high sure deviations of precipitation are found to be related to the SPI values going somewhere in the range of 1.50 and 1.99. The extremely high precipitation occasions coincide with SPI upsides of +2.0 or more. Be that as it may, for the station Hatwara, negative deviations of precipitation range between −39.9% in 1979 and −0.5% in 1994 (Fig. 14.3).

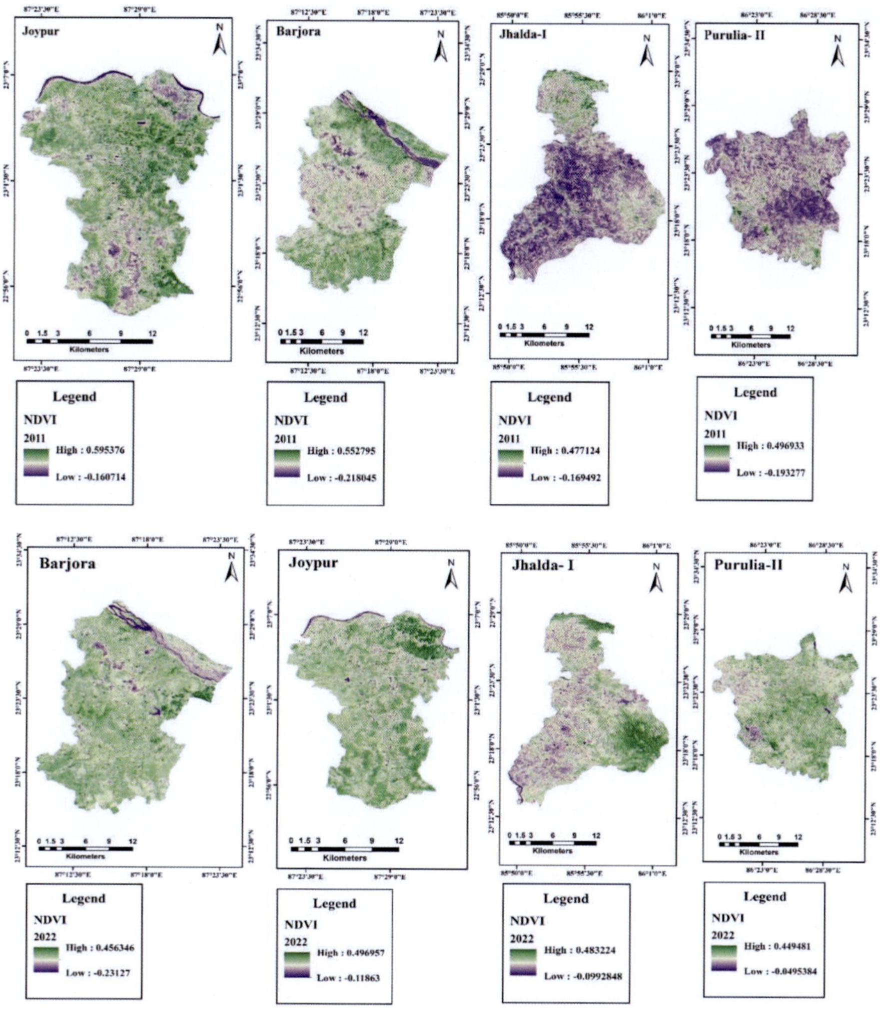

Fig. 14.3 NDVI maps of some selected CD blocks of Purulia and Bankura in 2011 and 2022

14.3.3 NDVI

The computed NDVI maps can truly explain the moisture status of the vegetation for the pre-monsoon month of May. The maps computed for some selected CD blocks in the Purulia and Bankura districts of West Bengal indicate a better moisture status in the vegetation (Fig. 14.3). The moisture potentialities within the vegetation are found to be declining in the year 2022, as indicated by the declining NDVI values (Fig. 14.3). The declining NDVI values can be interpreted as a result of the downward movement of the rainfall values.

Table 14.3 Difference between P (mm) and PET (mm)

Stations	December	January	February	March	April	May	June	July	August	September	October	November
Taldangra	-23.5926	-18.5471	-42.8849	-118.485	-255.925	-352.318	-126.613	68.084	72.85943	60.94829	-45.4209	-54.9129
Barjora	-21.5311	-17.576	-43.6694	-140.384	-291.936	-366.805	-155.701	61.656	87.012	52.72171	-24.3954	-55.9243
Hatwara	-21.0871	-16.664	-41.0931	-123.521	-253.077	-316.81	-109.344	76.37943	88.18971	78.82286	-26.1126	-49.8986
Jhalda	-22.4454	-16.1766	-40.9811	-118.099	-252.481	-309.613	-92.6671	95.29057	126.0957	83.70057	-22.9997	-42.4711

The typical month-to-month PET for the station Hatwara fluctuates from at least 28.6 mm in the period of January to a limit of 396.36 mm in the long stretch of May. The typical month to month PET qualities are viewed as 343.29 mm, 219.64 mm, 182.4 mm, and 157.91 mm for the long stretches of June, July, August, and September (rainy months), respectively. However, because the region as well as the stations gets higher measures of precipitation during storms and the sky is practically loaded with clouds, a lot of PET is not yet recognised during this season. The higher measure of insolation combined with the small cloud sum is responsible for the higher PET qualities during the pre-rainstorm. The month-wise contrasts of precipitation (P) and PET qualities for the station Hatwara show that during the non-rainy months, the PET qualities are viewed as higher than the P values (Table 14.3). Indeed, even in the rainy month of June, the precipitation amount is found to be less than the PET amount (P < PET = 109.34 mm). Dew is considered as a solid wellspring of dampness during the winter. Accordingly, the contrast between precipitation and PET during the winter is viewed as lower contrasted with the pre-rainy and rainy months. Assuming precipitation as water pay and PET as water loss, it is seen that 243.39 mm of precipitation moves as surface pursue off fulfilling the need for PET during the monsoon. In the event that a significant part of this sum (243.39 mm) is captured, the issue of water shortage can be settled to an impressive degree. However, for the station in general, the aggregate sum of PET is viewed as higher than the aggregate sum of precipitation (P < PET = 714.214 mm) (Table 14.3).

14.3.4 SDDI and PAR

During the monsoon period, the typical TA is viewed as higher than the typical TC because of slanted radiation demonstrating ample dampness in the plants; however, this present circumstance (TA > TC) is brief (Table 14.4). The dampness balance proportion of green beans changes totally in the pre-rainstorm period in contrast with the monsoon. Because of the continuous rain, the yields are sufficiently dampened during rainstorms. However, the inconsistent precipitation during the pre-rainstorm makes dampness stress inside the harvests more explicit.

The connection between various sorts of PARs and SDDI of the three harvests and their degree of importance have been worked out and introduced in tables (Table 14.4), which show areas of strength for a connection among SDDI and top occurrence PAR during summer with the computed R2 value of 0.770. The connection between these two is additionally observed to be huge at the 1% level. Additionally, during the monsoon, the connection between top occurrence PAR and

Table 14.4 Significance of correlation between PARs and SDDI for the three crops during pre-monsoon and monsoon

Season	PARs	R^2 value	t-value	p-value	Seasons	Tomatoes	Small vegetables	Green beans
Pre-monsoon	Top	0.77	4.844582	0.001868**	Pre-monsoon	0.421	0.253	0.457
	Bottom	0.455	2.416667	0.046318*		0.247	0.246	0.698
	Transmitted	0.4	2.162301	0.067379		0.302	0.359	0.584
	Intercepted	0.275	1.627981	0.147554		0.278	0.822	0.258
Monsoon	Top	0.647	4.281876	0.001607**	Monsoon	0.377	0.345	0.368
	Bottom	0.379	2.472256	0.032977*		0.356	0.678	0.47
	Transmitted	0.252	−1.83365	0.096597		0.453	0.577	0.680
	Intercepted	0.507	−3.20924	0.009344**		0.257	0.675	0.175

** Significant at 1% level; * Significant at 5% level

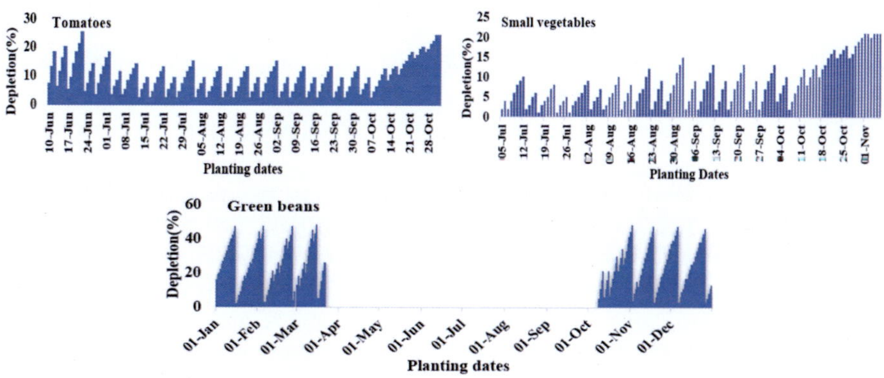

Fig. 14.4 Irrigation water requirement for some selected crops

SDDI is distinguished by areas of strength with the computed R2 and p upsides of 0.647 and 0.002 ($p < 0.01$), respectively. During the pre-monsoon stage, the changeability of radiation is viewed as a lot higher. Because of stable precipitation in the upper east, the base occurrence of PAR is fundamentally connected to the SDDI values (critical at the 5% level). The occurrence of a considerable amount of rainfall is responsible for the lower water demand during the monsoon, most particularly for the green beans (Fig. 14.4). However, for the other two crops, the water needed during the monsoon can be fulfilled by the heavy downpour during the season. During the non-rainy season, the dew deposited on the ground also acts as a reliable source of moisture.

14.3.5 Cobb-Douglas Production Estimation

To analyse the effect of various information assets (especially temperature and rainfall) on gross and net returns, the upsides of R2 are determined for each ranch size. For maize, the assessed coefficient of normal temperature (0.342) is viewed as

Table 14.5 Cobb-Douglas production estimation of different crops

Crops		Temperature (°C)	Rainfall (mm)
Maize	R square	1.000	
	Coefficients	0.342	2.685
	t Stat	13.109*	2.375
Aman paddy	R square	0.634	
	Coefficients	2.004	-1.254
	t Stat	1.070	-0.224
Potato	R square	1.000	
	Coefficients	-0.020	2.098
	t Stat	-0.546	82.376**
Cotton	R square	1.000	
	Coefficients	2.988	-0.303
	t Stat	60.019*	-0.247
Soybean	R square	0.989	
	Coefficients	-0.373	13.936
	t Stat	-3.628	9.267
Rose	R square	1.000	
	Coefficients	0.010	0.052
	t Stat	25.053*	6.173

Significant at **1% level and *5% level

certain and critical at the 5% level. The assessed coefficient of rainfall for potatoes is viewed as 2.098, and this worth is exceptionally huge at the 1% level. For maize and potatoes, almost 100 percent of the variety result can be explained by the relapse condition.

Cotton is considered a cash crop, and the review region is viewed as having potential for its growth. For the cotton likewise, almost 100 percent of the variety result can be explained by the relapse condition, dissimilar to the case with aman paddy, where only 80% of the variety result can be explained by the relapse condition. For the tomato crops, the coefficients of temperature and rainfall are determined as 0.134 and 0.205, respectively (Table 14.5). These qualities are viewed as significant at the 5% and 1% levels, respectively. In this situation, 100 percent of the variety result can be explained by the relapse condition. Dissimilar to tomatoes, the 't' values for temperature (−3.628) and rainfall (9.267) of soybean are not viewed as significant; however, the majority of the variety result of this yield can be explained by the relapse condition (Table 14.5).

14.4 Conclusion

For successful agricultural growth in an area, several key factors must remain in the region, such as sufficient rainfall, a good temperature regime, bright sunshine, and fertile soil. Although the area receives a significant amount of precipitation, the coarse-grained soils in the study area are characterised by a low water-holding capacity, and the top layer of the soil has little moisture retention capacity after very heavy rainfall. It moves more than surface runoff along terrain gradients. The rugged terrain, highly variable soils, and harsh climatic conditions are some of the factors that do not support the cultivation of long-lived crop varieties. Appropriate agricultural development in the region, therefore, must contain large amounts of runoff and focus on the introduction of new, short-lived cultivations; providing climate information to the agricultural community is also very important. Otherwise, the desired level of productivity cannot be achieved.

This study is of great importance in this sense, as it identifies the agricultural development of this region from an ecological point of view. Analysis of crop evaporation-to-precipitation ratios suggests that soybean, maize, and tomato are some of the crops that do not require irrigation water for cultivation. Even after heavy rain showers during monsoon and non-monsoon seasons, water migrates along the slopes of the area and exerts pressure on the top layer of soil in very short spans. Precipitation-to-PET ratios suggest that over 800 mm of excess precipitation would run off as overland flow during these three months of the monsoon (July, August, and September) after the PET demand had been met. In other non-monsoon months, the amount of PET is higher than the amount of precipitation. Therefore, much of this flow should be prevented in order to minimise water scarcity problems in the region. Water stored during the monsoons can be used for non-potable purposes during the dry season. Water conservation and runoff control in agricultural plots and the construction of dams and reservoirs are some of the techniques proposed for water conservation in the study area. Some CD blocks in the study area are practising water saving techniques, but they need to be implemented throughout the district, especially in the south-western part.

References

Aninagyei, I., & Appiah, D. O. (2014). Analysis of rainfall and temperature effects on maize and rice production in Akim Achiase, Ghana. *Scholars Academic Journal of Biosciences (SAJB), 2*(12B), 930–942.

Arora, N. K. (2019). Impact of climate change on agriculture production and its sustainable solutions. *Environmental Sustainability, 2*, 95–96.

Babar, S., Gul, S., Amin, A., & Mohammad, I. (2015). Climate change: Region and season specific agriculture impact assessment (thirty year analysis of Khyber Pakhtunkhwa i.e. 1980-2010). *FWU Journal of Social Sciences, 9*(1), 89–98.

Bera, S., Ammad, M., & Suman, S. (2017). Land suitability analysis for agricultural crop using remote sensing and GIS – A case study of Purulia district. *IJSRD International Journal for Scientific Research & Development, 5*(06), 999–1004.

Chakraborty, S., Pandey, R. P., Chaube, U. C., & Mishra, S. K. (2013). Trend and variability analysis of rainfall series at Seonath River Basin, Chhattisgarh (India). *International Journal of Applied Sciences and Engineering Research, 2*(4), 425–434.

Cherian, B., & Khanna, V. K. (2018). Impact of climate change in Indian agriculture: Special emphasis to soybean (Glycine Max (L.) Merr.). *Open Access Journal of Oncology and Medicine, 2*(4), 179–185.

Goswami, A. (2019). Identifying the trend of meteorological drought in Purulia District of West Bengal, India. *Environment and Ecology, 37*(1B), 387–392.

Ladan, S. I. (2014). An appraisal of climate change and agriculture in Nigeria. *Journal of Geography and Regional Planning, 7*(9), 176–184.

Liu, L., & Basso, B. (2020). Impacts of climate variability and adaptation strategies on crop yields and soil organic carbon in the US Midwest. *PLoS One, 15*(1), e0225433.

Lobell, D. B., & Gourdji, S. M. (2012). The influence of climate change on global crop productivity. *Plant Physiology, 160*, 1686–1697.

Mahato, A. (2014). Climate change and its impact on agriculture. *International Journal of Scientific and Research Publications, 4*(4), 1–6.

Mall, R. K., Singh, R., Gupta, A., Srinivasan, G., & Rathore, L. S. (2006). Impact of climate change on Indian agriculture: A review. *Climate Change, 78*, 445–478.

Manyeruke, C., Hamauswa, S., & Mhandara, L. (2013). The effects of climate change and variability on food security in Zimbabwe: A socio-economic and political analysis. *International Journal of Humanities and Social Science, 3*(6), 270–286.

Mishra, S. (2008). Dakshinbanger Paschimanchaler Samasya O Sambhobona. *Saar Samachar, 46*(1), 41–47.

Pareek, A., Dhankher, O. P., & Foyer, C. H. (2020). Mitigating the impact of climate change on plant productivity and ecosystem sustainability. *Journal of Experimental Botany, 71*(2), 451–456.

Salvo, M. D., Begalli, D., & Signorello, G. (2013). Measuring the effect of climate change on agriculture: A literature review of analytical models. *Journal of Development and Agricultural Economics, 5*(12), 499–509.

Sivakumar, M. V. K., Das, H. P., & Brunini, O. (2005). Impacts of present and future climate variability and change on agriculture and forestry in the arid and semi-arid tropics. *Climate Change, 70*, 31–72.

Soro, G. E., Noufé, D., Bi, T. A. G., & Shorohou, B. (2016). Trend analysis for extreme rainfall at sub-daily and daily timescales in Côte d'Ivoire. *Climate, 4*, 37.

Chapter 15
An Assessment of the Changing Environmental Factors of Estuarine Tidal Flats in Sagar Island

Tuli Sen and Ashis Kumar Paul

15.1 Introduction

The tidal flats usually get submerged at high tide and emerge at low tide. Horizontal tidal flat accretionary surfaces in the Hugli estuary are classified into three types: bank margin tidal flats, island margin tidal flats, and sand bars and char land tidal flats (Paul et al., 2017). Tides, storms, winds, waves, and sediment composition all have an impact on tidal flat processes. The Hugli estuary is one of the major distributary of the Ganga River system in West Bengal. It is 50 km wide near Sagar Island and extends 250 km up to Nabadwip. The Hugli Estuary has an impact on the dramatic changes in tidal flats on Sagar Island. Waves in the Hugli estuary are produced by trade winds and monsoon effects over the Bay of Bengal, with frictional temporal incidence on the open waterbody for the generation of wave heights on the sea face of the estuary up to 4 metres (Paul, 2022; Paul & Paul, 2022; Walker & Humpherys, 1983; Bhattacharyya et al., 2013; Mukhopadhyay, 2007). The waves moving along the southern coast of Sagar Island are classified as "medium energy waves" from January to February (spilling breaker) and "high energy waves" (plunging breaker) between August and September (Paul, 1989). This paper attempts to identify the tidal flat processes along the Hugli estuary channel and find out the impact of tidal processes on the channel morphology. The island is extended over the estuary section, lying only 6.5 metres above sea level (Paul, 2022), and is also accessible by ferry services across the Muriganga River. Because of their location in the dynamic environment, the island fringes are occupied by two types of tidal flats: (i) channel bank tidal flat and (ii) sea facing tidal flat.

T. Sen
Sagar Mahavidyalaya, Sagar Island, West Bengal, India

A. K. Paul (✉)
Department of Geography, Vidyasagar University, Midnapore, West Bengal, India
e-mail: akpauleastcoast@gmail.com

A. K. Paul, A. Paul (eds.), *Crisis on the Coast and Hinterland*,
https://doi.org/10.1007/978-3-031-42231-7_15

15.2 Methodology

The research was conducted using primary and secondary data, as well as their proper analysis in a geospatial and statistical environment, employing various methods of analysis using models and appropriate software. Weather data (temperature, rainfall, humidity, and solar) and cyclone data were obtained from the IMD, Pune. Tide gauge data from four different stations (Diamond Harbour, Gangra, Haldia, and Beguakhali at Gangasagar) were collected through the Tide Chart published by the Calcutta Port Trust (CPT) in 2006, 2010, and 2015. In addition, daily tidal data has been collected from the Gangasagar Light House point. Furthermore, an on-field seasonal primary survey was conducted at various times using photographic evidence to understand the nature of tidal flats. Identification of changing ecological status as well as other information about tidal flat processes is also carried out while taking landscape alteration, ecosystem nature, morphological variation of muddy and sandy flats, and tidal flat management into account.

15.3 Results and Discussions

15.3.1 Rise and Fall of Tides in the Hugli Estuary

Sagar may be because the river channel is much wider at the point of confluence with the sea than it is in Haldia. The seasonal variations between post-monsoon and dry lean flow phase (December) are 1.35 m and 1.61 m at Sagar and Haldia, respectively. The mean monthly tidal range at Haldia was 5.49 m in 2006 and 5.67 m in 2015, higher than at Sagar (4.90–4.97 m). A comparison of mean monthly tidal levels (MMTL) for the 3 years (2006, 2010, and 2015) of Sagar (3.06 m) and Haldia (3.19 m) depicted in the study. The mean high-water level (MHWL) difference between Haldia (6.12 m) and Sagar (4.69 m) is approximately 1.43 m, indicating a possible rise of the channel bed inland compared to the channel bed at Sagar's sea-face. Monthly mean highest high tide water at Sagar shows higher values during the monsoon and post-monsoon months of July to November (data from 2006, 2010, and 2015), making the island's low-lying areas vulnerable to flooding and breaches in embankments, roads, and so on.

15.3.2 Tidal Currents and Tidal Ranges in Sediment Transport and Accumulation

During neap tide, the low tidal currents allow a large portion of the sediment to settle on the bottom. Since the river discharge rate is high, sediment accumulates near the sea in a "turbidity region." The ebb and flow of the tides caused regular changes in the level of the sea along the coast, as well as the generation of tidal currents. Sediment movement in the Hugli estuary is complicated by reversing tidal currents, storm effects, and interactions with the longshore transport system. In the Rangafalla channel, the ebb tide is greater than the flood tide.

15.3.3 Tidal Ranges and Discharge in Estuaries

Flood and ebb tides occur twice daily, and the current in the tidal flat area changed direction every 6 hours. The spring tides, which occur during the HAT phase (Aug-Sept), produced the greatest rise and fall of water levels. The south-eastern lowland is fed by the Hugli and Muriganga rivers, through which tidewater flows deep inland, particularly in the spring tides (Paul, 2002; Paul et al., 2005; Paul & Kamila, 2018). The amplitude ranged from 4.88 m to 5.93 m in September 1984 (Paul, 2002) and 4.91 m to 6.03 m in September 2015 at the mouth of the Hugli River near Sagar Island, reaching 6.64 m at Haldia near Nayachar Island. Using the fluvial channel analogy, the mid-tide section of the estuary could be argued to be the hydraulically important channel where dominant discharge occurs. The Hugli Estuary's mouth is funnel-shaped due to the deposition of sediments and strong tide water movements towards the upstream sections. Waves or sediment concentration shape the island plumes and sholes at the entrances of Gangasagar creek, Muriganga creek, and Chemaguri Khal. Flood tide transports sediment into the inlet, resulting in the formation of a large flood tide delta. The data collection point for river discharge at various locations in the Hugli estuary is represented in this section (Fig. 15.1a). At Balari Semaphore, the ebb flux exceeds the flood flux, and sedimentation concentration rates are high between Balari Island and the north end of Nayachar. At the headward point of Hugli estuary, the discharge rate at Khejuri causes erosion of the Mandirtala tidal flat at Sagar Island. Figure 15.1b depicts the relationship between maximum velocity and tidal range at Hugli estuary. Since the ebb current is stronger at Beguakhali, erosion across the banks can be seen here. The same can be said for Sagar and Gangra points of the estuary sections. However, at Balari, the flood current speed is greater than the ebb current speed, resulting in faster sediment accumulation.

During the flood, the flood flux at the Rangafala Channel between Gangra Tower and Ghoramara Island is 699.44 × 106 m^3 and 815.63 × 106 m^3 during the ebb flux. This translates to a higher percentage of Flood Flux and ebb Flux than at Haldia channel and causing to a higher percent of flood flux (80.05%) and ebb flux (84.75%) at Rangafala Channel compared to Haldia channel's percent of flood flux (19.95%) and ebb flux (15.25%). The navigational capacity is superior in Rangafala channel than the Haldia channel. The distance between Balari Island and Nayachar Island is gradually getting narrowed. As a result, the sediment concentration is increased in smaller char land surface and causing lower draught available for navigation in the Haldia port (Fig. 15.1b).

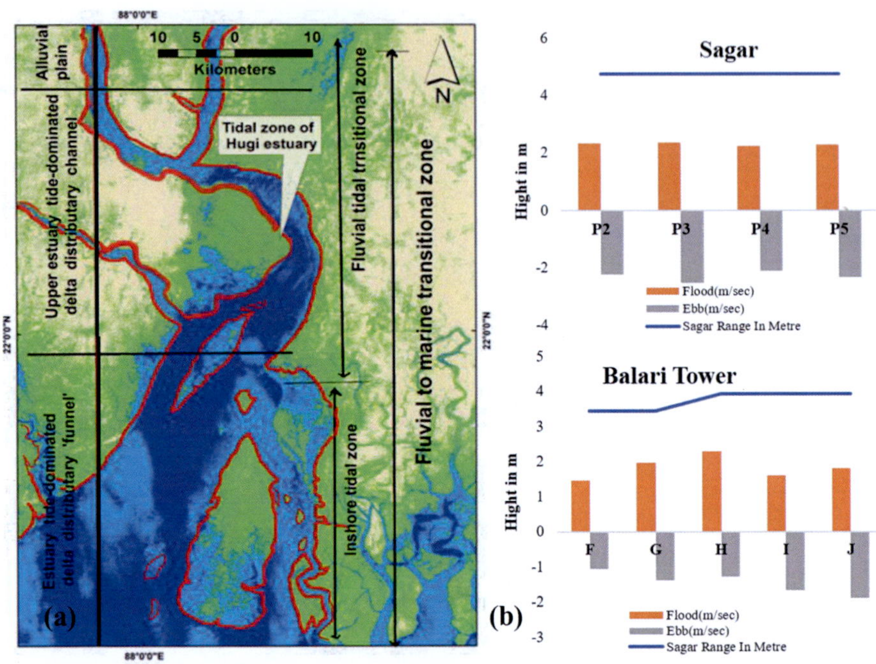

Fig. 15.1 (**a**) Red arc line showing horizontal extension of tidal flats of Hugli estuary and (**b**) relation of the maximum velocity and tidal range at Hugli estuary

15.3.4 Sediment Concentration, Salinity in Estuaries

Sediment concentration and salinity in estuaries are important in understanding erosion and accretion caused by sediment movement from one location to another. The average sediment concentration at Haldia Channel is higher than the average for all points. The concentration of sediment (1.61 gm/lt.) at point D is the highest of all. This point is prone to more sediment deposition, posing a problem for KPT's navigational movement (Figs. 15.2 and 15.4) along the Hugli estuary channel. The tidal flats of Beguakhali, Dhablat, Mandirtala, and Sagar have decreased due to sediment transport processes. In 2018, the West Bengal Government's Irrigation Department built a tetra-embankment on the bank of Boatkhali to protect the tidal flats. As a result, the western bank margin tidal flats of Sagar Island are squeezed and unable to support the tidal flat ecosystems but once dominated by mangroves, salt marshes, heaths, and sedge plants are affected by erosion at present. The gradual retreat of the tidal flat surfaces is also depicted by the unfavourable condition for sediment accumulation particular during the flood and ebb flux in the region.

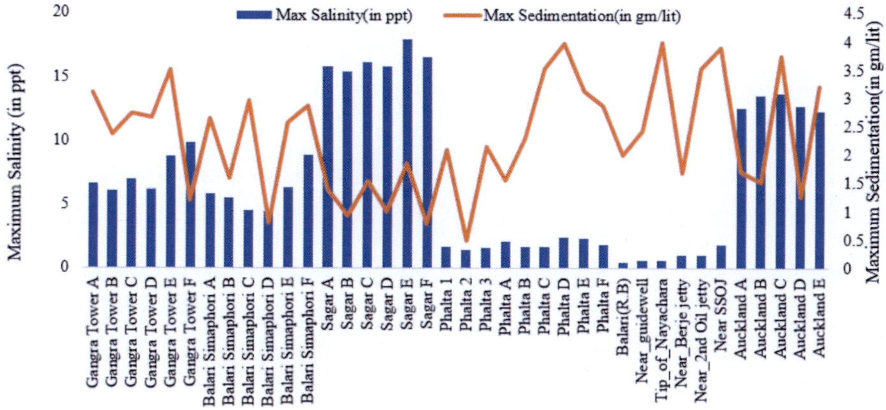

Fig. 15.2 Graphical representation of sediment concentration and salinity at Hugli Estuary

15.3.5 The Role of Waves in the Dynamics of Tidal Flats

The wind creates undulations on the surface of the water, which are known as waves. Waves consist of wave height (H), wavelength (L), and wave period (T). The time between the passes of two successive wave crests is referred to as the wave period. The quantitative representation of these parameters describes the wave climate of the tidal flats. At Sagar Island, the major types of wave breakers can be seen. Irregular bathymetry, the frequent location of bars, and the lower depth of Sagar Island offshore empower waves to travel at different angles, causing frictional momentum to be induced in the moving waves. At the same time, the tops of shorter waves spill over, dissipating wave energy on the sea face posing favourable condition for accretionary process in the tidal flat surfaces.

15.3.6 The Effect of Storms on the Tidal Flats

Storms are extreme geophysical events that pose a significant risk to human life and can cause significant damage to goods and the tidal environment. From a sociological standpoint, the severity of accepted risk is determined by who you are and what society you belong to at the time of the disaster. The poor and marginalised people who live in the tidal regime as well as the surrounding area suffer the most, and their misery is exacerbated by long-term effects (Smith, 1998). The majority of tropical cyclonic storms cross the coast at speeds ranging from 25 to 75 kilometres per hour. Between 1907 and 1966, there were 108 tropical cyclonic storms with winds of more than 25 km/h that crossed the Sundarbans coast. Cyclone damage potential has increased dramatically over the last few decades. The occurrence of cyclones at Hugli estuary is depicted graphically in Fig. 15.3 for depression of the severe

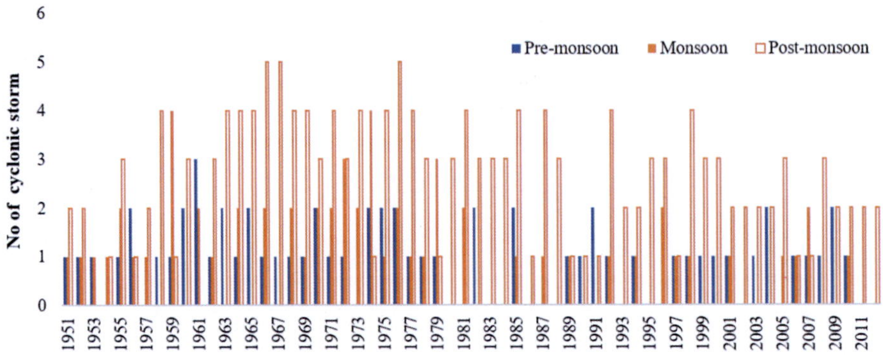

Fig. 15.3 Graph showing number of occurrences of Cyclonic storm at Hugli estuary (1891-2014)

cyclone, where the number of severe cyclones and cyclonic storms can be seen. During the post-monsoon season, the frequency of depression, cyclonic storm, and severe cyclonic storm occurrences at Hugli Estuary is highest among all seasons. The climate-induced coastal hazards and their consequences have produced the coastal squeeze effect on the intertidal habitats of the estuarine Island of the Hugli river mouth (Paul et al., 2023).

Storm damage to tidal flat vegetation causes severe erosion of coastal land. The storm surge has washed away unconsolidated dune sand, destroying a large portion of the vegetation in the area's coastal shelterbelt made by Casuarina plants. Mangrove trees, in general, suffer less damage from storms, and these trees protect the shoreline from being destroyed by storms. The tropical cyclone Amphan (2020), on the other hand, has a different effect on mangrove trees; it destroys a large portion of them. Sagar Island's shoreline marshes are well preserved because layers of over wash fan splays have been preserved beneath the present-day marsh depositional surface along the shoreline. The coarser sediments were deposited over the marsh beds as a result of the storms induced over wash process (Paul, 2022). The impacts of four different cyclones (2009, 2019, 2020, and 2021) of the previous decades were responsible for the transport of beach sands from the lower part of the intertidal zones on the sea face of the island. As a result, the beach lowering with clay bed outcrops occurred all along the tidal flats at the shore fringe of Sagar south.

15.3.7 The Fluvial Dynamics of the Hugli Downstream Section

The morphometric configuration of the Hugli estuary and Sagar Island is the result of ongoing fluvial sediment deposition in a succession of para-deltaic lobe progradation systems that evolved over the whole Holocene period (Allison et al., 2003; Goodbred & Kuehl, 2000) on the western and eastern troughs of the Bengal basin tectonic frame, influenced by eustatic, isostatic, and tectonic forces. During the monsoon months, freshwater flows increase and the salinity of the Hugli River moves southward. Salinity in the water decreases at the upper part and lower part of the Hugli estuary. The sedimentation and salinity of the Hugli estuary are inversely related (Fig. 15.4). During the monsoon seasons, the salinity near Balari Island is around 0.45 ppt, and sedimentation is around 2.00 g/lt. During the pre- and post-monsoon seasons, the volume of total tidal discharge at Diamond Harbour is 77,326 m^3/s. However, during the post monsoon months, the Hugli estuary is influenced by tide water movements, and for the time being, the tidal flats become stable with accumulation of tidal sediments (finer sands and silts) and transport of finer sediments towards upstream fringes.

15.3.8 *The Modern Sea Level Rise and Future of Tidal Flats*

Sea-level rise poses a real threat to coastal life. Increased storm surge intensity, flooding, and coastal damages are among the consequences. The present level of such data sets is a function of both past sea level changes and past crustal level

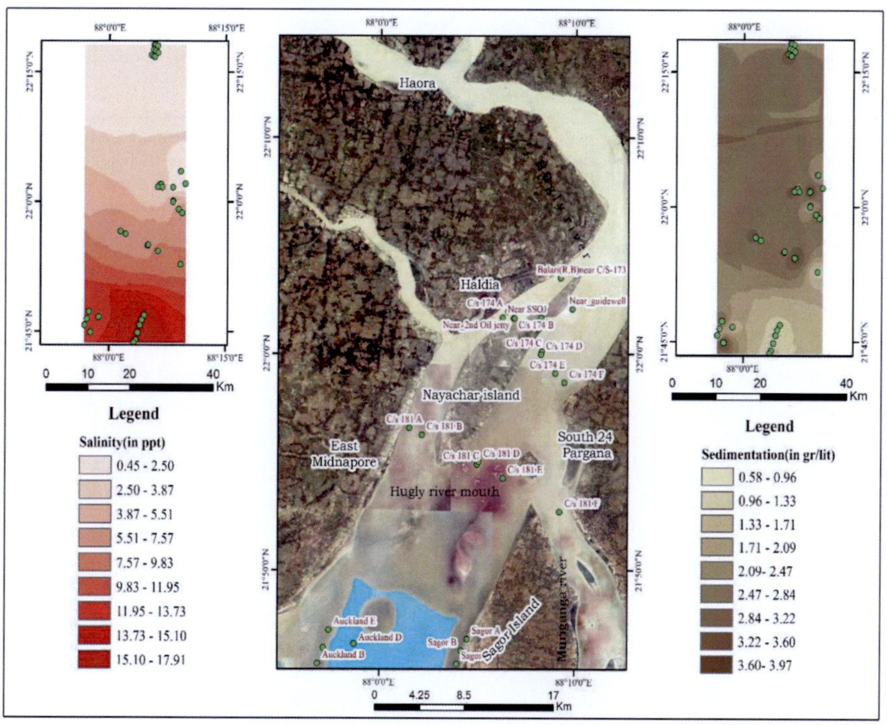

Fig. 15.4 Map showing sedimentation and salinity at Hugli Estuary

changes. In the old concept of eustasy, all sea-level changes occurred simultaneously around the world. All types of sea-level changes, in contrast to all types of crustal deformations, are referred to as "eustatic" (Mörner, 2004). West Bengal has eight PSMSL sites, all of which are located along the Hugli Estuary. Four stations can be filtered out: Sagar, Gangra, Haldia, and Diamond Harbour. Time-series analyses were conducted to determine the mean monthly and annual rates of change in sea level. Hugli estuary has significant intra-annual variance in local mean water due to freshets from the Indian summer monsoons, according to PSMSL monthly records. Between February and August, the short-term water level varies by 479 mm at the mouth of the estuary. The patterns of monthly fluctuations in sea level closely correspond to the northern Bay of Bengal coast's conventional seasonal divisions based on pre-monsoon (Feb–May) and monsoon (Jun–Sep) phases.

The MSL changes from 1937 to 2015, when all four stations are considered, show that while Sagar station shows a downward sea-level change, other stations, such as Gangra, Haldia, and Diamond Harbour, show the opposite change. The annual sea level trends of the Hugli estuary station. The sea level gradually rises above MSL as one travels north from the estuary's mouth at Sagar to Diamond Harbour. Satellite images show the accretion and erosion phases of estuarine islands located on various tidal sand ridges. Ghoramara island, for example, separated from Sagar between 1903 and 1904. Nayachar emerged in 1948 and gradually grew to become the estuary's second-largest island. The major impacts of sea-level rise are increased flooding and inundation of low-lying areas of tidal flats, shoreline retreat, and loss of land, and livelihoods. The Hugli estuary serves as a sediment sink, largely offsetting the effects of accelerated sea level rise on land loss.

15.4 Conclusion

The rise and fall of tidal currents, along with their range and sediment transport from the upper part of the Hugli estuary, and the sediment accumulation at various points, are observed. With the changing sea level, changing fluvial dynamics, tidal influences, and frequent cyclonic storms, these tidal flats left a significant stratigraphic signature after erosion. With the help of a texture analysis of the sediment, the composition of sand flats and mudflats was identified, which helped to understand the erosion and accretion of the tidal flats at different locations on those islands. The study also highlighted the increased sedimentation rate on the western bank of the Hugli estuary, causing the emergence of shoaling flats, sand bars, and extensive tidal flats. However, the effects of tidal current concentrations on Sagar Island's western shorelines have resulted in the degradation of estuary fringe tidal flats compared to Muriganga fringe tidal flats on Sagar.

References

Allison, M. A., Khan, S. R., Goodbred, S. L., Jr., & Kuehl, S. A. (2003). Stratigraphic evolution of the late Holocene Ganges–Brahmaputra lower delta plain. *Sedimentary Geology, 155*(3-4), 317–342. https://doi.org/10.1016/S0037-0738(02)00185-9

Bhattacharyya, S., Pethick, J., & Sarma, K. S. (2013). Managerial response to sea level rise in the tidal estuaries of the Indian Sundarbans: A geomorphological approach. *Water Policy, 15*(S1), 51–74. https://doi.org/10.2166/wp.2013.205

Goodbred, S. L., Jr., & Kuehl, S. A. (2000). Enormous Ganges-Brahmaputra sediment discharge during strengthened early Holocene monsoon. *Geology, 28*(12), 1083–1086.

Mörner, N. A. (2004). Estimating future sea level changes from past records. *Global and Planetary Change, 40*(1-2), 49–54.

Mukhopadhyay S. K. (2007). *The Hooghly Estuarine System, NE Coast of Bay of Bengal, India.* Workshop on Indian Estuaries, NIO, Goa.

Paul, A. K. (1989). *Coastal sediments of West Bengal. The observer, vol. XXXI* (pp. 1–10). Univ. of Calcutta.

Paul, A. K. (2002). *Coastal geomorphology and environment* (pp. 1–342). ACB Publication.

Paul, A. K. (2022, August). Dynamic behaviour of the estuaries in response to the phenomenon of global warming in the coastal ecosystems of West Bengal and Odisha, India. In *Transforming coastal zone for sustainable food and income security: Proceedings of the international symposium of ISCAR on coastal agriculture, March 16–19, 2021* (pp. 907–931). Springer International Publishing.

Paul, A. K., & Kamila, A. (2018). Studies on coastal morphometry with total station survey in shore face of Sagar Island to assess the impacts of sea level rise. In *38th INCA International Congress Hyderabad* (pp. 460–469).

Paul, A. K., & Paul, A. (2022). Adjustment of the coastal communities in response to climate variability and sea level rise in the Sundarban, West Bengal, India. In *Climate change, disaster and adaptations: Contextualising human responses to ecological change* (pp. 201–217). Springer International Publishing.

Paul, A. K., Chatterjee, S., & Paria, S. (2005). Morphodynamics and vulnerability issues of Nayachar Island in the Hugli Estuary section of Ganga Delta. *Indian Journal of Geography and Environment, 10*, 76–92.

Paul, A. K., Ray, R., Kamila, A., & Jana, S. (2017). Mangrove degradation in the Sundarbans. In *Coastal wetlands: Alteration and remediation* (pp. 357–392). Springer. https://doi.org/10.1007/978-3-319-56179-0

Paul, A. K., Paul, A., & Sardar, J. (2023). Susceptibility of the climate resilient landforms of the coastal tract of Odisha and West Bengal, India. *Journal of the Geological Society of India, 99*(6), 827–839.

Smith, J. A. (1998). Evolution of Langmuir circulation during a storm. *Journal of Geophysical Research: Oceans, 103*(C6), 12649–12668.

Walker, W. R., & Humpherys, A. S. (1983). Kinematic-wave furrow irrigation model. *Journal of Irrigation and Drainage Engineering, 109*(4), 377–392.

Chapter 16
Temporal and Spatial Changes in the Hugli Estuarine Environment: A Review of Nayachara Island

Ashis Kumar Paul, Phalguni Bhattacharyya, Anurupa Paul, and Joydeb Sardar

16.1 Introduction

The downstream section of the Bhagirathi-Hugli River, as the western-most distributary of the Ganga, was submerged slowly and steadily by the Holocene transgression (6000–7000 YBP) and gradually reshaped by the Late Holocene to recent sub-reccnt fluvio-marine depositional processes in the Bengal Basin (Allison et al., 2003). At the downstream end of Diamond Harbour, the Hugli flares up to the mouth and represents a perfect funnel-shaped estuary with sedimentary depositional sink modified by asymmetric tidal behaviour by faster movement of flood tides and slower movement of ebb tides along its course. The late Holocene advancement of the Ganga delta took place by a series of deltaic lobes towards south-southeast directions, with the role of active distributary channels as sedimentary environmental focie from time to time. In this connection, the Bhagirathi-Hugli delta lobe advanced towards the Bay of Bengal at the western boundary of the Ganga delta,

A. K. Paul (✉)
Department of Geography, Vidyasagar University, Midnapore, West Bengal, India
e-mail: akpauleastcoast@gmail.com

P. Bhattacharyya
Balarampur M N Vidyamandir, Bonhooghly, Narendrapur, Kolkata, West Bengal, India

A. Paul
Department of Remote Sensing & GIS, Vidyasagar University, Midnapore, West Bengal, India
e-mail: anurupapaul2017@gmail.com

J. Sardar
Department of Remote Sensing & GIS, Vidyasagar University, Midnapore, West Bengal, India

Centre for Environmental Studies, Vidyasagar University, Midnapore, West Bengal, India
e-mail: joydebsardar18@gmail.com

A. K. Paul, A. Paul (eds.), *Crisis on the Coast and Hinterland*,
https://doi.org/10.1007/978-3-031-42231-7_16

with the role of active sedimentary environmental focie (5200–5700 YBP) at the beginning stage of the Late Holocene deltaic evolution (Goodbred & Kuehl, 2000a, b).

The present study reveals the current depositional characteristics of the Hugli estuary, in which the modern hydrodynamic regimes and tidal sediment supply played major roles in the shaping, reshaping, and development of estuarine islands (Paul & Bandyopadhyay, 1986, 1987; Mukhopadhyay, 2007). Frequent cyclone breaks with storm surges and tidal surges transported shifting sands into the estuary section in the form of spits, bars, and shoals (Paul, 1996, 1997, 2000). The climate change reworking of the estuarine landscapes has produced a new trend of instability in the configurations of the younger and older islands within the morphodynamic setup. The islands are grouped into four major classes of morphodynamic setup in the downstream section of the Hugli estuary. They are known as: (i) the Haldia-Rangafala channel fringe estuarine complex, (ii) the Rangafala-Baratola channel fringe estuary complex, (iii) the Baratola-Muriganga channel fringe estuary section, and (iv) the shallow offshore complex at the confluence of the Muriganga and Hugli estuaries. In all cases, the local islands are forming and reforming into different shapes and sizes over a time scale under the impacts of energy levels, physiographic settings, hydrological behaviours, and sediment supply rates. Such dynamics will be more rapid and critical in the estuary sections with the impacts of sea level rise, climate change threats, and human activities in the river system (Paul, 2011, 2014, 2022; Paul & Kamila, 2018; Paul & Paul, 2022; Mukhopadhyay, 2007). The study also reveals that changing configurations of the islands, bank erosion in the estuary channel reaches, storm drift shifting sands, and tidal siltation provided sediment supply sources for the morphodynamic systems of the estuary sections of the region (Fig. 16.1).

16.2 Materials and Method of the Study

The downstream section of the Hugli estuary is delineated from the Survey of India (SOI) toposheets from Diamond Harbour to Sagar south to be considered the study area on the western margin of the lower Ganga delta (Fig. 16.1). Landsat 8 OLI and Google Earth Images (1985 and 2022) data are considered for the identification of the island configurations in the study. The estuary morphology of the Hugli downstream section is studied with the naval hydrographic chart and other bathymetric charts. Other aspects, such as data from different sources (India Meteorological Department (IMD), Indian National Center for Ocean Information Services (INCOIS), Kolkata Port Trust (KPT), etc.) and a field survey in the island territory, helped to carry out the final study.

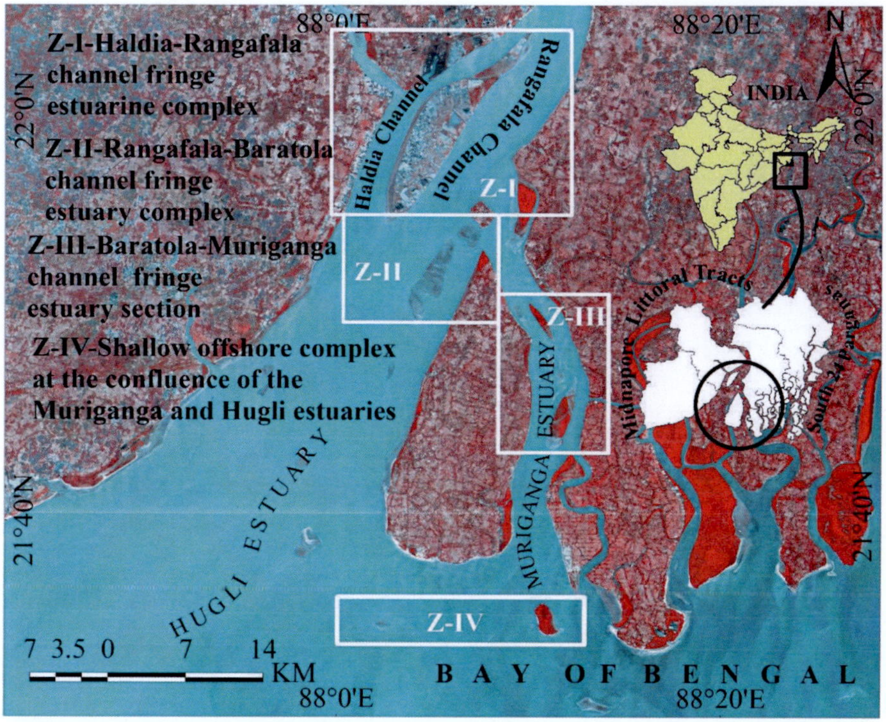

Fig. 16.1 Location of the study area and diversity in depositional landforms in Hugli estuary with four morphodynamic subsystems

16.3 Results and Discussion

The temporal and spatial changes in the estuarine landscapes depicted the system-subsystem sections in which the energy and materials interact with each other over a time to develop the depositional landforms. However, the interactions are guided by tidal energy variations, seasonal variation in downstream discharges, storm surge impacts into the arm of the sea, climate change reworking of the estuarine landscapes, coastal marine processes, and the rate of sea level rise in the funnel-shaped estuary complex of the Hugli River. The following sections will demonstrate the estuary subsystems as per the Hugli estuary channel, which reaches from Diamond Harbour point to the sea face of the Bay of Bengal.

16.3.1 Haldia Point Estuary Reach

Between the Rangafala channel and Haldia channel complex of the Hugli estuary, a number of islands and shoals are distributed in the form of Nayachar Island, Balari Island, Diamond sands, Haldia sands, and Nayachar south sand along a linear

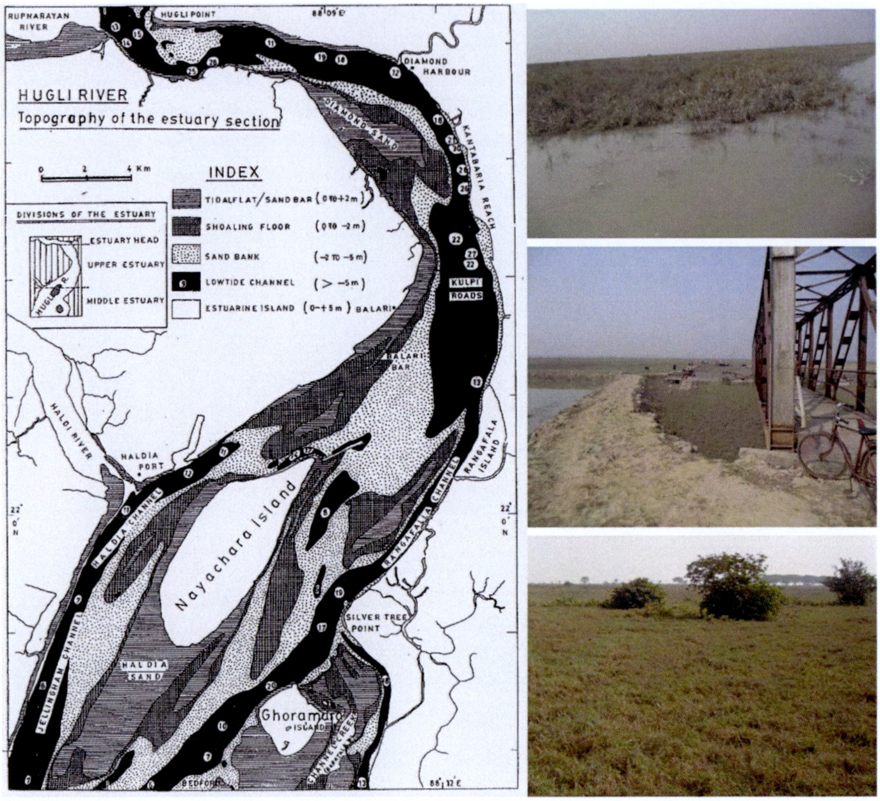

Fig. 16.2 The Haldia point and Silver Tree Point estuary reaches of the Hugli system with younger island surfaces of Nayachara Island colonised by halophytic grasses and mangroves

stretch (Paul, 2002). The subsystem is developed at the flared-up section of the estuary reach below Diamond Harbour Point and extended southward up to Nayachar South Sands. The linear depositional features are aligned by the downdrift currents of Rangafala Channel (eastern margin thalweg) and the updrift currents of Haldia Channel (western margin thalweg). Accretionary bars initiated at the inner bank of the Hugli estuary in the form of Diamond sands and gradually evolved into Nayachar, Balari, Haldia Sands, and Nayachar South Sands into the linear depositional subsystems guided by the hydrodynamic regimes of tidal currents, residual currents, storm surge waves, and sediment transport paths between the source and sink (Figs. 16.2 and 16.3a).

The ebb tide currents are stronger in the Rangafala channel compared to the Haldia channel, which represents a flood tide-dominated channel in the estuary reach. Thus, under the blocking effects of Balari and Nayachar islands, the Haldia channel arrests flood tide sediments with a rapid rate of siltation process to extend the Haldia sands on the eastern fringe of Nayachar island. The Nayachar South sands, on the other hand, were deposited by an uprushing tidal surge through the

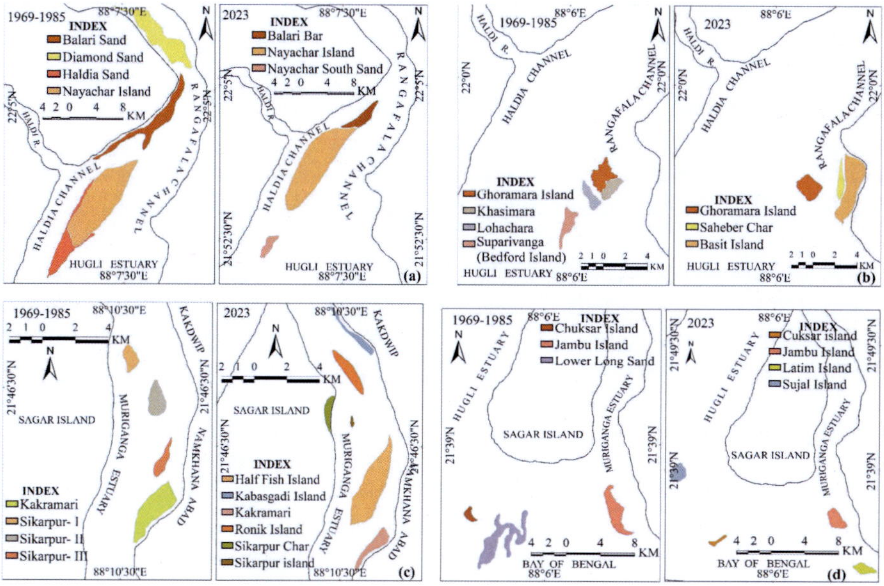

Fig. 16.3 Morphodynamic Subsystems (**a**) Haldia Point estuary reach with Nayachar island; (**b**) Silver tree point estuary reach with Ghoramara Island; (**c**) Muriganga-Baratola reach with Rohnik Island and Sagar Island; and (**d**) the sea facing distal islands subsystem with Jambu Island of the Hugli estuary from Diamond Harbours to the sea face of the Bay of Bengal

Hugli mouth and also by the accumulation of shifting sands during the storm surges in the estuary reach. Further, it was realigned by the downdrift currents of Rangafala channel and reshaped by the impacts of repeated cyclone landfalls when the Hugli downstream section acted as the passage of northward-advancing storm surges in the near past. Nayachar and Balari islands are covered by halophytic grasslands and mangroves (Paul et al., 2005, 2017) and drained by a network of tidal creeks. Both aggradation and progradation occur rapidly during the submergence of astronomic tides and occasional higher tides, but erosion occurs along the north-eastern margins of Nayachar when strong down-drift currents hit the island banks in the Rangafala channel section of the estuary.

16.3.2 Silver Tree Point Estuary Reach

During 1969 (SOI Toposheet), Ghoramara, Khasimara, Lohachara, Bedford, and Suparibhanga islands were aligned in NE-SW directions parallel to the projected bank of Silver Tree Point in the estuary reach. This is another subsystem of the Hugli estuary system, which represents the barrier between Rangafala downdrift currents and Baratola updrift currents in the region during the events of ebb tides and flood tides. Suddenly, the nucleus of the group Ghoramara island was severely affected by undermining the northwestern banks with direct hitting of Rangafala downdrift currents in the estuary reach. Active erosion has reduced the area from 19 km^2 to less than 6 km^2 and broken it into reduced Ghoramara, Bedford Shoal, and Shiber Char in the estuary reaches. Sliver Tree Point of the eastern bank of the Hugli estuary is affected by severe erosion caused by undermining the basement of the bank wall during the 1990s and the formation of an extensive char land at the bank margins. Baratola-Muriganga flood currents transported sediments through the tidal surges into the Shiber Char shoaling flat of the estuary reach. In the 1920s, a large part of the Khasimara-Lohachara group was covered by mangroves, tidal flats, and muddy tracts with tidal channel creeks (SOI Toposheet, 1925) in the estuary reach. Gradually, with a rapid rate of erosion, the islands were eroded due to the thalweg-shifting in Rangafala and the Baratola channel complex triggered by cyclone landfalls (1970, 1991, 1997, 1988–89) in the region of the Sundarban coastal tract. Lohachara-Khasimara-Ghoramara islands were reclaimed and inhabited by settlements (Paul, 2002) in the estuary reach. Eroded sediments of the subsystems are transported and accumulated in "Shiber Char" and "Bedford" in the form of shoaling flats in the region. Ghoramara was the nucleus of this island group in the subsystem, but the volume of sediments transferred and drifted from the centre to the secondary islands of smaller size in the form of accumulated bodies of shoals under the impacts of high energy events. Gradually, other islands grew by aggradation and progradation with an emerging colony of halophytic grasslands, mangroves, and other littoral vegetation on the basis of available soil nutrients, tidal drainage conditions, and the rate of fresh silt accumulation over time (Paul et al., 2017). Ghoramara will be further reduced in size by interplaying with increasing event frequencies of high energy phases and sediment transfers from island to island in the morphodynamic settings (Figs. 16.2 and 16.3b).

16.3.3 Baratola-Muriganga Estuary Channel Reach

The eastern branch of the Hugli estuary is separated by the location of Sagar Island and known as the Baratola-Muriganga channel section from north to south at the downstream course of 30 km length. The system includes three charlands fringed with the right bank and left bank of Muriganga (e.g., Kabasgadir char, Kakramari char, and Shikarpur char) and three mid-Channel Islands of elongated size along the lower most course of the estuary reach (i.e., Rohnik Island, Shikarpur Island, and Half Fish Island). Muriganga is a strong ebb-dominated channel in which islands and charlands are aligned with the downdrift currents (2.67 m/s, below 10 m depths over 5 m/s). The bank attached charlands and mid-Channel Islands are covered by mangroves and salt marshes with tidal creeks and mud flats. Bank erosion, island fragmentation, and island drifting occur in the high-energy events and supply sediments into the accumulated shoals in the estuary section. Temporary stability is achieved with long relaxation times (>20 years) between the high energy events frequency in the depositional environment. However, climate change-induced reworking of the estuarine landscapes will make them more dynamic with repeated landfalls of cyclones and an increasing rate of sea level rise processes. Increased tidal prisms in the Muriganga Estuary during the vents of surge elevation produced a major threat to the stability of the depositional environment. Thus, the Half Fish Island is broken into fragmented shoals, with bank margins at Mousuni Island and Sagar Island showing instability, and mangrove wetlands of the estuary fringe tidal flats getting trapped by overwashed sands and shifting sands in the estuary reach (Fig. 16.3c).

16.3.4 Sea Facing Distal Islands

The morphodynamic subsystem of the estuarine landscapes provides a critical condition in the sea face in which the distal islands are breaking into shoals, reducing into respective sizes, changing island configurations, and surprisingly drifting from their original sites in the Bay of Bengal due to the effects of storm surges, tidal surges, and cyclone landfalls in the previous decades. There are a number of islands and shoals in the region, and they are known as Jambu Island, Latim Island, Sagar Island, Lower Long Sand, Chuksar Island, and South Sand. They are rapidly interacting with each other by transferring the volume of sediments onto the shallow offshores, guided by the energy events and longshore sediment supply. Mangroves and tidal flat habitats are seriously affected in the distal islands due to their reduced sizes and fragmentations at present.

The basement structures of the island are exposed as base clay, sand sheet, compact sit flat, beach ridge, and sand dunes with vegetation (Paul, 1988, 1994). At the embayed shores surrounded by spits and beach ridges, mangroves often occupy the tidal flats of backwater settings in many occasions. Sandy shoals, on the other hand, were modified rapidly with rippled flats, sand waves, tidal pools, and beach plains by the littoral drifts and surging waves. During storms, the islands and shoals are submerged by seawater and wave reworked thus breaking into fragmented shoals or islands before compaction of the sediments. When the event frequency increases, the broken islands spread over a large area into a number of shoals and islands of low height surfaces on the sea face of the Hugli estuary. The western branch of the Hugli estuary to the west of Sagar Island is over 30 km wide and usually characterised by a flood-dominated channel that acts as the passage of landward-moving tropical cyclones from the northern Bay of Bengal (Fig. 16.3).

16.3.5 Energy Levels and Sediment Supply

The morphodynamics settings of the estuary reaches are affected by various energy levels at present. They are identified as anthropogenic controlled fluvial discharges and sediment loads, tidal asymmetry in the estuary, occurrences of tidal bores with high resonant waves, velocity variations in flood tide and ebb tide currents into the estuary, increased tidal prisms with surge water levels, climate change reworking with rising sea level, increased event frequency of cyclone landfalls and inundations, tidal surges during eastern winds, longshore current transport of sediment, and swell waves with steeper gradients into the system. Aggradation and progradation do not occur at similar rates under the impacts of various energy levels and diversity in morphodynamic settings at different estuary reach distances in the depositional environment. Multiple sediment supply sources feed the systems to operate through longshore transport, sea bed scouring in the storm surges, bank erosion materials, tidal sediments, and reworking of island sediments in the estuary system. The island groups in the morphodynamic settings interact and interplay with each other to adjust the various energy levels and reduced sediment supply to develop the signatures of landforms in the estuary reaches at present (Table 16.1).

Table 16.1 The configuration dynamics of the depositional landforms with island drift and island accumulation

Nayachara Group	Area-1969–1985 (sq.km)	Nayachara Group	Area-2023 (sq.km)	Ghoramara Group	Area-1969–1985 (sq.km)	Ghoramara Group	Area-2023 (sq.km)
Nayachar Island	35.4	Nayachar Island	48.2	Ghoramara Island	3.6	Ghoramara Island	3.54
Haldia Sand	10	Nayachar south sand	4	Khasimara	2.16	Basit Island	8.21
Diamond Sand	17.7	Balari Bar	6.44	Lohachara	2	Saheber Char	1.45
Balari Bar	13.7			Suparivanga	2.55		
Total area	76.8		58.64	Total area	10.31		13.2
Muriganga Group	Area-1969–1985 (sq.km)	Muriganga Group	Area-2023 (sq.km)	Jambu Group	Area-1969–1985 (sq.km)	Jambu Group	Area-2023 (sq.km)
Sirkarpur-I	1.1	Ronik Island	2.34	Jambu Island	7.92	Jambu Island	5.41
Sirkarpur-II	1.77	Kabasgadi Island	1.42	Chuksar Island	1.45	Chuksar Island	1.35
Sirkarpur-III	1	Sikarpur Char	1.13	Lower long Sand	3.4	Latim Island	3.37
Kakramari	4.1	Sikarpur Island	0.17			Sujal Island	5
		Half fish Island	8.7	Total area	12.77		15.13
		Kakramari	2.1				
Total area	7.97		15.86				

16.3.6 Estuary Process Variables

The wide, funnel-shaped estuary also acts as an arm of the sea in the downstream section of the Hugli River. Present-day processes involve multiple components of tidal, shallow marine, and fluvial environments, and they interplay and interact with each other within the physiographic settings of the estuary (Table 16.2).

16.3.7 Tidal Processes in the Estuary

Tidal influences with variable amplitudes (>6.0 m in Haldia and >5.0 m in Sagar) and ranges, flood tide-ebb tide velocities, asymmetry in tidal movements into the estuary reaches, events of tidal bores, tidal flood inundation levels into the islands and creeks, emerging tidal surge waves during the cyclonic winds (i.e., the Yass cyclone, 2021), and events of astronomic tides (HAT phase) interact to produce a significant energy level in sediment transport within the morphodynamic settings of the estuary reaches (Figs. 16.4 and 16.5).

16.3.8 Shallow Marine Environment at the Sea Face of the Estuary

Significant marine influences are felt on the sea face of the Hugli estuary with Muriganga-Baratola tidal river confluence. The wide estuary mouth with 47 km stretches of opening receives impacts from strong wind waves, swell waves, storm surge, longshore currents, tidal currents, and residual currents in different seasons. In the bathymetric chart of the Hugli mouth, it is observed that the sub-aqueous shelf deposits are jagged and aligned parallel to the estuary channel from north to south as alternate sand ridges and channels. They are actually modified by strong

Table 16.2 Events of Large Floods and High Magnitudes Cyclones in and around the Hooghly Estuary

Event of large flood	Year	Event of high magnitude cyclones	Year	Remarks
Large flood	1950	Severe cyclone	1942	Emerged sand bar
Largest flood	1978	Severe cyclone	1971	Mud deposition
Moderate flood	1984	Sundarban cyclone	1989	Vegetation cover
Moderate flood	1991	Sundarban cyclone	1991	Mangrove wetland extension
Moderate flood	2000	Cyclone Cedere	2007	Sand bar extension
Large flood	2008	Nargis cyclone	2008	Balari extension
		Aila cyclone	2009	Erosion and siltation

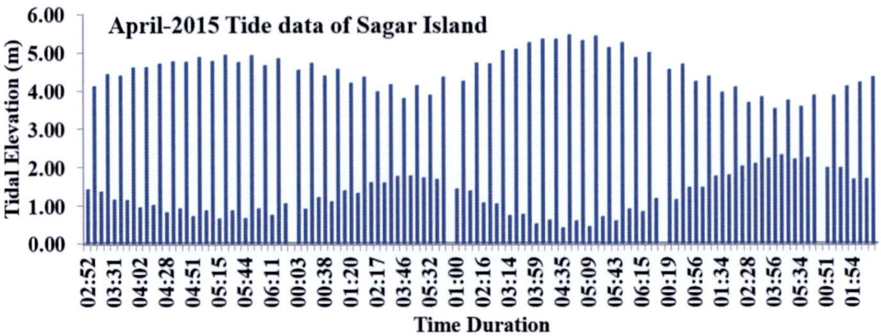

Fig. 16.4 The nature of monthly tidal cycles of Hugli estuary near Sagar Island. (on the basis of Tide Gauge data)

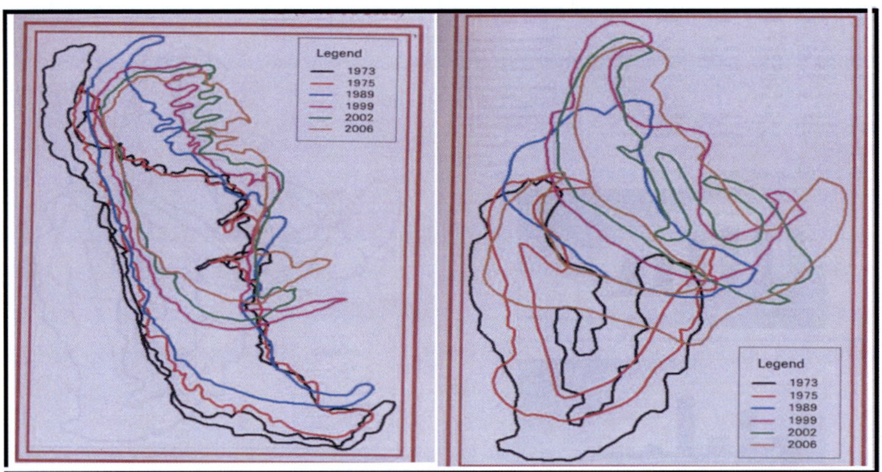

Fig. 16.5 Configuration dynamics of the distal islands (Jambu and Chuksar Island) in the Bay of Bengal with island drifting and reduction in sizes on the sea face of the Hugli estuary. (Using Landsat Images)

tidal currents and turbidity currents in the region, but possibly originated during the last glacial low stand (18,000 YBP) with the extension of fluvial valleys and natural levee deposits towards the sea. However, the early Holocene transgression (6000 YBP) submerged the extensive areas by the advanced sea and occupied the region of fluvial depositional features under the impacts of sub-marine processes and related modifications (Allison et al., 2003).

The distal islands and sandy shoals or bars on the sea face area, however, are influenced by the shallow marine environment with the activities of tides, currents, storms, littoral drifts, and the rate of present-day rising sea level (Fig. 16.5). Thus, the extensive tidal flats and mangrove wetland habitats of Jambu Island (6.36 km long and 1.70 km wide in 1985) are now reduced to a smaller island platform by erosion (3.2 km long and 2.06 km wide in 2023) and fragmentation in the sea face due to the impacts of dynamic marine processes. The Chuksar island in the sea face of the Hugli estuary to the south of Sagar Island was represented as a tiny island with backwater mangroves surrounded by wide sandy platforms and back dunes (1.7 km long and 0.74 km wide in 1985; Paul, 1990). There is no island in the region at present; it has drifted and broken into sandy shoals over wide areas of the shallow marine environment. They were mostly affected by the cyclone landfalls on the passage of Hugli Mouth in 1989, 2009, 2019, 2020, and 2021 with active storm surge elevations (2.5 m to 5.9 m, INCOIS) and inundation events.

16.3.9 Fluvial Regimes in the Estuary

Anthropogenic controlled fluvial discharges and sediment loads contributed a significantly reduced amount of discharges and fluvial sediments into the estuary reaches of the study area in the modifications of the island groups. Before 1975, a set of estuarine sand banks (Diamond Sands, Balari Sands, and Haldia Sands) was pointed towards the south and aligned parallel to the strong residual currents in the flared-up section of the Hugli estuary when fluvial sediments contributed to the formation of sandy platforms in the region. The high-magnitude flood frequency events in Hugli (1950, 1978, 2000, and 2008) contributed sediments to the island platforms, river beds, and sand banks but discharged the suspension loads into the sea face in the form of plumes of total suspended material (>80 mg/m^3, Paul, 2002). Aggradation and progradation were enhanced by fluvial sediment supply during the events of the annual floods before the anthropogenic controlled discharges and sediment transported into the island systems of the estuary sections. Presently, tidal influences pushed back the impacts of fluvial regimes back from the estuary sections in the form of fluvial backwater settings into the upstream sections of the Hugli River (Fig. 16.6). The estuary is acting as a significant mixing zone of freshwater discharges with the saltwater encroachments and pushing the saltwater into the sea during southwest monsoon months in the region. However, the similar estuary is dominated by tidal encroachments and also by marine influences during the dry months to control the morphodynamic systems in the modification of islands, shoals, bars, and bank margin depositional features.

16.4 Conclusion

The Hugli estuarine depositional features in specific physiographic settings of the four groups of islands and sandy shoals at the reach distances exchange sediment volumes with the interplay of energy and process variables under morphodynamic settings to modify their configurations over a time span in the region. The study reveals that the Nayachar group has changed to a relatively lesser extent in configuration in comparison to the Ghoramara group, Muriganga channel-reach island groups, and sea-facing island groups in the estuary system. The number of

Fig. 16.6 Mangroves and salt marsh creeks in Nayachar Island platform, Hugli Estuary

depositional units has changed from four islands to six islands (1985–2023) in the Muriganga estuary channel reach section between Sagar, Kakdwip, and Namkhana blocks. However, islands of the sea face at the confluences of Muriganga and Hugli demonstrated major changes due to climate change reworking with the storminess of the sea and an increasing rate of relative sea level change. On the other hand, the Ghoramara group of islands is lying at the junction between the impacts of a strong ebb-dominated channel (Rangafala) and the climate change reworking of estuarine landscapes to produce the dynamic depositional environment.

It is also believed that the rates of aggradation and progradation are relatively slower in the islands than the rates of storm frequency, climate change reworking, and saltwater inundations in the estuary system of the Hugli downstream section. All the major and older islands of the four morphodynamic settings are getting reduced in size (Sagar, Ghoramara, Jambu, and Nayachar islands) at present. Nayachar has expanded its linking with Balari Island, accumulation of sediments, vegetation cover surfaces, and extended tidal drainage systems with aggradation and rapid progradation processes in the estuary reaches. Human modifications of the island may invite a similar fate of size reduction like Ghoramara and Sagar Island by a rapid rate of erosion and an insignificant rate of aggradation.

References

Allison, M. A., Khan, S. R., Goodbred, S. L., Jr., & Kuehl, S. A. (2003). Stratigraphic evolution of the late Holocene Ganges–Brahmaputra lower delta plain. *Sedimentary Geology, 155*(3–4), 317–342.

Goodbred, S. L., Jr., & Kuehl, S. A. (2000a). The significance of large sediment supply, active tectonism, and eustasy on margin sequence development: Late Quaternary stratigraphy and evolution of the Ganges–Brahmaputra delta. *Sedimentary Geology, 133*(3–4), 227–248.

Goodbred, S. L., Jr., & Kuehl, S. A. (2000b). Enormous Ganges-Brahmaputra sediment discharge during strengthened early Holocene monsoon. *Geology, 28*(12), 1083–1086.

Mukhopadhyay, S. K. (2007). *The Hooghly Estuarine System, NE Coast of Bay of Bengal, India*. Workshop on Indian Estuaries, NIO, Goa.

Paul, A. K. (1988). Coastal Dune Systems of the Sundarban. *Geographical Review of India, Calcutta., 50*(1), 14–23.

Paul, A. K. (1990). *Chuksar Island at the head of Hugli estuary (In Bengali)* (Vol. 2, pp. 46–58). Visva Veeksha.
Paul, A. K. (1994). The Dune environment of West Bengal coastal plain. In K. R. Dikshit, V. S. Kale, & M. N. Kaul (Eds.), *India geomorphological diversity* (pp. 314–351). Rawat Publications.
Paul, A. K. (1996). Identification of coastal hazards in West Bengal and parts of Orissa. *Indian Journal of Geomorphology, 1*(1), 1–27. New Delhi, Academy and Law serial.
Paul, A. K. (1997). Coastal erosion in West Bengal. *Maeer MIT Pune Journal, iv*(15-16 (Special issue on coastal environmental management)), 66–84.
Paul, A. K. (2000). Cyclonic storm and their impacts on West Bengal coast. In V. G. Rajamanikam & M. J. Tooley (Eds.), *Quaternary Sea level variation* (pp. 25–57). New Academic Publishers.
Paul, A. K. (2002). *Coastal geomorphology and environment* (pp. 1–342). ACB Publication.
Paul, A. K. (2011). Coastal drainage problems in response to sea level rise: A study in MidnaporeBalasore coastal zones and Sundarban coastal tracts. *Indian Journal of Geology, 83*(1–4), 41–60.
Paul, A. K. (2014). Morphology of estuaries and tidal inlets: An emphasis on Hugli, Subarnarekha and Chilika systems along the Bay of Bengal shoreline. In *Proceedings of the national conference on modern trends in coastal and estuarine studies*, Tilak Maharashtra Vidyapeeth, Pune, Maharashtra India, 6–7 Feb 2014.
Paul, A. K. (2022, August). Dynamic behaviour of the estuaries in response to the phenomenon of global warming in the coastal ecosystems of West Bengal and Odisha, India. In *Transforming coastal zone for sustainable food and income security: Proceedings of the international symposium of ISCAR on coastal agriculture, March 16–19, 2021* (pp. 907–931). Springer International Publishing.
Paul, A. K., & Bandyopadhyay, M. K. (1986). Coastal sand dunes of Sagar Island, W.B., India. *Geographical Review of India, Calcutta, 48*(2), 41–48.
Paul, A. K., & Bandyopadhyay, M. K. (1987). Morphology of Sagar Island, a part of Ganga delta. *Geological Society of India, 29*(4), 412–423.
Paul, A. K. & Kamila, A. (2018). Studies on coastal morphometry with total station survey in shore face of Sagar Island to assess the impacts of sea level rise. In *38th INCA International Congress Hyderabad* (pp. 460–469).
Paul, A. K., & Paul, A. (2022). Adjustment of the coastal communities in response to climate variability and sea level rise in the Sundarban, West Bengal, India. In *Climate change, disaster and adaptations: Contextualising human responses to ecological Change* (pp. 201–217). Springer International Publishing.
Paul, A. K., Chatterjee, S., & Paria, S. (2005). Morphodynamics and vulnerability issues of Nayachar Island in the Hugli Estuary section of Ganga Delta. *Indian Journal of Geomorphology and Environment, 10*, 76–92.
Paul, A. K., Ray, R., Kamila, A., & Jana, S. (2017). Mangrove degradation in the Sundarbans. In *Coastal wetlands: Alteration and remediation* (pp. 357–392). Springer. https://doi.org/10.1007/978-3-319-56179-0

Chapter 17
Assessment of Hazards and Flood Risks in the Southwestern Sundarbans

Satyajit Dhara and Ashis Kumar Paul

17.1 Introduction

Many of the world coastal zones undergo continuous adjustment towards a dynamic equilibrium. The world's coast lines are now exposed to a variety of hazards by human activities. It has been estimated that 23% of the world's population lives both within 100 km distance of the coast and less than 100 m above mean sea level and population density in the coastal regions are about three times above the average (Small & Nicholls, 2003). The coasts are experiencing the adverse consequences of hazards related to climate change and sea level rise threats. The lack of resilience means an expression of the limitations of access and mobilization of the resources of human settlement and its incapacity to respond when it comes to absorbing the impact (Cardona, 2011). The low-lying coastal zones of West Bengal are inhabited by over 4.5 million population (2011 Census) particularly in the Indian Sundarbans.

The lower portion of the southwestern part of the Ganga delta where Hugli River entered the Bay of Bengal is known as Hugli estuary. The lower portion of the Saptamukhi River is known as Saptamukhi estuary. Due to the deposition of the sediments at the mouth of the river Ganga, large deltaic islands are developed in the region. The mouth of the Hugli River and Saptamukhi River jointly with many other rivers of the region comprises many island blocks. Hugli-Saptamukhi estuarine deltaic complex is one of the dynamic geomorphological and ecological areas of the West Bengal coast (Paul, 1990, 2000, 2002; Paul & Kamila, 2018).

The Indian Sundarbans is geographically distributed in the lower Ganga delta plain along with many other islands. The increased population in the low-lying

S. Dhara
Acharya Jagadish Chandra Bose College, Kolkata, West Bengal, India

A. K. Paul (✉)
Department of Geography, Vidyasagar University, Midnapore, West Bengal, India
e-mail: akpauleastcoast@gmail.com

A. K. Paul, A. Paul (eds.), *Crisis on the Coast and Hinterland*,
https://doi.org/10.1007/978-3-031-42231-7_17

coastal plains of West Bengal with the direct impact in human modifications of the landscapes, the natural delta building process, as well as the coastal ecosystem is very much affected. Presently, the global warming impacts with sea level rise threats, increased salinity in river and sea water, rate of coastal erosion and shoreline shift with many other factors make this area the most hazard prone and vulnerable tracts. Here the coastal communities are exposed to varieties of hazards. The present research work along the Hugli-Saptamukhi estuarine deltaic complex is carried out to identify the hazard exposure, and vulnerability relationship (Paul et al., 2016, 2017, 2018). If the intensity and frequency of hazards increase, the human community will be exposed to hazards. In this way the degree and the level of community vulnerability will also be increased. Therefore, the coastal hazards and community vulnerability are the two major aspects considered for the present research. Assessments of different types of hazards, its impact, and management are a big challenge for the planners to mitigate the coastal hazards and community vulnerability in the low-lying coastal tracts (Paul et al., 2005; Sahana et al., 2021; Paul, 2022, Paul & Paul, 2022).

17.2 Study Area

The coastal low-lying area has been reclaimed from the tidal water levels in the Sundarban mangrove forests during the early part of the nineteenth century.

This region is bounded by the Hugli River mouth in the west, the Saptamukhi River mouth estuary in the east, and the Bay of Bengal in the south (Fig. 17.1). The existing deltaic and estuarine tracts are drained by tidal creeks and tidal rivers and separated into islands. The regions under study include community development blocks and a few forested islands in the South 24 Parganas district of West Bengal.

17.3 Materials and Method

For the spatiotemporal analysis of the hazards, pre-field, field, and post-field methods are followed. CD block maps and hazard-related data are collected from the Census Department and Irrigation Department of West Bengal. The report on the district disaster management plan helped to identify the hazard-affected areas of the study; but the field data was collected through a questionnaire survey during the period of fieldwork in the region. The Survey of India (SOI) toposheets are mosaiced using the Earth Resources Data Analysis System (ERDAS) IMAGINE 9.2 software to carry out the study. The hazard map is prepared based on the hazard parameters for the region.

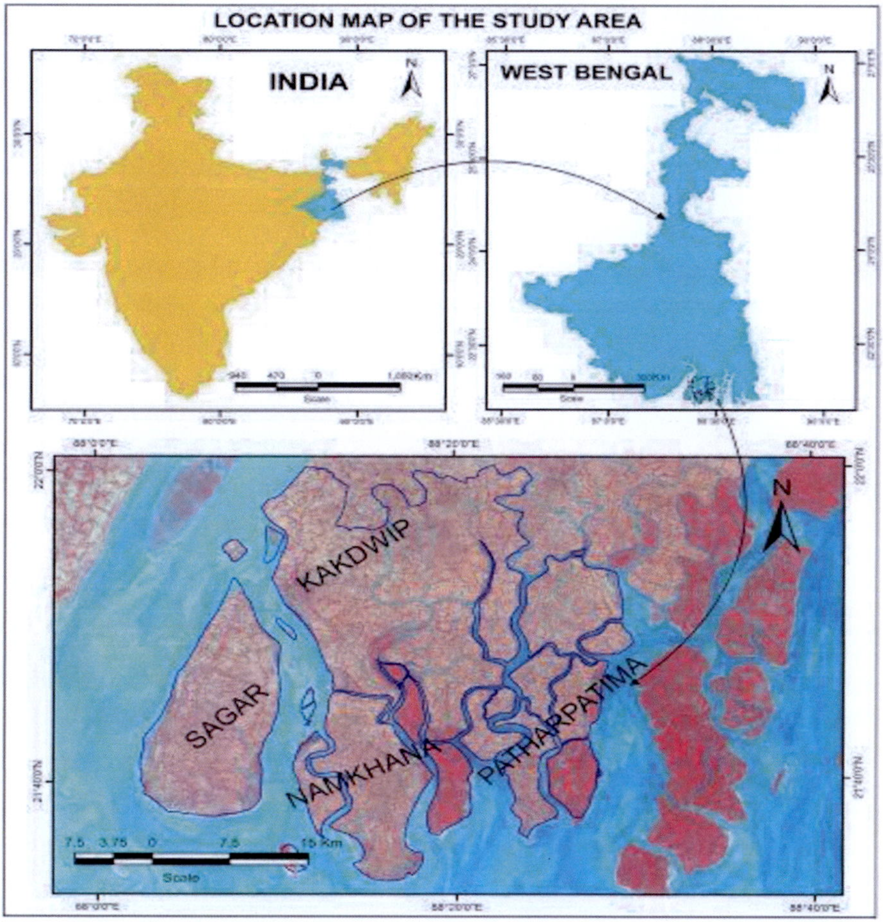

Fig. 17.1 The southwestern part of the Sundarbans with Hugli-Saptamukhi estuarine complex

17.4 Results and Discussion

17.4.1 Identification of Hazards

The study area is situated at the head of the funnel-shaped Bay of Bengal, and the southern part of this area is very much influenced by the coastal environment. The Sundarbans of the lower Ganga Delta Plain is located in the form of islands and tidal river mouths with tidal creeks and, thus, is very much exposed to a variety of hazards. The administrative blocks of the region occupy the deltaic islands of Kakdwip, Sagar, Namkhana, and Patharpratima in between the Hugli-Saptamukhi estuarine deltaic complex. The hazards that severely influence the region can be classified into two main classes: natural hazards and man-made hazards. The very common natural hazards affecting the entire coastal area of the study are cyclones with storm surges, tidal surges, coastal floods, and embankment breaching.

17.4.2 Cyclones

The cyclone is one of the most devastating types of natural hazards that turn into disasters in the coastal community blocks of West Bengal. In the previous decades, the frequency and damage potential of cyclones have increased to a great extent. In every year, from March to November, tropical cyclones affect the coastal blocks of three coastal districts in West Bengal.

The cyclone data from 1582 to 2021 represents the event frequencies and recurrence interval which are indicative of high cyclone vulnerability with reduction in recurrence interval for the case of the Indian Sundarbans (Fig. 17.2). The repeated cyclone hazards will aggravate the situation of human vulnerability in the region of the Indian Sundarbans in the coming years with rising sea level.

17.4.3 Coastal Flooding

Coastal flooding is a main problem for the dwellers of the southern part of the Indian Sundarbans. During the period of a strong tropical cyclone, a high tidal surge causes embankment breaching and the failure of sea walls. As a result, coastal flooding occurs in the low-lying area behind the breached embankments. More than six million people are at risk due to increased flooding and storm impacts (District Disaster Plan Report-DDPR, 2018). The high saline water table and poor drainage conditions of the region encourage the flood water to spread over large areas.

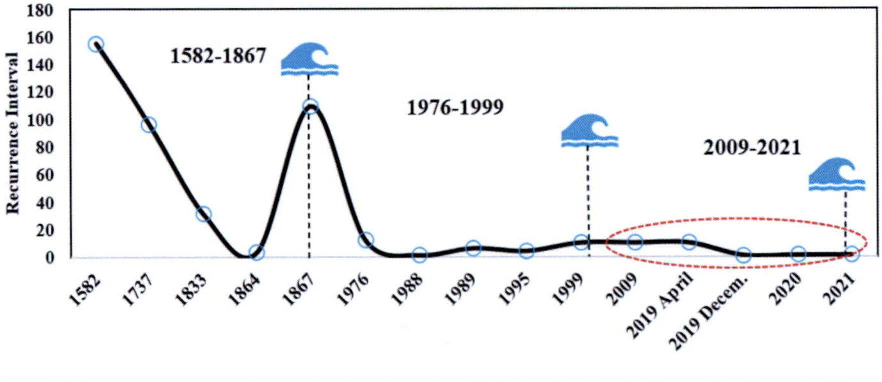

Fig. 17.2 The recurrence interval of cyclone landfalls in the Sundarbans showing significant reduction in the interval period indicating the hammering effects of energy levels into the environment

Sometimes coastal flooding occurs due to the coincidence of onshore strong winds with the spring high tides.

Water logging incidents happen when water overflows the river banks and enters the low area behind the embankments. During the rainy season, at some places, water stays in a stagnant condition for several days. During low tide, the local people cut the embankments and release the stagnant water into the river, and that causes a weaker condition of the embankments in the region. However, due to the absence of a sufficient number of sluice gates across the embankments, the drainage network is not in proper condition to release the excess water immediately. In Namkhana and Patharpratima blocks, such incidents are a very common phenomenon.

17.4.4 Impact of Hazards on Southwest Sundarbans

Natural disasters, especially hydro-meteorological hazards like cyclones and storm surges, frequently occur in the South 24 Parganas. The stretch of the Indian Sundarbans extends over an area of 9360 km^2 comprising 102 islands, of which 52 have human settlements. The coastal areas of the South 24 Parganas are inundated during storm surges and tidal surges as the area lies at the apex of the funnel shaped Bay of Bengal. Tropical cyclonic depressions lead to the occurrence of frequent cyclonic storms. Storm-induced wave surges causing tidal floods are another potential disaster for the local people. A much-talked-about disaster in recent times is the coastal erosion induced submergence of islands, especially in the Sagar block. The tidal movement has already engulfed Lohachara, and Suparibhanga and portions of Ghoramara Islands in the Sagar Block (Paul 2002). According to Ghosh et al. (2003, 2014), Ghoramara Island has already lost 75% of its area in the past 30 years. Environmental refugees are moving away to the adjacent islands and administrative blocks like Kakdwip and Diamond Harbour. The earlier researchers predicted that at the present rate of bank erosion, at least 30,000 people will be displaced from Sagar and another 2000 from the Namkhana block by the year 2020, and the western estuarine deltaic islands like Ghoramara will be completely disappeared (Paul 2022, Paul & Paul 2022). From the year 2001 to 2015, Sagar Island lost almost 7 km^2 than that of its previous area. Namkhana lost almost 9 km^2 of area, and Ghoramara Island lost 2 km^2 as well. According to the present estimation, the vulnerability to natural disasters like floods and cyclones is extremely severe (7 out of 13 blocks in the Sundarban area). Analysing the records of incidence of tropical cyclones in the Sundarbans from 1582 to 2014, we find that on average, strong cyclonic depressions occur from April to November (Disaster Report-DDRP South 24 Parganas, government of West Bengal, 2013).

17.4.5 Embankment Breaching

Embankments are very crucial for the existence of human settlements in the Indian Sundarbans. The total length of embankments in the Indian Sundarbans is 3500 km. Most of the embankments were constructed during British rule on pre-mature land to protect the human settlements in the region (Fig. 17.3). The earthen embankments are 150 years old and are being weakened day by day by the swirling currents that scour at their bases and by tidal surges coupled with strong winds and cyclones. During rainy season and Nor'wester events, embankment breaching takes place mostly in the reclaimed islands. The distributaries of the Ganges are fed by sea tides twice daily. The sea water enters more than 100 km inland through these estuaries and inundates the low-lying plains. Tidal effects, seasonal flooding, and storm impacts cause embankment breaching all along the river bank.

On May 25, 2009, the Aila cyclone hit the whole Sundarbans. The worst affected blocks were Kakdwip, Sagar, Namkhana, Patharpratima, Gosaba, Basanti, Sandeshkhali 1 and 2, Kultali, Haroa, Minakhan, and many other blocks of the coastal Bay of Bengal. High-speed winds, tidal surges, and storm surges damaged the embankment of this area, with 99.2 km of embankment damaged in the four blocks under Kakdwip subdivision, and the extent of the damaged embankment was

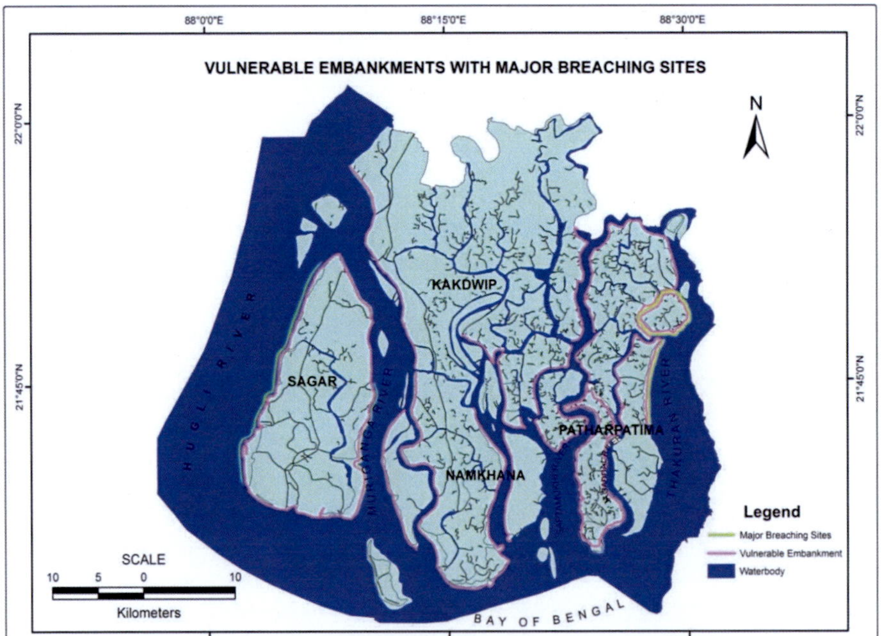

Source:SOI Toposheet,Irrigation Division (Joynagar,kakdwip,South 24 Pargana)

Fig. 17.3 The vulnerable embankments along the shorelines with active breaching sites registered in the southwestern part of the Indian Sundarbans after the impacts of cyclone Aila (2009)

highest at Patharpratima block. During the Aila cyclone in 2009, nearly a 778 km stretch of embankments were completely destroyed, and the saline water encroachment caused heavy damage to agricultural fields, fishing firm plots, and other man-made structures.

17.4.6 Coastal Erosion

During the storm events, the raised water level and high energy wave environment, or extreme wave events, affected the upper layer of shoreline sands along the inland river banks. Sundarbans is compared with polder land as well as reclaimed areas that are lying below the level of high tide water, which enters the interior part of the Sundarban delta plain twice daily through the tidal channels and rivers.

Many channels are choked with the high rate of siltation in the channel beds and are thus unable to hold the massive ingress of seawater in the cyclones. Partially constructed embankments and the weaker structure of the embankments are promoting bank erosion in many areas of the Patharpratima, Sagar, and Namkhana blocks. Some bank erosion is quite severe, and under natural conditions it is controlled by bank side vegetation loss (Fig. 17.4). In mangrove-dominated river banks, the roots of the plants hold the soil together and maintain the banks' stability. The local people are clearing the mangrove forests to increase the space for agriculture practices and to use wood as fuel. In this way, many river banks are affected as erosion of embankments takes place (Table 17.1).

All the lowlands of coastal West Bengal are fed by large estuaries, tidal inlets, and channels that help surges originating in the bay penetrate deep inland during larger cyclones. In May 1995, when a cyclonic storm coincided with the spring high tide, the bank receded by 8–10.5 m at Sagar Island (Shibpur and Dhablat areas). It is observed that damages to river banks and adjacent embankments caused by wave dashes are far less than the magnitude of destruction caused by currents at the bed of a river. High wind speeds caused by cyclones increase wave activity and change the current pattern of rivers. Thus, unconsolidated materials produced by fresh silt accumulation over the river banks are easily erodible by waves and the changing current pattern during cyclones. At Dhablat and Beguakhali villages of Sagar block, the amount of soil salinity before the Aila cyclone was lower than 3000 ppm, but after the Aila cyclone, the area was inundated by sea and river waters, which increased the amount of salinity in the soil up to 6000 ppm (moderately saline) at present.

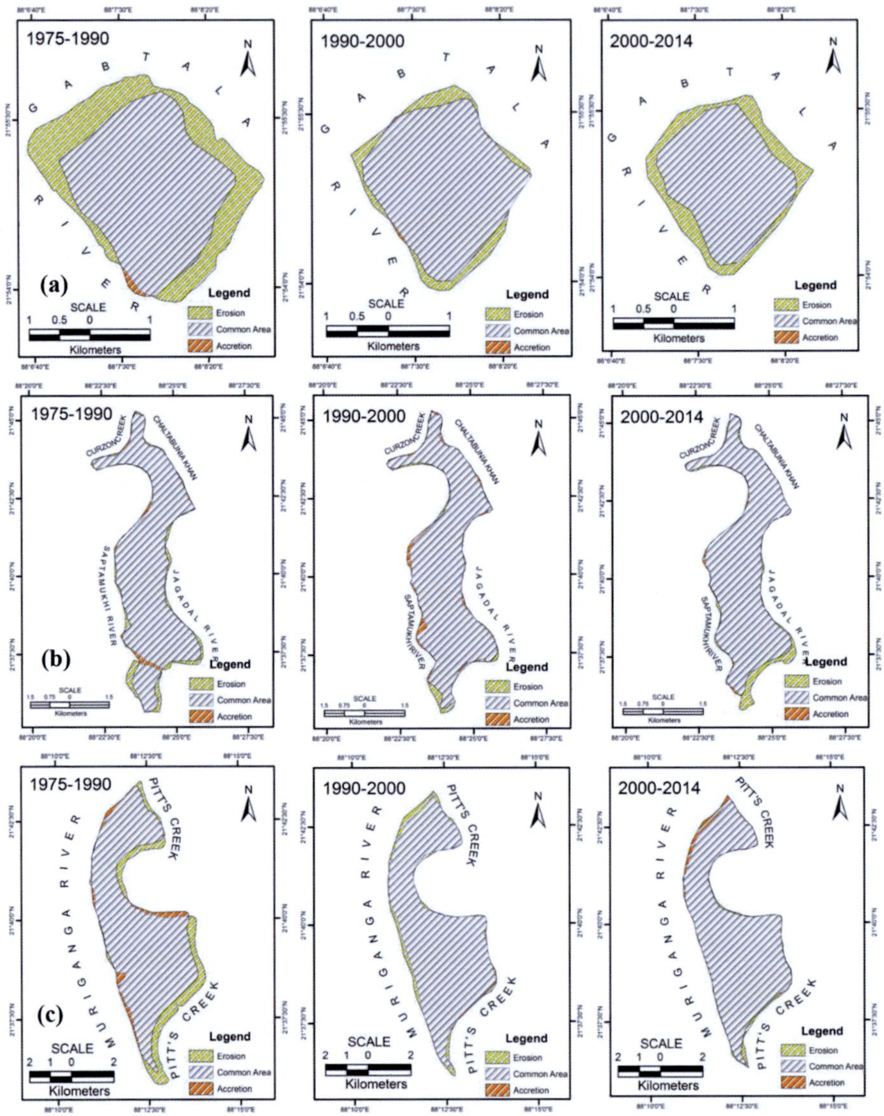

Fig. 17.4 Changing configurations of three major islands. (**a**) Ghoramara Island. (**b**) G-Plot & c. Mousuni Island of the southwestern Sundarbans represent the impact of cyclone hazards and the nature of energy levels in the deltaic and estuarine coastal regions

Table 17.1 The detection of net change area in the islands of the southwestern Sundarbans (1975–2014)

G-plot				Mousuni Island			
Year	Accretion (m/year)	Total area (km²)	Erosion (m/year)	Year	Accretion (m/year)	Total area (km²)	Erosion (m/year)
1975–1990	1.131	41.303	3.194	1975–1990	1.334	28	4.77
1990–2000	1.888	41.009	1.396	1990–2000	0.337	27	1.946
2000–2014	0.571	39.898	3.007	2000–2014	0.955	26	0.596
Ghoramara Island					Net Change Area (1975–2014)		
Year	Accretion (m/year)	Total area (km²)	Erosion (m/year)	Year	G-plot	Mousuni Island	Ghoramara Island
1975–1990	2.3	7.334	12.05	1975–1990	1.405 km²	2 km²	1.97 km²
1990–2000	1.55	4.5	2.35	1990–2000			
2000–2014	1.22	3.55	10.08	2000–2014			

17.5 Conclusion

The low-lying coastal areas of West Bengal have enhanced vulnerability to high-energy events such as tidal waves and storm surges due to the physical settings of the coast and increased human activities in the delta catchment as well as in the coastal hinterland areas due to drainage obstruction. Major physiographic changes took place in the coastal areas of the region due to the influence of the previous cyclones of 1942–43, 1978, 1988–89, 1995, and 2009. Large tracts of sandy beaches were reduced in many places (Sagar, Fraserganj, Bakkhali, and Gobardhanpur) at the threshold water level of storm periods (when the beach became submerged) due to erosion caused by quarrying, winnowing, and abrasion forces. The problem of bank erosion on both sides of rivers, estuaries, inlets, and tidal channels became severe after the landfall of last major cyclones in the coastal belts. Channel beds are sufficiently silted up, on the other hand, which encourages the thalweg shifting to cause bank erosion even during the period of smaller cyclones at present. The instability of foredunes is produced by high-energy events associated with major cyclones. Such type of sand dunes are moving over the tidal marshes (at Sagar Island, Fraserganj, Bakkhali, Henry Island, and the western portion of Gobardhanpur Village), where the marsh deposits are exposed on the foreshores and eroded under wave attack in the storms. Shore fringe mangrove wetland areas are destroyed in many places by direct attack from these waves. Tidal inlets and creek mouths are eroded on their bank margins and widening during storm events. The areas of some islands are gradually decreasing in the estuarine coast, and the Sagar Island, Ghoramara, Mousuni, and Tat Island are the best examples of such incidents.

Active erosion of the estuarine banks, local channel banks, and shoreline of the seaward side, shallower river beds in the coastal areas, and changes in water courses in the study area appear as a slow disaster on the sensitive alluvial coast. All such geomorphological and environmental incidents generated environmental refugees in the coastal villages of the Hugli-Saptamukhi Complex. The Aila cyclone of 25 and 26 May 2009 was one of the most destructive cyclones, displacing thousands of people and forcing them to flee their homes in the low-lying Sundarban coastal tracts and exposing human vulnerabilities. About 778 km of embankments were washed out and devastated during the Aila cyclone, making the low-lying areas significant vulnerable flood zones. Later, in 2019 and 2020, the Amphan and Bulbul cyclone landfalls inundated a large tract of Sagar Island, Namkhana Abad, and Tat Island in the southwestern Sundarbans.

References

Cardona, O. D. (2011). Disaster risk and vulnerability: Concepts and measurement of human and environmental insecurity. In *Coping with global environmental change, disasters and security: Threats, challenges, vulnerabilities and risks* (pp. 107–121). Springer.

Census Report. (2011). *District Census Handbook, South 24 Parganas, West Bengal*, PART XII-A. https://censusindia.gov.in/nada/index.php/catalog/1362/download/4462/DH_2011_1917_PART_A_DCHB_SOUTH_TWENTY_FOUR_PARGANAS.pdf

DDPR. (2013). *District disaster management plan, South 24 Parganas 2013–14* (pp. 1–112). https://wbiwd.gov.in/uploads/anual_flood_report/ANNUAL_FLOOD_REPORT_2013.pdf. Assessed: 2022.

DDPR. (2018). *District disaster management plan, South 24 Parganas 2018–19* (pp. 1–252). http://wbdmd.gov.in/writereaddata/uploaded/DP/DPSouth%2024Parganas%20%20%20%20%20%20%20%20%20%20%20%205803.pdf. Assessed: 2022.

Ghosh, T., Bhandari, G., & Hazra, S. (2003). Application of a 'bio-engineering' technique to protect Ghoramara Island (Bay of Bengal) from severe erosion. *Journal of Coastal Conservation, 9*(2), 171–178.

Ghosh, T., Hajra, R., & Mukhopadhyay, A. (2014). Island erosion and afflicted population: Crisis and policies to handle climate change. In *International perspectives on climate change: Latin America and beyond* (pp. 217–225). Springer.

Paul, A. K. (1990). Identification of coastal hazards in West Bengal and parts of Orissa. *Indian Journal of Geomorphology, 1*(1), 1–27.

Paul, A. K. (2000). Cyclonic storm and their impacts on West Bengal coast. In V. G. Rajamanikam & M. J. Tooley (Eds.), *Quaternary sea level variation* (pp. 25–57). New Academic Publishers.

Paul, A. K. (2002). *Coastal geomorphology and environment* (pp. 1–342). ACB Publication.

Paul, A. K., Chatterjee, S., & Paria, S. (2005). Morphodynamics and vulnerability issues of Nayachar Island in the Hugli Estuary section of Ganga Delta. *Indian Journal of Geomorphology and Environment, 10*, 76–92.

Paul, A. K. (2022, August). Dynamic behaviour of the Estuaries in response to the phenomenon of global warming in the coastal ecosystems of West Bengal and Odisha, India. In *Transforming Coastal Zone for sustainable food and income security: Proceedings of the international symposium of ISCAR on coastal agriculture* (pp. 907–931). Springer.

Paul A.K., & Kamila A. (2018). *Studies on coastal morphometry with total station survey in shore face of Sagar Island to assess the impacts of sea level rise 38th INCA International Congress Hyderabad* (pp. 460–469).

Paul, A. K., & Paul, A. (2022). Adjustment of the coastal communities in response to climate variability and sea level rise in the Sundarban, West Bengal, India. In *Climate change, disaster and adaptations: Contextualising human responses to ecological Change* (pp. 201–217). Springer.

Paul, A. K., Jana, S., Kamila, A., Bal, A., Sultana, F., Soren, S., & Mondal, D. (2016). Problems of some erosion affected areas of Sagar Island and proposal for mangrove Ecotourisim development sites in a sustainable manner. *Eastern Geographer, XXII*(1), 569–582.

Paul, A. K., Ray, R., Kamila, A., & Jana, S. (2017). Mangrove degradation in the Sundarbans. In *Coastal wetlands: Alteration and remediation* (pp. 357–392). Springer. https://doi.org/10.1007/978-3-319-56179-0

Paul, A. K., Kamila, A., & Ray, R. (2018). Natural threats and impacts to mangroves within the coastal fringing forests of India. In *Threats to mangrove forests* (pp. 105–140). Springer.

Sahana, M., Rehman, S., Paul, A. K., & Sajjad, H. (2021). Assessing socio-economic vulnerability to climate change-induced disasters: Evidence from Sundarban biosphere reserve, India. *Geology, Ecology, and Landscapes, 5*(1), 40–52.

Small, C., & Nicholls, R. J. (2003). A global analysis of human settlement in coastal zones. *Journal of Coastal Research*, 584–599.

Chapter 18
Perils of Premature Reclamation: Case Studies from the Indian Sundarbans

Kanailal Das and Karabi Das

18.1 Introduction

The Sundarbans is the largest mangrove ecosystem in the world, and it is now under severe threat from resource exploitation and recurrent hazards and disasters. It has been confirmed from the dating of peat materials that the area was covered by an extensive mangrove patch. This region was transformed into an open tidal basin, which gradually got filled up by sediments accommodated in the generated space caused by subsidence. Further displacements of the Holocene shoreline (5000–3000 YBP) were recorded in the region through aggradations and progradation. Recent alluvium to very thick Tertiary sediments can be found in the Bengal Basin (Allison et al., 2003). The stratigraphic records found at Calcutta show peat beds and decayed woods containing the remains of *Heritiera* minor (Sundari), which are found at 9 m depth at Fort William and at 10 m depth at Sealdah. The peat beds with tree stumps of mangroves were also found at Canning, at the head of the River Matla (Paul, 2002; Paul et al., 2017).

The deltaic alluvial plain of the Sundarbans gently slopes towards the south and southeast (Paul, 1998, 2002; Paul et al., 2016) direction to the Bay of Bengal. The elevation of the plain ranges from 3.0 to 6.0 m above sea level and is characterised by low-lying depressions. The intertidal areas of the Sundarbans are bounded by the Dampier-Hodges line with the river Hugli in the west, the Bay of Bengal in the south, and Kalindi-Raimongal distributaries in the east. To the north, Dampier-Hodges connects Kakdwip to Basirhat. The major distributaries of the area include the Hugli system, the Matla-Bidyadhari system, and the Ichhamati

K. Das (✉)
Parameswar Mahavidyalaya, Namkhana, West Bengal, India
e-mail: kanailaldas17@gmail.com

K. Das
Department of Geography, Dr. Kanailal Bhattacharyya College, Howrah, West Bengal, India

A. K. Paul, A. Paul (eds.), *Crisis on the Coast and Hinterland*,
https://doi.org/10.1007/978-3-031-42231-7_18

Kalindi-Raimongal system, influenced by the macro-tidal environment of a tidal range of more than 4 m, and the deltaic low-lying areas experience tidal inundations. The tides enter the channels, spread over the channel banks, and inundate the inter distributary areas. The depth of the sediments remains maximum near the river banks, and with the increase in distance, the fresh silt content becomes less towards the inner part of the island (Paul, 1985, 2002).

Tidal amplitudes are found to range from 4.88 to 5.93 m (Sept. 1984) at the mouth of the Hugli River near Sagar, and this ranges from 4.41 to 6.80 m at Garden Reach near Calcutta. Whenever the spring tide coincides with cyclones, they lead to an unusual increase in the height of high tide (Fig. 18.1a). The climate of the Sundarbans is humid tropics, associated with wet summers and dry winters. Temperature ranges from 30–39 °C during summer and 10–18 °C during winter. Nor'westers are frequent during the pre-monsoon summer season and associated with the occurrences of thunderstorms and hailstorms. More than 80% of the annual rainfall takes place in the rainy season due to the southwest monsoon. During the southwest monsoon, the average wind speed ranges from 30 to 50 km/h during storms, the wind velocity exceeds 100 km/h. During winter, the weather becomes calm and the wind speed goes down to 10 km/h all along the coasts of West Bengal. The study highlights the problems of land reclamation and resultant drainage decay in the Indian parts of the Sundarbans (Paul & Paul, 2022; Paul, 2022). It is also estimated that there are various types of vulnerabilities in the low-lying Sundarban

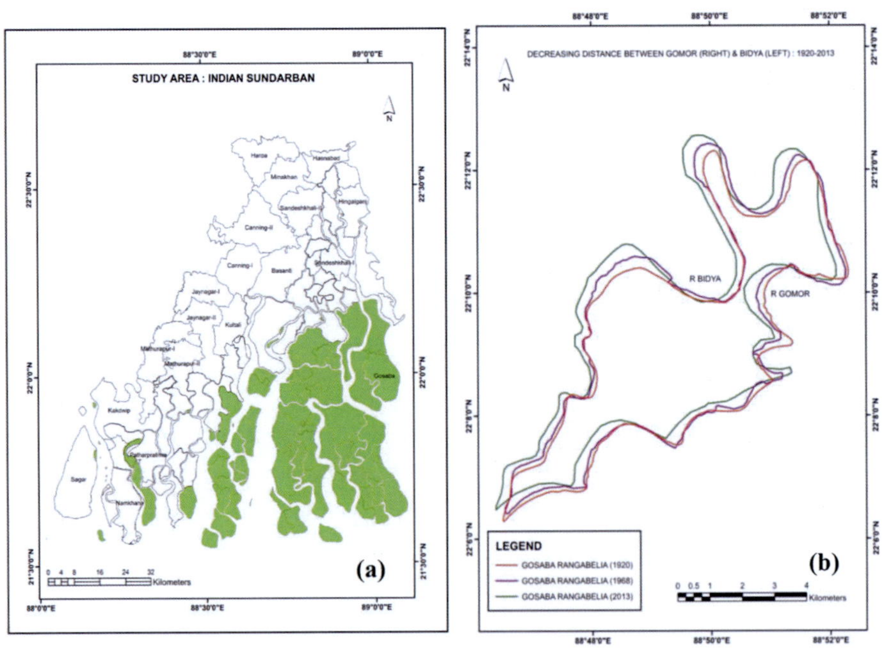

Fig. 18.1 (**a**) The reclaimed tracts and forested parts of the Sundarban Islands in India and (**b**) decreasing distance between rivers Bidya and Gomor

coastal tracts, in consideration of some mitigating strategies. The research is carried out in the reclaimed parts of the Indian Sundarbans to demonstrate the present-day impacts of the pre-mature land reclamations.

18.2 Methodology

The Survey of India (SOI) toposheets of 1920 and 1968–1969 were procured from the Survey of India in consideration with the Google Earth imagery of 2021 to identify the drainage discontinuities and drainage decay of the Indian Sundarbans. The temporal changes are documented with the overlays of the vector data using Quantum Geographic Information System (QGIS) software. Various types of vulnerability were estimated using the aspect data on population at risk, evacuation gaps, and the topographic characters from the report of the District Disaster Management Plans, South 24 Parganas, for the present study. For the calculation of the locational vulnerability and insularity index of the study area, the length of riversides is measured using QGIS software with the Google Earth imagery of 2021.

18.3 Results and Discussion

18.3.1 Settlement History of the Sundarbans

Discontinuation of the settlement in the historical phases is probably responsible for the occurrences of natural disasters such as cyclones, subsidence, flooding, and earthquakes. The reclamation of the forest area of the Sundarbans started with the activities of the British East India Company for the expansion of trade and commerce. In accordance with the plan of Claude Russell, the then Collector General, 1771, the land area was leased to the landlords of the Sundarbans for seven consecutive years without rent on the assessment survey of the land. The next major attempt to clear forest cover for settlement and economic activities was made by Mr. Tillman Henckell, the Magistrate of Jessore in 1783, who submitted a proposal of granting a lease of forest to the people to convert the forests of the Sundarbans into agricultural land. In the Sundarbans, three types of lands were reclaimed: (i) land reclaimed by broader zamindars; (ii) land reclaimed by the Patitabadi talukdars in excess of the quantity for which they paid revenue; and (iii) land reclaimed by persons without title (Ascoli, 1921; Pargiter, 1934).

These steps taken for the reclamation of these lands were: (i) demarcation of the lot surrounded by the streams; (ii) embankment construction along the bank of the stream surrounding the lot; (iii) subsequent construction of strong dams across the mouths of smaller streams to keep the salt water out; and (iv) after completion of the above work, clearing the forest, digging tanks, and constructing huts for the

Table 18.1 Various stages of land reclamation in the Sundarbans from 1770 to 1951

Time span	Areas reclaimed
1770	Haroa, Bhangar, Kulpi, Hasnabad, Canning, Baruipur, Jaynagar, Patharpratima
1770–1873	Hasnabad, Haroa, Bhangar, Kulpi, Minakhan, Canning, Jaynagar, Mathurapur, Sagar
1873–1939	Whole of Sandeshkhali, Kakdwip, residual portions of Sagar and Canning, greater parts of Namkhana, Patharpratima, Basanti, Mathurapur, Kultali, Hingalganj, Gosaba
1945–1951	Some additional settlements came up at Hingalganj, Gosaba, Basanti, Kultali, Mathurapur, Patharpratima, and Namkhana. Rehabilitation of East Pakistan refugees in portions of Hingalganj, Gosaba, Satjelia, Kultali, Patharpratima, Namkhana, Sagar, Mousuni Island, and Jharkhali (Basanti)
After 1951	Satjelia Island, Marichjhanpi (restored mangrove forest, 1979)

Modified after Mukherjee (1969, 1984)

introduction of cultivation (Table 18.1). In this connection, John F. Richards showed in 1990 that total wetlands, comprising both tidal mangrove swamps and surface water, between the 24 Parganas (India) and Bakarganj and Khulna (Bangladesh), declined by 2750 km^2 between 1880 and 1940. And as the pace of reclamation quickened between 1940 and 1980, about 5230 km^2 of wetlands got reduced. The remaining 10,000 km^2 of the Sundarbans came under government status by 1947, of which 60% went to Bangladesh and 40% to India after independence (Paul, 2002).

18.3.2 Problems of Premature Reclamation

As a result of reclamation, small creeks were blocked and tidal water was not allowed to enter the islands, thus most of the areas of the estuary that would submerge under 25 or 50 cm of shallow water could not get inundated further. For more than 200 years, these embankments were successful in preventing the entry of tidal water and sediments, but as a result, the settlements remained at a lower level than the level of the water's surface. The sediments eroded from the southern islands got deposited on the riverbed itself as the spill areas of rivers got reduced. Similarly, the unplanned reclamation for expansion of human settlements has resulted in disruption of the land-building process. The estuary tries to breach the embankments to regain the lost spill areas and starts to deposit sediments on the bed itself to reduce the depth in the restoration of the estuarine equilibrium. As a consequence of the reclamation of premature lands, subsidence, alteration in aquifers, poor drainage conditions, ponding effects of storm water, and saltwater breaches increase the hazards in the polderised landscapes (Paul, 1991, 2002). The natural subsidence was compensated by the vertical accretion of tidal silt and the overflow of silt accumulated in wetlands or swamps. However, subsidence occurs without compensatory siltation over protective areas, but wave dashing and tidal currents are restoring the wetlands through the erosion of beaches, coastal dunes, and river banks, as well as through embankment failure and saltwater inundation. The present-day sea level rise and diversity in land subsidence rates have caused many environmental problems for the reclaimed Sundarbans (Hazra et al., 2002). The Rangabelia-Uttardanga Island Unit of the Gosaba block has registered a rapid rate of erosion due to the channel bank retreats between rivers Bidya and Gomor, which have been decreasing from 731.55 m (1920) to 360.56 m (2020) in the 100-year time gap.

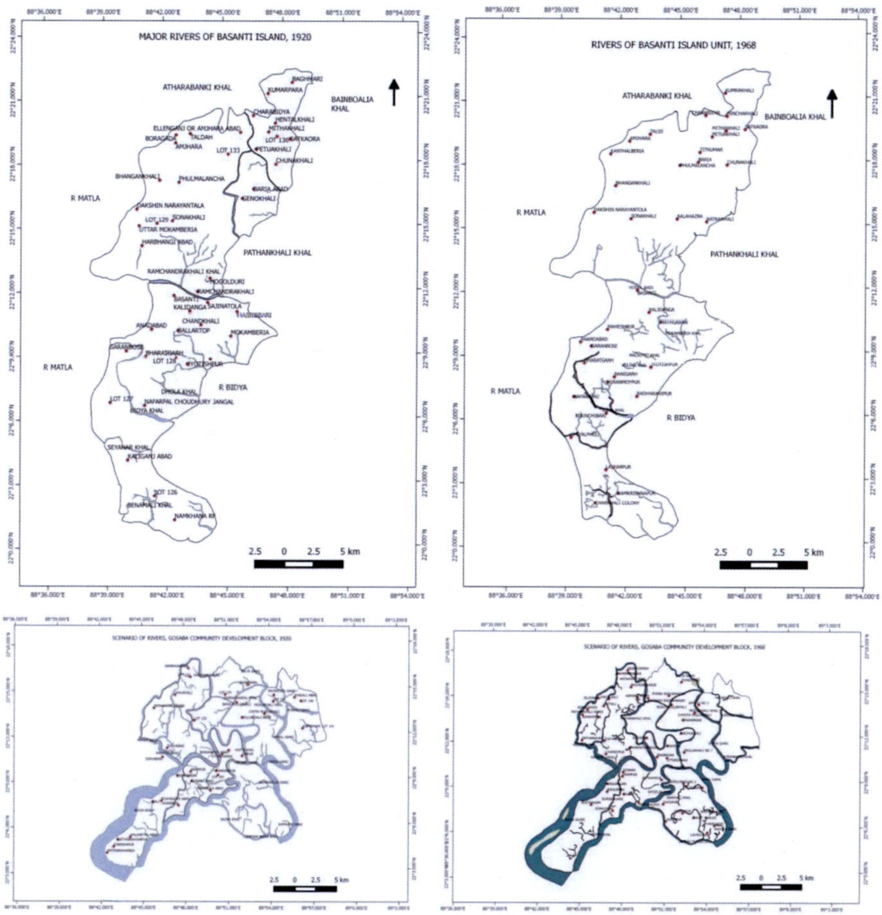

Fig. 18.2 Drainage decay in the temporal span from 1920 to 1968 (SOI Toposheets) in Basanti and Gosaba blocks of the Indian Sundarbans

18.3.3 Drainage Decay and Discontinuity

Owing to both natural and anthropogenic causes, the scenario of river decay has become persistent. The funnel-shaped macrotidal Hugli Estuary is susceptible to time velocity asymmetry, which has increased in-channel sedimentation (Figs. 18.1b and 18.2). As the intra-island rivers get silted up, the big rivers keep on eroding both the banks and widening the channels. This scenario was observed in the case of the river Muriganga. Degeneration of the creek joining rivers Matla and Durgaduani near Bijoynagar, Bali, was seen between 1920 and 1968, which further deteriorated in 2021. All the intra island rivers from Sagar Island to Gosaba-Basanti Sector have degenerated between 1920 and 1968 and have further deteriorated in 2021. River decay has been observed due to both natural and anthropogenic stresses, and these cases of river decay have accelerated fluvial erosion at present (Paul & Kamila, 2018).

18.3.4 Vulnerability

Physical vulnerability involves whatever in the built environment is at risk of being affected. The construction of resistant structures in hazard-prone areas shall decrease the physical vulnerability (Paul, 1996, 2000). Social vulnerability is the measure of the societal, political, and cultural factors that control the vulnerability of a group or the population that is economically vulnerable. Environmental vulnerability indicates the health and quality of the environment, which play an important role in increasing or decreasing the vulnerability (Paul, 1990). Physical vulnerability is assessed, taking into account the population likely to be exposed to various hazards. For this, the population at risk was considered in the study area. Mouza wise physical vulnerability has been calculated for the community development blocks of Gosaba and Basanti. This has been calculated by finding out the population that can be evacuated at the time of any hazard in that area using the data procured from the district disaster management plans of South 24 Parganas. This population has been subtracted from the total population to find the population at risk (Fig. 18.3bi–bii). The population at risk has been divided by the total population to determine physical vulnerability (Bandyopadhyay, 1997; Das, 2015). Most of the community development blocks, located in the coastal fringes of the Sundarban Biosphere Reserve, have ranked high in vulnerability as a result of high exposure and sensitivity with a low level of adaptation. The coastal fringes are badly affected by loss of assets, and low health status, less access to safe drinking water, low income, and low infrastructural development are the causes of high vulnerability. Patharpratima, Gosaba, Namkhana, and Kultali recorded very high exposure, but Patharpratima, Namkhana, Kultali, Gosaba, and Basanti ranked high in sensitivity. However, Hasnabad, Sandeshkhali II, Namkhana, Gosaba, Patharpratima, Hingalganj, and Kultali showed low adaptation (Paul, 1990; Sahana et al., 2021).

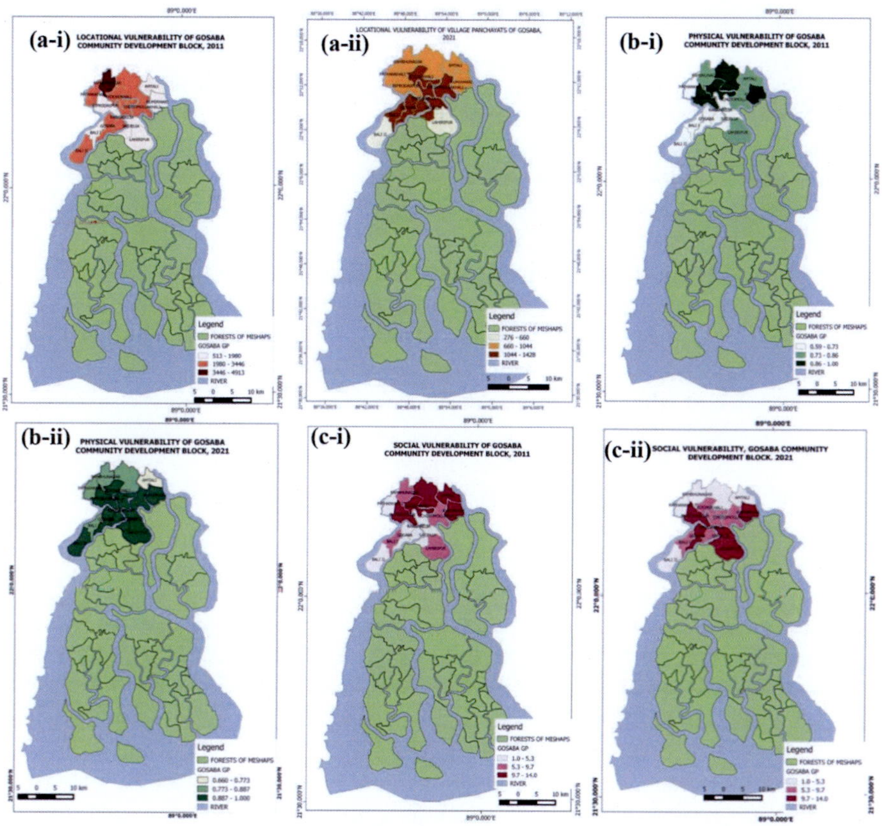

Fig. 18.3 (**ai–aii**) The locational, (**bi–bii**) physical, and (**ci–cii**) social vulnerabilities assessed for the Gosaba Community Development Block of the Indian Sundarbans

The physical vulnerability maps prepared for the year 2011 show the case of Gosaba, in which this measure is highest for the three villages where the physical vulnerability is 1 (one), leaving the total population completely at risk. This proves that low infrastructural development in these villages cannot combat the risks faced by these areas. In another way, the village of Pathankhali scores a zero level of risk (the lowest), which shows Pathankhali is armed with the best possible resources to combat the hazards experienced, as the Pathankhali College and other concrete buildings can be used for evacuation, which decreases its vulnerability. In the case of Basanti, four villages out of 37 have the highest physical vulnerability, which is showing as a 1 (one) risk score. This indicates that Basanti's major problem is a low level of infrastructural development, which is a hindrance towards combating hazards. Physical vulnerability for the year 2021 shows that the lowest vulnerability was observed in Amtali, moderate vulnerability was observed in five villages, and the highest vulnerability was observed in nine villages in the Gosaba block. In case of physical vulnerability (Fig. 18.3bi–bii), Kumirmari ranks the highest, the lowest physical vulnerability was observed in Radhanagar and Taranagar village panchayat areas.

The strategic location of a place casts its influence on the vulnerability of the residents. The site and situation of any settled area are thus very important when it comes to consideration of the hazards faced and the degree of vulnerability of the residents. Gosaba and Basanti, two blocks of the Indian Sundarbans, are constantly in the grip of natural hazards as the location of these two blocks is influenced by the rivers. Thus, the lengths of the river sides are computed, and the areas of the mouzas have been taken into account to find the insularity index. After that, the insularity index was multiplied with the population density of the mouzas to find the locational vulnerability index of the mouzas (Das, 2015). The locational vulnerability was mapped according to the village panchayats of Gosaba block for the year 2011 (Fig. 18.3ai–aii), in which the lowest locational vulnerability was observed in six villages. Moderate locational vulnerability was observed in eight villages in Gosaba. The highest locational vulnerability was found at Sambhunagar village panchayat. In the case of locational vulnerability in the year 2021, Lahiripur and Bali II village panchayats have recorded low vulnerability. Moderate vulnerability was recorded by seven village panchayats. Similarly, the highest vulnerability was recorded by the village panchayats of Bali-I, Gosaba, Rangabelia, Satjelia, and Chotomollakhali villages.

The social vulnerability score (Fig. 18.3ci–cii) was given according to the population that can be evacuated at the time of hazards. In the case of social vulnerability in 2011, the lowest vulnerability was observed in five villages; moderate vulnerability was observed at Sambhunagar, Bali-I, Chotomollakhali, and Lahiripur villages. In the case of social vulnerability in 2021, the lowest vulnerability was observed in six villages, and moderate vulnerability was observed at Kochukhali, Chotomollakhali, Rangabelia, and Bali-I villages. Highest vulnerability was observed at Bipradaspur-Satjelia, Gosaba, Lahiripur, and Kumirmari villages, but in the case of social vulnerability, Kumirmari ranks the highest both in 2011 and 2021.

18.4 Conclusion

The Indian part of the Sundarbans is truly vulnerable at present and is experiencing frequent cyclones (Aila in 2009, Bulbul in 2019, Amphan in 2020, and Yaas in 2021), subsequent saline water intrusion, and water logging. As a result of this increasing vulnerability, the people of the Indian Sundarbans are demanding a permanent solution in the form of concrete embankments in order to combat fluvial erosion and inundation. There are two types of embankments in the Sundarbans: river dykes and sea dykes. Even concrete embankments could not save Dhablat, near the confluence of Baratala on Gangasagar Island. High and wide embankments along Baliara on Mousuni Island have been washed out at a regular interval. It is realised that very high concrete embankments cannot arrest erosion for long but can protect settlements from erosion for 2–4 years in the region. Thus, where the embankments have survived, the channel fringe areas are aggrading and are colonised with mangroves.

Based on the above study, a few recommendations in the following way may be useful for the revival of the drainage systems and island aggradation characters: (i) the Sundarban river fringe areas were aggrading nature before the construction of the embankments; (ii) the alternative concrete embankments that can withstand erosive forces and can be cost effective for river bank management; (iii) the channel fringe area should restore the mangrove-dominated buffers, and the island interior region needs restoration of the aggradation process by the Tidal River Management (TRM) method; (iv) the excavation of creeks will revive the natural flow, which will reduce or prevent the water logging in the island interior sites; (v) the historical Gherpukur structure is needed for each island to rescue and rehabilitate the disaster-affected people in the low-lying islands; and (vi) the knowledge of local people should be applied in the construction of embankment structures and other protective structures along the island banks.

References

Allison, M. A., Khan, S. R., Goodbred, S. L., Jr., & Kuehl, S. A. (2003). Stratigraphic evolution of the late Holocene Ganges–Brahmaputra lower delta plain. *Sedimentary Geology, 155*(3–4), 317–342.

Ascoli, F. D. (1921). *A revenue history of Sundarbans from 1870 to 1920*. Bengal Secretariat Book Depot.

Bandyopadhyay, S. (1997). Natural environmental hazards and their management: A case study of Sagar Island, India. *Singapore Journal of Tropical Geography, 18*(1997), 20–45.

Das, K. (2015). *Assessment of coastal vulnerability in the deltaic reclaimed Sundarban for disaster management, a case study at Gosaba – Basanti Island Units, of South 24 Parganas District*. West Bengal.

Hazra, S., Ghosh, T., Dasgupta, R., & Sen, G. K. (2002). Sea level and associated changes in the Sundarbans. *Science and Culture, 68*(9–12), 309–321.

Mukherjee, K. N. (1969). Nature and problem of neo-reclamation in the Sundarbans. *Geographical Review of India, 31*(4), 1–20.

Mukherjee, K. N. (1984). *History of land reclamations in the Sundarban* (pp. 1–14). Indian Journal of landscape Ecology and Ekistics.

Pargiter, F. E. (1934). *A revenue history of the Sundarbans from 1765 to 1870*. Bengal Government Press.

Paul, A. K. (1985). *The development of accumulation forms and erosion in the coastal tracts of Balasore, Medinipur and 24 Parganas districts of Orissa and W.B* (pp. 19–32). Indian Journal of Landscape Systems and Ecological Studies.

Paul, A. K. (1990). Identification of coastal hazards in West Bengal and parts of Orissa. *Indian Journal of Geomorphology, New Delhi, Academy and Law serials, 1*(1), 1–27.

Paul, A. K. (1991). Effective management strategies for the coast of West Bengal. *Journal of Geographical Review India Calcutta, 53*, 60–74.

Paul, A. K. (1996). Identification of coastal hazards in West Bengal and parts of Orissa. *Indian Journal of Geomorphology, 1*(1) New Delhi, Academy and Law serial, 1–27.

Paul, A. K. (2000). Cyclonic storm and their impacts on West Bengal coast. In V. G. Rajamanikam & M. J. Tooley (Eds.), *Quaternary Sea level variation* (pp. 25–57). New Academic Publishers.

Paul, A. K. (2002). *Coastal geomorphology and environment* (pp. 1–342). ACB Publication.

Paul, A. K. (2022, August). Dynamic behaviour of the estuaries in response to the phenomenon of global warming in the coastal ecosystems of West Bengal and Odisha, India. In *Transforming coastal zone for sustainable food and income security: Proceedings of the international symposium of ISCAR on coastal agriculture* (pp. 907–931). Springer.

Paul A.K. & Kamila A. (2018). *Studies on coastal morphometry with total station survey in shore face of Sagar Island to assess the impacts of sea level rise 38th INCA International Congress Hyderabad* (pp. 460–469).

Paul, A. K., & Paul, A. (2022). Adjustment of the coastal communities in response to climate variability and sea level rise in the Sundarban, West Bengal, India. In *Climate Change, disaster and adaptations: Contextualising human responses to ecological Change* (pp. 201–217). Springer.

Paul, A. K., Jana, S., Kamila, A., Bal, A., Sultana, F., Soren, S., & Mondal, D. (2016). Problems of some erosion affected areas of Sagar Island and proposal for mangrove Ecotourism development sites in a sustainable manner. *Eastern Geographer, XXII*(1), 569–582.

Paul, A. K., Ray, R., Kamila, A., & Jana, S. (2017). Mangrove degradation in the Sundarbans. In *Coastal wetlands: Alteration and remediation* (pp. 357–392). Springer. https://doi.org/10.1007/978-3-319-56179-0

Paul. A. K. (1998). *Morphoecological management of the Sundarban, India*. Symp. Vol. Geomorph Env. Manage. Allahabad Geog. Soc. Allahabad (pp. 668–682).

Richards, J. F. (1990). Agricultural impacts in tropical wetlands: Rice paddies for mangroves in South and South East Asia. In M. Williams (Ed.), *Wetlands- a threatened landscape* (pp. 140–180). Balckwell.

Sahana, M., Rehman, S., Paul, A. K., & Sajjad, H. (2021). Assessing socio-economic vulnerability to climate change-induced disasters: Evidence from Sundarban biosphere reserve, India. *Geology, Ecology, and Landscapes, 5*(1), 40–52.

Chapter 19
Inventory of Landforms and Geomorphosites for the Promotion of Geotourism in South Andaman Island, India

Swati Ghosh, Ashis Kumar Paul, and Dipanjan Das Majumdar

19.1 Introduction

The landforms of the Andaman and Nicobar Islands are very much attractive to the visitors for their geomorphological significances and scenic beauty along the Andaman Sea and Bay of Bengal. South Andaman is an administrative unit of district in which the main Andaman ridges and valleys and islands of the Andaman Sea and the Bay of Bengal are isolated from the main land area. So far, 39 islands of the Andaman Group have allowed visitors to explore their natural beauty and geomorphological features. Every year about 4.5 lakh tourists visit the island destination sites to explore the multiple recreational activities. The south Andaman Island system comprises island-fringed coral reefs, shore platform geomorphology, shore cliff features of geological and geomorphological significances, island platform, terraces with vegetation cover, hill ridges and highland surfaces, valley flats topography, limestone caves and other karstic features, waterfalls, and short flowing streams with their riparian environment (Paul, 1998, 2005, Paul et al., 2017, 2018a, b, 2021; Acharya et al., 2022; Gee, 1927).

S. Ghosh
Department of Geography, Dum Dum Motijheel College, Kolkata, West Bengal, India

A. K. Paul (✉)
Department of Geography, Vidyasagar University, Midnapore, West Bengal, India
e-mail: akpauleastcoast@gmail.com

D. D. Majumdar
Department of Geography, Netaji Satabarshiki Mahavidyalaya,
Ashoknagar, West Bengal, India

A. K. Paul, A. Paul (eds.), *Crisis on the Coast and Hinterland*,
https://doi.org/10.1007/978-3-031-42231-7_19

The present study attempted to explore the specific coastal features and island interior features of geomorphological significances for the promotion of tourism attractions and activities. Currently the geomorphological features and their natural heritage significance as well as the environmental values of the landscapes are accepted by the visitors and tourism department for promotion of destination sites in the island systems environment for making awareness among the tourists to understand the significance of the oceanic island environment. There are many attractive geomorphosites in South Andaman district recognised by visitors and selective eco-tourists. They are grouped as (i) coral reef tourism, (ii) shoreline geomorphological features (e.g., sea beaches, natural arch, sea caves, karstic tours, shore platform mangroves, vertical sea cliffs, beach rock features, etc.), and (iii) interior island features (e.g., waterfalls, limestone caves, mangrove creeks, evergreen forests, and hills and terraces, etc.). These geomorphic sites are well connected with Port Blair by roads and waterways with different forms of nature tourism, adventure tourism, and with developed recreational infrastructures.

Multiple processes like geological structures, rate of weathering activities, tectonics setup, marine processes tidal influences, significant wave height, and past sea level changes are the major factors for shaping and reshaping the landscapes in the tropical island environment. Such dynamic landform features produced by natural processes with vegetation covers and marine environment have generated their heritage value, environmental value, and scenic beauty for creating attractions to the visitors, tourists, excursionists, and others. Currently the areas of geomorphological significances are the major destination of the tourist from abroad. The study also highlighted the assessment of geomorphological values of the landscapes in South Andaman districts.

19.2 Materials and Method of the Study

South Andaman district is extended from Humphrey Strait (northern boundary) to extreme south end of Little Andaman (10° Channel) island, and the region is fringed by the Andaman Sea towards east and Bay of Bengal towards west (Fig. 19.1). The area is characterised by topographic diversities of longitudinal ridges and valleys, coastal embayments, islands, and fringing coral reefs. There are a number of short flowing streams across the coastal landscapes and along the main island interior landforms as longitudinal streams. The majority of the river system longitudinal profiles show the presence of two to three knick points along their courses. However, the island interior longitudinal streams have a relatively wider valley flat surface. The elevation of the South Andaman district ranges from 20 to 365 m (Mount Harriet) in height with significant relief features (345 m).

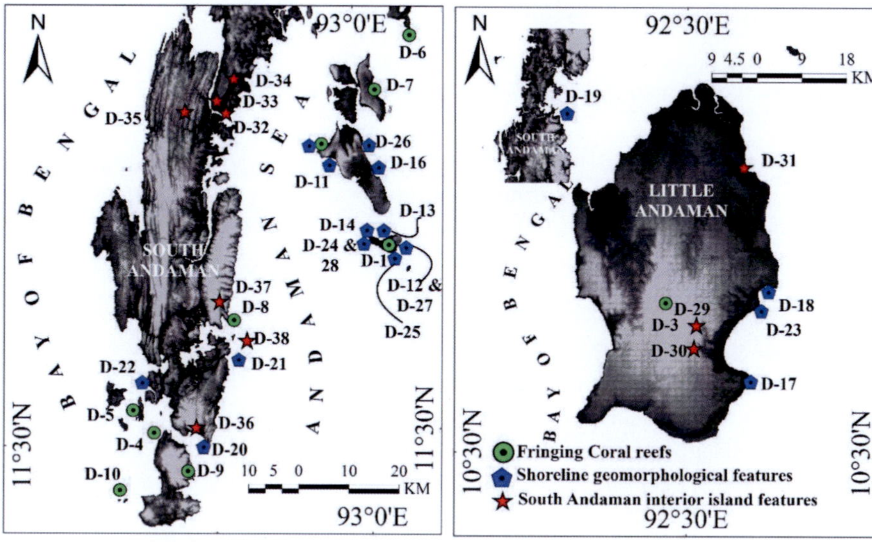

Fig. 19.1 Location of the destination for geomorphosites in South Andaman district, India

The study is conducted on the basis of survey of India (SOI) toposheets, shuttle radar topography mission (SRTM) digital elevation model (DEM), Google Earth Image, and repeated field survey during the research work in South Andaman (2016–2018). The geomorphosites have been explored and identified considering with their natural heritage landscape significances, and assigning the weightage for ranking the scientific values (Sce), ecological values (Eco), cultural values (Cult), and aesthetic-landscape (Est) values to estimate final geomorphic values of each destination site in the present study (Codrea et al., 2022; Kubalíková, & Kirchner, 2016). The entire geomorphosites are grouped into three different categories of geomorphological significances such as; (i) coral fringe coast (10 sites), (ii) shore fringe geomorphic features (12 beaches, 6 other features), and (iii) island interior landscapes (10 sites) in south Andaman district. Each criterion has been assigned a weight according to its importance for the South Andaman geomorphosites within this study. They have assigned the geomorphic values between 0 and 10, where 0 is the lowest value and 10 is the highest. Further, the integrated score for estimated values under four criteria is achieved in the study and a total of 38 destination sites are ranked as Rank-I, Rank-II, Rank-III, and Rank-IV for final geomorphic value. A framework of geomorphosite databases is prepared in the present study with their geomorphological values (V_{GEO}) following Eq. 19.1:

$$V_{GEO} = (\text{Sce} \times 0.01) + (\text{Eco} \times 0.30) + (\text{Cult} \times 0.15) + (\text{Est} \times 0.45) / 4 \quad (19.1)$$

19.3 Results and Discussion

19.3.1 Prospects of Tourism in South Andaman

Andaman Islands are considered to be a visitor's paradise because of several natural resources. During the early centuries the situation of the island was grim from a tourist's perspective. Compared to its actual potential, the development of the tourism industry in the island was disheartening. The South Andaman Island was isolated till the early twentieth century, although after the tsunami in December 2004 it gained popularity as a tourist destination. The island has a locational advantage due to the proximity of international maritime routes and hence has a great potential to attract tourists both from domestic and foreign locations. The island is situated close to South East Asian countries. This island is considered to be located in such an advantageous position wherein many of the international trade and tourism routes get initiated and terminated. Thus, the island has immense potential for tourism. The island offers three main categories of attractions for the tourists (Fig. 19.2b–g).

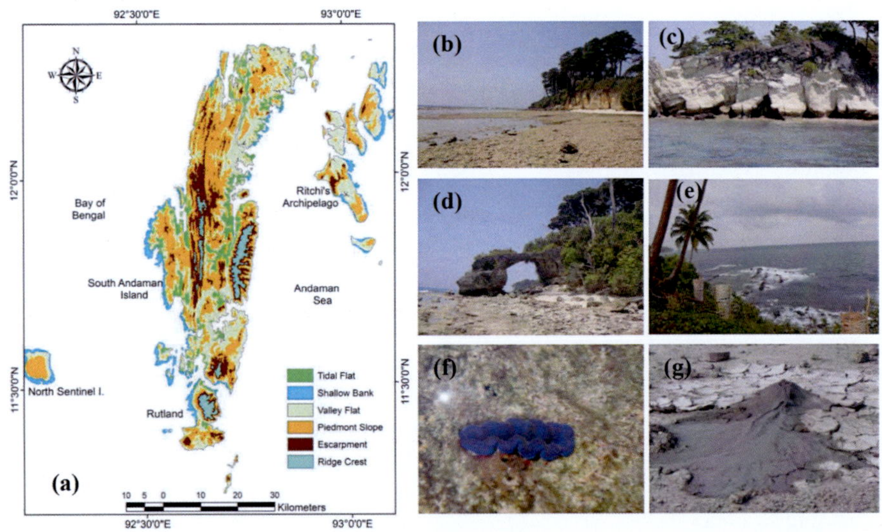

Fig. 19.2 (**a**) Terrain classification depicted the physiographic units of the South Andaman and the attractive geomorphosites of the South Andaman; (**b**) Cliff at Laxmanpur beach of Neil Island; (**c**) Notches and sea caves at the base of limestone cliff at elephant beach of the Havelock Island; (**d**) Natural arch at Laxmanpur-II beach at Neil Island; (**e**) Ophiolitic exposure at the south eastern part of South Andaman Island near Corbyn's cove island; (**f**) Living coral at Joly Buoy island; and (**g**) Mud volcano at Baratang island, South Andaman

Historical places: The Cellular Jail had been set up for the freedom fighters and prisoners prior to 1898. This island has got significant historical importance, as the first prison was constructed in this island. The Ross Island, which was then the administrative headquarters of the Britishers and for the Japanese troops, holds historical significance. Humphreygunj is another place of importance from the point of view of freedom fighters, wherein a large number of Indian patriots were shot dead by the Japanese. *Natural Beauty:* The breath-taking natural beauty of some islands like Jolly Buoy, Cinque Island, Red Skin Island, Havelock Island, Neil Island, and beaches like Corbyn's Cove, Chidiya Tapu, Radhanagar, Ramnagar, Laxmanpur, and Sitapur area are the major attractions for nature lovers. *Museums and Parks:* Several museums like Anthropological Museum, Samudrika Naval Marine Museum, Zoological Survey of India, Marine Museum, and parks viz., Mahatma Gandhi Marine National Park, Mount Harriet Park, Chidiya Tapu, are places of interest in South Andaman Island.

Thus, the above study depicts that there is a lot of scope in the field of tourism which is yet to be explored and developed in order to enrich the economic and social conditions. Although it has to be enhanced, retaining all the rules and regulations (as per coastal regulation zone (CRZ) notifications) deployed by the government for this group of Islands. The ecological balance should not be disturbed in any way to bring in additional facilities for the tourists. Andaman group of islands are a well-admired place by many such tourists who have travelled the place. But in order to make this place more attractive and bring in the list of wonders, proper resource management and publications will be planned in the near future.

19.3.2 Tourist Arrival in South Andaman

Andaman is a relatively new tourist destination in India. Earlier there had been a very less number of tourists who had shown interest for this place to visit the island. A statistic published by IP&T (Information, Publicity, and Tourism) and the A & NI Administration in 1980 reveals that the total number of tourists who visited the island was less than 10,000. Over the years the number steadily increased and crossed almost one hundred thousand (1,00,000) in 2004. Although due to the advent of the tsunami in the year 2004, the number of tourists who visited the place dropped to 32,381. The number increased to 1,95,369 tourists in the subsequent years.

19.3.3 Seasonality of Tourist Arrival

While considering the pattern of tourist arrival in the island, one of the interesting aspects to be studied is the season of tourist arrival. According to Butler and Poria (2001), there are two basic sources of the phenomena of seasonality – "natural" and

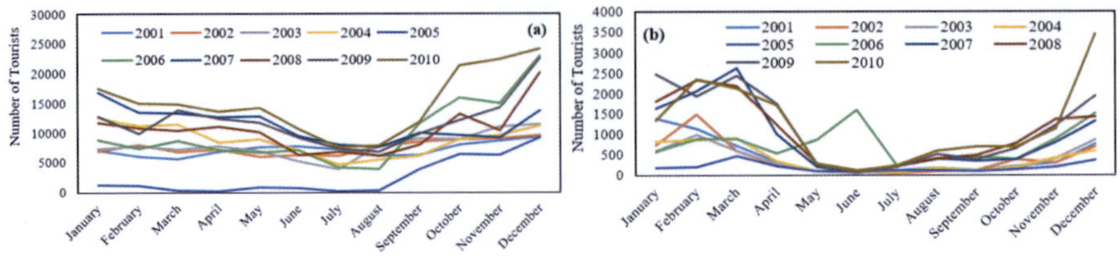

Fig. 19.3 The inflow of (**a**) domestic and (**b**) foreign tourists in South Andaman (2001–2010). (Source: Tourism Department of Andaman)

"institutional". The natural seasonality is the result of variability in climatic conditions like temperature, rainfall, daylight, etc., in the destination. Seasonality is an important issue in the Andaman Island because the tropical regions like the Indian subcontinent experience extreme temperatures, high rainfall, and humidity which results in the reduction of tourist demand at certain times of the year. Institutional seasonality according to Butler is the result of human decisions that can be due to a combination of factors like religious, cultural, ethnic, social, etc. In A & NI, the monthly tourist arrival data shows a great impact of seasonality during the calendar year (Fig. 19.3a). An analysis indicates that there is no high variation in the arrival of domestic tourists in different months of the year, though a marginal increase is found to take place in the months of November, December, and January. This is mostly due to the comfortable climatic condition that prevails in the island during that time and due to the public holidays. The number of incoming foreign tourists across the year varies significantly. The foreign tourists visited the island mostly during the winter time, that is from mid-October to mid-March, while there is a sharp decline observed during summer to monsoon (Fig. 19.3b).

19.3.4 Tourist Places and Tourism Activities

The Andaman is well known for its natural beauty, but the tourist spots are concentrated in certain specific areas. Below are the places that are well known for their tourist attractions in South Andaman. Tourism holds a significant role in boosting up the economic conditions of any place. This has a direct positive impact on the social and economical conditions of the place. After a generic survey was conducted, it was thus concluded that tourists have a strong fascination towards relaxing and swimming on the beaches present in the island. They have strongly recommended the scenic beauty of the place. Besides it has been analysed that tourists like to avail various water sports activities like snorkelling and scuba-diving. Red Skin, North Bay, and Jolly buoy are famous for corals where tourists are taken to enjoy snorkelling. Glass-bottomed boats are also available to view the corals from the transparent glasses. In Havelock and Neil Islands, often foreigners are found engaged in snorkelling without any assistance of the localities. Wandoor provides such a facility wherein tourists are taken from Havelock by a dongle. Professional Association of Diving Instructors (PADI) are available for assistance. Moreover, it has been observed that mostly foreigners across the world have a passion of trekking through the natural trails near Havelock and Neil Islands. Apart from international tourists, domestic tourists have shown interest by exploring the scenic beauty of the nearby places across the island. Cellular Jail, Corbyn's cove, Anthropological Museum, Samudrika Museum, Mahatma Gandhi Marine National Park, and Mini Zoo are considered to be few places of attraction.

19.3.5 Tourism Industry & Economy

The island is a reservoir of natural resources that influence the economy of the island. The major sources of revenue of South Andaman Island are agriculture, handicrafts industry, hotel, restaurant business, and transport business. But the interesting fact is that all these industries are well connected to the tourism industry of the islands. Handicraft industry, which is one of the major sources of income for the people in Port Blair, is a major attraction for tourists. The products are made up of shells which have a high demand across the world. People living in the coastal areas indulge themselves in the practice of shell crafting. The shells are collected, cleaned, and processed suitably followed by shaping different designs to ornament gift items as well as decorative items which can be used to adorn houses. Timber and wood works are another major section of handicraft industry. The island is a home of several kinds of timber species. These timbers are cut into different shapes and sizes to make furniture, mats, and other products which has drawn the attention of several tourists.

19.3.6 Estimated Geomorphic Values of the Geomorphosites

All the landscape systems of South Andaman district are categorised into three different types of geomorphological significances. They are identified as: (i) ten destination sites under coral reef coasts; (ii) eighteen destination sites under shoreline geomorphological features; and (iii) remaining ten destination sites under island interior landscapes for the assessment of scientific values, ecological values, cultural values, and aesthetic-landscape values in the study (Table 19.1). As per the scientific values and ecological values the highest weightage value is assigned for eight destination sites of island interior landscapes; and considering the cultural values and aesthetic-landscape values the highest weightage value is assigned for six destination sites for coral reef coasts and six destination sites of the interior landscapes in South Andaman (Fig. 19.4). A total of 38 destination sites are distributed in the three major landscapes of the oceanic island systems. Among them, seven destination sites of coral reef coasts are located along the island fringes of the Andaman Sea, whereas, only three destination sites are located in the Bay of Bengal. The fifteen destination sites of shoreline geomorphological significances are represented by the Andaman Sea coastal fringe areas of South Andaman district, but the remaining three numbers of destination sites are distributed in the Bay of Bengal fringe coastal areas and especially on little Andaman Island. However, the island interior geomorphic and ecological features are mostly distributed in the main island landscapes complex of the South Andaman district.

Considering the above geomorphic significances with values of four criteria, the final geomorphic values are estimated for the 38 destination sites in the study. The result shows that the Rank-I category geomorphosites include four destinations, island interior landscapes. The rank-II category geomorphosites include eight destinations of coral fringed coasts, island interior landscapes, and shore fringe geomorphological importance. However, the rank-III category of destinations include 14 geomorphosites and Rank-IV category destinations include 12 number of

Table 19.1 Assessment of geomorphic values for the 38 geomorphosite destinations in South Andaman district based on scientific values, ecological values, cultural values, and aesthetic-landscape values. The final integration scores and their ranking to identify the category of geomorphosite in the emerald islands

		Geomorphological Values (V_{GEO})					
Major category of the landscapes	Geomorphosites	Scientific value (Sce)	Ecological value (Eco)	Cultural value (Cult)	Aesthetic-landscapes value (Est)	Score	Rank
Fringing coral reefs	Neil Island (D-1)	6	6	5	4	1.1025	Rank-IV
	Havelock Island (Elephant beach point) (D-2)	6	5	5	5	1.14	Rank-IV
	Little Andaman (D-3)	7	5	6	7	1.405	Rank-III
	Jolly Buoy Island (D-4)	8	7	6	6	1.445	Rank-III
	Redskin Island (D-5)	8	8	7	8	1.7825	Rank-II
	South Button (Havelock Island) (D-6)	7	7	5	6	1.405	Rank-III
	Henry Lawrence Island (D-7)	7	7	6	8	1.6675	Rank-II
	North Bay Island (D-8)	6	6	5	4	1.1025	Rank-IV
	Rutland Island (D-9)	7	8	7	8	1.78	Rank-II
	Twin Brothers Island (D-10)	7	6	7	7	1.5175	Rank-III

(continued)

Table 19.1 (continued)

Major category of the landscapes	Geomorphosites	Geomorphological Values (V_{GEO})				Score	Rank
		Scientific value (Sce)	Ecological value (Eco)	Cultural value (Cult)	Aesthetic-landscapes value (Est)		
Shoreline geomorphological features	**Beaches**						
	Radhanagar Beach (D-11)	8	6	5	6	1.3325	Rank-III
	Sitapur Beach (D-12)	5	4	4	5	1.025	Rank-IV
	Bharatpur Beach (D-13)	6	5	4	4	0.99	Rank-IV
	Laxmanpur Beach-I (D-14)	5	5	3	4	0.95	Rank-IV
	Elephant Beach (D-15)	5	6	4	5	1.175	Rank-IV
	Kala Pathar Beach (D-16)	4	6	3	5	1.135	Rank-IV
	Hut Bay Beach (D-17)	6	7	4	4	1.14	Rank-IV
	Buttler Bay Beach (D-18)	6	8	5	5	1.365	Rank-III
	Karmatang Beach (D-19)	7	8	5	5	1.3675	Rank-III
	Munda Pahar Beach (D-20)	6	7	5	6	1.4025	Rank-III
	Corbyn's Cove Beach (D-21)	7	5	4	5	1.105	Rank-IV
	Wandoor Beach (D-22)	7	5	4	5	1.105	Rank-IV
	Other features						
	Kalapathar limestone topography (D-23)	8	7	6	4	1.58	Rank-III
	Laxmanpur-II natural arches (D-24)	8	8	6	6	1.88	Rank-II
	Ramnagar shore platform mangroves (D-25)	5	8	3	6	1.625	Rank-II
	Vijaynagar shore platform mangroves (D-26)	5	8	3	5	1.5125	Rank-III
	Sitapur white claystone cliff (D-27)	6	5	3	5	1.335	Rank-III
	Neil Island beach rock platform (D-28)	5	5	2	3	1.025	Rank-IV

(continued)

Table 19.1 (continued)

Major category of the landscapes	Geomorphosites	Geomorphological Values (V_{GEO}) Scientific value (Sce)	Ecological value (Eco)	Cultural value (Cult)	Aesthetic-landscapes value (Est)	Score	Rank
South Andaman Interior Island features	Whisper Wave Waterfalls (D-29)	9	9	7	6	2.04	Rank-I
	White Surf Waterfall (D-30)	9	9	7	8	2.265	Rank-I
	Dugong Creek (D-31)	9	9	6	8	2.2275	Rank-I
	Baratang mangrove creek (D-32)	8	7	6	7	1.9175	Rank-II
	Baratang limestone cave (D-33)	9	8	7	8	2.19	Rank-I
	Baratang mud volcano (D-34)	9	5	3	4	1.365	Rank-III
	Jarawa Reserve Forest (Hill Forest) (D-35)	8	8	6	7	1.9925	Rank-II
	Munda Pahar Forest (Chidiya Tapu) (D-36)	7	7	4	5	1.57	Rank-III
	Mount Harriet peak (D-37)	8	8	5	7	1.955	Rank-II
	Ross Island (D-38)	7	5	6	4	1.3825	Rank-III

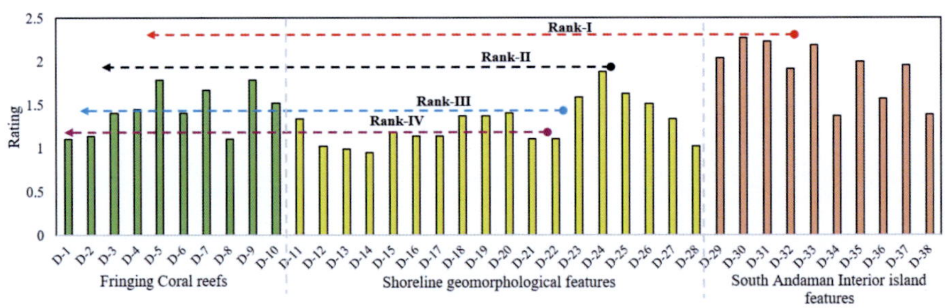

Fig. 19.4 Estimated ranking for geomorphosite destinations based on scientific values, ecological values, cultural values, and aesthetic-landscape values in South Andaman district

geomorphosites in the South Andaman district. The rank-I geomorphosites are connected with easy access routes and they are geomorphologically most significant with their four significant values. The Rank-II geomorphosites category destinations are accessible and explored by the adventure tourism process along the island fringes, coral reefs, and island interior landscapes. Geomorphosites of Rank-III and Rank-IV categories are located far away from Port Blair and their access is relatively not easy in comparison to Rank-I and Rank-II destination sites. The SRTM DEM and SOI toposheets helped to identify the different terrain characters of south Andaman district. As many as six major terrain units are identified after the consideration of contour patterns and generation of a Digital Elevation Model for the study area (Fig. 19.2a). The Google Earth image is used for the validation of geomorphic characteristics for each destination site in South Andaman district. A repeated field survey method helps to monitor the features and tourism significance of each geomorphosite. The 38 geomorphosites are identified and plotted in a framework scheme to estimate the geomorphic values. They are classified into three major groups with estimation of scientific values, ecological values, cultural values, and aesthetic-landscape values.

19.4 Conclusion

The natural landscape of South Andaman district attracts a large number of domestic tourists and foreign tourists in the emerald islands. The domestic tourist is involved in the tourism process for a few days (7 10 days) under a package tour guided by different tourism agents and tour operators in the island. However, the foreign tourists visit the different geomorphosites of isolated island landscapes and stay for a month or more in the eco-huts. They actually rely on the aesthetic-landscape values of the geomorphosites and various adventure tourism practices with coral watching, sun bathing, and nature hunting in the islands.

The tourism department of Andaman promotes eco-tourism in multiple islands where the minimum tourism recreation infrastructure are available. The assessment of geomorphological values of the various destination sites will provide a very good data base, a framework of geomorphosites for managing the travel and tourism process by the coastal managers and administrators in a sustainable way. Environment regulations of 2009 are advocated in favour of tourism lobby in the Andaman group of islands. The island protection zone is strictly followed for the construction of touristic recreational infrastructures as per the size of the island (smaller island and big island). Therefore, there are a lot of opportunities to promote eco-tourism, nature tourism, and geo-tourism activities. The assessment of geomorphosites along with their geomorphological values will help in the promotion of the tourist activities in this manner.

References

Acharya, A., Mondal, B. K., Bhadra, T., Abdelrahman, K., Mishra, P. K., Tiwari, A., & Das, R. (2022). Geospatial analysis of geo-ecotourism site suitability using AHP and GIS for sustainable and resilient tourism planning in West Bengal, India. *Sustainability, 14*(4), 2422.

Codrea, P. M., Bilașco, Ș., Roșca, S., Irimuș, I. A., Iuliu, V., Rusu, R., et al. (2022). The integrated assessment of degraded tourist Geomorphosites to develop sustainable tourism: A case

study of Grădina Zmeilor Geomorphosite, North-West Region, Romania. *Applied Sciences, 12*(19), 9816.
Gee, E. R. (1927). The geology of the Andaman and Nicobar Islands, with special reference to Middle Andaman Island. *Record of the Geological Survey of India, 59*, 208–232.
Kubalíková, L., & Kirchner, K. (2016). Geosite and geomorphosite assessment as a tool for geo-conservation and geotourism purposes: A case study from Vizovická vrchovina Highland (eastern part of The Czech Republic). *Geoheritage, 8*, 5–14.
Paul, A. K. (1998). *Problems of groundwater contamination and water scarcity in Andaman and Nicobar Islands DELF, New Delhi* (pp. 13–33).
Paul, A. K. (2005). *Tsunami: An assessment of the disaster over the nations in the Indian Ocean* (pp. 1–125). India ACB Publications.
Paul, A., Bandyopadhyay, J., & Paul, A. K. (2017). Monitoring vertical cliffs, embayments and shore platform morphology in parts of Neil Island, South Andaman using spatial information technology. In *38th Asian conference on remote sensing – Space applications: Touching human lives*. ACRS.
Paul, A., Bandyopadhyay, J., & Paul, A. K. (2018a). Geomorphological mapping and environmental zoning approach to coastal management in Havelock Island, South Andaman, India. In *39th Asian conference on remote sensing: Remote sensing enabling prosperity* (pp. 346–357). ACRS.
Paul, A. K., Kamila, A., & Ray, R. (2018b). Natural threats and impacts to mangroves within the coastal fringing forests of India. In *Threats to mangrove forests* (pp. 105–140). Hazards, Vulnerability, and Management.
Paul, A., Sardar, J., & Bandyopadhyay, J. (2021). *Assessment of Geo-Environmental aspects for Geo-Tourism site selection of South Andaman District, India, ISSN: 2348 – 6740, Volume 8 , Issue No. 1 - 2021 Pages 1-12.*
Poria, Y., Butler, R., & Airey, D. (2001). Clarifying heritage tourism. *Annals of Tourism Research, 28*(4), 1047–1048.

Chapter 20
Historical and Geomorphological Background of Ancient Khejuri-Hijili and Tamralipta Ports in South Bengal

Mihir Pradhan and Ashis Kumar Paul

20.1 Introduction

The study area is formed under the deltaic floodplain environment of the Hugli Estuary. Later, it is modified by marine and riverine processes. Physiographically, it lies between 2 and 3 meters above the mean sea level, thus, it is considered as coastal low land of the estuarine bank. Khejuri is largely exposed to cyclones that frequently lash out at the estuary during the summer. In the thirteenth century, Tamluk was a seaport of north-east India near the confluence of the river Rupnarayan, the river Hugli, and the Bay of Bengal, which existed till the eleventh century. However, Balasore near the mouth of the Budhabalanga in Orissa in the seventeenth century was situated at the mouth of the Bay of Bengal and Subarnarekha delta. Khejuri enjoyed locational advantages for trade and commerce, which consequently helped it to emerge as a port and trade centre (Sarkar, 1977). The area under consideration has been differently named and called by the foreign and native people at different times. In most cases, such names owe their origin to the natural resources that were available locally. On the southern bank of the river Rupnarayan, there was a port at present Tamluk, mentioned as "Tamalites" in Ptolemy's geography in the historical past. It was an important seaport in ancient times, but it lost its importance towards the end of the tenth century due to a lack of sufficient navigable depth caused by gradual silting process in the river beds. Then the Hijili rose to prominence as a seaport (Pradhan, 2003, Bowrey, 1904). Characteristically, Khejuri-Hijili is formed of by Gangetic Deltaic alluvium. In very few places there are gentle undulating surfaces with wavy rolling topography which are slightly elevated above the

M. Pradhan
Kalagachia Jagadish Vidyapity, Khejuri, West Bengal, India

A. K. Paul (✉)
Department of Geography, Vidyasagar University, Midnapore, West Bengal, India
e-mail: akpauleastcoast@gmail.com

A. K. Paul, A. Paul (eds.), *Crisis on the Coast and Hinterland*,
https://doi.org/10.1007/978-3-031-42231-7_20

alluvial plain. In the early stages of its formation, this tract was slightly elevated above sea level and was intercepted by numerous streams. But the Khejuri-Hijili sectors were silted up and devastated by the consecutive storms and flood events in 1807, 1831, 1833, etc., and lastly in 1864 (Gastral et al., 1866). Because of its coastal location, frequent tidal spills caused logging of saline water in this area, which harmed crop production for years and made the area prone to malaria (Hunter, 1997). These were the main causes of the deterioration of this area. Later, the island was covered with natural mangroves and littoral vegetations. Gradually, it became the abode of fishermen through many geomorphological, social, economic, political, and historical phases; it had come to the present state.

The study area, popularly known as the coastal plain, consists of an extensive low-lying tract of fluvio-marine deposits. The entire alluvial tract has evolved through the seaward advancement of delta fan deposits by Haldi and other rivers that originated from Chhotonagpur plateau as rainfed tributaries of the Hugli River at the Holocene regressive phases of the sea, sub-tidal to inter-tidal deposits between the repeated marine transgression-regression phases within 6000 and 500 YBP, and finally by estuarine deposits of the Hugli, Haldi, and Rasulpur rivers (Mukhopadhyay, 1972). A few borehole records of lithological logs from Haldia and the adjacent region indicate the presence of a thick clay blanket to a maximum depth of 425 feet (or 127.5 m) with some ventricular pockets of granular materials. The upper surface of the clay layer is produced by intertidal and estuarine floodplain deposits of the Hugli, Haldi, etc., rivers in the recent to sub-recent past. Estuarine tidal flood deposits are still active around the channel margins, channel beds, and within the unprotected tidal spill basins. The entire surface of Khejuri, Hijili, and Haldia ranges from 2 to 2.4 m in height above the mean sea level. The low land surface is dissected by several tidal flood plains with creek networks including the Hijili tidal canal, and the Rasulpur Haldi estuaries. Many abandoned channels, paleochannel beds, tidal flats, natural levees, and islands represent the lowland complex of the region by nature (Paul, 2002). Recently, the tidal channels have been choked up in the inner parts by tidal siltation and due to anthropogenic activities. Reclamation of low land from an active tidal spill basin with regular protective earthen embankments has prevented the region from tidal inundation and natural siltation at present.

20.2 Materials and Method

The sources of the data for this study may be classified into two categories comprising with archaeological tools and literary documents. The combined testimony of these two categories of sources has helped us to reconstruct the history of Tamralipta and Khejuri-Hijili in particular. Literary sources, including the frequent references to Tamralipta in the literary works of various categories, help us to throw sufficient light on Tamralipta as a city-port and a noted centre of Buddhist learning, and later to Khejuri, a port on the sea mouth estuary. On the other hand, the archaeological sources not only help us trace the antiquity of the study areas to a hoary past but also throw light on many interesting points of socio-cultural history of importance. Thus, on the basis of all these sources, an honest and humble attempt has been made for the first time in this work to depict a comprehensive history of Tamralipta janapada in the ancient period.

The diary of William Hedges during his agency in Bengal (1887) is a unique source of early history of Bengal as well as the study area. Some land revenue recorders (Hunter, 1996, 1997; O'Malley, 1995; Ascoli, 1917) have documented many historical events and general accounts of this region clearly. Economic activities depend on physical and nonphysical environments. These environments, such as agriculture, marine resources, fishing, etc., have been broadly explained by the block, district, and state planning organizations. The background of this lowland coast is described through historical, geological, and instrumental records, and the physical environmental settings are discussed by researchers (Paul, 2002). Finally, these collected data have been analysed to determine how much economic support influenced the resources that will afford the coastal people of the study area to maintain their livelihoods after the sequential deterioration of the port-centric area. Finally, with the repeated filed works remains of early structures, monuments, and office buildings have been attentively observed, and written reports, research articles, etc., have been carefully studied for sequencing the socio-economic history of the area.

20.3 Results and Discussion

Ancient Tamralipta or modern Tamluk is located on the right bank of the river Rupnarayan and the river Hugli confluence in Purba Medinipur district, West Bengal. It is frequently mentioned in Jaina, Buddhist, and Brahmanical Sanskrit works. It must have been in existence before the birth of Christ. Ptolemy (circa 150 A.D.), a Greek geographer, also noticed it in the geography, calling it TAMALITES and placing it along the right bank of the river Ganges. Tamralipta comes from two Sanskrit words: "Tamra" means copper, and "Lipta" means plenty (Bhowmik, 1991). In early times, copper from the Chhotonagpur plateau region was collected and exported through this port, so it bears this name. It first emerged as a port at which merchants and others embarked for Ceylon (modern Sri Lanka) and the Far East. According to the Chinese pilgrim Fa-Hian (405–11 A.D.), it was in the sea mouth, and he also resided there for 2 years in the Buddhist monasteries. Later on, another Chinese pilgrim, Hiuen Tsiang, in the seventh century A.D., said that it lay near an inlet of the sea and was 10 Li (about 3.2 km) in circuit. He witnessed a Buddhist sputum erected by Maurya Emperor Ashoka in 268–232 BCE (O'Malley, 1995). Though there were ups and downs in trade relations between India and China, there is specific evidence of brisk trade between these countries since the first century B.C. This trade relationship was maintained through the port of Tamralipta, and it was fostered by the demand for luxury goods at the Chinese court. On the basis of the longer record of Tamralipta as an international port, it is believed that the major part of India's foreign trade to the Far East was carried through the port of Tamralipta during the first century A.D. (Ray, 1979). Missions to the Chinese court during the Kushana period between 89 and 161 A.D. must have sailed down the Ganga and left for China from the port of Tamralipta. The sending of an embassy to the yellow gate (the Chinese court) in the third century A.D. by a king of Tan Mei is also suggestive, as it is believed by some that Tan Mei is an equivalent of Tamralipta (Beal, 1869; Sen, 1988).

Tamralipta has ever been remembered as one of the greatest city-ports of Southern Asia and a great centre of Buddhist culture and learning. After a few centuries, it gradually lost its importance as a seaport owing to the silting up of the channel that formerly connected it with the Bay of Bengal. This channel survived until the middle of the sixteenth century and is shown in the earliest European maps. Now Tamluk is famous as an old town. It was constituted as the first municipality of the undivided Midnapore district in 1864 (O'Malley, 1995). Though the proto-historic human settlement at Tamralipta is evident from the archaeological findings, the detailed political history of Tamralipta janapada (the kingdom) in the ancient period is not clearly explained. Tamralipta is often referred to in literature both as a city-port and a distinct janapada, sometimes as a separate one and sometimes as a part of a neighbouring kingdom or janapada. During this period, this place was known by various names like Tamalite, Damalipta, Tamalipta, Tamalini, Tamalipti, Vishnugriha, Stambapura (stambapu), Tamrallpti, Velakula Tamalika Tamraliptaka, and Tamraliptlka, as evident from both foreign and indigenous sources (Bhowmik, 1991; Campos, 1956).

The Rupnarayan River Estuary is a sub-system of the Hugli Estuarine Complex. This estuary is gradually mixed with saltwater as it approaches the outer estuary near Hugli point, while the surplus seawater of the Hugli embayment moves in and out of the estuary with each tidal cycle. Its estuary is progressively wider at the lower section, but it is slightly reduced at the point of intersection with the Hugli embayment. Several islands and linear bars are found in the river channel, while accretions in the shape of grass-covered chars are common in the meso-tidal estuary. A large portion of intertidal flat usually gets exposed along the channel banks and island banks at every tidal cycle. The wide mud flats and reed swamps are ideally favoured by a large volume of tidal exchange and rapid siltation or surface accretion. In recent time, tidal shoals, linear bars, and meandering courses have also been regularly distributed near the Tamluk area (Paul, 2002).

20.3.1 Port-Centric Activities

In the Bowrey chart of 1669 A.D., present-day Tamluk is mentioned as "TAMALEE" on the right side of the River Rupnarayan, and the Khejuri-Hijili twin islands are distinctly plotted near the Hugli estuary as "CASNREE" "HIJILEE", respectively (Fig. 20.1a). Tamralipti, modern Tamluk, is mentioned in the Indian epic, the Mahabharata, the Vrihat Samhita, and the Ceylonese chronicle, the Mahavamsa. In

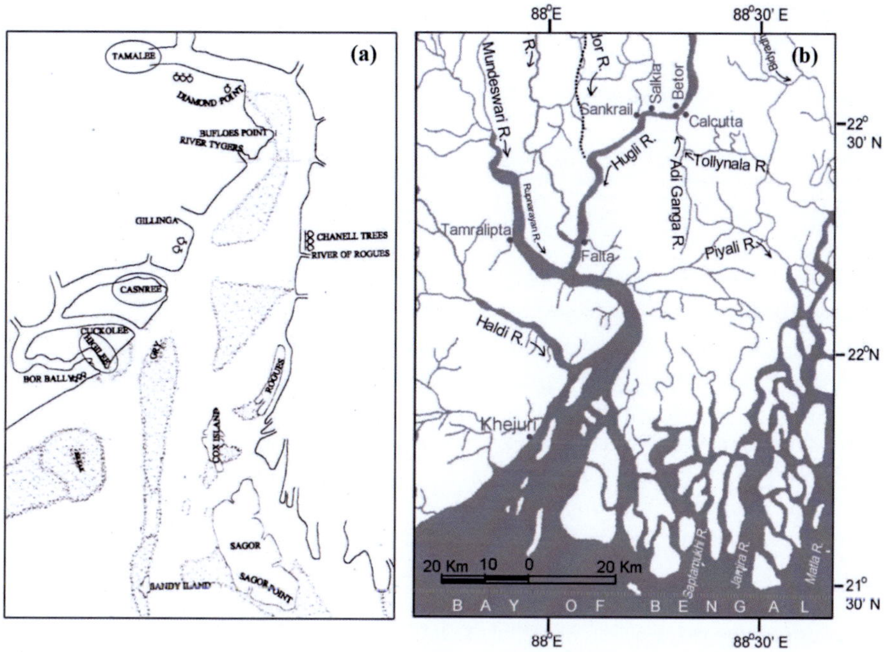

Fig. 20.1 (**a**) Location of Tamluk and Khejuri Ports after modified by Thomas Bowrey Chart-1669 (**b**) Old historical ports of Bengal. (From Karan, 2002)

the Periplus of the Erythran Sea (first century A.D.), there is a reference to a port on the mouth of a river in Bengal that indicates the Tamralipti. It was the capital of ancient Shuma in the eighth century of the Christian era and formed a part of the Magadha Kingdom under the Mauryas. Due to the process of land formation near the river mouth, Tamluk, once situated on the seashore, is now 90 km from the sea mouth. By the tenth century, the channel on which Tamluk was situated and which afforded facilities for navigation was silted up, but it continued to be an emporium of commerce until the sixteenth century. For well-nigh 2000 years, this far-famed port of the east was visited by foreign merchants from China to the Mediterranean; here lay in anchor hundreds of Bengal Ships, which carried to distant shores the merchandise, culture, religion, and art of Bengal through the epochs (Mukherjee, 1938). From the medieval Bengali literature like Mangal Kavya, we may derive the information about the articles of trade (Bhattacharya, 1970). Bengali merchants, at a later period, gradually assumed the role of distributors and established themselves in the coastal trade through Tamluk port, which extended as far as Ceylon. At that time, trade was carried out through a barter system. A list of articles of trade under that system is given below (Table 20.1 and Fig. 20.1b).

Khejuri, a part of the Kanthi coastal tract (regionally extended from the Rupnarayan River to the Subarnarekha delta around the Hugli estuary), is associated with several beach ridges and sand dunes and intervening tidal flats at the south

Table 20.1 List of trade articles through Tamluk port

Exports from Bengal	Trade types	Imports in Bengal
Deer	In exchange of	Horse
Glass	In exchange of	Emerald
Sea-salt	In exchange of	Rock-salt
Ape	In exchange of	Elephant

west of this area. The area is formed under the deltaic floodplain environment of the Hugli Estuary. Later, the area was modified by tidal processes. This coastal low land tract lies between 2 and 8 meters above the mean sea level. The beach ridges and sand dunes of this area are partly obliterated, segmented, and cut by tidal waves, anthropogenic activities, and weathered in a subaerial environment. These coastal features signify that the area was once in a coastal and estuarine environment (Paul, 2002). Khejuri is also prone to flood and storm hazards because of its estuarine location. So, the flood-prone areas were once protected by earthen embankments. By 1587, Hijili had become an important trade centre. From the account of *Ralph Fitch,* it is reported that the then-Hijili was a great heaven where every year many ships would come from Negapattam, Sumatra, and Malacca loaded with rice, cotton-cloth, wool, sugar, pepper-corn, and other victuals. Ralph Fitch, an English traveller, described the place as having "very much rice and cloth made of grasses, which they call yerva (a Latin word meaning grass), which is like silk." (Hedges, 1883). It is now inhabited by fishermen, as are also Ingellie and Kedgerrie, two neighbouring islands on the west side of the mouth of the Ganges.

20.3.2 Occupational Activities and Their Changes under Colonialism

In the second quarter of the seventeenth century, the Dutch began to conduct trading activities there. The English appeared as rivals in the later half of that century. A list (Table 20.2) of imported and exported goods is given below. The introduction of colonial rule over the area in the seventeenth century brought about dramatic changes in the occupation of the local community. Earlier, the local people used to produce paddy, vegetables, fish, salt, etc., according to their needs. But after the arrival of the foreign traders, the local people were compelled to be engaged in the production of salt, indigo, green vegetables, sea fish, pigs, goats, etc., to meet the foreigners' needs. They also became wage workers in the ships and sloops and import-export activities run by the foreigners (Table 20.3).

Table 20.2 List of trading articles after being modified by Karan (2002)

Goods brought by colonizers	Sale of goods by colonizers
From Malacca – cloves, mace, muslins.	Rice, butter, sea fish, cows, buffaloes, sugar, vegetables, spices, wood, and others ship-building materials, jute, oil and oil seeds, cotton, silk, scents made from flower, prickle and sweetmeat made by them from Bengal's fruits, areca nut, wax, potassium nitrate, straw, bamboo, opium, embroidered clothes, hogs, long pepper, hen, etc.
From Sumatra – cloves nutmegs, muslins.	
From Borneo – camphor, ivory	
From Ceylon – cinnamon	
From Malabar – pepper	
From China – silks, gilt furniture	
From Maldives – Sea-shells used for ornaments and coins	
From Coromondal – coast-big sea shells called as Chan quos	
From Solor and Timor – sandal-woods (red and white colour)	
From Batavia (modern Jakarta) – silver, sugar	
From their own country like fruits, flowers, garments, etc.	

Table 20.3 Occupational changes in Khejuri and Tamluk areas after being modified by Karan (2002)

Past occupation	Present occupation	Remarks
Indigo cultivation	Rice paddy cultivation	Past occupation fully controlled by East India company on the basis of the availability of local resource for fulfilment of their own demand. Present occupation controlled under the independent administration. Past occupation also selected by land resources. Present occupation is controlled by both land and water resources.
Rice cultivation	Fish farming	
Collection of non-timber resources from tidal forest	Brick manufacturing	
Salt processing	Fish catch from estuary	
Marginal occupation (fish catch from the different wet lands)	Agricultural labour	
Daily labour	Marginal occupation	
Seasonal vegetables cultivation	Crab potting	
Other occupation related to trade and commerce	Collection of non-living resources (silt collection)	
	Molasses processing from date-palm tree	
	Salt process (marginal)	
	Vegetable production	
	Trawler artisan	
	Engaged with commercial prawn culture in agricultural land	

20.4 Conclusion

The port of Tamralipta was mentioned by Pliny in the first century A.D. and by Ptolemy, a Greek philosopher, in the second century A.D. At that time, it was famous as a kingdom extending from the Hugli River in the east to the Subarnarekha River in the west. The local Sarak communities of the Jain religion were engaged to collect copper from the nearby eastern part of the Chhotonagpur plateau. They transported these through waterways to Tamralipta for sailing towards Ceylon, Subarnabhumi (modern Indochina and Indonesia regions), etc. Just after the abolition of "Tamralipta Bandar" and other old ports of Bengal, Khejuri turned into a full-fledged sea port. From the sixteenth century A.D., Khejuri turned into a port when a navigation route had been maintained touching the area. But the thalweg shifting towards east (about 8 km within 1768–1918 A.D.) due to continued silt deposition and the tectonic position of the Bengal basin caused deterioration of the port.

In 1679, British initiatives and activities turned Khejuri from a coastal halt station to a full-fledged port. Though the Portuguese traders started commercial activities long before, during the sixteenth century A.D. a navigation route touched the area. From the beginning of the nineteenth century, the rapid shifting of the main estuarine channel towards the east caused deterioration in this port. The river and the estuary also offer livelihood options to the local people. The coastal location provides a special importance to the area. Huge water resources can be utilized to meet the increasing demand of people. But development activities without much attention to environmental issues have resulted in a great pressure on natural resources. So, people's awareness, participation, and response are musts in every step of management policies. The huge manpower can also be used for proper utilization of resources both on land and in water. Agriculture and its by-products can be practised in an environmentally friendly manner with proper management policies. The involvement of skilled labour in agro-industries can open up new prospects in this area.

References

Ascoli, F. D. (1917). *Early revenue history of Bengal and the fifth report 1812* (pp. 1–14). Clarendon Press.

Beal, S. (1869). *Travels of Fah-Hian and Sung-Yun fromChinato India (400A.D. & 518 A.D.)* (pp. 140–148). Trubner and Co..

Bhattacharya, A. (1970). *Bangla Mangalkavyer Ithihas* (5th ed., pp. 170–179). CAL.

Bhowmik, M. (1991). *History, culture and Antiquities of Tamralipta*. Unpublished Ph.D. thesis of Calcutta University (pp. 3–31).

Bowrey, T. (1904). *A geographical account of the countries round the Bay of Bangal-1669–1679* (S. R. C. Temple, Ed., pp. 12–36). Hakluyt Society/Cambridge/MDCCCCV.

Campos, J.J.A. (1956). *History of the Portuguese in Bengal*. Janaki Prakashan (pp. 1–43, 112–168 and 204–228).

Gastrell, L., Col, J. E., & Blanford, H. F. (1866). *Report on the Calcutta cyclone of the 5th October-1864 government of Bengal* (pp. xx–xxv and 127–132).
Hedges, W. (1883). *On the early History of the Company's Settlement in Bengal and on Early charts and topography of the Hugli River* (R. Barlow, Trans.) (pp. 33–69). Hakluyt Society.
Hunter, W. W. (1996). *The annals of rural Bengal* (Vol. 3, pp. 272–304). West Bengal District Gazetteer/Government of West Bengal.
Hunter, W.W. (1997). *A statistical accounts of Bengal*. West Bengal District Gazetteer/Government of West Bengal (Reprint Edition, pp.6–24 and 167–227).
Karan, M. (2002). *Rachana Sankalan (Hijlir-Masnad-e-Ala, Khejuri Bandar*. Published by Khejuri Itihas Sanrakshan Parshat, Samskriti Samsad.
Mukherjee, R. K. (1938). *The changinh face of Bengal- A Study in Riverine Economy*. University of Calcutta (Reprint Edition 2009, pp. 35–49).
Mukhopadhyay, S. C. (1972). Techtonic landforms of the South Western fringe area of the Bhagirathi-Hoogly Basin. In K. G. Bagchi (Ed.), *The Bhagirathi-Hoogly basin (Proceedings of the interdisciplinary symposium)* (pp. 39–48). Calcutta University/Department of Geography.
O'Malley, L. S. S. (1995). *Bengal district gazetteers Midnapore District*, new edition. Higher Education Department/Government of West Bengal (pp. 124–141 and 221–244).
Paul, A. K. (2002). *Coastal geomorphology and environment* (pp. 1–342). ACB Publication.
Pradhan, M. K. (2003). *'Hijlinama' Contai House Owner's*. Association Publication (pp. 7–29, 51–61 and 152–188).
Ray, N. R. (1979). Tamralipta and Ganges: Two port cities of ancient Bengal and connected considerations. *Geographical Review of India, Calcutta, 41*(3), 205–222.
Sarkar, S. J. (1977). *The history of Bengal-Muslim period (1200–1757) Patna Janaki Prakashan* (pp. 368–370).
Sen, G. P. (1988). Tamralipta- commerce and culture. In B. Chattopadhyay (Ed.), *Culture of Bengal through the ages some aspects* (pp. 14–34). University of Burdwan.

Chapter 21
An Assessment of the Impacts of Marine Litter in the Coastal Regions of West Bengal and Odisha on Flora, Fauna, and Humans

Anurupa Paul, Joydeb Sardar, Ashis Kumar Paul, Sk Saharukh Ali, Punam Debnath, Abantika Dey, Suparna Banerjee, Avik Saha, and Shrabani Mukherjee

21.1 Introduction

The Bay of Bengal fringe coasts along the regions of West Bengal and Odisha are highly contaminated due to the concentration of marine litter generated by human activities at present. Several activities like tourism and recreation, marine fishing, agricultural runoff, urban sewage materials, runoff of the untreated waste from aquaculture ponds, trading and commerce, shipping ports, cultural activities along the shores, and dry fish processing have generated litter in the region. The wetlands, sea beaches, sand dunes, tidal inlets, mangrove habitats, shorelines, lagoon fringes, backwaters, and halophytic grasslands are lined by accumulated litter in many places along the coast (Paul, 2002, 2022; Paul et al., 2017, 2018; Bugoni et al., 2001; Ryan, 2015). Agitated sea waters accumulate foaming waters along the shorelines of West Bengal and Odisha during the monsoon months (particularly during July and August) after the mixing of agricultural runoff and untreated fish pond waters with the sea waters. Both the frequencies and types of litters are estimated in the study along the selective coastal stations. However, the religious activities related to the gatherings at Gangasagar, Puri, and Chandrabhaga released the highest amount of litter on the seashores. Materials such as plastics, nylons,

---

A. Paul (✉) · A. Dey · S. Banerjee · A. Saha · S. Mukherjee
Department of Remote Sensing & GIS, Vidyasagar University, Midnapore, West Bengal, India
e-mail: anurupashis@gmail.com

J. Sardar
Centre for Environmental Studies, Vidyasagar University, Midnapore, West Bengal, India
e-mail: joydebsardar18@gmail.com

A. K. Paul · S. S. Ali · P. Debnath
Department of Geography, Vidyasagar University, Midnapore, West Bengal, India
e-mail: akpaul_geo2007@mail.vidyasagar.ac.in

A. K. Paul, A. Paul (eds.), *Crisis on the Coast and Hinterland*,
https://doi.org/10.1007/978-3-031-42231-7_21

polythene sheets, cloths, leathers, geo-fibre bags, and polypropylene bags create cover surfaces on the primary sand dunes, which are dominated by a colony of creepers, heaths, and grasses along the high tide lines. The soil fauna and colony expansion by plant communities suffer mostly in the strandlines of low-lying coasts, particularly in Sagar Island, Burahbolong river mouth, Rushikulya river mouth bars, and Paradeep shoreline. There are other materials like broken glasses, glass bottles, bulbs, thermocol, foams, plastic bottles, paper, cups made out of paper lined with plastic or wax films, tin, cans, and thick nylon ropes that produce problems in mangrove regeneration by restricting the germinated seeds from coming into contact with the soil surface in the coastal wetlands. The natural bioturbation processes by sandy mio-fauna (Uca sp.) were affected in many cases due to the accumulation of plastic in the high tide line around the sandy shores in West Bengal and parts of northern Odisha.

Marine litters are categorized in a scientific framework on the basis of types, sources, user communities, and areas of accumulation to achieve the sustainable goals in conservation and management of the shoreline of the region. The sensitive ecosystems of the coast, their environmental values, and their environmental rights are completely ignored by the user communities of natural resources through the dumping of litter and wastes into the emerald nature. A guideline is prepared to remove the litter and wastes from the sensitive habitats for the conservation of the coastal marine environment of the region of West Bengal and Odisha.

21.2 Study Area

In the current study, 16 stations along the coasts of West Bengal and Odisha were chosen to conduct research on the concentration and impact of marine litter (Fig. 21.1). Regionally, the coastal-marine environment of the two administrative states consists of deltaic wetlands, tidal rivers and creeks, estuaries and lagoons, beach ridge sand dunes, shoreline beaches, mangroves, saltmarshes, spits, and barrier bars. The Bay of Bengal fringe coasts under a high-energy environment act as a buffer against storms and erosion and filter the hazardous waste from the toxic effects on the sensitive environment.

21.3 Materials and Methods

Numerically, the types and sources of litter are estimated along the 16 shoreline stations at a specific distance and width of shoreline transects. The total number of litter items per transect length and transect width from each station is calculated to obtain the Clean Coast Index (CCI) in the study. The clean coast index is based on the following Eq. 21.1: (Alkalay et al., 2007; Chambault et al., 2018; Fernandino et al., 2016; Hardesty et al., 2016; Ross et al., 1991):

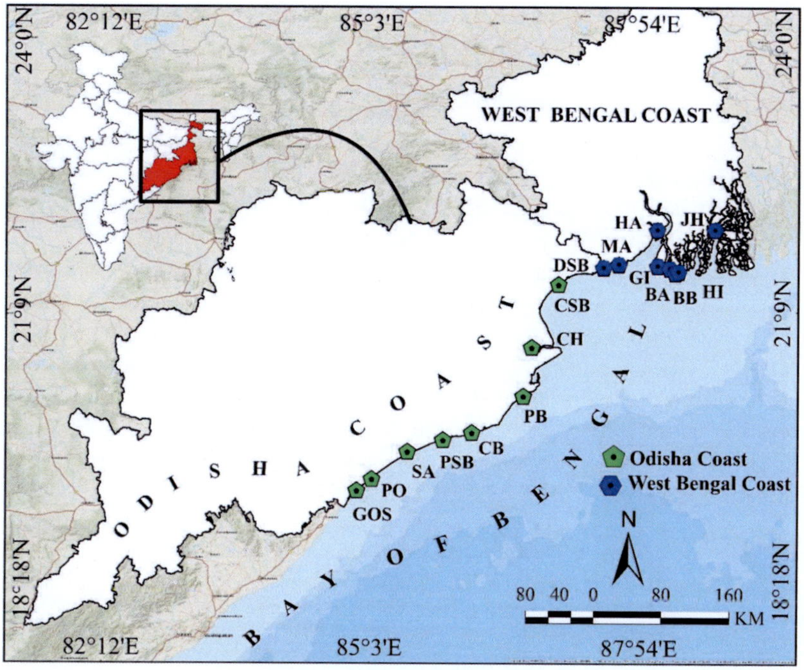

Fig. 21.1 The study area with location of 16 coastal sections for the assessment of impacts of the marine litters

$$CCI = (\text{Total no.of plastic items or marine litters}) / (\text{Total sampled area}) \times K \quad (21.1)$$

The total number of litters per m^2 area and coefficient $K = 20$ (the constant K value was set at 20 in the Mediterranean environment) are considered for the calculation of the index (Li et al., 2016; Munari et al., 2016; Carson et al., 2013; Okuku et al., 2021). The shoreline sections are classified based on very clean to dirty according to the scale provided in the present output of the index. However, the values of the CCI scale are classified as very clean (0.00–0.80), clean (0.80–1.60), moderate (1.60–2.40), and dirty (>2.40) in this calculation after computing the data.

21.4 Results and Discussion

The accumulation of litter reveals the role of the user community, which uses the natural resources of the locality and throws the debris or waste into the beaches, dunes, backshores, and wetlands to litter the coast. The types of user communities are classified as 27, and the four coastal stations in West Bengal and Odisha represent 14–16 types of users of the natural resources. However, nine coastal sections of

Odisha and West Bengal show 10–12 types of user communities, and the remaining three coastal sections are dominated by 6–8 user communities of natural resources. Potagada, Chandrabhaga, and Henry's Island of the region are used by only a selective community (Table 21.1). The areas of the three coastal stations are however noted for their ecotourism significance, so they survive as less littered environments along the shorelines.

Several sensitive coastal habitats of the shoreline sections are affected by the concentration of marine litter. They are categorized as sandy, clayey, and watery habitats of the alluvium coast. The high-energy marine environment in many sections of the coast drifted the litters into mangrove wetlands, saltmarshes, lagoons, and beach ridges. Problems of anchorage and rooted conditions caused by mangrove seeds are experienced in the wetland soils due to the dumping of litter along the deltaic shores of the Sundarban and Bhitarkanika. Several bioturbated mudbanks and beach plains of Mandarmani, Chandipur, Talsari, Puri Marine Drive, and Gopalpur were obstructed by thick deposits of litter or debris from waste.

21.4.1 Clean Coast Index (CCI)

The 16 coastal sections of Odisha and West Bengal are selected for the estimation of litters along the shorelines. In the study, the types of litter are identified at a specific distance and width along the shorelines to obtain the frequency of items per square metre area. After calculating the computed data on the frequency of litter items per square metre area, the area was again classified as very clean (five stations), clean (another five stations), moderate (five stations), and dirty (only one station) coast (Fig. 21.2, Table 21.2).

The dirty and moderate classes of coasts require immediate measurements by operating cleaning processes through the activities of NGOs and local people for different coastal sections. The shorelines of West Bengal and Odisha, on the other hand, are influenced by the seasonal direction of longshore currents and littoral drifts, and fed by a series of estuaries, embayments, and backwaters. The debris of litters is accumulated and reworked into the marshes, swamps, beaches, dunes, and estuarine islands by the littoral drifts during storms and southwest monsoon currents in the region. Sometimes they are exported into the sensitive ecosystems of the coast fringe wetlands and play a significant role in the degradation of soil conditions by trapping the mangrove seeds and by blocking the seepage of waters with nutrient loads into the soils. Thus, patches of open space are found into the shore fringe halophytic grasslands and swampy tracts as a result of the concentrated accumulation of debris or drifted litter.

Table 21.1 Types of marine litters accumulated in the coastal sections of Odisha and West Bengal

Location	Types of litters	Accumulation areas of sensitive environment	User communities
Gopalpur on sea	Tourists, urban areas, sea defences, fishermen, picnicking, beach stalls, fishing harbours, shipping ports, dry fish processing platforms, and foresters	Beach, wetland, dune	Tourists, fishermen
Potagada	Ships and trawlers, tourists, sea defences, fishermen, recreation-boating and agriculture firms,	River bank	Tourists, fishermen
Satapada	Ships and trawlers, industries, tourists, fishermen, beach stalls, cultural festival, agriculture fields, dry fish processing platforms, festival and fair grounds, lime-factory	Lagoon fringe area	Tourists, locals, and trade and commerce
Puri Sea Beach	Urban areas, tourists, sea defences, fishermen, placer mining, agriculture firms, hotel and restaurants, agriculture fields, clinical debris, hawkers, cremation grounds/animal Caracas and lime-factory	Beach and dune	Tourists, religious gathering, fishermen, trade and commerce, and visitors
Chandrabhaga Beach	Tourists, fishermen, beach stalls, agriculture firms, hotel and restaurants, agriculture fields, eco-tourism sites, and hawkers	Beach and dune	Tourists and religious gathering
Paradeep Beach	Ships and trawlers, urban areas, tourists, fishermen, salt processing, beach stalls, shipping ports, ferry services, cultural festival, agriculture fields, hawkers, lime-factory	Beach and dune	Tourists, fishermen ports, industrial activities, and mining
Chandvali	Ships and trawlers, urban areas, tourists, fishermen, eco-tourism sites, beach stalls, shipping ports, cultural festival, agriculture fields, hotel and restaurants, foresters, festival and fair grounds, hawkers, and lime-factory	River bank and forest	Tourists, locals, trade and commerce, festival, ferry services
Chandipur Sea Beach	Ships and trawlers, tourists, fishermen, foresters, picnicking, beach stalls, fishing harbours, ferry services, hotel and restaurants, agriculture fields, dry fish processing platforms, and cultural festival	Beach, dune, and estuary	Tourists, locals, fishermen
Digha Sea Beach	Ships and trawlers, urban areas, tourists, sea defences, fishermen, foresters, picnicking, beach stalls, agriculture firms, fishing harbours, cultural festival, hotel and restaurants, shipping ports, dry fish processing platforms, irrigation departments, clinical debris, and hawkers	Beach, dune, and wetlands	Tourists, locals, fishermen, trade and commerce, and harbours

(continued)

Table 21.1 (continued)

Location	Types of litters	Accumulation areas of sensitive environment	User communities
Mandarmani	Ships and trawlers, tourists, fishermen, foresters, salt processing, picnicking, irrigation departments, agriculture firms, hotel and restaurants, agriculture fields, dry fish processing platforms, and hawkers	Beach, dunes, and wetlands	Tourists, locals, fishermen, and hotel business
Haldia	Ships and trawlers, urban areas, tourists, sea defences, fishermen, hotel and restaurants, agriculture firms, fishing harbours, shipping ports, ferry services, hotel and restaurants, agriculture fields, irrigation departments, clinical debris, and industries	Estuary, wetlands, and flood plain	Industry port, visitors, and locals
Gangasagar	Tourists, fishermen, foresters, picnicking, cultural festival, hotel and restaurants, dry fish processing platforms, irrigation departments, eco-tourism sites, festival and fair grounds, and cremation grounds/animal Caracas	Beach, dune, wetlands forests	Festival, religious gathering, fishermen, tourists, and foresters
Baliara	Ships and trawlers, tourists, sea defences, fishermen, picnicking, beach stalls, fishing harbours, ferry services, hotel and restaurants, foresters, and eco-tourism sites	River bank, island, forest, and beach	Tourists and locals
Bakkhali	Ships and trawlers, tourists, sea defences, fishermen, foresters, picnicking, beach stalls, agriculture firms, fishing harbours, cultural festival, hotel and restaurants, irrigation departments, hawkers, and cremation grounds/animal Caracas	Beach, dune forest, and mangroves	Tourists, hotel business, foresters, aquaculture farming, and fishing harbours
Henry	Tourists, fishermen, foresters, agriculture firms, agriculture fields, eco-tourism sites	Beach, wetlands forests, mangroves	Tourists, fishing, aquaculture
Jharkhali	Ships and trawlers, tourists, fishermen, foresters, picnicking, beach stalls, ferry services, hotel and restaurants, irrigation departments, eco-tourism sites, and cremation grounds/animal Caracas	Mangroves forests, estuary	Tourists, locals, and fishermen

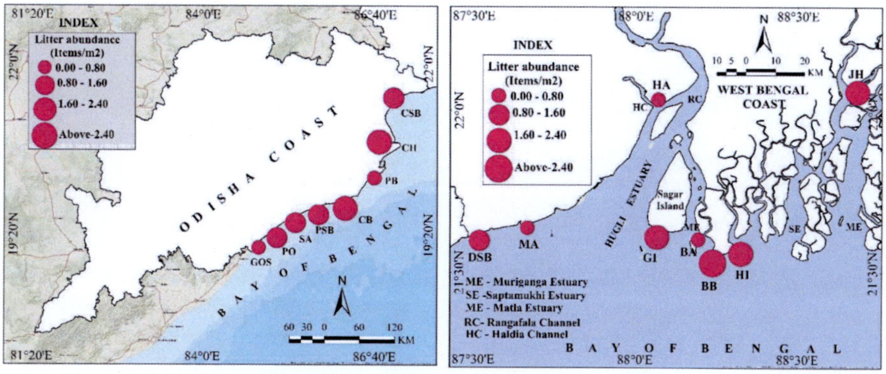

Fig. 21.2 The result of litter abundance along the shoreline sections of Odisha and West Bengal

21.4.2 Sources and Types of Marine Litters

Chandvali, Digha, Haldia, and Bakkhali coastal sections represent a maximum accumulation of 14–16 types of litter. The areas are visited by tourists, visitors, and travel operators and communicated by transport services. Locals, the fishing community, authorities of ports and harbours, large-scale gatherings by cultural festivals and industries, as well as activities of trade and commerce produce varieties of litter that are mostly nondegradable items. Among them are plastics, glasses, foams, fibres, hard leathers, cloths, polythene, nylons, medicinal waste, cigarette buts, thermocol, electronic cables, geo-tube bags, coconut shells, animal Caracas, etc. There are nine coastal sections in which 10–12 types of litter are found. They include as Gopalpur, Satapada, Puri, Paradeep, Chandipur, Mandarmani, Gangasagar, Baliara, and Jharkhali in the region. As a result of the estimation of the litters, it is observed that the places are mostly visited by tourists, visitors, and people involved in the operation of transport services, travel groups, trade and commerce, and hotel and restaurant businesses. The religious gatherings, cultural festivals, fishermen's communities, beach placer mining activities, shipping ports, and fishing harbours generate debris and litter in the coastal sections of Odisha and West Bengal (Fig. 21.3).

However, some sections of the coastal region, that is, Potagada, Chandrabhaga, and Henry's Island, are visited by selective tourists for their ecological significance. The study reveals that six to eight types of litter are found in the coastal sections, particularly due to eco-tourists, local fishing activities by rural fishermen communities, aquaculture-related activities, gathering during festivals, and maintaining the cleaning process by local authorities (Fig. 21.4, Table 21.1). The types of litter collected from the beaches, mangrove forests, sand dunes, and river banks included as plastics, nylons, bags, glasses, medicinal wastes, coconut shells, cloths, rope, tins, gas lights, fishing nets, floats, plastic bottles, body spray bottles, electric bulbs, bricks, and burnt wood. Thus, the user communities of waste materials may be categorized as visitors, tourists, fishermen, and people relying on natural resources in the region.

Table 21.2 The final output of CCI values and their rankings for the coastal sections of Odisha and West Bengal

Location	Beach Id	Total number of marine debris items per transects	Length of the transect (m)	Width of the transect (m)	CCI	CCI Rank
Gopalpur on sea	GOS	5444	1816	171	0.36	Very Clean
Potagada	PO	2118	706	70	0.86	Clean
Satapada	SA	1590	318	80	1.25	Clean
Puri Sea Beach	PSB	7448	1862	86	0.94	Clean
Chandrabhaga Beach	CB	4081	800	50	2.05	Moderate
Paradeep Beach	PB	4050	1350	296	0.21	Very Clean
Chandvali	CH	1270	254	50	2	Moderate
Chandipur Sea Beach	CSB	3976	994	70	1.15	Clean
Digha Sea Beach	DSB	9756	1626	115	1.05	Clean
Mandarmani	MA	2611	544	125	0.77	Very Clean
Haldia	HA	2940	980	83	0.73	Very Clean
Gangasagar Island	GI	6594	942	75	1.87	Moderate
Baliara	BA	2655	1062	85	0.59	Very Clean
Bakkhali Beach	BB	3500	500	52	2.69	Dirty
Henry's Island	HI	1488	372	35	2.29	Moderate
Jharkhali	JH	1220	610	20	2	Moderate

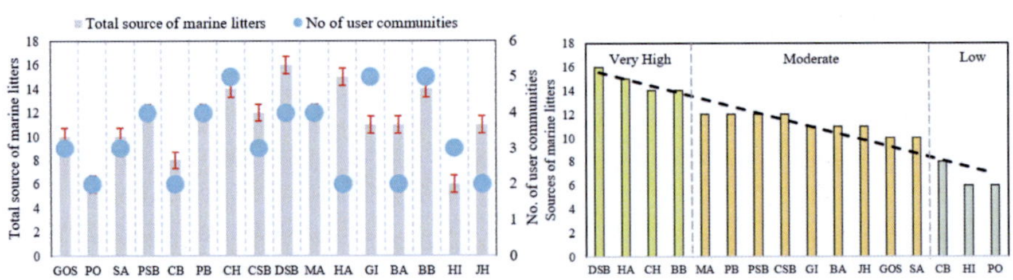

Fig. 21.3 The section-wise total sources of marine litter produced by user communities along the coasts and their rankings

21.4.3 Suggested Management Practices

The litters from the marine fronts, local sites, and drifted materials from distant areas accumulate in the coastal fringe beaches, sand dunes, forests, and inlet channel mouths at a rate that depends on the number of visitors, the location of beach stalls, hotels, and restaurants, urban areas, local activities, and large-scale gatherings during festivals in the coastal sections. Similarly, the winds, tides, waves, currents, and storm surges transport and rework the waste and play significant roles as carriers of litter fragments from distant land to the sensitive habitats of the low-lying coasts. The attractions of natural sites may be lost to visitors in the near future as a result of debris accumulation. Thus, a framework of management is recommended in the work to inform the coastal managers and administrators of their supporting data base when designing the management practices for any coastal section of the

Fig. 21.4 (**a**) Glass bottle litter transferred by boat from Mousuni Island; (**b**) Multiple items of litters accumulated on the halophytic grassland of Bakkhali; (**c**) Litter debris produced by fishing activities in the Petuaghat harbour; (**d**) Broken trawls and their fragments sunk into the mudflat after the storm in the fishing harbour of West Bengal; (**e**) Plastics and thermocol plates drifted into the mangrove swamp floor at Haripur section of Saptamukhi estuary; (**f**) Unwanted litters accumulated at the base of an erosive sand dune at Digha; (**g**) Litters along the high tide line drifted and accumulated on Fedric Island; (**h**) Beach stalls and related litters on the Puri sea beach of Odisha; and (**i**) The items of litter thrown away into the forest fringe shoreline of Chandrabhaga after the landfall of a cyclonic storm

region. Management issues, programme implementation authorities, funding management, and policy interventions are considered in the framework of management for the clearing and removal of wastes or litter fragments from the intertidal habitats (Table 21.3).

The municipality authorities of the coastal urban sections should use the beach cleaning equipment for removing beach pollution such as glasses, syringes, plastics,

Table 21.3 A framework of management practices suggested for marine litters in the coastal marine environment

Management Issue	Programme Implementation	Funding Management	Policy Interventions
Beach cleaning management	Local authority, local government, NGOs, and institutions	Public funding, private funding, and environment tax	CRZ Act, ICZM & EPA
Dune cleaning assessment	Local authority, forest department and institutions	Public funding, private funding, and environment tax	CRZ Act, ICZM & EPA
Wetland cleaning activities	Local authority, NGOs, forest department, and individuals	Public funding, private funding, and environment tax	CRZ, ICZM & Wetland Acts
Visitor management	Environment volunteers, NGOs, and local authority providing with wayfinding signs: identification, directional, informational, and regulatory	Public funding, environment tax, and ecosystem service bond	Environmental Zoning Approach, Municipality Act & Regular cleaning process by the volunteers
Urban waste management	Municipality, Local Development Board, local government, and institutions	Private funding from hotelier association, and public funding	CRZ & Municipality Acts
Aquaculture ponds related waste management	Fishery department using treatment plants	Public funding, and private funding	CRZ & Wetland Acts
Agricultural waste disposals management	Agriculture department using treatment plants	Public funding, private funding for large land holdings	Environment Protection Act
Navigation and trawling related waste management	Port authority and harbour authority, NGOs, and coast guards	Funding from port and harbour authorities, and coast guard funding	Environment Protection Act & CRZ Act
Fishermen colony related waste management	Environment management authority, State Planning Board	Public funding and environment bonds	Environment Protection Act & CRZ Act

cigarette buts, and other unwanted debris. The beach tech cleaning machine is generally used to remove oil from beach sands after oil spill disasters, but the port authorities and estuary management authorities such as Ganga Action Plan (GAP) also may be involved in the coastal pollution cleaning process for the region. The involvement of NGOs will support the government authorities in implementing the programmes at the local level for cleaning and removing the litter pollution from the wetland habitats of the coastal region. Coastal Regulation Acts (2018–2019) and Environment Protection Acts (1986) should be strong enough to protect sensitive coastal habitats from pollution and litter impacts.

Awareness among tourists, visitors, excursionists, people related to trade and commerce, fishing activities, placer mining, and activities involving untreated waste dumping into the wetlands is immediately needed to restore the natural habitats from the impacts of marine litter. NGOs may be involved in this purpose to raise awareness of environmental issues among visitors and others about the importance of coastal cleaning processes. Environment bonds and Environment Service Taxes are also needed to be implemented over the user communities of coastal marine resources in the near future for the protection of the environment. Coastal urbanization has expanded along and across the wetland habitats, shorelines, and fringe forests of Puri, Gopalpur on the sea, Paradeep, Digha, Haldia, and Diamond Harbours sections of Odisha and West Bengal. Environment friendly urban waste management should be implemented in such coastal urban units immediately to reduce the release of sewage effluents into the sea, wetlands, and tidal creeks in untreated conditions. Beach litter pollutants are often explored at the sites of sewage outlet channels across the shorelines in many cases. Thus, waste treatment by natural filtering of slurry products from sewage effluents may be conducted in the urban centres using the sand pits across the beach ridges of the coast.

Agricultural washout deposits with nutrient loads and aquaculture ponds released nutrient loads into the coastal lagoons, back waters, estuaries, and mangrove fringe wetlands, causing eutrophication and producing accumulated litter along the shorelines on the dynamic coast. Such litters are explored from the lake Chilika beds, estuary fringe tidal flats, and Bhitarkanika and Sundarban mangrove fringe islands after the landfall of tropical cyclones in the regional settings. Treatment of effluents is also needed immediately to protect the rate of litter accumulation on the shoreline at present. As per the guidelines of MOEF, an EIA is needed at every 2-year interval period for the assessment of urban fringe marine water quality in the coastal sections and the beach pollution cleaning quality of the region.

21.5 Conclusion

The following conclusion is drawn on the basis of the above study. Marine litters are now considered a big environmental issue in the coastal management. The environment lobbying is surpassed by the hoteliers lobby, the industrial lobby, the trade and commerce lobby, and the traditional customary rights of the coastal poors in regulating the environment for the conservation of coastal habitats. In the study, 16 coastal stations in Odisha and West Bengal states are considered for assessment of the impacts of marine litter. On the basis of the coastal cleanness index, they are classified into four categories (very clean, clean, moderate, and dirty), with three West Bengal stations and two Odisha stations representing the very clean coast. However, one coastal station of the Indian Sundarbans in West Bengal shows a dirty coast as pert as the output of CCI for the entire study area. The types and sources of litter are also estimated for each coastal section to identify the user communities of natural resources in the coastal marine environment. It is observed that tourists, visitors, fishery sectors, urban managers, people of trade and commerce, hotel and restaurant workers, and coastal poors are using the resources of the coast and polluting the coastal marine environment.

The seasonal drifting of pollutants and storm-drift marine litters encroaches into the halophytic grasslands, mangrove wetlands, sea beaches, and sand dunes of the coastal sections to hinder the growth and regeneration of flora and fauna. An appropriate management framework is prepared in the study to minimize the rate of litter accumulations in the sensitive coastal habitats of the tropical environment.

References

Alkalay, R., Pasternak, G., & Zask, A. (2007). Clean-coast index—a new approach for beach cleanliness assessment. *Ocean and Coastal Management, 50*(5–6), 352–362. https://doi.org/10.1016/j.ocecoaman.2006.10.002

Bugoni, L., Krause, L., & Petry, M. V. (2001). Marine debris and human impacts on sea turtles in southern Brazil. *Marine Pollution Bulletin, 42*(12), 1330–1334.

Carson, H. S., Lamson, M. R., Nakashima, D., Toloumu, D., Hafner, J., Maximenko, N., & McDermid, K. J. (2013). Tracking the sources and sinks of local marine debris in Hawai 'i. *Marine Environmental Research, 84*, 76–83.

Chambault, P., Vandeperre, F., Machete, M., Lagoa, J. C., & Pham, C. K. (2018). Distribution and composition of floating macro litter of the Azores archipelago and Madeira (NE Atlantic) using opportunistic surveys. *Marine Environmental Research*. https://doi.org/10.1016/j.marenvres.2018.09.015

Fernandino, G., Ellif, I. C., Silva, R. I., Brito, T. S. A., & Bittencourt, C. S. P. (2016). Plastic fragments as a major component of marine litter: A case study in Salva-dor, Bahia, Brazil. *Journal of Integrated Coastal Zone Management, 16*(3), 281–287. https://doi.org/10.5894/rgci649

Hardesty, B. D., Lawson, T., Van Der Velde, T., Lansdell, M., & Wilcox, C. (2016). Estimating quantities and sources of marine debris at a continental scale. *Frontiers in Ecology and the Environment, 15*(1), 18–25. https://doi.org/10.1002/fee.1447

Li, W. C., Tse, H. F., & Fok, L. (2016). Plastic waste in the marine environment: A review of sources, occurrence and effects. *Science of the Total Environment, 566–567*, 333–349. https://doi.org/10.1016/j.scitotenv.2016.05.084

Munari, C., Corbau, C., Simeoni, U., & Mistri, M. (2016). Marine litter on Mediterranean shores: Analysis of composition, spatial distribution and sources in north-western Adriatic beaches. *Waste Management, 49*, 483–490. https://doi.org/10.1016/j.wasman.2015.12.010

Okuku, E. O., Kiteresi, L., Owato, G., Otieno, K., Omire, J., Kombo, M. M., Mwalugha, C., Mbuche, M., Gwada, B., Wanjeri, V., Nelson, A., Chepkemboi, P., Achieng, Q., & Ndwiga, J. (2021). Temporal trends of marine litter in a tropical recreational beach: A case study of

Mkomani beach, Kenya. *Marine Pollution Bulletin, 167*, 112273. https://doi.org/10.1016/j.marpolbul.2021.112273

Paul, A. K. (2002). *Coastal geomorphology and environment* (pp. 1–342). ACB Publication.

Paul, A. K. (2022, August). Dynamic behaviour of the estuaries in response to the phenomenon of global warming in the coastal ecosystems of West Bengal and Odisha, India. In *Transforming coastal zone for sustainable food and income security: Proceedings of the International Symposium of ISCAR on Coastal Agriculture, March 16–19, 2021* (pp. 907–931). Springer International Publishing.

Paul, A. K., Ray, R., Kamila, A., & Jana, S. (2017). Mangrove degradation in the Sundarbans. In *Coastal wetlands: Alteration and remediation* (pp. 357–392). Springer. https://doi.org/10.1007/978-3-319-56179-0

Paul, A. K., Kamila, A., & Ray, R. (2018). *Natural threats and impacts to mangroves within the coastal fringing forests of India* (Vol. 25, pp. 105–140). Springer Nature, Coastal Education and Research Foundation, JCR. https://doi.org/10.1007/978-3-319-73016-5

Ross, J. B., Parker, R., & Strickland, M. (1991). A survey of shoreline litter in Halifax Harbour 1989. *Marine Pollution Bulletin, 22*(5), 245–248. https://doi.org/10.1016/0025-326X(91)90919-J

Ryan, P. G. (2015). A brief history of marine litter research. In *Marine anthropogenic litter* (pp. 1–25). Springer. https://doi.org/10.1007/978-3-319-16510-3_1

Chapter 22
Groundwater Contamination due to Saline Water Encroachment in Coastal Andhra Pradesh, with Particular Emphasis on the Morphological Units

Anurupa Paul, Ashis Kumar Paul, Joydeb Sardar, Bubay Bid, and Subhajit Mandal

22.1 Introduction

The younger coastal plains of Holocene to Recent-subrecent sediments in Andhra Pradesh are geomorphologically and lithologically prone to saltwater ingression into the coastal aquifers like other states of the eastern coast of India. As a result of the composition of unconsolidated alluviums, over extraction of groundwater for irrigating the delta plains as well as coastal plains, spread of coastal urbanization, reduction in freshwater flows into the downstream sections of coastal plain river systems, climate change reworking of the coastal landscapes by sea level rise threats, coastal erosion, and impacts of storm surge inundations, the groundwater is contaminated by such events of saline water encroachments. The sand-silt-clay and gravel beds with deltaic and marine alluviums provide favourable conditions for percolation and seepage flow at the land-water interface if the region is over pumped for groundwater uses and the gradient of the hydraulic head is reduced in the aquifer along the younger coastal plain of the coastal state.

The study demonstrates lithological and morphogenetic regions, rainfall diversity zones, land use and land cover types, sea level rise threats, storm surge inundation vulnerabilities, and groundwater quality of a few sections of the coastal. Using

A. Paul (✉) · B. Bid · S. Mandal
Department of Remote Sensing & GIS, Vidyasagar University, Midnapore, West Bengal, India
e-mail: anurupashis@gmail.com

A. K. Paul
Department of Geography, Vidyasagar University, Midnapore, West Bengal, India

J. Sardar
Centre for Environmental Studies, Vidyasagar University, Midnapore, West Bengal, India
e-mail: joydebsardar18@gmail.com

A. K. Paul, A. Paul (eds.), *Crisis on the Coast and Hinterland*,
https://doi.org/10.1007/978-3-031-42231-7_22

geospatial techniques, existing secondary data on groundwater conditions, other physical components of the coastal zones, and surface runoff characteristics with drainage systems and climate change threats, a spatial index of saltwater encroachments is prepared to delineate the boundaries of saltwater contamination zones into the coastal aquifers of the state. Such a method of study as well as the database will provide the ingredients for sustainable water uses and the management of contamination zones in the coastal belts of regional settings. Similarly, the method of spatial indexing for saltwater contamination would also be applicable for other coastal regions along the eastern shoreline of India. Coastal urbanization will be threatened by such aquifer contamination from saltwater encroachments influenced by sea level rise threats and over-extraction of groundwater resources in the near future. Many researchers studied on various aspects of groundwater quality in coastal urban areas (Rao et al., 2002), saltwater intrusion into coastal aquifers through model studies (Datta et al., 2009), sources of saltwater intrusion in Koleru lake (Harikrishna et al., 2012), investigation of saline water intrusion in Godavari delta (Naidu et al., 2013) using an integrated approach, effect of aquaculture practices in the groundwater in east Godavari district (Penmetsa et al., 2013), groundwater contamination types in Srikakulam (Rao et al., 2011) because of saline water intrusion, types of saltwater intrusion pathways using remote sensing techniques (Dhakate et al., 2016), seawater intrusion in Visakhapatnam (Mamatha et al., 2022), and investigation of groundwater quality using geostatistical techniques (Sangadi et al., 2022) for the Andhra Pradesh coast.

To reduce the contamination level of groundwater quality, the following measurements may be recommended for the coastal parts of the state. Restricted use of groundwater in the coastal urban centres, construction of artificial recharge pits, sharing of freshwater discharges into the downstream sections of the coastal streams, and rainwater harvesting should be practised immediately to reduce the rate of saltwater ingression into the coastal groundwater aquifers.

22.2 Materials and Method

The present work is carried out in the coastal tracts of Andhra Pradesh to find out the physical, environmental, and human factors as per their importance in the assessment of groundwater contamination potentials with saltwater ingression (Fig. 22.1a). A total of 10 factors are assessed for the selective coastal sections with the weighted sum method to set the integrated scores according to the importance of the factors in the regional settings. All the factors are analysed using Shuttle Radar Topography Mission (SRTM) DEM, Landsat 8 Operational Land Imager (OLI) satellite data, Google Earth Images, and other aspect data from India Meteorological Department (IMD), Indian National Center for Ocean Information Services (INCOIS), the Intergovernmental Panel on Climate Change (IPCC), the Survey of India, the Central Ground Water Board (CGWB) report, and the Central Water Commission (CWC) handbook for calculating the scores to identify the potential contamination zones.

The future prediction of the saltwater ingression limit into the groundwater aquifers is estimated in this work using the Analytical Hierarchy Process (AHP) method

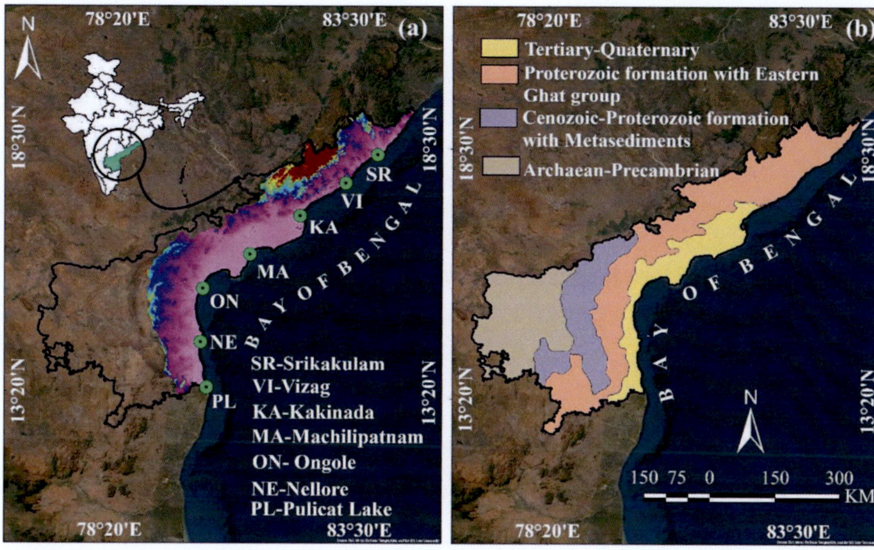

Fig. 22.1 (**a**) Coastal Andhra Pradesh and (**b**) Coastal geology of the state of Andhra Pradesh

and the fuzzy membership method to develop index values for zonal concentrations of saltwater contamination in the groundwater. However, the validation of such a prediction model is conducted using the receiver operating characteristic curve (ROC), and their accuracy level is tested with the AUC and AHP methods for the output of significance. The calculation shows that the AUC is more accurate at the 85.5% level than the AHP method is at the 81.0% for application in the region. Finally, the result of the AHP method and fuzzy membership method denotes the future trend and regional diversity in saltwater encroachment in Andhra Pradesh coastal tracts.

The multi-criterion decision matrix (AHP method) is used in weighted overlay tools to identify the hotspot areas of the saltwater ingression zones in the present study area considering the following Eq. 22.1. A few researchers have applied this equation in investigating the groundwater potential zones and their mapping in other areas of the country (Parimala & Lopez, 2012):

$$S = \sum_{i=1}^{n} W_i \times X_i \tag{22.1}$$

where, S refers to the suitability index for each pixel map. W_i is the assign weight of the ith layer and X_i score of the ith each criteria layer. n refers to the number of suitability layer. The future prediction of the model is calculated using the following Eq. 22.2:

$$Gp = \frac{1}{K} \sum k^{th} v^{\text{response}} \tag{22.2}$$

where, Gp refers to the groundwater prediction and k indicates the separate trees in the method. The performance of the model accuracy has been evaluated in the validation of the saltwater ingression using the following Eq. 22.3:

$$\text{Accuracy} = \frac{TP + TN}{TP + TN + FP + FN} \tag{22.3}$$

where, TP refers to the true positives, TN indicates true negatives, FP refers to the false positives (error type I), and FN refers to the false negatives (error type II). TP and TN are the numbers of pixels that are correctly grouped.

22.3 Results and Discussion

This section represents ten factors and their characteristics in influencing the groundwater contamination potentials by saltwater ingression. The future potential areas and their characteristics are also calculated to highlight the environmental crisis of the region related to saltwater encroachments.

22.3.1 Coastal Geology and Geomorphology

Coastal Andhra Pradesh comprises four major groups of rocks that extend gradually from the shoreline to the hinterland areas. They belong to (i) the Quaternary-Tertiary group of rocks constituting thick alluviums, laterites, and sandstones along the Bay of Bengal coastal zones; (ii) the crystalline sediments, on the other hand, represent thick deposits of sediments in the Cuddapah basin from the upper Precambrian to early Precambrian group of rocks; (iii) the Deccan traps with basalt and other intrusive rocks in combination with Intertrappean bed materials belong to the Cretaceous. Infratrappean beds of Rajahmundry areas comprise limestone and sandstones that underlie the lava trap flows. (iv) However, the Archaean-Precambrian group of rocks includes Granite, Gneiss, Schists, the Bended Gneissic Complex, Khondolites, and Charnockites in the western margin of Andhra Pradesh, particularly along the eastern ghat exposures. Among the geologic units, the Tertiary-Quaternary groups are considered potential aquifers of the groundwater quality (Figs. 22.1b and 22.2a).

In geomorphology, the coastal state may be classified as (i) rocky coast; (ii) coastal plain and delta plain alluviums; (iii) basinal structure of the west central highlands; and (iv) elevated ridges of south western tracts. The coastal plains and delta plains of Godavari R., Krishna R., and Pennar R. sections extend along the Bay of Bengal shoreline towards the east, and the rocky coast, backed by hills and valleys, extends in Vishakhapatnam, Vijaynagaram, and Srikakulam districts of the state towards the northeast. Topographically, the coastal tracts

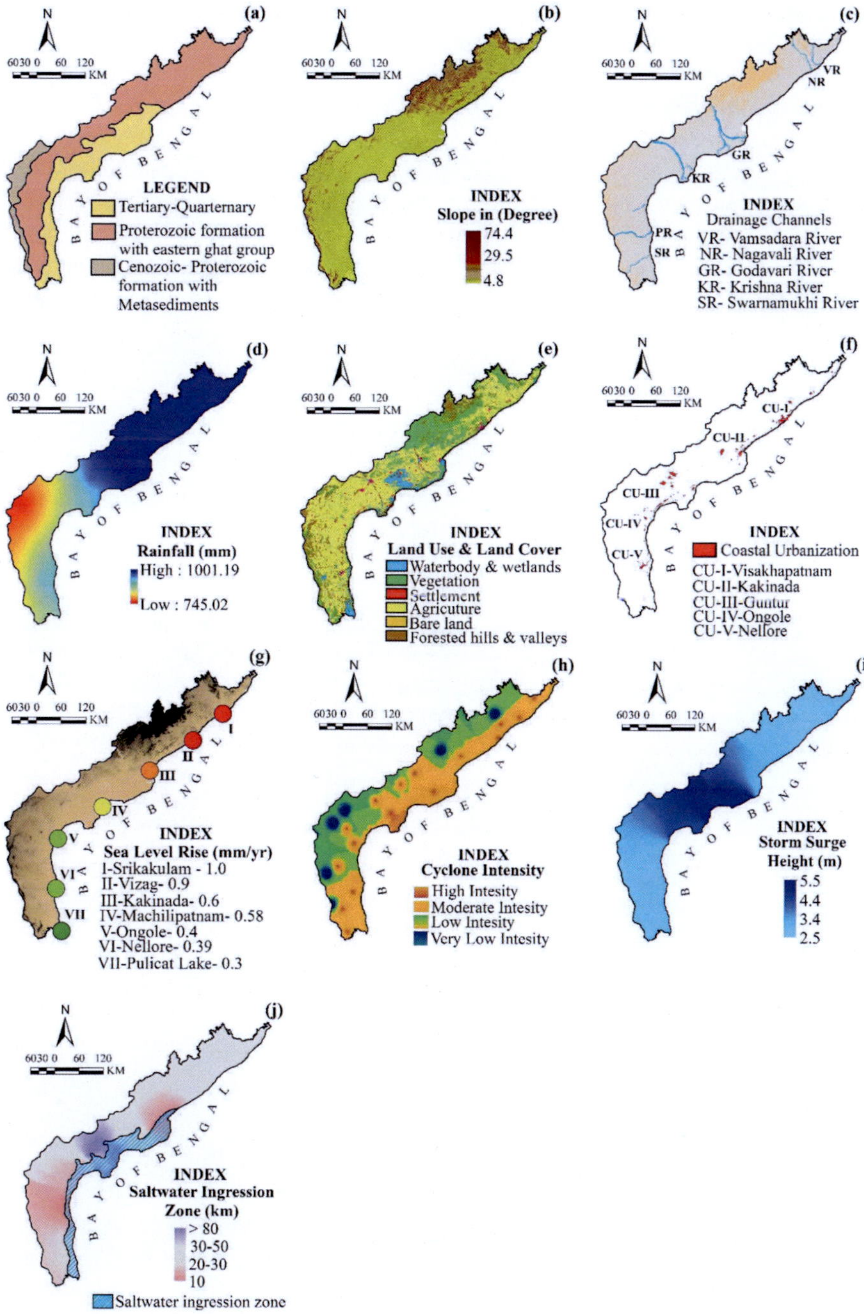

Fig. 22.2 Various aspects of physical, environmental, and human factors of the coastal Andhra Pradesh showing (**a**) Coastal Geology, (**b**) Slope, (**c**) Drainage, (**d**) Rainfall, (**e**) Land Use and Land Cover, (**f**) Coastal urbanization, (**g**) Rates of sea level rise rate, (**h**) Cyclone intensity, (**i**) Storm surge, and (**j**) Saltwater ingression zone

are backed by highland surfaces elevated up to 800 m in height and slope towards the east and southeast directions. The flat, low-lying delta plain surface is composed of fluvio-marine depositional alluviums with beach ridges, levee deposits, paleo channels, and estuarine deposits. Geomorphological surfaces from the west to the east have been documented from the Polarimetric Phased Array L-band Synthetic Aperture Radar (PALSAR) DEM. The hills, ridges, highlands, basins, and plain lands are descending at the steps from west to east and cut by drainage systems following the regional surface gradients of the coastal state. The river systems of Godavari, Krishna, Pennar, Vamshadhara, Gosthani, and Nagavali are flowing to the Bay of Bengal, cutting transversely to the shoreline from west to east. Wider delta plains are produced by the Godavari, Krishna, and Pennar rivers in the southeastern tract and lie 10–20 m in heights above the mean sea level. Several palaeo channels, beach ridges, and levees provide the conditions for transporting sea waters inland in the region of coastal plain alluviums (Figs. 22.2b and 22.3).

Fig. 22.3 Sandy and Graveliferous materials of the coastal tracts prone to saline water intrusion into the groundwater in Visakhapatnam and Bheemunipatnam areas along the Bay of Bengal shoreline

22.3.2 Coastal Drainage Features

Freshwater flow into the downstream sections declined rapidly along the drainage systems due to human-controlled water and sediment discharges and reduced rainfall due to climate change impacts in the previous decades. Several studies by the researchers and the CWC database of river discharges in the peninsular drainage channels demonstrated a 40% reduction after 1960 (Das, 2021; Das et al., 2022). Due to a significant reduction of fresh water flow in the lower courses, the river beds are silted up and filled with the deposition of bars, shoals, and islands (Fig. 22.2c). The encroachment of tidal flows into the upstream channels transported sea waters and occupied the floodplain wetlands across the channels in the region. Deltaic distributary channels, Paleo channels, and low-lying wetlands are usually influenced by marine environments and saltwater flooding, particularly during the landfall of cyclones and the impacts of storm surges. As the gradient of downstream sections is very low in the coastal plain drainage channels, the saltwater encroachment advanced landward sections at high sea levels. The six major river systems of the coastal tract are used as passage for transporting saltwaters into the landward sections of the state.

22.3.3 Land Use and Land Cover Types

The delta plains and coastal plains of Godavari R., Krishna R., and Pennar R. are intensively cultivated by multiple cropping patterns using groundwater irrigation. A number of settlement sites and urban centres have emerged in the coastal plains in expanses of wetlands and alluvium tracts in the west Godavari, east Godavari, Krishna, Guntur, Nellore, and Prakasham districts. Similarly, urbanization processes expanded in Vishakhapatnam, Vijaynagaram, and Srikakulam districts of the highlands coast of Andhra Pradesh. The municipality's towns and other urban water supplies are also overly dependent on the pumping of groundwater in the region. Various land use practices like salt manufacturing pans, aquaculture ponds, and placer mining sites occupy the coastal fringes of estuary banks, lagoons, backwaters, bay side areas, sand dunes, beach ridges, and delta plains. Coastal wetland habitats such as mangroves, saltmarshes, tidal flats, spits, beaches, sand dunes, and river mouths are under stress due to an overdependence on natural land cover resources. Such a rapid rate of land use conversion and over-extraction of the groundwater is causing fluctuating aquifer levels in the region (Figs. 22.2e and 22.3). Kakinada and Visakhapatnam cites have occupied the coastal fringe landscapes and modified their habitats.

22.3.4 Variation in Rainfall

There is a significant variation in average annual rainfall distribution patterns between northeastern districts and other eastern and southeastern districts of the coastal state. The national average annual rainfall map shows that Srikakulam, Vijaynagaram, Visakhapatnam, West Godavari, East Godavari, Krishna, and parts of Nellore districts received above 1000 mm of rainfall during the year 2021. Guntur and other areas of Nellore district received 900–1000 mm of rainfall. However, the coastal districts of Guntur, Prakasam, Chittoor, and the eastern parts of Kadappa received 800–900 mm of rainfall during this time. The coastal districts were affected by the seasonal variation of rainfall, for example, the high rainfall (62.6–174.1 mm) in the southwest monsoon months (May–November) and low rainfall in the northeast monsoon months (December–April), when only 6.5–23.2 mm of rainfall were recorded in the region (Fig. 22.3d). The amount of such rainfall plays a significant role in recharging the groundwater and surface runoff towards the Bay of Bengal following the surface gradient.

22.3.5 Population and Urbanization

Population concentration, density, and coastal urbanization play a significant role in the consumption of water from the groundwater aquifers. Intensity of cropping and prolific urban growths, as well as urban sprawling and irrigation of the productive alluviums, need water.

Thus, over-extraction or overpumping of groundwater is practised to support the population and urbanization in the coastal plains of Andhra Pradesh. Major urban centres of the region include Vishakhapatnam, Kakinada, Machilipatnam, Guntur, Nellore, Chirala, Srikakulam, etc., and population density ranges from 500 to 700 persons per square kilometre in the rural settlements (2011 Census of India) of the coastal tracts (Fig. 22.2f). As the Godavari delta, Krishna delta, and adjacent coastal plains are noted for agricultural activities, the river waters and groundwaters are heavily used for irrigating the lands to feed the enormously growing population of the region. During the months of February, March, and April, the region receives very less amount of rainfall, and over-extraction of groundwater is the only solution to supply the agricultural needs.

22.3.6 Climate Change Threats

Dynamic stream flow and reductions in freshwater flow or river runoff in the Godavari, Krishna, and Penner River basins are caused by climate change impacts and human activities. During the past five decades, the changes have been associated with variations in precipitation in the catchment areas of the river basins of the Godavari and Krishna (Das, 2021). However, the average annual temperature is increased in the northern districts of Andhra Pradesh than it was during the previous two decades (2001–2021). Similarly, the coastal districts of the state were affected by a series of cyclone landfalls with encroaching storm surges in the low-lying coastal plains from 1970 to 2022. The increased SST of the Bay of Bengal, the reduction in stream flow into the coastal zones, and the change in salinity gradients caused the intensification of cyclones in the Bay of Bengal in the previous decades (Fig. 22.2h). The low-lying areas liable to saltwater inundations during the tidal surges and storm surges promoted salt encrustation over the coastal plain alluviums, and vertical translocation by rainwater solution and percolation injected the saline waters into the groundwater aquifers. Such processes of saltwater ingression are enhanced by repeated cyclone impacts on the coast caused by climate change threats.

22.3.7 Sea Level Rise Effects

The IPCC working committee report 2021, and the tide gauge data of Kakinada port and Visakhapatnam port reveal the current status of sea level rise rate in the Bay of Bengal fringed Andhra Pradesh coast, which is ranging from 0.93 to 1.03 mm/year. Thus, the increasing rate of sea level rise threats will enhance the saltwater ingression into the coastal aquifers across the freshwater–saltwater interface zones into the coastal tracts (Fig. 22.2g). The alternate sand layers trapped between the clay blankets are vulnerable to saltwater ingression, with an increasing rate of sea level rise threat along the coastal plains in Andhra Pradesh.

22.3.8 CGWB Report (2014)

The Central Ground Water Board (CGWB) of India, by their experimental wells along the Andhra Pradesh coast, prepared the report on groundwater status in 2018. From the status report, it is observed that southeastern and eastern districts of the coastal plain are affected by saltwater ingression into the aquifers up to 10 km inland from the shoreline position (Fig. 22.2j). The groundwater quality index is prepared by assigning weightage to the physiographic environmental, hydrologic, lithologic, and human factors of the coastal tracts for the estimation of integrated scores and to find out the landward extension of the saltwater contaminated zones. The potential for contaminated groundwater aquifers due to encroachment and ingression of saltwater from the sea may extend further inland in the near future.

22.3.9 Surge Water Heights and Tidal Inundations

The factors of surge heights and tidal inundations are considered as potential reasons for ingression of saltwaters into the coastal aquifers. Surge water heights and tidal inundations vary from place to place on the basis of coastal topography, intensity of surges, tidal bores, storm waves, and the location of funnel-shaped estuaries along the coast. Occurrences of the surge heights and tidal inundation encroachments are estimated by the researchers (Babu et al., 2015; CGWB, 2014) in coastal Andhra Pradesh. The surge height is relatively higher in Machilipatnam, Ongole, and Kakinada (3.0–5.5 m) than the areas of Kalingapatnam, Vishakhapatnam, and Nellore (2.6–2.8 m). However, the landward encroachment of tidal inundation zones is very high in Godavari, Eluru, Machilipatnam, and Guntur (50–80 km inland) in comparison to the areas of Pulicat, Nellore, Ongole, and Kakinada (10–30 km inland) in the coastal plains (Fig. 22.2i).

The composition of the coastal plain alluviums, paleo channels, beach ridge Cheniers, and marshy wetlands allows and invites the saltwaters to percolate into the aquifers if the groundwater piezometric pressure is depleted in the region of surge and tidewater inundation.

22.4 Assessment of Potential Areas of Groundwater Contamination

The shorelines of coastal Andhra Pradesh are categorized into four different sections for the assessment of potentially contaminated aquifers by encroachment or ingression of saltwater. They are classified on the basis of topographic contrasts and physiographic settings of the coast, for example, (i) northeastern coastal tracts; (ii) Godavari Krishna delta plains; (iii) southeastern coastal plains; and (iv) the Penner Delta and Pulicat Lake Province. A total of ten factors are considered, including the landward extension of the 2014 contaminated coastal aquifers (CGWB, 2014), and weightage are assigned from 0 to 10 for each parameter as per their importance to identify their role in determining the saltwater ingression limit in the coastal tracts (Table 22.1).

Similarly, the future encroachment limit of saltwater into the coastal aquifer is also predicted in this work using the AHP method and the fuzzy membership method. The SRTM Dem (2022) and Landsat 8 (2022) data are used in the assessment of Land Use Land Cover (LULC), drainage characteristics, physiographic settings, slope, and finally rainfall diversity (2021) data, which are integrated to get the current status of the saltwater ingression limit in comparison with the CGWB (2014) report on groundwater contamination with saltwater encroachment into the aquifers of the Andhra Pradesh coastal plains.

Thus, the current status as well as future prediction of the shifting of groundwater contamination zones towards inland with saltwater ingression are estimated in

Table 22.1 An integrated approach of ten parameters with four different coastal sections of Andhra Pradesh using weightage sum method to calculate the integrated score to identify the saltwater contamination potential zones

Parameter	Northeastern coastal tracts (A)	Godavari-Krishna delta plains (B)	Southeastern coastal plain (C)	Penner delta and Pulicat lake (D)	Total Score	Integrated Score
Coastal geology	4	9	7	8	28	7
Coastal geomorphology (slope and elevation)	3	7	5	6	21	5.25
Coastal drainage features	4	9	6	8	27	6.75
Land use and land cover types	5	9	6	6	26	6.5
Variation in rainfall	6	6	4	5	21	5.25
Population and urbanization	6	7.5	6	6.5	26	6.5
Climate change threats	3	9	6.5	7	25.5	6.375
Sea level rise effects	3	4	3.5	3.5	14	3.5
Surge water height and tidal encroachment	2.5	7.5	6.5	6	22.5	5.625
Saltwater ingression (CGWB)	3	9.5	7.5	8	28	7
Total score	39.5	77.5	58	56		
Integrated score	9.875	19.375	14.5	14		

this study, and it will provide additional information to the coastal managers and administrators for implementing sustainable groundwater management (Fig. 22.4a, b). The CGWB (2014) data is used to validate our present work based on geospatial techniques and field surveys in Andhra Pradesh coastal zones from time to time (2003–2018). Result of the Integrated Scores assessed on ten environmental factors for the four physiographic settings of Andhra Pradesh demonstrated very high scores (Rank-I) for the Godavari-Krishna delta plains (19.38); a high to moderate score (Rank-II) for the southeastern coastal plain (14.50) and the Penner delta with Pulicat lake region (14.00); and finally, a low score (Rank-III) for the northeastern coastal tracts (9.88) of the region fringed with the Bay of Bengal (Fig. 22.5a, Table 22.1). Thus, the result shows that there is a strong similarity between such an assessment of potential groundwater contamination zones and the actual impacts of saltwater ingression into the coastal aquifers of the region studied by the Central Ground Water Board during 2011–2014 (CGWB, 2014) in Andhra Pradesh.

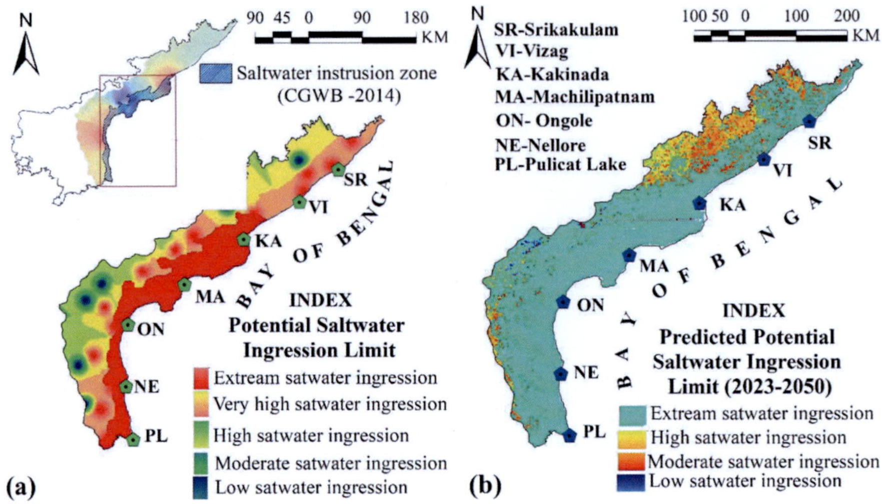

Fig. 22.4 (**a**) The current status (2022) of saltwater intrusion zones in the coastal aquifers estimated with the AHP method and (**b**) The future predictive potential zones of saltwater intrusion investigated using Fuzzy Membership method applied for coastal Andhra Pradesh to find out the landward shifting of the contaminated zones

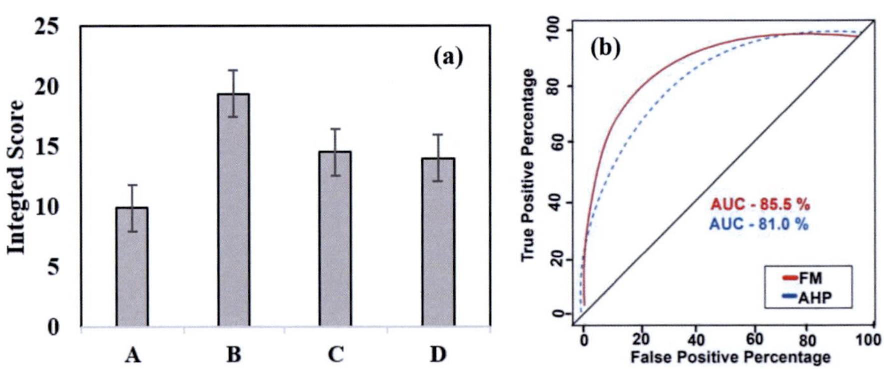

Fig. 22.5 (**a**) The error bars showing the 5% error for the assigned weightage value and their diversity in the four different coastal sections to investigate the potential zones of saltwater intrusion and (**b**) The ROC curves represent the accuracy levels for the validation of AHP and Fuzzy Membership methods in the study

In other ways, the result of the AHP method and fuzzy membership method applied for assessing the current status and future prediction of saltwater contamination zones indicated areas of Kakinada, Machilipatnam, Ongole, Nellore, and Lake Pulicat under the extremely high saltwater encroachment potential of Andhra Pradesh (Fig. 22.4a, b). Inland shifting of potential contaminated zones is reflected in the AHP method of the present study in comparison with the information received from the CGWB (2014) report for the coastal zones of the state. Result of the fuzzy membership method applied for the study shows, on the other hand, that Vishakapatnam, Vizayanagaram, and Srikakulam regions are of very high potential, and western portions of the Delta Plain are of high to moderate potential for future ingression rates of saltwater into the aquifers. However, a few low patches of potential contamination zones are also demarcated in the western highlands of crystalline rocks and also along the western margins of the Eastern Ghat granulite terrains of Andhra Pradesh in the near future.

Finally, the validation of such a prediction model using the receiver operating characteristic curve (ROC) prepared with pixel-by-pixel calculation shows the accuracy validation for the AUC in the fuzzy membership method as 85.5% and the accuracy validation for the AHP model as 81.0% in the present study (Fig. 22.5b). From the accuracy point of view, the predictive model is highly justified with such a positive relationship between false positive percentage and true positive percentage ratio, and the fuzzy membership method is more significant than the AHP in the model's accuracy.

22.5 Conclusion

The coastal groundwater contamination caused by saltwater ingression is a major environmental issue in Andhra Pradesh along the three coastal sections fringed by the Bay of Bengal. Physiographically, the regional settings for the coastal tracts can be grouped as northeastern tracts of highland coast with rocky and alluvial materials; Godavari-Krishna delta plains backed by highlands with Quaternary alluviums; southeastern tracts of low land coast with the Penner delta and Lake Pulicat region; and elevated ridges of southwestern tracts. The unconsolidated alluviums of the coastal plains and semi-consolidated weathered materials of highland areas are liable to saltwater ingression if the saltwater–freshwater interface is affected by physical, environmental, and human factors. The CGWB estimated the groundwater quality of coastal Andhra Pradesh in 2014 using the data of groundwater monitoring wells and the application of geoinformatics to delineate the landward boundaries of saltwater contaminated aquifers, but spatial variation was not illustrated broadly in this output of information.

The present study highlighted a weighted sum approach in the integration of ten major factors as per importance in the coastal sections of four physiographic settings using SRTM DEM, Landsat 8 satellite data, Google Earth Images, and other aspect data on rainfall, cyclones, sea level rise rates, storm surge penetration distance, tidal encroachments, and CGWB water quality data at the first step. In the second step, the groundwater contamination zones due to saltwater ingression into the aquifers is estimated up to 2022 to explore the current status using satellite data with the AHP method. This study represents the further landward encroachments of the groundwater-contaminated zones in comparison with the CGWB (2014) study based on ground monitoring wells. Affected areas of contaminated potentials are ranked into extreme to very high contaminated zones, high to moderately contaminated zones, and low contaminated zones as per the spatial diversity of saltwater ingression in the physiographic settings of Andhra Pradesh.

Finally, the saltwater ingression into the groundwater aquifers is predicted using the fuzzy logistic approach with a future predictive model. This approach identified the newer sites of groundwater contamination with saltwater ingression in the inner parts of entire coastal Andhra Pradesh. However, the CGWB (2014) data is used in the study for validation of contaminated sites, and ROC is applied to improve the accuracy of such model studies to identify the groundwater quality sites affected by saltwater ingressions. The data set and methods of study will provide attractive information for sustainable groundwater management in the coastal tracts of Andhra Pradesh.

References

Babu, A., Narayana, P. S., & Prasad, G. (2015). Study of geoinformatics for East Coast of India-along the AP coast (Andhra Pradesh) India. *International Journal of Innovative Research in Science, Engineering and Technology, 4*(11), 1–15. https://www.ijirset.com/upload/2015/november/84_24_Study.pdf

CGWB. (2014). *Report on status of ground water quality in coastal aquifers of India*. Ministry of Water Resources Government of India. Assessed 2014.

Das, S. (2021). Dynamics of streamflow and sediment load in Peninsular Indian rivers (1965–2015). *Science of the Total Environment, 799*, 149372. https://doi.org/10.1016/j.scitotenv.2021.149372

Das, S., Kandekar, A. M., & Sangode, S. J. (2022). Natural and anthropogenic effects on spatio-temporal variation in sediment load and yield in the Godavari basin, India. *Science of the Total Environment, 845*, 157213. https://doi.org/10.1016/j.scitotenv.2022.157213

Datta, B., Vennalakanti, H., & Dhar, A. (2009). Modeling and control of saltwater intrusion in a coastal aquifer of Andhra Pradesh, India. *Journal of Hydro-Environment Research, 3*(3), 148–159.

Dhakate, R., Sankaran, S., Kumar, V. S., Amarender, B., Harikumar, P., & Subramanian, S. K. (2016). Demarcating saline water intrusion pathways using remote sensing, GIS and geophysical techniques in structurally controlled coastal aquifers in Southern India. *Environmental Earth Sciences, 75*, 1–19.

Harikrishna, K., Ramprasad, D. N., Venkateswara, T. R., Jaisankar, G., & Venkateswara, V. R. (2012). A study on saltwater intrusion around Kolleru Lake, Andhra Pradesh, India. *International Journal of Engineering and Technology, 4*(3), 133–139.

Mamatha, K., Kumar, K. S., Srinivas, N., Gopamma, D., & Ziauddin, A. (2022). Preliminary study on seawater intrusion in coastal aquifers of Visakhapatnam. In *Advances in behavioral based safety: Proceedings of HSFEA 2020* (pp. 113–125). Springer Nature Singapore.

Naidu, L. S., Rao, V. V. S. G., Sarma, V. S., Prasad, P. R., Rao, S. M., & Rao, B. M. R. (2013). An integrated approach to investigate saline water intrusion and to identify the salinity sources in the Central Godavari delta, Andhra Pradesh, India. *Arabian Journal of Geosciences, 6*, 3709–3724.

Parimala, M., & Lopez, D. (2012). Decision making in agriculture based on land suitability-spatial data analysis approach. *Journal of Theoretical and Applied Information Technology, 46*, 1.

Penmetsa, A. R. K. R., Muppidi, S. R., Popuri, R., Golla, S. B., & Tenneti, R. (2013). Aquaculture and its impact on ground water in East Godavari District Andhra Pradesh, India–a case study. *International Research Journal of Environment Sciences, 2*(10), 1–5.

Rao, N. S., Rao, J. P., Devadas, D. J., & Rao, K. S. (2002). Hydro-geochemistry and groundwater quality in a developing urban environment of a semi-arid region, Guntur, Andhra Pradesh. *Geological Society of India, 59*(2), 159–166.

Rao, G. V., Naidu, C. K., & Mouli, S. C. (2011). Contamination of groundwater in Srikakulam coastal belt due to salt water intrusion. *International Journal of Engineering and Technology, 3*(1), 25–29.

Sangadi, P., Kuppan, C., & Ravinathan, P. (2022). Effect of hydro-geochemical processes and salt-water intrusion on groundwater quality and irrigational suitability assessed by geo-statistical techniques in coastal region of eastern Andhra Pradesh, India. *Marine Pollution Bulletin, 175*, 113390.

Part III
Assessment Through Remote Sensing & GIS Approaches

Chapter 23
Geomorphological Analysis of the Coral Fringed Coasts of Andaman and Nicobar Islands Using Geospatial Techniques

Anurupa Paul and Ashis Kumar Paul

23.1 Introduction

The Andaman and Nicobar Islands form an island arc along the convergent boundary of two ocean plates, denoting the location of land masses under active plate tectonic stress. The accretionary prisms in the form of linear ridges emerged above the sea as a north-south island barrier, separating the Bay of Bengal to the west and the Andaman Sea to the east. The islands in the Andaman group are more continuous than those in the Nicobar group, which are discontinuous with large gaps between them towards the south occupied by seas. The longest archipelago of the Indian subcontinent represents different settings of fringing coral reefs in the Bay of Bengal and Andaman Sea. A transect across the island group from west to east represents the tectonic framework with the settings of different activities. The Indian Ocean plate is pushing into the Thai-Burma Microplate along the north-south axis of the subduction zone to give rise to the Andaman Trench towards the western margin of the accretionary prisms, or Andaman Ridge. Similarly, the archipelago of Andaman is evolved within the accretionary prisms, with emerged blocks separated by the frontal Andaman thrust belt, Jarawa thrust, eastern margin fault, and diligent fault from west to east along the north-south axis into ridges and basins. The continuous pushing effect of the Indian Ocean plate has created a depression, or forearc basin, between the main ridge of Andaman and the eastern margin volcanic arc of the system in the Andaman Sea. However, the back arc basin is located to the east of the volcanic arc in the Andaman Sea spreading centre. Previous studies by

A. Paul (✉)
Department of Remote Sensing & GIS, Vidyasagar University,
Midnapore, West Bengal, India
e-mail: anurupashis@gmail.com

A. K. Paul
Department of Geography, Vidyasagar University, Midnapore, West Bengal, India

A. K. Paul, A. Paul (eds.), *Crisis on the Coast and Hinterland*,
https://doi.org/10.1007/978-3-031-42231-7_23

researchers (Rodolfo, 1969; Khan & Chakraborty, 2005; Curray, 1991, 2005; Curray et al., 1979; Sharma & Srinivasan, 2007) reveal that the Oligocene is considered the period of pronounced upliftment in the island arc system in which the geological evolution of Andaman and Nicobar took place.

From the outcrop of rocks, it is revealed that the Andaman ridges and Nicobar Islands are underlain by ophiolites, intermediate volcanics, older sedimentaries, and crystalline limestone to the upward growth of shale-limestone series, Andaman Flysch, calcareous mudstone, siltstone, chalk, limestone to recent coral rags, beach sands, and shell limestone in the top. They are separated, however, by a series of unconformities with formations ranging from the Late Cretaceous to the Oligocene, Miocene, Pleistocene, and Recent Sub-Recent Stages. The stratigraphic succession on the sea cliffs represented such formations and exposures in different places of the archipelago. Physiographically, the Andaman and Nicobar Islands can be grouped as: (i) inner parts of the main Andaman ridge; (ii) scattered islands; (iii) coastal fringes; (iv) shallow offshore banks with fringing coral reefs; (v) and straits, channels, or bays. Unique features are associated with the physiographic units of the archipelago in India. Various processes like ancient sea level variations, weathering, and denudations; a tropical climate with high temperatures and high amounts of rainfall; tectonics with uplift, subsidence, and an active stress zone of the subduction boundary; phytokarstic processes; fluvial, coastal, and marine processes; and volcanic activity all played significant roles in the modification of the island configurations.

A scheme of classification of geomorphological features is prepared on the basis of order and scale in the study. Coral reef areas and their settings are also identified in the island groups using geospatial techniques. The ecology and geomorphology of coral reefs are studied in selective islands of South Andaman after a series of filed works and consideration of Landsat 8 images of the region. Finally, the seismotectonic effects of the 2004 Great Andaman earthquake in geomorphology using geospatial techniques demonstrated the role of processes in the modification of island configurations in the archipelago in the past and present.

23.2 Study Area

The present study areas of the island groups extend from 6° 45′ N to 13° 40′ N along the Bay of Bengal and Andaman Sea in India. However, the main Andaman ridge runs from Landfall Island to Little Andaman Island with a continuous landmass and scattered islands. The Nicobar group is separated into many major and minor landmasses by east-west passages of water bodies in the northern Indian Ocean. They extended from Car Nicobar to Great Nicobar Islands in the regional setting of an archipelago (Fig. 23.1).

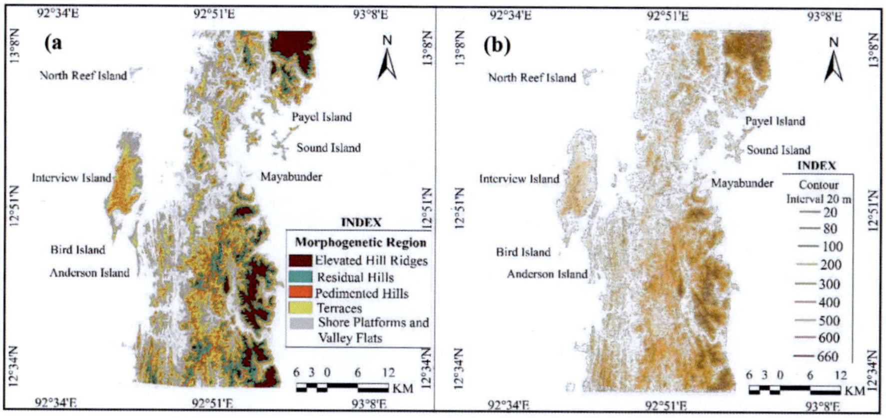

Fig. 23.1 Morphogenetic regions and contour patterns of Andaman Islands (North and Middle Andaman) (**a**) There are six morphogenetic regions in the North and Middle Andaman; (**b**) The linear hill ridges of the eastern part and north-eastern part are showing the elevated land surface by the contour patterns

23.3 Methodology

The materials used in the study include the Survey of India's Toposheets (RF: 1:50,000), existing literature on the Andaman and Nicobar group of islands, the Landsat 8 image, the Shuttle Radar Topography Mission (SRTM), Digital elevation model (DEM), the Advanced Land Observing Satellite (ALOS), Phased Array Type L-Band Synthetic Aperture Radar (PALSAR) DEM, and ground truthing information obtained during fieldwork in the islands. The islands are physiographically classified on the basis of topographic diversity. The nature of landform units was represented by various cross-profiles of the two island groups from the DEM-based contour plans. Location and regional settings of coral reefs have been identified using Landsat 8 classification methods for the study areas. Both the active reefs and raised reefs are explored with geospatial techniques in the study.

23.4 Results and Discussion

Various aspects of the archipelago are dealt with in this section of the study. Tectonics, geologic settings, physiographic diversity, multiple processes of the oceanic environments, geomorphological uniquenesses, organic activities of the fringing coral reefs, and sea level changes appeared as major factors in shaping the landscapes of the Andaman and Nicobar Islands. Geospatial techniques are used to express the geomorphological characters of the coral fringed coasts of the region in this section.

23.4.1 *Tectonic Framework*

As per the tectonic settings of the region, the Andaman and Nicobar group of islands belongs to the collision coast, because it is situated in the active plate margins of island arc setting. On the basis of tectonics, Inman and Nordstrom (1971) classified the world's coasts into three major types to explain their origin and resultant landforms. Relative upliftment and subsidence took place in the coastal parts and adjacent hinterlands due to the location of the island system along the thrust belt related stress zones between the subducted and abducted plate margins of the Indian Ocean plate and the Burma microplate.

Using a cross-sectional form at 11° North latitude, the schematic features represent the framework of the outer arc, fore arc basin, volcanic arc, and back arc basin positions of the Andaman and Nicobar region. The Andaman and Nicobar ridge runs parallel to the outer arc section, areas of the West Basin, the invisible bank, and parts of the Andaman Sea appear in the fore arc basin, and the active and dormant volcanoes (Barren and Narcondam islands) of the region are located in the volcanic arc. However, the back arc basin contains the central Andaman basin and the east basin. In the regional framework of tectonic settings, there are three major faults along the island system (Curray, 2005) which are named the Jarawa fault, the Diligent fault, and the Eastern Margin fault. The continuation of EMF is absent in the Nicobar group towards the south of the system. The accretionary prisms of the outer arc region are sliced by the eastward dipping faulting, tilting, and folding that give rise to the several longitudinal hills and valleys in the Andaman and Nicobar Island systems. The tectonic framework of the archipelago shows the vulnerability of seismic activities under such a plate tectonic stress zone. Like the great earthquake of 2004, many earlier earthquakes in the region depicted evidence of seismo-tectonic activities that played a role in the vertical changes of the island surfaces. According to one estimate based on terrace dating in Interview Island, Middle Andaman, the uplift rates of reef terraces range from 1.73 mm per year to 6.45 mm per year (Rajendran et al., 2008), demonstrating the impact of neo-tectonics on the archipelago's coastal fringes. The seismo-tectonic impact of the 2004 great earthquake, on the other hand, shows significant uplift of Andaman (1–1.5 m) blocks and subsidence of the Great Nicobar (1–3 m) block. A few researchers also estimated the post-earthquake readjustment of the island blocks (Das & Pramanik, 2011) in the island systems of the regional tectonic setting.

23.4.2 *Geologic Settings*

The geologic formations of the Andaman and Nicobar Islands were studied earlier by many researchers based on geological investigations and stratigraphic classifications (Oldham, 1885; Tipper, 1911; Gee, 1927; Chatterjee, 1984, 1967; Karunakaran et al., 1968a, b; Srinivasan, 1986). However, following the rock outcrops, the geologic settings of the archipelago are studied by Bandyopadhyay and Carter (2017) from the North Andaman to the Great Nicobar Islands. The basement formations from the Late Cretaceous to the Oligocene periods are occupied by rocks of older sedimentaries, ophiolites, crystalline limestones, volcanics, and Andaman Flysch, separated by a series of unconformities. Relatively younger formations of Miocene-Pleistocene and Recent-Sub-Recent groups dominated by calcareous mudstone, siltstones, chalk, shell limestones, coral rags, and beach sands occupy parts of the south Andaman, Little Andaman, Car Nicobar, Great Nicobar, and other islands. Erosional activities, exposure of rock outcrops, and denudational characteristics vary from north to south on the basis of rock types, plate tectonic stresses, and exposure of the shorelines to wave attacks. Steep sea cliffs and karstic landforms are produced in the limestone rocks, marls, chalks, and calcareous mudstones in the islands of Nicobar group and parts of South Andaman.

The fringing coral reefs after deformations with concentration of wave energy and tsunami activities transported reefal debris and coral rubbles into the shore platforms to accumulate and participate in the reef flat morphodynamics. Lithification, cementation, and accumulation of debris and beach sands with precipitation of calcium carbonates and aragonites have produced the beach rocks at the high tide shorelines. Field observations on a number of islands demonstrated that most of the visible exposures of bedrock from the basement of the shore platform to the crestal parts of sea cliffs were affected by deformations during the great earthquake of 2004. They were affected by faulting, jointing, rupturing, collapsing, and also by the impact of slope failure with direct wave attacks at the cliff base after deformations. Larger reef boulders were thrown out onto the shore platforms from the oyster banks and coral banks by the tsunami energy after the event of earthquake (Fig. 23.2).

Fig. 23.2 Measuring the dip of the bedding plane on the rock outcrop of the shore cliff at Havelock Island (Archipelago formation)

23.4.3 *Physiographic Diversity*

A scheme of landform classification is prepared on the basis of origin, elevation, form, and shape in different orders. Using the SRTM DEM, ALOS PALSAR DEM, Google Earth Image, field survey, and existing literature, the diversity of physiographic settings is identified in the Andaman and Nicobar group of islands. The study reveals that there are four different types of physiographic settings in the archipelago. The uniqueness of the collision coast with the dynamic stress zone, lithological structure, and exposure to wave attacks, denudational activities, and climate characters of the region were represented by second-order landforms of each physiographic unit. As many as seventeen subunits of major landform units and multiple numbers of small-scale features indicate the physiographic diversity of the region. The cross-sectional forms of the island sections in North Andaman, Middle Andaman, and Great Nicobar Island revealed similar types of major landforms and subunits.

Six morphogenetic regions have been identified from the contour plan prepared by ALOS PALSAR DEM. The similar features are also extracted from Interview Island, North Reef Island (North Andaman), and Great Nicobar Island using the PALSAR DEM at local scale. Physiographically, the coastal fringe landforms are relatively diverse and deformed due to the impacts of tectonics, exposure to wave attack, lithological complexities, and active denudational processes. The available morphogenetic features identified from the embayed shores of western coast and eastern coast reveals two different stages of evolution in Andaman and Nicobar Islands. From the model of evolution of embayed coasts (Strahler, 1960) it is observed that the Bay of Bengal fringe coast represented the early youth stage and the Andaman Sea fringe coast on the eastern side of the archipelago represented the late youth stage in the evolution of landforms. The western shores are fringed with series of bays, coves, embayments, headlands, promontories, steep sea cliffs, islands, ria type estuaries, crenulated or broken shorelines, erosive banks, and luxuriant coral reefs with emergent features. However, there are six major bays or wide embayments fringing accretionary features along with headlands, sea cliffs, coves, beaches, estuaries, straight shoreline, broken shoreline in few places, islands, and both mangroves and fringing coral reefs in Andaman Sea coast. The tectonics and paleoclimatic processes, including sea level changes, are very important to the geological settings of the oceanic coasts, which have been operating in the region for a long time in the modification of the shorelines. Similarly, coastal types of the Great Nicobar Island also demonstrate signatures of the early youth stage on the western margins and the late youth stage on the eastern margins. The western coast of the island is heavily broken with submerged features (Fig. 23.3).

The inner parts of main Andaman ridge are characterized by appearance of a series of folded and east ward tilted longitudinal ridges. They are separated by a number of longitudinal valleys occupied by drainage channels and also connected with parallel channels coming down from the western and eastern hills into the valleys. Lower parts of the longitudinal valleys are submerged and sediment filled. The

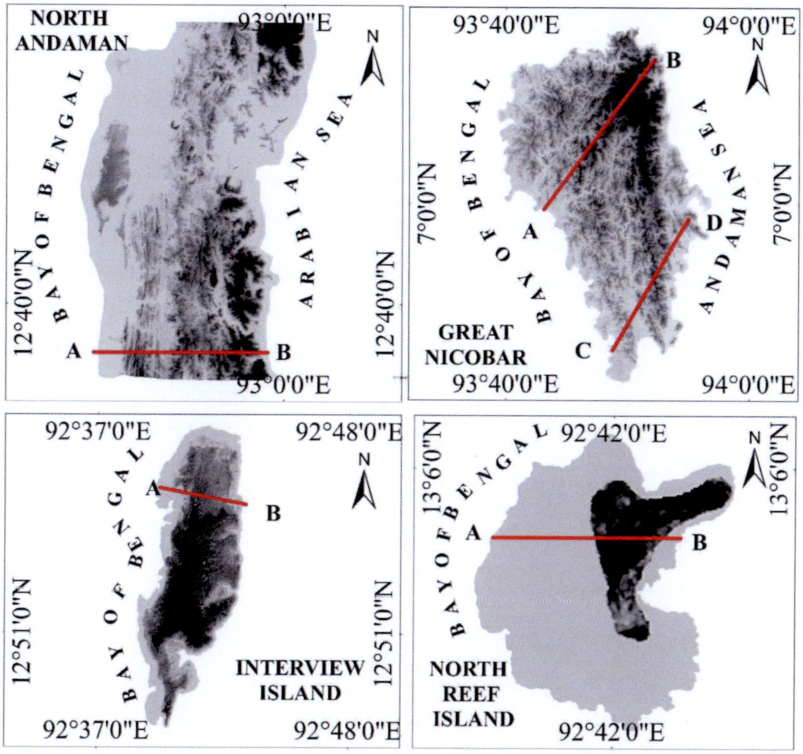

Fig. 23.3 The ALOS PALSAR DEMs for North and Middle Andaman and Nicobar Islands at the regional scale and Interview Island and North Reef Island at the local scale depict the physiographic diversities

average elevation of the hills ranges from 300 to 700 m though there are several peas in North Andaman (Saddle Peak, 732 m), Middle Andaman (Mt. Diavolo Peak, 295 m), South Andaman (Mt. Harriet Peak, 565 m), Little Nicobar (Mt. Deobon, 435 m), and in Great Nicobar (Mt. Thailler Peak, 642 m) islands. Limestone caves are found in the jointed and fractured hills of the archipelago through active geochemical processes in the thicker limestone beds. The Middle Andaman limestone caves of Baratang represent a truly unique development of the limestone dissolution process and cave geomorphological features. However, the sea caves and Karstic caves are well developed in the shell limestone, chalk, and white claystones of Ritchie's Archipelago, Little Andaman, and Car Nicobar Islands.

Intermontane valley flats are visible in between wide gaps of longitudinal hill ridges in middle and north Andaman. The valley flats are also found on Havelock Island and Great Nicobar Island, where the valley fill surfaces with alluvium are cut by meandering drainage channels and also provide favourable sites for potential groundwater aquifers. Betapur channels of such valley flat surface of Middle Andaman represents a unique character of alluvial channel. The valleys, flat

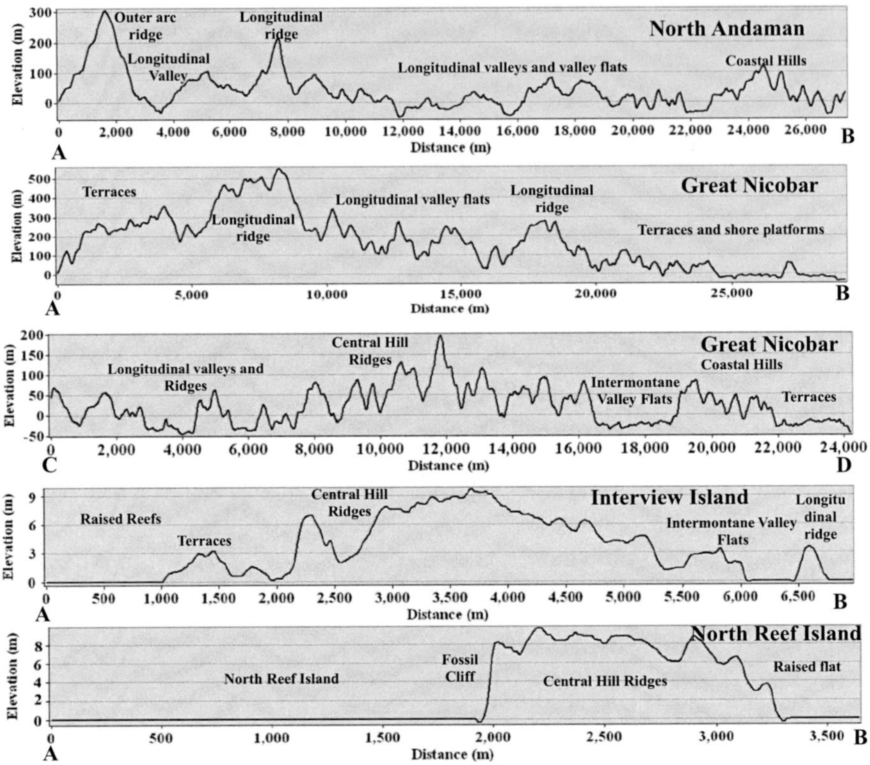

Fig. 23.4 The cross-sectional forms of topography depict the diversity in morphogenetic regions

surfaces of the areas, and Havelock Island are sites of settlements and cultivation of rice-paddy crops (Fig. 23.4).

On the outer arc ridge, a series of islands are located parallel to the coast of the Andaman group, extending from north to south at a regular gap or distance. There are about (30) thirty islands extended as emerged parts of the submerged ridge in the region of the Bay of Bengal. Similarly, on the Andaman Sea, slightly away from the main group of Andaman Island coasts, a series of islands (>26 in number) run longitudinally from north to south as emerging portions of a submerged ridge. Both the groups of islands in the west and east are noted for extensive fringe corals, reef flats, shore platforms, and wave-cut sea cliffs. Other groups of islands are located on the Andaman Nicobar ridge, that is, Rutland Island, Little Andaman, Cinque Island, Brother Islands, Car Nicobar, Little Nicobar, Kamorta, Katchall, Trinkart, Chowra, Teressa, Bompoka, and Great Nicobar Island, from 11° North to 6° 45′ North in the northern Indian Ocean. They are also fringed with coral banks, wave-cut sea cliffs, shore platforms with rock outcrops, and central highlands with significant drainage channels. Among them are the Galathea drainage channel of Great Nicobar, Bumila Creek, and Dugong Creek of Little Andaman; Mita Nallah and Phanda Nallah of Rutland Island, which are significant.

There are other groups of islands in the volcanic arc of tectonic settings of the archipelago. They are known as Narcondam Island, Barren Island, and Tillalangchong Island, and are located over the submarine rises in the Andaman Sea. However, Narcondam is a dormant volcano, and it rises abruptly from the mean sea level as a tower-shaped hill at 13° 26′ N latitude and 94° 15′ E longitude. There is no crater on the island, but the peaked hill has runoff channels, water divides with sharp ridges, eroded slopes with depressional surfaces, and forested lands surrounding the hill slopes. The shorelines are jagged and marked by narrow shore platforms, promontories, sea cliffs, and other wave-cut and wave-built features. Three ash layers are delineated stratigraphically (13° 25′ 54.73′ N and 94° 15′ 25.51′ E) from the exposure of the rock outcrop with a wave-cut sea cliff. The island fringe sloping surfaces are cut by deeper valleys of the runoff channels.

Barren island (12° 17′ 2.68″ N and 93° 51′ 41.24″ E) on the other hand reveals the most active volcano in Andaman Sea areas of the Indian subcontinent. Three to four craters have been identified in the dome-shaped hills of the island. The active crater and its surroundings are accumulated by volcanic ash materials, volcanic debris, and lava sheets, with a seaward projected lava delta of recent origin. Other parts of the island's surfaces are vegetated and marked by valleys formed by runoff channels. The shore fringe areas are characterized by relatively wide shore platforms, wave cut sea cliffs, headlands, bays, and coves. Fringing coral reefs are located on the shallow offshore banks of the island. The volcano erupted frequently in the decades of the 1990s, 2000, and 2010. The last catastrophic eruption took place in 1991, when the flora and faunal habitats were abundant but reduced due to the impact of lava flows on the island.

Finally, various types of submarine features are identified from the bathymetric chart (General Bathymetric Chart of the Oceans (GEBCO)), navigational charts, and Survey of India's Toposheets of the oceanic region (Fig. 23.5). The significant features of this category include submarine rises, submarine troughs, submarine ridges, channel straits, and gaps of water space between islands. Deeper channel

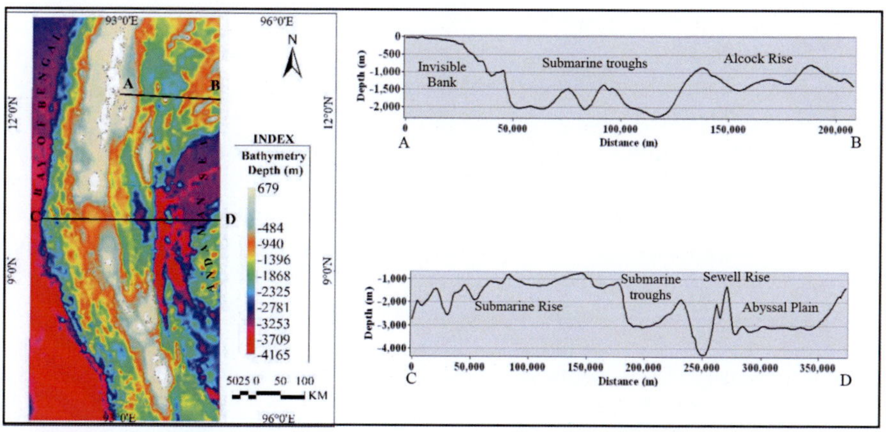

Fig. 23.5 The GEBCO data depicts the submarine features with their cross-sectional forms at A-B section and C-D section across the Andaman Sea

straits stretching from east to west separated the Andaman Island blocks at Mayabunder, Baratang, Port Blair, and Rutland. Those straits are known as Karen Strait, Kadamtala Strait, Hadoo Point Strait, and Rutland Strait in the Andaman Group of Islands. Ten Degree Channel water space separated Little Andaman from Car Nicobar Island in the south which is an underwater gap between Andaman and Nicobar ridge. The passes or underwater gaps and channel straits are extending from east to west cutting across the longitudinal submarine ridges, troughs, and submarine rises at the bottom of the sea. As per the bathymetric contour patterns, it is observed that submarine rises are large, flat-topped surfaces of considerable variety with aseismic ridges above the abyssal floor. Alcock Rise is surrounded by −2000 m contour to −500 m contours over which the Barren and Narcondam islands are standing in Andaman Sea. There is also another Swell Rise (−2000 to −500 m) in the bottom of the sea in the form of a submarine plateau towards the south of the Alcock Rise along the volcanic arc. Fossil coral debris and a thinner cover of sediments are lying over the submarine rise. Thus, physiographically, the Andaman and Nicobar Islands have four different varieties of landforms which are bearing the contrasting features in the archipelago.

23.4.4 Multiple Processes

Various processes which are acting in the region include tectonic stresses, vertical change in the surfaces, shore line exposures to wave attack, ocean currents, variation in tidal range from 6° 45′ N to 13° 26′ N along the island systems, sediment supply and transport along the shorelines, coastal climates, and the biogeomorphological activities in the shores and hinterlands. Since the last glacial low stands (21,000 YBP) the sea level changes took place rapidly and gradually slowed down the rate of change to achieve the present position. Coral reefs have adjusted themselves to such changes in sea level in the past. Other processes like calcification rates by coral reefs, sea grasses, and algae mats vary from place to place in the island environment of the tropics and play significant roles in the lithification and cementation of sediments. Karstic processes by biological activities, dissolution process, and salt weathering process played their role rapidly in the limestone terrains. Catastrophic events like seismo-tectonic activities (i.e., the 2004 great earthquake) and tsunami impacts also produced morphological changes, particularly along the shorelines, sea cliffs, and coral reefs of the islands. Relatively younger islands of Archipelago series with Pleistocene-Holocene and Recent-Sub-recent formations are affected by sediment supply and transport along the shore platforms and river mouths after the impacts of the high-intensity storm rainfall during atmospheric depressions and particularly during the occurrences of east coast and west coast swell waves in the region. During heavy rains, the slope wash process transported the majority of sediments from coastal hills and terraces to the shorelines of Havelock Island, Neil Island, and Little Andaman. A large part of the islands of South Andaman were also inundated by sea waters during the storm surge that

Fig. 23.6 Volcanisms and seismo-tectonic activities: (**a**) Mud volcano eruption at Baratang Island; (**b**) Active volcano at Barren Island; (**c**) Subsidence at Indira Point (Light House under the sea); and (**d**) Uplifted coral reef in Interview island during 2004

occurred in the cyclone Lahar (2013). Signatures of such multiple processes are observed along the coastal fringes of the islands during the field visit (Fig. 23.6).

23.4.5 Geomorphological Features

The Andaman and Nicobar Islands group is a natural museum of unique geomorphological features. The shore platforms, sea cliffs, marine terraces, "Ria"-type estuaries, and coral reefs are unique features in the archipelago. Islands are surrounded by horizontal to slightly seaward inclined platforms that have been shaped over time by wave cut processes with the sea level attained in its current position. They are backed on the windward side by sea cliffs and fronted by reef flats or seaward inclined banks connected with fringing coral reefs in tropical oceanic islands. However, marine terraces on the back shores represent emergent features, and "Ria"-type estuaries across the coastline indicate submergent features in the regional settings of the islands (Fig. 23.6).

As a result of seismic effects and tectonic upliftments on the outer arc region, like the net effect of 2004 great earthquake, the terraces developed along the coastal fringe environment of several islands. They consisted of limestone, formed from coral and shell sand, and overlying as cap rock at different elevations on significant surfaces at Interview Island, North Reef Island, Anderson Island, Flat Island, North Sentinel Island, and parts of Little Andaman along the Bay of Bengal. The vertical changes of such coastal terraces are estimated by researchers (Rajendran et al., 2008) through the age determination of terraces of different altitudes. Though the upliftment rates vary from 1.33 mm/yr^{-1} in Little Andaman to 2.80 mm/yr^{-1} in South Andaman and 2.45 mm/yr^{-1} in North Andaman, the vertical changes are attributed to the effects of the subduction earthquake in the region. Coastal terraces were also identified in the John Lawrence group of islands in the Andaman Sea during the field survey. Coral fragments and shell sands are capped by the elevated surfaces at 2–3 different levels on Neil Island and Havelock Island.

Shore platforms are sites of carbonate machines in which sediments are derived from the coral reefs, wash deposits from the land margin limestone terrains, precipitated carbonates from the sea waters during high tides, and by degassing of bicarbonates at the low tide. Biological activities on the platforms by algae mats, sea grasses, and rock-boring animals supply a significant volume of bio-clasts into the system. Finally, the morphodynamics system of the shore platforms is run by the energy inputs from tidal pumping, wave run-ups and breakers, wave abrasion and salt weathering, and longshore currents as well as cross-shore currents. Weathering, erosion, transportation, and accretion keep the system active for the processing of shore platform sediments under sea level cycles with the upliftment rate of the limestone islands. However, cementation and lithification of materials by carbonates and cements (aragonites, calcium carbonates, dolomites, etc.) will produce beach rocks on the high tide level through the seepage waters and degassing of bicarbonates in the region. Sediments are also supplied into the shores by mass wasting deposits from the landslide of sea cliffs. Various tafoni features and karst features are found on the shore platform nearby the algal mats over limestone surfaces. Rock pools, tide pools, and other water bodies provide various micro habitats for patch corals, sea grasses, algae mats, and other marine animals on the shore platforms. During the 2004 earthquake event, tsunami-genic drift of eroded boulders from coral banks and oyster banks was also supplied, along with rubble and coral debris, into the shore platforms of the Andaman Islands. The seaward sides of the shore platforms are affected by wave breakers, and spurs and grooves features are well developed along this zone of the windward side of the islands. Two sea-level cycles have been identified in the development of shore platforms on Havelock and Neil Islands.

Geomorphological features are well developed on the windward side of the island shoreline. The vertical sea cliffs with different elevations are found in parts of Little Andaman, Neil Island, Havelock Island, John Lawrence Island, and Car Nicobar Island on the windward sides (Paul et al., 2017, 2018). They consist of chalk, white claystone, shell limestone, and coralline fossiliferous limestones of different groups. Notches form in the relatively softer rocks at the basement as a result of wave erosion, and fissure caves at the jointed rocks are extended in many places as a result of seepage flow from the landward side and concentration of wave energy on the seaward sides in the island margin flanks. Sea caves in the thick limestone bed at the base represent developed cave openings, cave roofs, and cave interiors in the exposure rocks of sea cliffs. Wave splashing has produced tafoni marks on the limestone rocks at the base of the cliffs on many islands. Relatively softer rocks of the headlands and promontories are affected by dissolution processes and wave erosion processes to give rise to the development of natural arches, and after roof collapsing of the arches they produce stacks on the erosive basement. There are lines of seepage pits on the cliff wall at the exposures of permeable and impermeable layers, and seasonal springs develop along this line of seepage points on the island during monsoon months. When the crestal parts of the cliff become unstable in the region, rock fall and debris fall occur at the base of the cliff. During the great earthquake (2004), such instability developed with the formation of joints, widening of fractures, reactivation of the fault line, and collapsing processes in the sea cliffs of the Andaman and Nicobar Islands. The cliff recession rate is recorded with a temporal assessment of satellite data using Landsat-8 for a few islands in South Andaman (1993–2018). The outcome of such a study reveals a significant rate of recession within the span of a 25-year time period, ranging from 0.78 to 2.26 m per year, which is indicative of the integrated role played by tectonics, weathering, lithology, exposure to wave attacks, and catastrophic events along the tropical coast.

Fringing coral reefs are relatively better developed in the Bay of Bengal fringe shoreline of Andaman and Nicobar Islands affected by west coast swell waves at the windward side. The coral reefs are located at a depth between 10 and 50 m in Havelock and Neil Island and at depths between 20 and 50 m in Little Andaman along the seaward slopes of Andaman-Nicobar ridge. The classified image from Landsat-8 shows the variation of coral reefs in the region, and GEBCO data from the Coastal Bathymetry shows the depth of their habitat. Ria coasts are well represented by the Andaman and Nicobar group of islands. The lower parts of river valleys and river mouths are affected by submergence due to seismo-tectonic activities in the past events. Subsidence occurred in South Andaman and Nicobar Islands during the 2004 earthquake, when the larger drainage mouths of Katchall Islands, Kamorta Island, and Great Nicobar Island subsided and submerged sea water occupied the river mouths, supplying marine sediments over the region's subsided river beds and degraded fringe forests. Such estuaries are developed in a transverse direction to the shoreline, but rias are also developed in the longitudinal valleys of the North and Middle Andaman, in which the river mouths were subsided and submerged in the ancient past during rising sea levels during marine transgression during the Holocene epoch (Figs. 23.7 and 23.8).

In many places, the subsided basin of the lower estuary is filled by marine sediments over the submerged forests, and they are now flooded by sea waters during

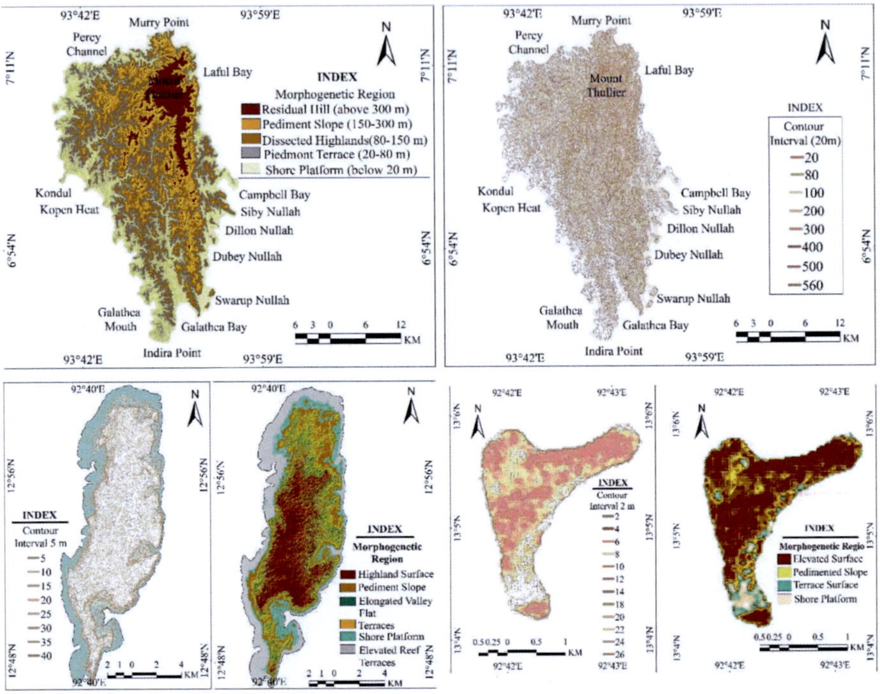

Fig. 23.7 Based on the ALOS PALSAR DEM, morphogenetic regions and contour patterns depict the geomorphological features at the local scale in (**a**) The Great Nicobar Island showing five morphogenetic regions; and two major linear ridges with valley flats and shoreline configurations by contour patterns, (**b**) The Interview Island of Middle Andaman showing six micro morphogenetic regions with shore parallel terraces; and surface configurations of the island by contour patterns, and (**c**) The North Reef Island of North Andaman showing four morphogenetic regions with reef terraces; and micro surface configurations by the countour patterns

high tides, swell waves, and storm surges. Some of the rias are uplifted in the Andaman Islands, and their mouths were temporarily closed by the dumping of marine sediments. Such uneven characters of vertical change of the surfaces produced a complex nature of ria coast in the region at present (Paul, 1998, 2005).

Fig. 23.8 (continued) Havelock Island; (**d**) Whisper waterfalls along the knickpoint of inland drainage channel across the limestone cliff in Little Andaman; (**e**) The tower karst of shale limestone with jointed block at Little Andaman; (**f**) Radhanagar sea beach at Havelock island with thicker deposits of coralline sands and silica sands; (**g**) Baratang limestone cave with a developed stalactite hanging from the roof; (**h**) Moving across the Dugong creek fringed with mangrove forests of estuarine setting at Little Andaman (**i**) Seaward projected lava delta sheets developed from the active volcanic crater after eruptions of the Barren island in Andaman sea during 1990s (**j**) A natural arch is developed due to marine erosion of the softer rocks at the headland of Neil Island

Fig. 23.8 Geomorphic features: (**a**) Submerged coral reef with restricted zooxanthella in Neil Island; (**b**) A promontory with a mark of light house over a limestone terrain in Havelock Island; (**c**) Shore platform fringed with mangrove and fronted by accretion of coral rubbles and debris in

23.4.5.1 Coral Reefs and Ecology and Morphology

Coral reefs are located on the islands of outer arc ridge in the Bay of Bengal, islands of the fore arc ridge in Andaman Sea, islands of the volcanic arc ridge in Andaman Sea, and in the shelves of the embayed shores of the archipelago as stronger bioshield. The zooxanthellae, other algal mats, and coral reefs reciprocally interact with each other to maintain their luxuriant growth and health conditions in the tropical sediment-free, transparent sea water. Other favourable conditions such as attachment surfaces, sun light

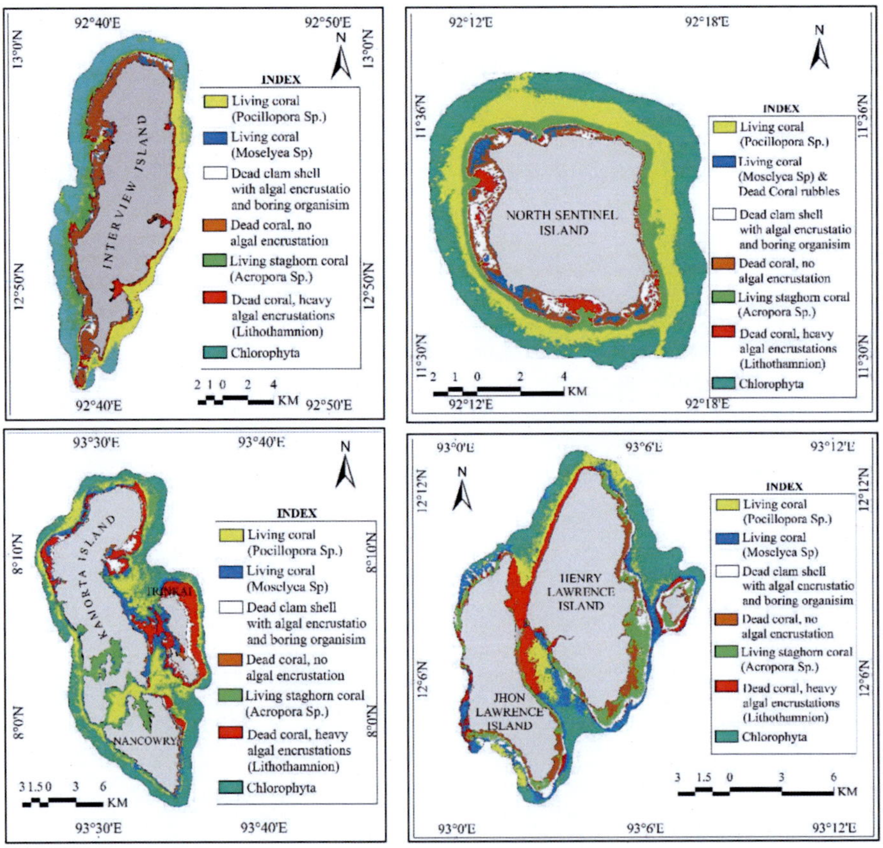

Fig. 23.9 Result of Infrared band of Landsat 8 OLI depicts coral reefs of Andaman and Nicobar Island (**a**) Interview Island with raised coral reef in the Bay of Bengal; (**b**) North Sentinel Island with fringing coral reef in the Bay of Bengal; (**c**) Kamorta, Nancowry, and Trinkat islands of Andaman Sea with fringing coral reefs in Nicobar Group, and (**d**) John Lawrence and Henry Lawrence of Ritche's Archipelago in Andaman Sea with fringing coral reefs of Andaman Group. Coral reef classes of seven categories identfied through the geospatial techniques in the four significant coral reef islands

penetration, low turbidity, high sea water temperature (25–29 °C), moonlit nights, tidewater fluctuations, sea waves, swell waves, and storm waves aided coral growth in the Andaman and Nicobar ridge of the northern Indian Ocean (Fig. 23.9).

There are 135 coral species in the Andaman and Nicobar Islands (Pillai, 1983), and among them the table corals, finger corals, brain corals, and fan corals are considered as land-building corals. Coral growth is determined by the rate at which the branching and sub-branching of corals grow vertically with concentrations of precipitated calcium carbonates from seawater. Usually, 100% coral cover areas accrete the calcification rate of 10 kg $CaCO_3$ m^2/yr.; 100% algal pavement areas get 4 kg $CaCO_3$ m^2/yr. and the bare sand and rubble cover surfaces accrete only 1 kg $CaCO_3$ m^2/yr. in the coral reefs. Thus, reef development and their luxuriant growth provide a high rate of calcification for the stability of the bioshield. The Andaman reef growth also consists of five processes, which are viewed in the reef structures like other areas (Scoffin, 1987; Wood, 1998, 1999). The primary framework of coral reefs consists of the living tissue of branching corals over the coral limestone; the secondary framework is produced by encrustation and epibiota; boring processes are performed by micro and macro boring animals of bivalves; and internal sediments are supplied by sands, shells, and reefal debris; finally, the cementation process is run by early cements precipitated into the voids to develop reefs fringing the shallow submarine banks of the islands. Coral reefs attempt to adjust to fluctuating sea levels through give-up, progradation, catch-up, back-stepping, and keep-up processes after such structural development (Neumann & Macintyre, 1985; Hubbard et al., 1997). All stages of coral reefs are found in the Andaman and Nicobar group of islands as the region experiences frequent sea level changes in the geological past. The emerged reefs and drowned reefs are viewed in Interview Island, North Reef Island, Neil Island, and Great Nicobar Island (Fig. 23.10).

Morphologically, the fringing coral reefs of Redskin Island, Jolly Buoy Island, Maly Island, Hobbday Island, Boat Island, and North Sentinel Island on the windward sides (Bay of Bengal fringe coast) are characterized by seaward inclined reef banks, shallow reef flats, and coral beach sands. Ritchie's Archipelago, a group of islands in the Andaman Sea, consists of shallow reefs and reef banks with patch reefs in the rock pools of shore platforms. The steeply inclined banks of Andaman's channel straits are covered in luxuriant coral growths. During swell waves and storm surges in the islands, the fringing coral reefs supply organic sediments in the form of dead coral debris to the shore platforms and coralline sands to the shoreline beaches along the High Tide Line (HTL). The coral reefs are good indicators of climate change records along the coastal fringes. Coral bleaching is being studied by Earth System Science Organisation (ESSO)/National Institute of Ocean Technology (NIOT) in South Andaman (North Bay).

Fig. 23.10 Patch corals in living condition in the rock pools within the shore platforms in Neil: (**a**, **b**) *Montipora tuberculosa* and (**c**, **d**) *Psammacora obtusangura* with sediment accretion due to tsunami event in 2004

23.4.6 Coral Reef Problems

The coral reefs of any location have to keep their natural growth upward from the attachment surface against multiple odds in a tropical oceanic environment. To assess the actual growth of the coral reefs in such active collision coasts, measurements are needed on the upliftment rate of the basement surface, subsidence rate of the bottom topography, rate of sea level rise, rate of sea level fall, rate of degradation of the coral reefs, and the rate of natural growth of corals. Coral reefs grew against all odds in the geological past to get light and temperature of sea water in a favourable depth condition for survival. However, the natural growth of the coral reefs may not adjust to the rapid rise and rapid fall of sea levels, and in that case, they may not survive under such stress factors. Raised coral reefs and submerged coral reefs have been produced in the past due to the dynamic nature of land and sea levels. The archipelago of Andaman and Nicobar has also experienced such impacts over the coral reef growth. Thus, raised coral reefs, submerged coral reefs, and living or active coral reefs are viewed in the regional settings of the group of islands. This type of problem was experienced by the island system during the 2004 great earthquake and tsunami waves, when a large part of the coral reefs of Interview Island, North Reef Island, Flat Island, Anderson Island, and North Sentinel Island were raised above the present sea level and came under subaerial modifications. Similarly, due to land subsidence, parts of shallow reefs in Great Nicobar Island and Kamorta Island were submerged and degraded through sediment accretion.

23.4.7 The Great Earthquake-2004 and Tsunami Event

The coseismic deformation on the exposed land due to upliftment and subsidence along the outer arc ridge during the great earthquake revealed vertical changes in the coastal features along the Bay of Bengal and Andaman Sea (Table 23.1). The earthquake caused a total uplift of 1–1.5 m above high tide level in North Andaman and a subsidence of 3–3.5 m below HTL in Great Nicobar Island. Various emergent and submergent features have been recorded in the islands. The vertical changes of the region may be categorized in the following ways:

A large part of the mangroves of the intertidal region emerged and moved towards the supratidal region in North Andaman, Wandoor, Little Andaman, and Middle Andaman. The coral reefs of North Andaman have raised above the HTL along the outer arc ridge. Shoreline configurations have changed with erosion and accretion processes after the earthquake and tsunami impacts. Because of their location in the stress zone of tectonic settings, the rock outcrops exposed on the sea cliffs are jointed and fractured.

Land subsidence has created depressional basins that have been occupied by seawater inundations on Katchall, Kamorta, and Trinkart Islands. Headlands are detached to isolate the smaller size light houses in the island. At Indira Point, the famous lighthouse is now lying on the sea bed after the event of Great earthquake 2004. Campbell Bay and Hut Bay were flooded by tsunami waves after the vertical changes in low land coast (Paul, 2005). Following the tsunami's impact in the South Andaman Islands, coral rubbles, ridges, and coral sands were dumped on shore platforms (Fig. 23.11).

Table 23.1 Types of vertical changes in Andaman and Nicobar Islands after the great earthquake, 2004

Types of vertical change	Features development	Region affected
Upliftment on the outer arc ridge	Reef terraces, raised beach features, raised part of longitudinal valleys, collapsing sea cliffs, raised river mouths	North Andaman, Middle Andaman, western part of South Andaman, and Little Andaman
Subsidence in Nicobar Islands	Ria coast, submergent river mouths, submergent forests, submerged coral reefs, salt water encroachments and inundations	Parts of South Andaman, Nicobar Islands, Great Nicobar

Fig. 23.11 The effects of earthquake and tsunami (2004): (**a**) rock outcrop and the sea wall collapsed in Neil Island and (**b**) Tsunami damage in the Erstwhile village of the Car Nicobar Island

23.4.8 Geo-Tourism Prospects

There are areas of geomorphological and geological significance in the Andaman and Nicobar group of islands that can attract tourists from different corners of the country and from abroad. The most significant geomorphosites in the archipelago are natural arches, sea caves, sea cliff outcrops, water falls, limestone caves, coral reefs, sea beaches, mangrove creeks, historical settlements of Ross Island, rain forests, mud volcanoes, and active volcanoes. Approximately 39 islands are now open to visitors for tourism and recreational purposes (Paul et al., 2021). Tourists in the region enjoy coral watching, sunbathing on the seashores, sea bathing, sailing in the sea, diving into the sea, and boating with water sports (Fig. 23.12). The interisland ships transport the visitors from Port Blair to different islands. At the moment (2022), over 5 lakh tourists from India's main land visit the islands each year. Geotourism sites are well connected by roads and sea routes, with Port Blair providing transportation for the tourists. The rain forests, mangrove forests, and littoral forests of the island landscape are other attractions in the region.

23.5 Conclusion

Using geospatial techniques, the above study represented the geomorphological characteristics and nature of the coral reef fringed coasts of the Andaman and Nicobar group of islands. Tectonically, the archipelago demonstrated as an active collision coast along the subduction plate boundary between two ocean plates. Thus, the island systems are experienced by tectonic zones of the outer arc ridge, fore arc basin, volcanic arc, and back arc basin from west to east. Geomorphological diversity is also reflected all along the group of islands in different tectonic zones. The SRTM DEM and ALOS PALSAR DEM are used to generate contour plans at

Fig. 23.12 Natural arch and limestone caves in the seaward projected promontory in Neil Island

regional and local scales for the representation of the archipelago's physiographic settings and geomorphology. Landsat 8 OLI data is used for the classification of the coral reefs of the islands, and using the GEBCO data, various submarine features have been identified in the region.

The patterns and forms of the topography helped to identify the broad physiographic settings as coastal fringe landforms, inner parts of the main Andaman ridge and linear valleys, scattered islands in the seas, and submarine features of the Bay of Bengal and Andaman Sea. They are also compared with the tectonic settings to identify their role with other subaerial processes in the evolution of the landforms. There are types of emergent, submergent, and composite shoreline features with signatures of landform diversities influenced by seismo-tectonics, exposure to wave attacks, lithology, weathering rates, and the tropical climate of the region. Karstic features are well developed in the limestone terrains of the outcrops. By classifying the Landsat 8 OLI images, over six types of coral reefs are identified on a few islands. The great earthquake and tsunami events of 2004 caused significant damage to reef systems by reducing algal cover and raising reefs in various sections of the islands.

Submarine topography is classified in the region using the bathymetry as submarine rise, submarine troughs, shelves, volcanic islands, and sea gaps between the Andaman Nicobar ridge, which is well connected with the Sumatra ridge and over which the volcanic islands of the Andaman Sea are located. The island interior geomorphology and drainage characteristics are also explored in the present study using the ALOS PALSAR DEM to identify the surface expressions after shaping and reshaping the landscapes under different processes. Such a study will help to create a database for the sustainable use of natural resources and their appropriate management.

References

Bandyopadhyay, P. C., & Carter, A. (2017). Chapter 2 Introduction to the geography and geomorphology of the Andaman–Nicobar Islands. *Geological Society, London, Memoirs, 47*(1), 9–18. https://doi.org/10.1144/M47.2

Chatterjee, P. K. (1967). Geology of the main islands of the Andaman Sea. In *Proceedings Symposium on Upper Mantle Project* (pp. 348–360). Geophysical Research Board, National Geophysical Research Institute.

Chatterjee, P. K. (1984). The Invisible Bank fault and geotectonics of the Andaman Nicobar Islands. *Quarterly Journal Geological Mineral Metallurgical Society India, 56*, 28–40.

Curray, J. R. (1991). Possible greenschist metamorphism at the base of a 22 km sediment section, Bay of Bengal. *Geology, 19*, 1097–1100.

Curray, J. R. (2005). Tectonics and history of the Andaman Sea region. *Journal of Asian Earth Sciences, 25*(1), 187–232.

Curray, J. R., Moore, D. G., Lawver, L. A., Emmel, F. J., Raitt, R. W., Henry, M., & Kieckhefer, R. (1979). Tectonics of the Andaman Sea and Burma: Convergent margins. In J. Watkins, L. Montadert, & P. W. Dickerson (Eds.), *Geological and geophysical investigations of continental margins* (Vol. 29, pp. 189–198). American Association Petroleum Geologists, Memoir.

Das, D. P., & Pramanik, K. (2011). Evaluation of impact of earthquake and tsunami on the coastal morphology of Andaman-Nicobar Island using multi-sensor temporal satellite data: Indian Jour. *Indian Journal of Geosciences, 65*, 9–22.

Gee, E. R. (1927). The geology of the Andaman and Nicobar Islands, with special reference to Middle Andaman Island. *Record of the Geological Survey of India, 59*, 208–232.

Hubbard, D. K., Gill, I. P., Burke, R. B., & Morelock, J. (1997). Holocene reef backstepping—southwestern Puerto Rico shelf. In *Proc 8th Int Coral Reef Symp* (Vol. 2, pp. 1779–1784).

Inman, D. L., & Nordstrom, C. E. (1971). On the tectonic and morphologic classification of coasts. *The Journal of Geology, 79*(1), 1–21.

Karunakaran, C., Ray, K. K., & Saha, S. S. (1968a). A revision of the stratigraphy of Andaman and Nicobar Islands, India. *Bulletin of the National Institute of Sciences of India, 38*, 436–441.

Karunakaran, C., Ray, K. K., & Saha, S. S. (1968b). Tertiary sedimentation in the Andaman–Nicobar geosyncline. *Journal Geological Society of India, 9*, 32–39.

Khan, P. K., & Chakraborty, P. P. (2005). Two-phase opening of Andaman Sea: A new seismotectonic insight. *Earth and Planetary Science Letters, 229*(3–4), 259–271.

Neumann, A. C. & Macintyre, I. (1985). Reef response to sea-level rise: keep-up, catch-up, or give-up. In *Proceedings of the Fifth International Coral Reef Congress Tahiti, 27 May-1 June 1985 volume 3: Symposia and seminars (A)* (pp. 105–110). Antenne Museum-EPHE.

Oldham, R. D. (1885). Notes on the geology of the Andaman Islands. *Records of Geological Survey of India, 18*(3), 135–145.

Paul, A. K. (1998). *Problems of groundwater contamination and water scarcity in Andaman and Nicobar Islands*. DELF, New Delhi, Published by Ashutosh College, Calcutta-26. (pp. 13–33).

Paul, A. K. (2005). *Tsunami: An assessment of the disaster over the nations in the Indian Ocean* (pp. 1–125). India ACB Publications.

Paul, A., Bandyopadhyay, J. & Paul, A. K. (2017). Monitoring vertical cliffs, embayments and shore platform morphology in parts of Neil Island, South Andaman using spatial information technology. *38th Asian Conference on Remote Sensing - Space Applications: Touching Human Lives, ACRS.*

Paul, A., Bandyopadhyay, J. & Paul, A. K. (2018). Geomorphological mapping and environmental zoning approach to coastal management in Havelock Island, South Andaman, India. *39th Asian Conference on Remote Sensing: Remote Sensing Enabling Prosperity, ACRS* (pp. 346–357).

Paul, A., Sardar, J., & Bandyopadhyay, J. (2021). Assessment of geo-environmental aspects for geo-tourism site selection of South Andaman Districts, India. *Journal of Coastal Sciences, 8*(1), 1–12. ISSN 2348-6740.

Pillai, C. G. (1983). Coral reefs and their environs. *CMFRI Bulletin, 34*, 36–43.

Rajendran, K., Rajendran, C. P., Earnest, A., Prasad, G. R., Dutta, K., Ray, D. K., & Anu, R. (2008). Age estimates of coastal terraces in the Andaman and Nicobar Islands and their tectonic implications. *Tectonophysics, 455*(1–4), 53–60.

Rodolfo, K. S. (1969). Bathymetry and marine geology of the Andaman Basin, and tectonic implications for Southeast Asia. *Geological Society of America Bulletin, 80*(7), 1203–1230.

Scoffin, T. P. (1987). *An introduction to carbonate sediments and rocks. Reef growth* (pp. 77–88). Blackie.

Sharma, V., & Srinivasan, M. S. (2007). *Geology of Andaman-Nicobar: The Neogene*. Capital Publ.

Srinivasan, M. S. (1986). Neogene reference section of Andaman-Nicobar: Their bearing on volcanism, sea-floor tectonism and global sea-level changes. In *Indian chapter IGCP Projects 195 and 197. Meeting* (pp. 295–308).

Strahler, A. N. (1960). *Physical geography*. Wiley.

Tipper, G. H. (1911). The geology of the Andaman Islands, with reference to the Nicobars. *Geological Survey of India, Memoir, 35*, 195–213.

Wood, R. (1998). The ecological evolution of reefs. *Annual Review of Ecology and Systematics, 29*(1), 179–206.

Wood, R. (1999). *Reef evolution*. Oxford University Press on Demand.

Chapter 24
Mangroves in Cyclone-Battered Sundarbans, India: A Geoinformatics-Based Multi-temporal Study

Ashis Kumar Paul, Anurupa Paul, Joydeb Sardar, Ratnadeep Ray, Khadija Khatun, Sukumar Chand, Rimpa Maji, and Sk Saharukh Ali

24.1 Introduction

Forest damages caused by notorious cyclones (1988, 2007, 2009, 2019, and 2020) of the previous decades have exposed the capacity and inability of the mangrove buffers in the Indian Sundarbans. The buffer width varies from 149 km (SE) to only 30 km (SW) in the coastal fringe forests and is separated by the larger tidal river mouths of Ichhamati-Raimangal, Gosaba, Matla, Thakuran, and Saptamukhi in the physiographic settings (deltaic and estuarine) of the Indian Sundarbans. Mature mangroves were distributed on the seafront island shores, inner deltaic platforms, and the levee surfaces of the tidal channels of the coastal settings and inland deltaic settings (Paul, 2002). Other areas are occupied by younger mangroves, stunted mangroves, and saltmarsh vegetation in the lower part of the Ganga delta. The thicker and more compact mud banks, nutrient-rich muds, and active tidal flats with saltwater inundations harbour the quality mangroves. Rapid aggradation,

A. K. Paul · S. S. Ali
Department of Geography, Vidyasagar University, Midnapore, West Bengal, India
e-mail: akpauleastcoast@gmail.com

A. Paul (✉) · K. Khatun · S. Chand · R. Maji
Department of Remote Sensing and GIS, Vidyasagar University, Midnapore, West Bengal, India
e-mail: anurupashis@gmail.com

J. Sardar
Centre for Environmental Studies, Vidyasagar University, Midnapore, West Bengal, India
e-mail: joydebsardar18@gmail.com

R. Ray
Department of Earth Sciences and Remote Sensing, JIS University, Kolkata, India

A. K. Paul, A. Paul (eds.), *Crisis on the Coast and Hinterland*,
https://doi.org/10.1007/978-3-031-42231-7_24

autocompaction of tidal sediments, freshwater mixing or brackish water quality, and stability of the substrate alluviums support ideal environmental gradients for holding the mangrove regeneration processes in the deltaic islands.

Using the satellite data (Landsat 4 and 5, 7 and 8), the areal coverage index, Normalised Differential Vegetation Index (NDVI), Enhanced Vegetation Index (EVI), and Leaf Area Index (LAI) are assessed for mangrove forests in the region in pre- and post-storm situations to identify the nature of cyclone-battered mangrove forests in the Indian Sundarbans. The damage types are observed in the islands during and after the occurrences of a few cyclones (e.g., Aila, Bulbul, and Amphan) in the Indian Sundarbans. In the field of surveying and monitoring vegetation, they are identified as (i) loss of mangroves by erosion, (ii) top breakage of mangrove trees by wind storms, (iii) overwash sand lobes encroachments and forest damages, (iv) loss of soil nutrients by removal of top soils, (v) burnt-out trees by salt sprays and moisture sucking effects, (vi) dieback of mangroves by salt encrustation of pan surface clay beds, and (vii) damages due to twisting of trees by strong wind speed. Observations also show that 45% of mangroves in the Indian Sundarbans were damaged by the multiple effects of cyclones. The habitat restoration of mangroves with green solution management through repeated plantations of mangrove saplings in the favourable tidal flats, regular monitoring of the plantation sites, and participation of the locals in the management process will extend the composition of mangrove buffers in the Indian Sundarbans and reduce the damages caused by cyclones.

Mangroves of the Indian Sundarbans are also studied on different aspects of environmental issues as cyclonic impacts (Paul, 2000b, 2002; Ghosh & Naskar, 2012); forest types and quality (Paul, 1996a; Hati et al., 2021; Mondal & Paul, 2022); ecological diversity of mangrove forests (Choudhury et al., 1984; Ghosh et al., 2015; Ray et al., 2014); climate change impacts on the mangrove forest in the Indian Sundarbans (Paul, 1996b; Raha, 2014; Chakraborty et al., 2014; Paul & Paul, 2022); and physiographic diversity of Indian Sundarbans with vegetation characteristics (Paul & Bandyopadhyay, 1985, 1988; Paul, 2000a, 2014, 2022; Majumdar et al., 2012) by the different researchers.

24.2 Materials and Methods

The cyclone-battered Sundarbans are studied using geoinformatics for identification of forest damages to highlight the nature of changes in the isolated forest patches of Lothian Island and Dhanchi Island units in the Saptamukhi estuary and Thakuran estuary. Landsat data series (2000, 2010, 2015, and 2021) of the Sundarban deltaic islands, field monitoring after the storms, and other aspect data related to cyclone damages are used to carry out the research in the Indian Sundarbans. The spectral index of satellite data is analysed in software tools for recording the changes in forest resources after each storm in the Sundarbans. Final output maps of classification images, vegetation index calculation, and leaf area concentration zones during and after the events of the storm will highlight the types of forest damages in the study areas.

Based on K-Means clustering unsupervised classification algorithm the mangrove forests of Lothian and Dhanchi Island are classified for assessment of temporal changes from 2000 to 2021. However, the EVI, NDVI, and LAI are calculated for the island forests to estimate the leaf morphology and health conditions under the impacts of cyclonic storms. The following Eqs. 24.1, 24.2, and 24.3 (Rouse et al., 1974; Green et al., 1997; Huete et al., 2002) are used in calculation to express the above parameters:

$$\mathrm{EVI} = G \frac{(\mathrm{NIR} - \mathrm{Red})}{(\mathrm{NIR} + C_1 \times \mathrm{Red} - C_2 \times \mathrm{Blue} + L)} \tag{24.1}$$

where, L refers to the canopy background adjustment for correcting the nonlinear matrix function, differential NIR, and red radiant transfer through a canopy; C_1 and C_2 refer to the coefficients of the aerosol resistance term (the blue band to correct for aerosol influences in the red band); and G is a gain or scaling factor. The coefficients adopted for the MODIS EVI algorithm arc $L = 1$, $C_1 = 6$, $C_2 = 7.5$, and $G = 2.5$:

$$\mathrm{NDVI} = \frac{(\mathrm{NIR} - \mathrm{Red})}{(\mathrm{NIR} + \mathrm{Red})} \tag{24.2}$$

$$\mathrm{LAI} = 12.74 \times \mathrm{NDVI} + 1.34 \tag{24.3}$$

24.3 Study Area

Indian Sundarbans harbour diverse character of mangroves, which are distributed in south-eastern zones, estuarine complexes, and southwestern zones in between Saptamukhi River and Ichhamati-Raimangal River. However, the study is conducted at the local level, particularly in Lothian Island and Dhanchi Island of the Indian Sundarbans. Storm damage plays a significant role in successional modification in mangroves. Being located in the mouth of the Saptamukhi tidal estuary, the Lothian Island represents the signatures of storm surges and tidal surges in the south western Sundarbans. Dhanchi Island, on the other hand, bears multiple effects of cyclonic storms along the Thakuran River estuary reaches in the middle part of the Indian Sundarbans (Fig. 24.1).

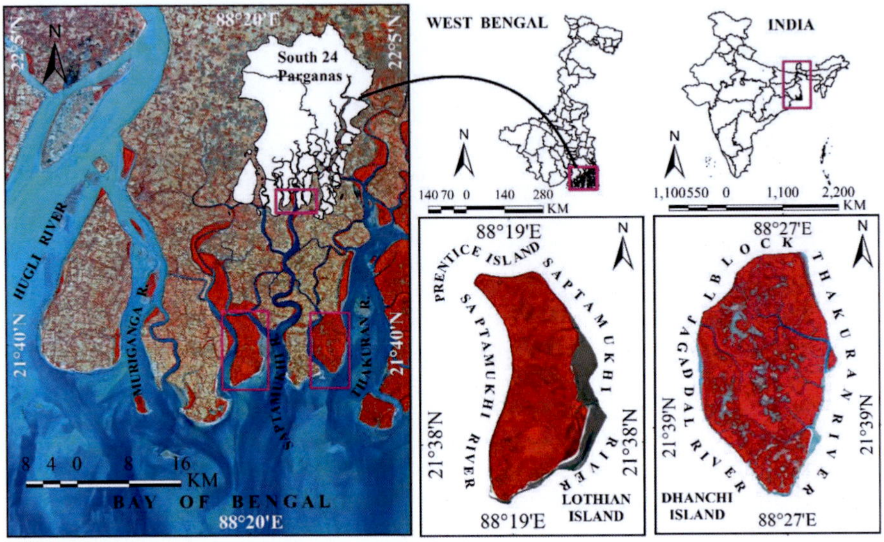

Fig. 24.1 Southwestern part of Indian Sundarbans with estuarine location of Lothian and Dhanchi Island

24.4 Results and Discussion

The section will demonstrate the declining trend of mangroves on Lothian Island and Dhanchi Island in multiple ways of impacts on the cyclone batter environment. The cyclonic wind velocities, wind-driven wave steepness, associated longshore currents, wind-driven tidal surges, storm surges, and saltwater inundations played an important role in the modifications of mangroves during the previous cyclones (1988, 1995, 2007, 2009, 2020, and 2021). Mangrove losses resulting from erosion of the island banks, shifting sands by washover deposits, hypersalinities, and storm damage to trees are discussed using geoinformatics and field surveys (Fig. 24.2).

24.4.1 Storm Induce Land Erosion

Changing configurations of the island shorelines and tidal channel banks due to land erosion have reduced the areal coverage of mangroves on Lothian Island and Dhanchi Island (Figs. 24.2 and 24.3). Mangrove-dominated island banks on the sea face are affected by the removal of top soil and sheet erosion due to breaking waves and longshore current energies concentrated in the forest fringe landscapes. Tidal surges driven by strong wind speed along the estuary banks and other tidal river banks undermined the bank margin soils and removed the trees with collapsing banks after the storms. About 1.77 km^2 area of forest lands are eroded in Lothian

Fig. 24.2 Mangrove Forest of Dhanchi and Lothian Island: Levee forest along the tidal creeks, shoreline mangroves damaged by the cyclonic storms, channel bank mangroves affected by erosion in the storms, and tidal swampy forest

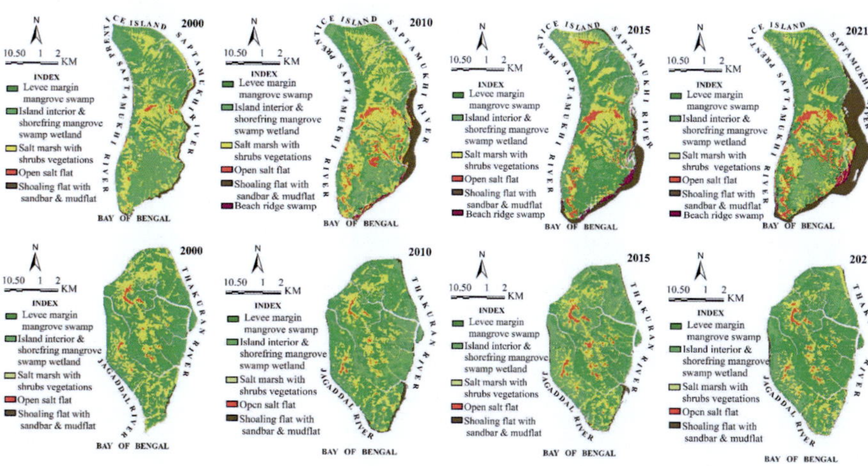

Fig. 24.3 Classification of mangrove ecosystem diversity on a temporal scale from 2000 to 2021 using the Landsat imagery for Lothian and Dhanchi island in the Indian Sundarbans

Island, and 2.55 km^2 of forest lands are eroded in Dhanchi Island, between 2005 and 2022, when maximum impacts were registered during the events of cyclones. The mangrove coverage area is reduced to 1.55 km^2, particularly due to erosion between

2006 and 2022, though afforestation was introduced after 'Aila' cyclone (2009) in the region.

The mangrove forest of Dhanchi and Lothian Island are classified on the Landsat images from 2000 to 2021 to identify the different categories of mangrove ecosystems and their dynamics (Figs. 24.3 and 24.4, Table 24.1). The study

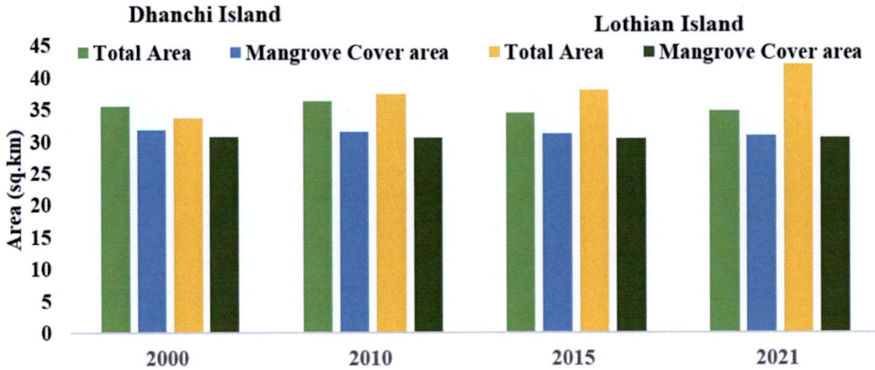

Fig. 24.4 Changing coverage of mangrove forest area in Dhanchi and Lothian Island

Table 24.1 The classification of mangrove habitat types and their changes during 2000, 2010, 2015, and 2021 for the Lothian and Dhanchi Islands

Sl. No.	Mangrove Habitats	Area in 2000 (sq. km)	Area in 2010 (sq. km)	Area in 2015 (sq. km)	Area in 2021 (sq. km)
Mangrove Zonation in Lothian					
1	Beach ridge swamp	0	0.29	0.73	0.57
2	Island Interior and Shorefring Mangrove Swamp Wetland	13.21	5.82	6.24	7.78
3	Levee margin mangrove swamp	6.24	10.15	12.70	11.54
4	Open Salt Flat	0.79	2.70	2.62	3.81
5	Salt marsh with shrubs vegetations	11.27	14.35	10.79	10.69
6	Shoaling flat with sandbar and mudflat	1.16	2.43	2.89	6.78
7	Tidal creeks and waterbody	1.03	1.59	1.92	0.82
Mangrove Zonation in Dhanchi					
1	Island Interior and Shorefring Mangrove Swamp Wetland	1.80	8.10	12.47	6.18
2	Levee margin mangrove swamp	19.24	15.33	10.60	17.48
3	Open Salt Flat	0.59	0.43	1.09	0.75
4	Salt marsh with shrubs vegetations	10.69	7.98	8.05	7.24
5	Shoaling flat with sandbar and mudflat	0.28	1.005	0.73	0.52
6	Tidal creeks and waterbody	2.83	3.43	1.52	2.56

Fig. 24.5 Loss of shore fringe mangrove forests due to (**a**) transgressive sand sheets and (**b**) wind damages with the signatures of relict forest or submerged forest at the sea front positions resultant from the impacts of cyclones

represents loss of the mangrove vegetations over the temporal scale in the region of the Indian Sundarbans resulting from cyclone batterer environment and storminess of the sea

24.4.2 Shifting Sands by Washover Deposits

The narrow shoreline beaches and low-height beach ridges of these islands were unable to hold the storm surges of cyclones 'Aila' and 'Amphan', thus sediment-laden waters encroached into the forest belts by advancing overwashed sands. Mangroves were trapped by such overwash sand deposits after the cyclone Amphan. A large tract of sand shoals was extended towards the south-eastern fringe forest in Lothian Island after the events of previous cyclones (Fig. 24.5). Tidal surges driven by winds during 'Yass' cyclone (2021) pushed the sands into the forests through overwashed sand fan lobes in the modification of mangroves. Southern parts of Dhanchi Island were also affected similarly by mangrove damage from shifting sands.

24.4.3 Hypersalinities

The storm surge sea waters inundated the low-lying surface pans with concave upward-shaped mud banks, and stagnant sea waters held for a long time in the depressions after the landfalls of repeated cyclones in the region of Lothian and Dhanchi islands. As the cyclone passed over the region, the muddy surface of depression basins became encrusted with salts due to evaporation in the region. Such hypersalinities of the surface soils do not allow the mangrove to grow or regenerate the mangrove seeds. The dieback of mangroves is registered at the pan fringes, possibly due to increased soil salinity after evaporation. Thus, a series of salt pan areas emerged in the coastal parts of Lothian and in a large part of Dhanchi Island. Thus, the gradual development of saline blanks has reduced mangroves in the Sundarbans.

24.4.4 Storm Damages to the Mangrove Trees

The types of damage to mangroves are observed in the Sundarbans during the field visits for monitoring forest status after the events of the Aila cyclone (2009) and the Amphan cyclone (2020). Storm surge elevations of the notorious cyclones were registered at 3.5–4.0 m (2009) above the ground level in the south-eastern parts of the Indian Sundarbans around the Malta–Raimangal–Ichhamati complex (Gayathri et al., 2015). Similarly, the elevation of the surge level was registered as 2.5–3.5 m on the shorelines fringed with mangroves in the western part of the Indian Sundarbans during the landfall of tropical cyclone 'Amphan' in 2020 (INCOIS, 2020), around the Matla-Hugli estuarine complex. Such devastating surge heights produced bank erosion, bank instability, and overwash encroachments with external inputs of sand over the forest fringe environment (Paul et al., 2017, 2018) in parts of the Indian Sundarbans. Chunks of the forest fringe lands were collapsed with uprooted trees along the tidal river banks after undermining the valley sides with strong currents generated in the surge waters in the cyclone-affected region. Wood logs of the mature mangrove trees were degraded by saltwater inundations and aggradations of sediments along the river banks after the events of cyclones, and many logs were also drifted into the downstream with the energy levels of return currents. However, on the sea face along the shore fringe mangroves, removal of top soils and overwash sand deposits destroyed mature mangrove trees after the events of tropical cyclone landfalls in the region. Mangrove loss is also documented from the emerging salt pans within the forest lands after they hold the surge waters into the surface depressions. Gradually, with the evaporation process, the clay beds were encrusted with salt crystals and destroyed mangroves by 'die-back' processes in the emerging hypersaline patches or saline blanks within the forest lands as a result of surge water penetration into the island interiors of the Sundarbans (Paul et al., 2018) during the cyclones (Fig. 24.5).

Trees lined with top breakages along the sea face represent the impact of wind storms in Dhanchi Island and Lothian Island of the southwestern shorelines in the Sundarbans between the Saptamukhi and Thakuran complexes. As they are damaged on the island platform surface, the basement layer of swamp clays has been removed by surface etching during storm water level currents and wave abrasion processes in the region. Thus, emerging stiff clay beds in unfavourable conditions do not support the mangrove regeneration process along the unstable shorelines by anchoring the viviparous mangrove seeds that drift on the sea face during storms. Mature mangrove trees were also twisted, uprooted, bended, and burned alive during strong wind velocity, salt sprays, and prevailing dry winds in the vicinity of the wall of circular movement of cyclonic winds after the landfall in the forest fringe environment. Using geoinformatics in the study, we demonstrate the Normalised Difference Vegetation Index (NDVI), Enhanced Vegetation Index (EVI), and Leaf Area Index (LAI) values that indicate the status of mangrove forests spatially and temporarily after and before the events of tropical cyclone landfalls in the Sundarbans (Figs. 24.6 and 24.7). Such degraded areas of forests are reflected in the LAI output maps of the two islands under study. The study highlights the reduction of NDVI, EVI, and LAI values for the cases of Lothian and Dhanchi islands in a temporal scale from 2000 to 2021. In Lothian Island the NDVI was 1.0 in 2000 and it is reduced to 0.96 in 2021; EVI on the other hand, showed 0.64 in 2000 and it is reduced to 0.57 in 2021; and LAI is also reduced to 13.48 in 2021 from 14.08 in 2000. Similarly, in Dhanchi island the NDVI showing reduction of 0.96 from 0.98, EVI showing a slight increase of 0.69 from 0.63 over the temporal scale, and the LAI is significantly reduced into 13.65 from 13.74 due to the impact of cyclonic storms by this time period (Figs. 24.6 and 24.7).

24.4.5 Climate Change Reworking and Forest Management

Climate change acceleration with sea level rise and storminess of the sea affected the deltaic islands of the Indian Sundarbans by reworking coastal landscapes with marine influences (Paul, 1996b; Paul & Paul, 2022; Raha, 2014; Chakraborty et al., 2014). The rate of aggradation and the rate of progradation of the islands were balanced by the supply and accumulation of sediments as well as the uninterrupted continuation of natural mangrove regeneration processes on the island platforms (Table 24.2 and Fig. 24.8). As the energy levels, anthropogenic controlled discharges of the river basin, and marine impacts have changed, the equilibrium is affected in sedimentation, and the reworking of landscapes further aggravates the situation of mangrove habitats in the Indian Sundarbans.

Attempts have been made to restore the vibrating ecosystem through different approaches to forest management. They are identified as (i) ecological engineering methods; (ii) mangrove afforestation for wild life conservation; (iii) improving ecological buffers in the biosphere reserve; and (iv) emerging concepts of the environmental value of the landscapes. Habitat restoration and carbon sequestration are other functions of mangrove forests currently realised by environmentalists, researchers, locals, and coastal managers in the region for managing the mangrove buffers. To minimise the impacts of cyclones and the coastal squeeze in the Sundarbans, people are involving themselves in the mangrove afforestation programmes.

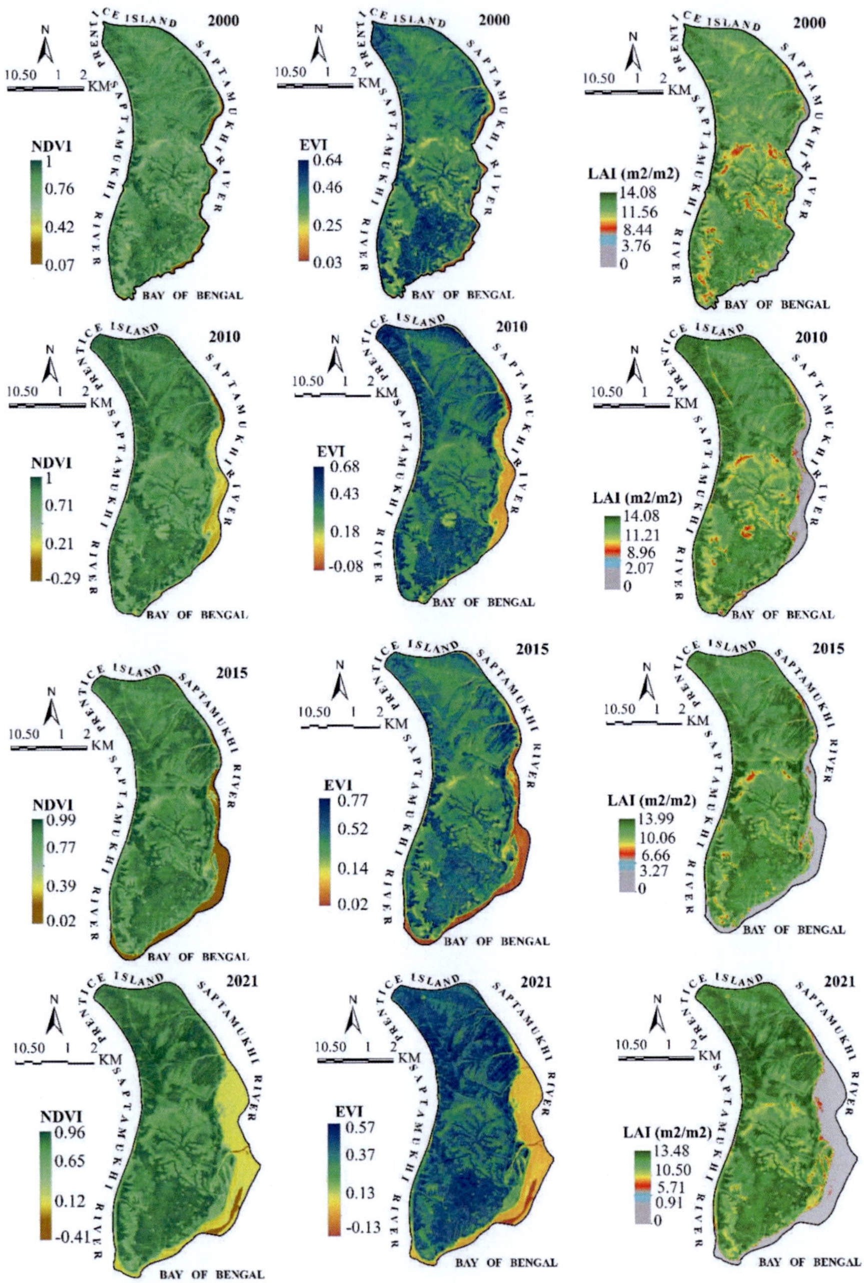

Fig. 24.6 Estimated spectral indices signature for NDVI, EVI, and LAI in Lothian island of Indian Sundarbans from 2000 to 2021

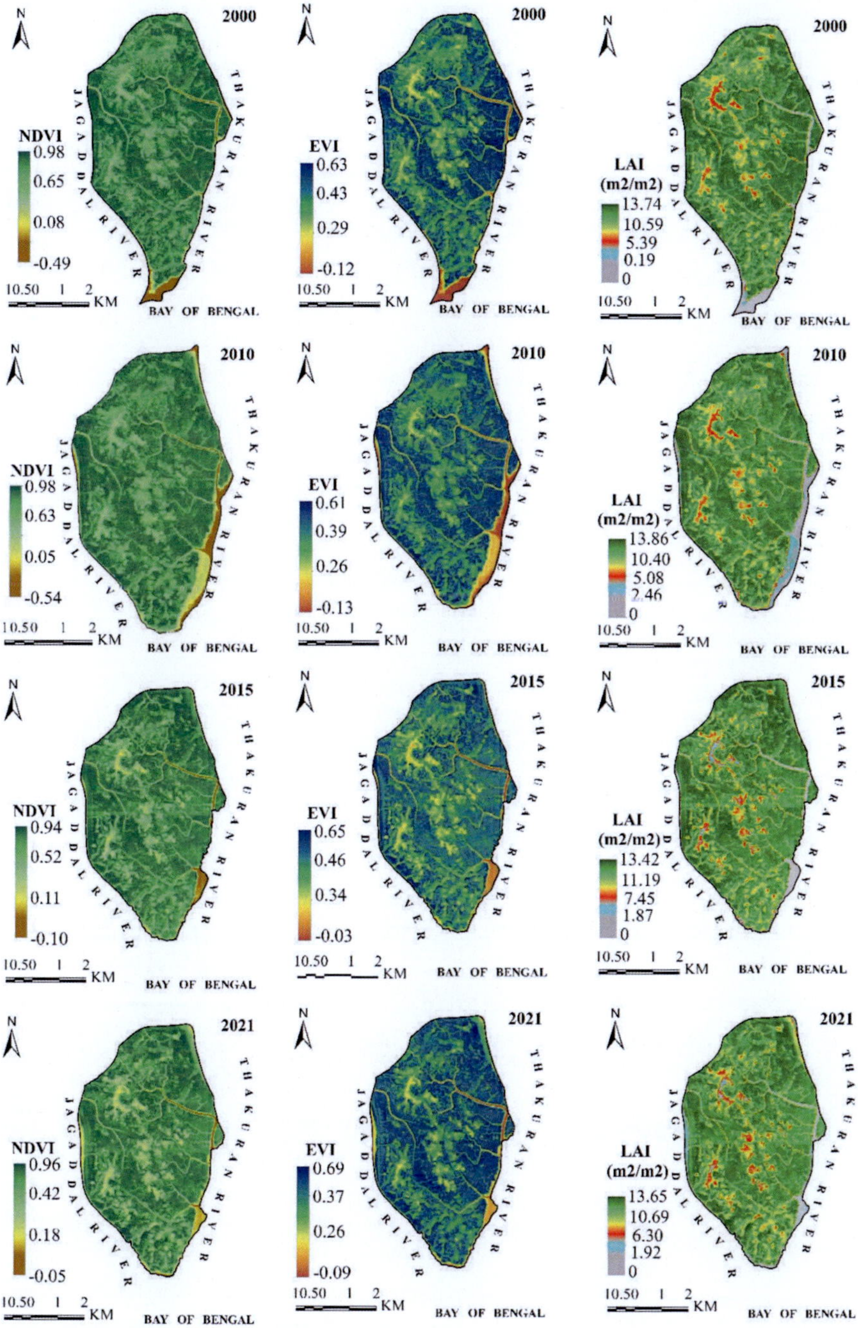

Fig. 24.7 Estimated spectral indices signature for NDVI, EVI, and LAI in Dhanchi island of Indian Sundarbans from 2000 to 2021

Table 24.2 The structure of forest habitats in physiographic settings and their degradation levels as a result of cyclonic storms in the previous decades along with the impacts of climate change reworking of the landscapes of estuarine setting in the Indian Sundarbans

Physiographic Settings	Forest Quality	Sedimentary Environment (Sediment Input and Output)	Tidal Environment (Inundation Frequency)	Water Quality in Pre-monsoon (Water Salinity in PPT)	Energy Levels
Fringe forest of the shorelines along the tidal rivers	Dynamic forest	Ag + Pg + Rg	Frequent	16–17	Tidal currents and tidal surges
Adjoining forest of the salt pans	Dwarfed forest	Dg	Low	25–28	Ponded water
Levee forest along the creeks	Living forest	Pg + Ag	High	17–18	Tidal flood current
Tidal swamp forest in the low gradient flats	Living forest	Ag	High	18–19	Tidal flood current
Littoral fringe forest	Buffer forest	Tg	Moderate	20–21	Storm surges
Relict forest along the seashores	Dead forest	Rg	Frequent	>21	Storms, winds, waves, longshore currents

Key: *Ag* Aggradation, *Pg* Progradation, *Dg* Degradation, *Tg* Transgressive sands, *Rg* Retrogradation

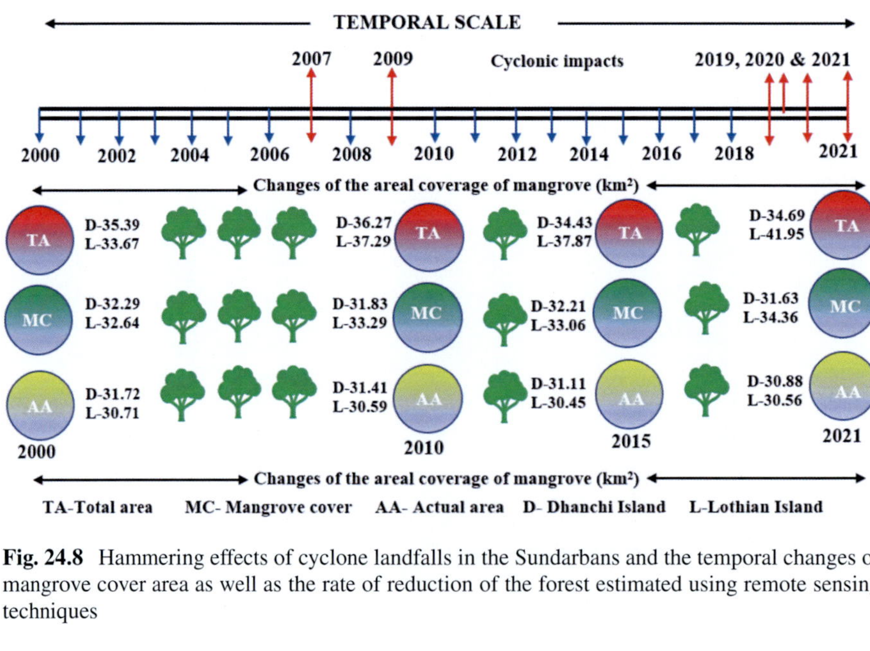

Fig. 24.8 Hammering effects of cyclone landfalls in the Sundarbans and the temporal changes of mangrove cover area as well as the rate of reduction of the forest estimated using remote sensing techniques

24.5 Conclusion

The Indian part of the Sundarbans was battered by eight notorious cyclones from 1988 to 2021 at an average recurrence interval of 4 years, which destroyed the mangrove forests in the coastal fringes. Climate change reworking with high rates of event frequencies from cyclones and sea level rise impacts has demonstrated the significant loss of mangroves in the Indian Sundarbans. The forest cover area changes for Lothian Island and Dhanchi Island represent spatiotemporal variations using image classification (Landsat 8 OLI) of satellite data (2000 to 2021) to detect the sites of degradation. Post-cyclone damages to the forest resources are also identified with the use of Leaf Area Index (LAI) values calculated for the study areas in the present work. The estimated values of NDVI, EVI, and LAI for the Lothian and Dhanchi island represent significant loss of vegetations and quality of vegetations along the shore fringes, channel banks, salt pans, and the littoral tracts. In Dhanchi island the LAI values are significantly reduced from 13.74 (2000) to 13.65 (2021). Similarly, the LAI is also reduced from 14.08 (2000) to 13.48 (2021) in Lothian Island. This measurement indicates the loss of forest quality in estuarine islands at the shore face and channel banks in the Indian Sundarbans under cyclone batter environment.

Cyclonic storms played a significant role in the zonation and plant succession of mangroves over a long time in the Sundarbans. Island interior zoned species have been exposed along the tidal river banks as a result of channel shifting caused by erosion and the loss of primary mangroves from the eroded mud banks after the events of cyclones. Signatures of relict forests are visible along the shore fringe forests, where exposures of vertical tree stumps with older pneumatophores are found on the erosive mud banks of the Bay of Bengal fringe islands. Storm-damaged trees are also found above the high tide line on Dhanchi Island of the Thakuran-Matla Estuarine Complex. Expansion of saltpans or saline blanks in the interior parts of the islands indicates the damages to the forest stands due to emerging hypersalinity. Salt flats and salt marshes represent stunted or dwarfed vegetation on the salt-affected surface in the inner parts of the islands. The immediate effects of cyclonic storms are seen in the forest fringe environment, with the signatures of partially damaged trees in the southwestern Sundarbans. Application of geoinformatics in the generation of databases will provide information for forest management in the deltaic islands. Ecological engineering methods, such as constructing drainage ditches to connect the tidal rivers for supply of moisture, can restore forests in favourable conditions, but mangrove afforestation in the degradation sites will restore habitats with monitoring the status of planted trees in the Sundarbans.

References

Chakraborty, S. K., Dutta, S. M., Ghosh, P. B., Ray, R., & Paul, A. K. (2014). Impact of global warming on Sundarbans mangrove ecosystem, India: Role of different assessment tools from ecosystem monitoring to molecular markers. In *Proceedings of the international conference on Green India: Strategic Knowledge for Combating Climate Change–Prospects and Challenges* (pp. 181–200). Pondicherry University. Excel India Publishers.

Choudhury, A., Bhunia, A. B., & Nandi, S. (1984). Preliminary survey on macrobenthos of Prentice Island, Sundarbans, West Bengal. *Records of the Zoological Survey of India, 81*(3–4), 81–92.

Gayathri, R., Bhaskaran, P. K., & Sen, D. (2015). Numerical study on Storm Surge and associated Coastal Inundation for 2009 AILA Cyclone in the head Bay of Bengal. *Aquatic Procedia, 4*, 404–411.

Ghosh, N., & Naskar, K. R. (2012). Aila cyclone, mangrove ecosystem and anthropogenic factor- a socio-economic scan. In A. K. Samanta & S. Panda (Eds.), *Some aspects of coastal vegetation in India, including Andamans and Sundarbans* (pp. 109–111). Ramnagar College.

Ghosh, A., Schmidt, S., Fickert, T., & Nüsser, M. (2015). The Indian Sundarban mangrove forests: History, utilization, conservation strategies and local perception. *Diversity, 7*(2), 149–169.

Green, E. P., Mumby, P. J., Edwards, A. J., Clark, C. D., & Ellis, A. C. (1997). Estimating leaf area index of mangroves from satellite data. *Aquatic Botany, 58*(1), 11–19.

Hati, J. P., Samanta, S., Chaube, N. R., Misra, A., Giri, S., Pramanick, N., et al. (2021). Mangrove classification using airborne hyperspectral AVIRIS-NG and comparing with other spaceborne hyperspectral and multispectral data. *The Egyptian Journal of Remote Sensing and Space Science, 24*(2), 273–281.

Huete, A., Didan, K., Miura, T., Rodriguez, E. P., Gao, X., & Ferreira, L. G. (2002). Overview of the radiometric and biophysical performance of the MODIS vegetation indices. *Remote Sensing of Environment, 83*(1–2), 195–213.

INCOIS (2020). Rapid intensification of super cyclone amphan fueled by marine heatwaves. *IIOE-2 Newsletter, 6*(6), June, 2022. https://incois.gov.in/documents/IIOE-2/Publications/IIOE-2-DOC_NL_243.pdf

Majumdar, D. D., Bera, S., Ray, R., & Paul, A. K. (2012). An assessment of diversities in halophytic grassland and mangrove swamp in Nayachar Island, Hugli estuary, West Bengal. In *Some aspects of coastal vegetation in India, including Andamans and Sundarbans* (pp. 112–123). Ramnagar College.

Mondal, B. K., & Paul, A. K. (2022). Application of participatory rural appraisal and geospatial techniques for analysing the dynamics of Mangrove Forest and dependent livelihood in Indian Sundarban. In *Conservation, management and monitoring of forest resources in India* (pp. 409–455). Springer International Publishing.

Paul, A. K. (1996a). Degradation of coastal vegetations in West Bengal. *India Journal of Landscape and Ecological studies, 19*(1), 39–50.

Paul, A. K. (1996b). Identification of coastal hazards in West Bengal and parts of Orissa. *Indian Journal of Geomorphology, 1*(1), 1–27.

Paul, A. K. (2000a). Physiography of the mangrove swamps- A study in the Sundarban (W.B). In *Sundarban Mangals* (pp. 152–170). Naya Prakash.

Paul, A. K. (2000b). Cyclonic storm and their impacts on West Bengal coast. In V. G. Rajamanikam & M. J. Tooley (Eds.), *Quaternary sea level variation* (pp. 25–57). New Academic Publishers.

Paul, A. K. (2002). *Coastal geomorphology and environment* (pp. 1–342). ACB Publication.

Paul, A. K. (2014). Morphology of estuaries and tidal inlets: An emphasis on Hugli, Subarnarekha and Chilika systems along the Bay of Bengal shoreline. In *Proceedings of the national conference on Modern Trends in Coastal and Estuarine Studies,* Tilak Maharashtra Vidyapeeth, Pune, Maharashtra India, 6–7 Feb 2014

Paul, A. K. (2022, August). Dynamic behaviour of the estuaries in response to the phenomenon of global warming in the coastal ecosystems of West Bengal and Odisha, India. In *Transforming coastal zone for sustainable food and income security: Proceedings of the international symposium of ISCAR on Coastal Agriculture, March 16–19, 2021* (pp. 907–931). Springer International Publishing.

Paul, A. K., & Bandyopadhyay, M. K. (1985). The role of mangroves in deltaic morphology: A study in Prentice and Lothian Islands, Sundarbans, W.B. In The Mangroves, L. J. Bhosle (Ed.), *Proceeding of national symposium on Biology, Utilization and Conservation of Mangroves* (pp. 218–221).

Paul, A. K., & Bandyopadhyay, M. K. (1988). Morpho-ecological dynamics of the mudflats in the Sundarban, West Bengal. *Geographical Review, 49*(1), 1–17.

Paul, A. K., & Paul, A. (2022). Adjustment of the coastal communities in response to climate variability and sea level rise in the Sundarban, West Bengal, India. In *Climate change, disaster and adaptations: Contextualising human responses to ecological change* (pp. 201–217). Springer International Publishing.

Paul, A. K., Ray, R., Kamila, A., & Jana, S. (2017). Mangrove degradation in the Sundarbans. In *Coastal wetlands: Alteration and remediation* (pp. 357–392). Springer. https://doi.org/10.1007/978-3-319-56179-0

Paul, A. K., Kamila, A., & Ray, R. (2018). Natural threats and impacts to mangroves within the coastal fringing forests of India. In *Threats to mangrove forests: Hazards, vulnerability, and management* (pp. 105–140). Springer.

Raha, A. K. (2014). Sea level rise and submergence of Sundarban Islands: A time series study of estuarine dynamics. *Journal of Ecology and Environmental Sciences, 5*(1), 114–123. SSN: 0976-9900 & E-ISSN: 0976-9919.

Ray, R., Paul, A. K., & Basu, B. (2014). OIF based indeces oriented ecological classification using LANDSAT TM digital data–a case study on Beluchary and Dhulibasan Island Groups, Sunderban, West Bengal, India. *International Journal of Remote Sensing Applications, 4*(1), 56–60.

Rouse, J. W., Haas, R. H., Schell, J. A., & Deering, D. W. (1974). Monitoring vegetation systems in the Great Plains with ERTS. *NASA Spec. Publ, 351*(1), 309.

Chapter 25
Multivariate Analysis of Coastal Vulnerabilities for Tamil Nadu Coast Using Remote Sensing and GIS Techniques

Debabrata Ghorai and Ashis Kumar Paul

25.1 Introduction

Coastal zone is the intersecting area where land, climate, and marine process occur. This zone is very dynamic in nature and extremely important to coastal countries for their future, particularly regarding the status of its natural resources which provide life support and economic development opportunities to coastal settlements (Clark et al., 1992). Nowadays the coastal zone is facing many problems mainly due to coastal erosion, pollution, developments, coastal habitats loss along with natural disasters consisting of cyclones, tsunamis, floods, land erosion by shoreline shifting, and salinity ingress (Nayak, 2000, 2002; Ajai et al., 2013). These disasters are increasing in the coastal zone due to human interference in the coast such as development activity, tourism activity, research activity (agricultural practice, medicine, etc.), and exploitation of natural resources. The sustainable management of the coastal zone to balance the coastal ecosystem is highly required from the impact of the above disasters. Perhaps, composite vulnerability assessment can be studied with the use of historical records of disasters along with various physical and socio-economic variables over the coastal region. There are a small number of studies on composite vulnerability/risk assessment that have been done previously by considering a combination of few disasters (Unmesh & Narayanan, 2009; Bhadra et al., 2010; Mahapatra, 2015; Das, 2012; Patnaik & Narayanan, 2005; Luers et al., 2003; Moss et al., 2001; Kaly et al., 2002; Downing et al., 2001; Iyengar & Sudarshan, 1982; Bhadra et al., 2010; Paul & Chatterjee, 2010). Therefore, the present study

D. Ghorai (✉)
Department of Research and Development, Dvara E-Registry, DBS House, Hyderabad, India
e-mail: ghoraideb@gmail.com

A. K. Paul
Department of Geography, Vidyasagar University, Midnapore, West Bengal, India

A. K. Paul, A. Paul (eds.), *Crisis on the Coast and Hinterland*,
https://doi.org/10.1007/978-3-031-42231-7_25

analyses the composite vulnerability of coastal taluk of Tamil Nadu and Puducherry coast using above disasters.

The Intergovernmental Panel on Climate Change (IPCC, 1996) defines the vulnerability as “the text to which climate change may damage or harm a system”. Vulnerability assessment which includes a combination of various factors that determine the degree to which someone’s life, livelihood, property, and other assets are put at risk by a discrete and identifiable event in nature and in society (Mahapatra, 2015; CGWB, 2014). Social vulnerability deals with those demographic and socioeconomic factors that increase the impacts of hazard on exposed populations (Tierney et al., 2001; Center, 2002). Composite vulnerability and risk are estimated for the coastal zone, which is associated with multiple disasters due to natural and man-made activity that may cause severe damage or loss to the society. Composite vulnerability index is calculated using various factors relating to coastal disasters and socioeconomic activity. Normalizing and assigning the weights to these composite vulnerability factors are considered for the estimation of composite index scores or values along the coastal zone. A total of 37 coastal districts of Tamil Nadu and Puducherry are considered for the present study.

25.2 Materials and Methodology

Composite vulnerability study is highly required from the threats of multiple hazards, which are uncertain and cannot be predicted for sustainable coastal zone management to reduce the pressure on the coastal resources and balance the coastal biodiversity. The vulnerability of a system is depending on the degree of its exposure, that is, how frequent and to what extent it will be exposed to the impact, its sensitivity (as different systems have different sensitivities to the impact), and its adaptive capacity or resilience (Saxena, 2013; Seenipandi et al., 2019). The composite vulnerability of coastal taluks of Tamil Nadu and Puducherry coast to multihazard (floods, cyclones, tsunamis, and salinity ingress) impacts has been assessed using 30 parameters towards exposure, sensitivity, and adaptive capacity. The analytical hierarchy process (AHP) method is followed to assess the flood vulnerability (rainfall, relative elevation, slope, drainage density, soil type, land use, and land cover); cyclone vulnerability (elevation, wind speed, and land use and land cover), and tsunami vulnerability (elevation, slope, and distance from the shoreline and land use and land cover) at the district level (NRCS 2016; SAC 2012). The GALDIT model is used to assess the salt water intrusion vulnerability (G-groundwater occurrence-aquifer type, A-aquifer hydraulic conductivity, L-level of groundwater height above sea level, D-distance from shoreline, I-impact of existing status of seawater intrusion, and T-thickness of aquifer being mapped) in the districts of Tamil Nadu coasts. Composite vulnerability index (iCVI) for the entire coastal taluk in Tamil Nadu and Puducherry are expressed as (Eq. 25.1):

$$\text{iCVIk} = 0.5 \times \left\{ \left(1/t \right) \Sigma wnynk \right\} 1/\left(m+1 \right) \tag{25.1}$$

where, iCVIk is the integral composite vulnerability index, *t* is the total number of parameters, *w* is the weight of the exposure parameter which is estimated by adopting AHP (Saaty, 1980), *y* is the indefinite integral expressed by the following equation (Eq. 25.2), *n* is the number of exposure indicator, *k* is the number of regions/taluks, and *m* is the minimum number of parameters among exposure, sensitivity, and adaptive capacity component:

$$y = \int (sx - a) = mdx = (m(sx - a)m + 1) / (s(m + 1)) + c \tag{25.2}$$

where, *x* is the normalized score of exposure parameter, *s* is the sensitivity index (SI), *a* is the adaptive capacity index (ACI), and *c* is the constant of integration which is considered as exposure index (EI) for this study. The EI, SI, and AI index values were estimated using the following formula as follow in Eqs. 25.3 and 25.4:

$$I = \sum ej (\text{for } j = 1 \text{ to } p) \tag{25.3}$$

where, *I* is the index of the component, *e* is the normalized score of the parameter which can be estimated by the following formula (Eqs. 25.4 or 25.5) based on the functional relationship towards vulnerability, and *p* is the number of parameters:

$$ei = (Xij - \text{Mini}\{Xij\}) / (\text{Maxi}\{Xij\} - \text{Mini}\{Xij\}) (\text{when positive relation}) \tag{25.4}$$

$$ei = (\text{Maxi}\{Xij\} - Xij) / (\text{Maxi}\{Xij\} - \text{Mini}\{Xij\}) (\text{when inverse relation}) \tag{25.5}$$

where, *X* is the value of parameter, *i* is the number of regions/taluks, and *j* is the number of parameters. Exposure parameter normalized score was derived using Eq. 25.4. It is clear that all these scores will lie between 0 and 1. The value 1 corresponds to that region with maximum vulnerability and 0 corresponds to the region with minimum vulnerability. Using Eq. 25.3, estimate the exposure index (EI), sensitivity index (SI), and adaptive capacity index (ACI) scores, which are divided into three categories, namely, low, moderate, and highly vulnerable regions.

25.3 Results and Discussion

25.3.1 Flood Vulnerability

Flood vulnerability zone divided into five vulnerability zones such as very high (4.33–4.99), high (3.85–4.33), moderate (3.39–3.85), low (2.90–3.39), and very low (1.64–2.90) flood vulnerability zone. Analysis of the flood vulnerability zone shows that very high to very low flood vulnerability zone areas constitute 26.40%, 24.97%,

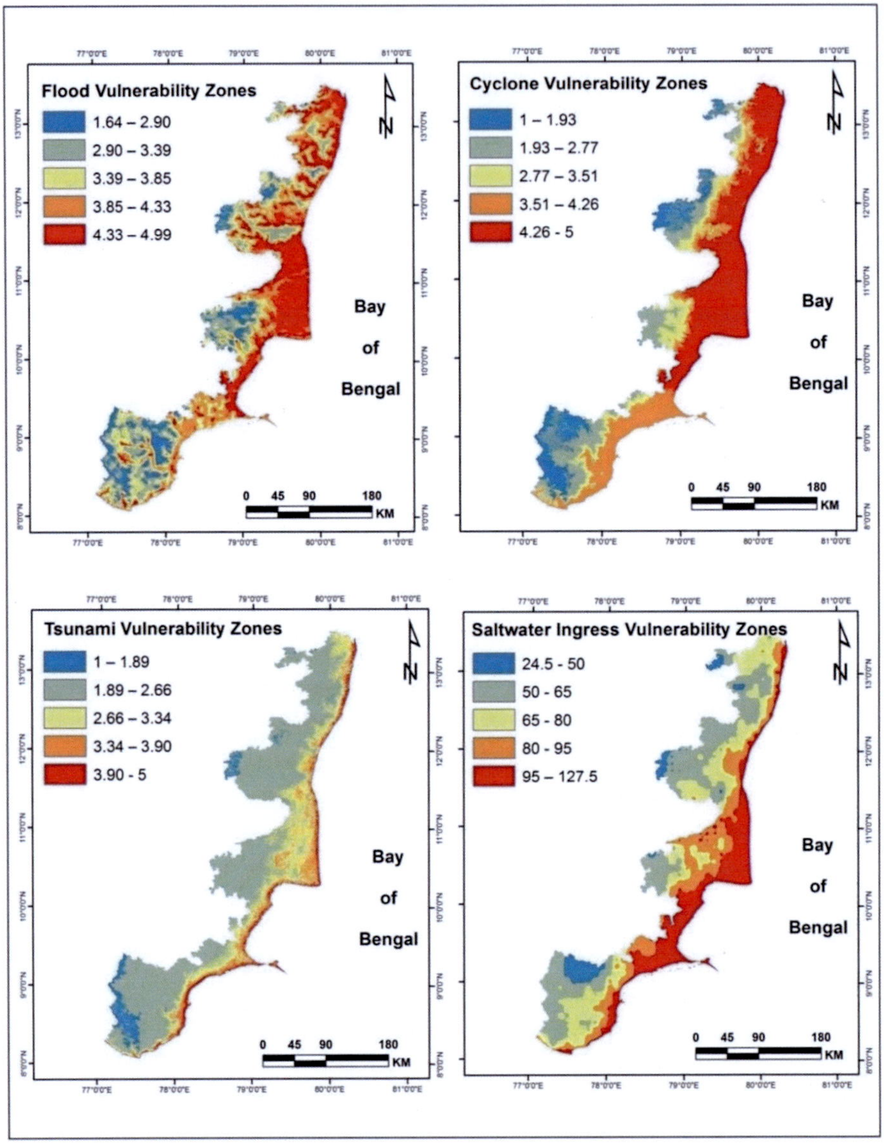

Fig. 25.1 The vulnerability factors (flood, cyclone, tsunami, and saltwater ingression) of the coastal districts in Tamil Nadu and Puducherry

19.33%, 18.62%, and 10.67%, respectively, of the study area (Fig. 25.1). It is observed that the highest flood vulnerable zone area comes under very high flood vulnerable zone followed by high flood vulnerable zone. As per the assessment of flood vulnerability, the districts of Thiruvallur, Chennai, Kancheepuram, and

Villupuram have the highest flood vulnerability zone. However, Cuddalore, Nagapattinam, Thiruvarur, and Karaikal districts show the flood vulnerability zone under very high flood risk. These districts are part of Cauvery delta, which falls under low-lying topography, which may be the reason for very high flood vulnerability in these districts. Similarly, Thanjavur, Pudukkottai, Ramanathapuram, Thoothukudi, Tirunelveli, Kanyakumari, and Puducherry districts are experiencing high flood vulnerability zones. The highest high flood vulnerability zone is observed in Vilathikulam (419.73 km^2) taluk followed by Tindivanam (395.20 km^2), Ramanathapuram (366 km^2), Kadaladi (329.25 km^2), and Pattukkottai (321.46 km^2) taluks.

25.3.2 Cyclone Vulnerability

Cyclone vulnerability zone divided into five vulnerable zones such as very high (4.26–5), high (3.51–4.26), moderate (2.77–3.51), low (1.93–2.77), and very low (1–1.93) cyclone vulnerability zone. The analysis of the cyclone vulnerability zone shows that very low to very high cyclone vulnerable zone areas constitute 12.18%, 17.96%, 12.72%, 20.16%, and 36.98% of the area, respectively. It is observed that the highest cyclone vulnerable zone area comes under very high cyclone vulnerability zone followed by high cyclone vulnerability zone. The map of cyclone vulnerability shows that almost all the northern to central coastal districts of Tamil Nadu and Puducherry come under highest very high cyclone vulnerability zones ranging from 57.12% to 99.54% of the area (Fig. 25.1).

25.3.3 Tsunami Vulnerability

Tsunami vulnerability zone is divided into five vulnerable zones such as very high (3.90–5), high (3.34–3.90), moderate (2.66–3.34), low (1.89–2.66), and very low (1–1.89) tsunami vulnerability zones. The analysis of tsunami vulnerability zones shows that very low to very high tsunami vulnerable zone areas constitute 5.08%, 62.91%, 18.41%, 9.68%, and 3.91% of the area, respectively. It is observed that the highest tsunami vulnerable zone area comes under the low tsunami vulnerability zone followed by a moderate tsunami vulnerability zone. The estimated tsunami vulnerability zone shows that almost entire Tamil Nadu coastal districts come under low tsunami vulnerability category ranging from 39.23% to 84.99%, as tsunami water can inundate few kilometres towards land from shoreline. However, Chennai and Puducherry districts show high tsunami vulnerability zone which cover 42.82% and 40.09% of the area (Fig. 25.1).

25.3.4 Saltwater Intrusion Vulnerability

Saltwater ingress vulnerability zone has been divided into five vulnerable zones such as very high (95–127.5), high (80–95), moderate (65–80), low (50–65), and very low (24.5–50) saltwater ingress vulnerability zones. Salinity intrusion is mainly observed along the coastal districts with certain degree of extent towards land from shoreline. Using the GALDIT model salinity ingress zones were estimated. The analysis of salinity ingress vulnerability zone shows that very high to very low salinity ingress vulnerability zones are constituted of 21.74%, 18.70%, 25.05%, 29.73%, and 4.77% of the area, respectively (Fig. 25.1). It is observed that the highest saltwater ingress vulnerable zone area comes under the low vulnerability zone followed by a moderate vulnerability zone. The highest high salinity ingress zone is in Karaikal, which is 100%, followed by Nagapattinam (90.31%) and Ramanathapuram (77.22%).

25.3.5 Exposure Index (EI)

Total eight hazard zones were used to calculate EI. The value of the EI was between 0.28 and 4.66. It was divided into three vulnerability classes, namely, low (0.28–1.50), moderate (1.50–3.00), and highly vulnerable area (3.00–4.66), respectively (Fig. 25.2). It is supposed that the lower range of EI values represents low vulnerability and the higher range of values indicates high vulnerability. The EI map shows that 25 coastal taluks mostly from central Tamil Nadu coast (Puducherry to Ramanathapuram) are highly vulnerable due to large coastal length, wider low-lying area, and severe historical disasters along with a high level of salinity ingress. The low vulnerable taluks are Ottapidaram, Radhapuram, Kalkulam, and Vilavancode because of the presence of rocky marine terrace, elevated sand dunes with teri sand, and marine cliff which are working as a defence against natural disasters especially storm surge, giant sea waves, tsunamis, and cyclones. The topography of these taluks is highly elevated with a small shoreline length which might be another reason for not having much impact of exposure vulnerability (Fig. 25.2).

25.3.6 Sensitivity Index (SI)

A total of nineteen parameters were used from coastal resources, land resources, and demographic variables to calculate SI. The range of the SI was between 3.33 and 8.28 and it was classified into three vulnerability classes such as low (3.33–5.00), moderate (5.00–6.50), and high (6.50–8.28) vulnerability (Fig. 25.2). It was approximated that the lower range of SI values indicates low vulnerability and the upper

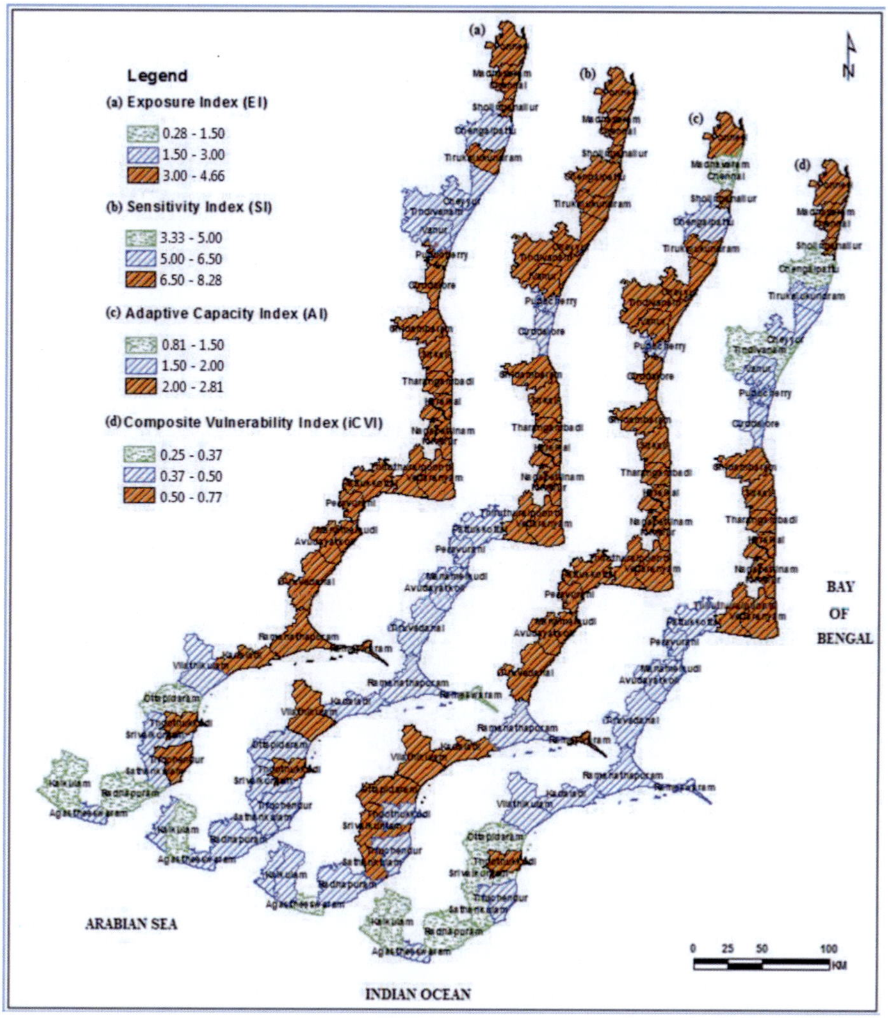

Fig. 25.2 The composite vulnerability index calculated for Tamil Nadu and Puducherry coastal districts using geospatial tools

range of values indicates high vulnerability. The map of SI revels that a total 19 coastal taluks mainly from the northern to the central part of the Tamil Nadu coast are highly vulnerable because of a large number of coastal resources, land resources, and human settlement density along with highest population. The low vulnerable coastal taluks are Rameswaram and Kalkulam and the remaining 16 coastal taluks are medium vulnerable which are situated on the central to the southern Tamil Nadu coast.

25.3.7 Adaptive Capacity Index (ACI)

The ACI was calculated using three variables. The range of ACI was 0.81–2.81, and it was further divided into three vulnerability classes, namely, low (0.81–1.50), moderate (1.50–2.00), and high (2.00–2.81) vulnerability (Fig. 25.2). The ACI map shows that a total 26 coastal taluks which are situated on the northern to the central coast of Tamil Nadu are highly vulnerable because of a large number of illiterate people and less density of road/railway network, as well as because not much coastal embankment development has been done. On the other hand, Madhavaram, Chennai, Agastheeswaram coastal taluks are low vulnerability and the remaining eight coastal taluks mostly from the southern Tamil Nadu coast are medium vulnerable (Fig. 25.2).

25.3.8 Composite Vulnerability Index (iCVI)

The iCVI was calculated using Eq. 25.5, which is a newly developed model or formula prepared for composite vulnerability analysis on the coast to identify the multiple disaster vulnerability regions. The iCVI ranges between 0.25 and 0.77, wherein a lower value indicates low vulnerability and a higher the value indicates high vulnerability. This index was further classified into three vulnerability classes such as low (0.25–0.37), moderate (0.37–0.50), and high (0.50–0.77) vulnerable classes (Fig. 25.2). The high vulnerability coastal taluks are Ponneri, Madhavaram, Chennai, Sholinganallur, Chidambaram, Sirkali, Tharangambadi, Nagapattinam, Kilvelur, Vedaranyam, Thiruthuraipoondi, Thoothukkudi, and Karaikal. These taluks are having highest low-lying areas that are liable to floods due to cyclones, tsunamis, and storm surges with heavy rainfall. Apart from these, highly sensitive components and low adaptive capacity make all these coastal taluks highly vulnerable. The low vulnerable coastal taluks are Chengalpattu, Tindivanam, Ottapidaram, Srivaikuntam, Sathankulam, Radhapuram, Kalkulam, and Vilavancode. The reason for low vulnerability for these taluks could be the presence of a rocky/cliff coast, rocky marine terrace, highly elevated sand dunes, dense plantation/forest coverage, and hilly topography with a small extent of shoreline which may result in a low to medium value of EI, SI, and ACI. The remaining 16 coastal taluks are medium vulnerable which are situated on the central to southern coast of Tamil Nadu and few are situated on the northern coast of Tamil Nadu. A rank to all the coastal taluk has been given based on the iCVI score which shows that Chennai is the highest vulnerable region followed by Vedaranyam, Sholinganallur, Nagapattinam, and Thoothukkudi taluk. The safest taluk is Kalkulam followed by Vilavancode, Radhapuram, Tindivanam, and Ottapidaram.

25.4 Conclusions

The present study highlights the use of geospatial technology in the assessment of coastal vulnerability. A composite vulnerability assessment approach of the coastal taluk of Tamil Nadu and Puducherry has been established and many recommendations for sustainable management of coastal resources have been suggested. The composite vulnerability zone for other coastal taluks of Tamil Nadu and Puducherry comes under moderate category. This study also described the integrated coastal resource management for mangroves, coral reefs, and land use, wherein the management approaches have been discussed. During this research, it is identified that, space technology has immense influence in the decision-making process. It encompasses data or information generation of land use, natural resources, climate, and urban systems for better management of resources. It also helps to protect ourselves with other coastal habitats from the impact of natural calamities like tsunami, flood, cyclone, etc., by estimating the probability of occurrence with the associated information and preparing the contingency for the management of the coastal zone.

References

Ajai, B. M., Unnikrishnan, A. S., Rajawat, A. S., Bhattacharya, S., Ramakrishnan, R., Kurian, N. P., Hameed, S., & Sundar, D. (2013). Demarcation of coastal vulnerability line along the Indian coast. *Journal of Geomatics, 7*(1), 25–31.

Bhadra, A., Bandyopadhyaya, A., Hodam, S., Yimchungru, C. Y., & Debbarma, R. (2010). *Assessment of vulnerability of Arunachal Pradesh (India) to floods*. http://www.iwra.org/congress/resource/2916898.pdf

Center, H. (2002). *Human links to coastal disasters*. The H. John Heinz III Center for Science, Economics and the Environment.

CGWB. (2014). *Report on status of ground water quality in coastal aquifers of India*. Central Ground Water Board (CGWB), Ministry of Water Resources, Government of India.

Clark, J. R., Garcia, S. M., & Caddy, J. F. (1992). *Integrated management of coastal zones* (Vol. 327, pp. 93–95). FAO. https://www.fao.org/docrep/003/t0708e/T0708E00.htm#TOC

Das, S. (2012). The role of natural ecosystems and socio-economic factors in the vulnerability of coastal villages to cyclone and storm surge. *Natural Hazards, 64*(1), 531–546.

Downing, T. E., Butterfield, R., Cohen, S., Huq, S., Moss, R., Rahman, A., et al. (2001). *Vulnerability indices: Climate change impacts and adaptation. Policy Series 3*. United Nations Environment Programme.

IPCC. (1996). Climate change 1995: Impacts, adaptations and mitigation of climate change. In *Intergovernmental panel on climate change, summary for policy makers*. World Meteorological Organization.

Iyengar, N. S., & Sudarshan, P. (1982). A method of classifying regions from multivariate data. *Economic and Political Weekly, 17*, 2048–2052.

Kaly, U., Pratt, C., & Howorth, R. (2002). A framework for managing environmental vulnerability in Small Island Developing States. *Development Bulletin (Canberra), 58*, 33–38.

Luers, A. L., Lobell, D. B., Sklar, L. S., Addams, C. L., & Matson, P. A. (2003). A method for quantifying vulnerability, applied to the agricultural system of the Yaqui Valley, Mexico. *Global Environmental Change, 13*(4), 255–267.

Mahapatra, M. (2015). *Vulnerability of Gujarat coast due to sea level rise – An investigation based on remote sensing & GIS techniques* [Ph.D. Thesis, Nirma University], (pp. 382–481).
Moss, R. H., Brenkert, A. L., & Malone, E. L. (2001). *Vulnerability to climate change: A quantitative approach*. Pacific Northwest National Laboratory (PNNL-SA-33642). Prepared for the US Department of Energy (pp. 155–167).
Nayak, S. (2000). Critical issues in coastal zone management and role of remote sensing. In *Subtle issues in coastal management* (Vol. 2000, pp. 77–98). Indian Institute of Remote Sensing.
Nayak, S. (2002). *Use of satellite data in coastal mapping (invited paper)*. Indian Cartographer, CMMC-01 (pp. 147–156).
NRCS. (2016). *United States Department of Agriculture, saturated hydraulic conductivity, natural resources conservation service soils*, http://www.nrcs.usda.gov/wps/portal/nrcs/detail/soils/survey/office/ssr10/tr/?cid=nrcs144p2_074846. Accessed on Aug 2016.
Patnaik, U., & Narayanan, K. (2005). *Vulnerability and climate change: An analysis of the Eastern Coastal Districts of India*. Paper presented at Human Security and Climate Change workshop, 21–23 June, Oslo.
Paul, A. K., & Chatterjee, S. (2010). A manual for coastal risk assessment in Bay of Bengal coast. In *4th session of the IAG working group on geomorphological hazards (IAGEOMHAZ) & international workshop on geomorphological hazards*. Department of Science and Technology, Govt. of India Ministry of Earth sciences, Govt. of India Tamil Nadu State Council for Science and Technology.
Saaty, T. L. (1980). *The analytic hierarchy process: Planning, priority setting, resources allocation*. McGraw-Hill.
SAC. (2012). *Coastal Zones of India*. Space Application Centre (SAC), ISRO, Ahmadabad, Sponsored by Ministry of Environment & Forests (MoEF), Govt. of India, http.www.sac.gov.in
Saxena, S. (2013). *Vulnerability assessment of Cuddalore Coast, Tamil Nadu, India: The science policy interface* [Ph.D. Thesis, Faculty of Civil Engineering, Anna University]
Seenipandi, K., Nainarpandian, C., Kandathil, R. K., & Sellamuthu, S. (2019). Seawater intrusion vulnerability in the coastal aquifers of southern India—an appraisal of the GALDIT model, parameters' sensitivity, and hydrochemical indicators. *Environmental Science and Pollution Research, 26*, 9755–9784.
Tierney, K. J., Lindell, M. K., & Perry, R. W. (2001). *Facing the unexpected: Disaster preparedness and response in the United States*. Joseph Henry Press.
Unmesh, P., & Narayanan, K. (2009). *Vulnerability and climate change: An analysis of the Eastern Coastal Districts of India*. MPRA Paper No. 22062, posted 14. April 2010 01:06 UTC. Online at http://mpra.ub.uni-muenchen.de/22062/

Chapter 26
Managing the Coastal Squeeze and Wetland Loss in Sagar Island in a Sustainable Framework Using Geospatial Techniques

Joydeb Sardar, Anurupa Paul, Kushal Nayak, Subhankar Naskar, Soumen Dey, Ipsita Mallick, Jatisankar Bandyopadhyay, and Ashis Kumar Paul

26.1 Introduction

The coastal parts of Sagar Island along the sea face retreated landward sufficiently by losing the sea beaches, mudbanks with relict mangroves, sand dunes, tidal creeks, backshore saltmarshes and mangrove swamps, village ponds, rice paddy fields, grasslands, earthen embankments, terrestrial vegetation, and settlements during the previous decades (1984 to 2022) due to erosion and advancement of the sea (Paul, 2000, 2002, 2022; Paul & Paul, 2022; Paul et al., 2018). The rate of shoreline retreat assessed using temporal images from 1985 to 2019 shows lands lost by 1000 m in the island's south-eastern corner and 500 m in the island's south-western fringe coast, at rates of 29 m/year and 16 m/year, respectively. Such loss of coastal lands and landscape ecology, as well as the diversity of wetlands, over a 38-year period indicates coastal squeeze events in response to the sea level rise scenario in the region of the estuarine coast on Sagar Island.

During the previous decades, on average, two cyclones attacked the Sundarban shore fringes with high wave energy, storm surges, tidal surges, and wind storms, causing the shorelines of Sagar Island to erode rapidly (Paul, 1997). As the island is located at the estuary front of the Hugli downstream section, the wetlands have been lost between realigned embankments and the advanced high tide line of the region.

J. Sardar
Centre for Environmental Studies, Vidyasagar University, Midnapore, West Bengal, India
e-mail: joydebsardar18@gmail.com

A. Paul (✉) · K. Nayak · S. Naskar · S. Dey · I. Mallick · J. Bandyopadhyay
Department of Remote Sensing and GIS, Vidyasagar University, Midnapore, West Bengal, India
e-mail: anurupashis@gmail.com

A. K. Paul
Department of Geography, Vidyasagar University, Midnapore, West Bengal, India

A. K. Paul, A. Paul (eds.), *Crisis on the Coast and Hinterland*,
https://doi.org/10.1007/978-3-031-42231-7_26

As a result of this coastal squeeze, a significant amount of landed property, settlements, tidal creeks, mangroves, saltmarshes, sand dunes, sea beaches, and engineering structures were lost along the different sections of the coast. In view of the above problems, the section-wise squeeze and diversity of losses are estimated for a sustainable framework of management to restrict the retreat of land margins. After estimating the coastal squeeze at the local level, embankment realignment techniques and new resource-use policies will be implemented.

26.2 Study Area

Sagar Island is a deltaic estuarine island that represents a historically reclaimed portion of the Indian Sundarbans. The shore fringe buffers of coastal wetlands are isolated and restricted by the protective cover of embankments after land reclamation. As a result of erosion, the wetland buffers have been narrowed in front of the protective embankments along the island fringe shoreline (Fig. 26.1).

26.3 Materials and Methods of the Study

Changing surface configurations, estimation of the shoreline beach widths of the three different physiographic settings of the triangular-shaped island, rate of shoreline changes along the seaward side, Hugli estuary side, and Muriganga

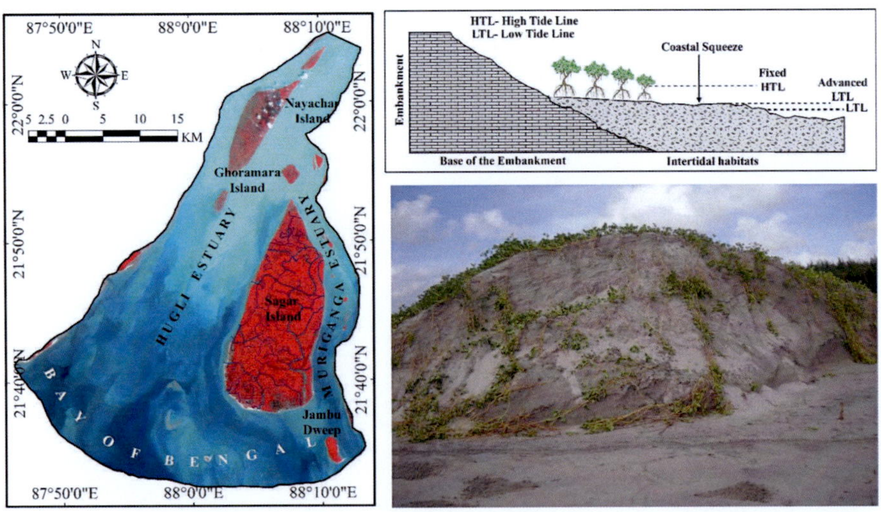

Fig. 26.1 Location of Sagar Island in Hugli estuary: Diagram depicts the coastal squeeze; coastal sand dunes lost from the intertidal habitat zone in Sagar Island

estuary side of the island, and spatiotemporal changes of the mangrove wetlands along the shore fringe areas are studied using the Landsat 8 OLI data, Google Earth Images, and field survey techniques to validate the output of such a study. The coastal squeeze could not ensure the coastal resiliencies in the vulnerable tract, thus a sustainable framework for the coastal squeeze is suggested in the present study.

26.4 Results and Discussion

The sea beaches, tidal flats, salt marshes, mangroves, sand dunes, and tidal creeks of the intertidal region of Sagar Island provided an ecological buffer, economic resources for the coastal poors, recreational ground, environmental benefits, and carbon sequestration. Historically, as a result of land reclamation (1810–1941), the habitats have suffered widespread losses in the inner parts and along the blocked shorelines by long earthen embankments. Climate change and sea level rise as well as the reduction of freshwater flows and sediment discharges through the estuary prevented the existence of intertidal habitats in the region. The vertical accretion rate and seaward erosion rate are influenced in the face of sea level rise, and that can prevent the long-term existence of intertidal habitats, particularly due to a lack of sediment supply. As the high-water mark is being fixed by the defence structure and, in response to sea level rise, the lower water line is migrating landward, coastal squeeze is emerging as a limiting factor in the survival of intertidal wetland habitats at present (Fig. 26.2).

26.4.1 Living Shorelines

The nature-based adaptation solution is immediately needed to adapt to sea level rise along the intertidal shores of the island. The present study has identified the more habitat-friendly options in the region to make the coast a living shoreline. During the 1980s, the shorelines were fringed with wide sea beaches backed by elevated sand dunes, tidal creeks with mangrove fringe banks, back shore sandy tracts with primary sand dunes, back shore mangroves and saltmarsh tracts, halophytic grass lands, relict mud banks with vertical mangrove tree stumps, and tidal flats (Paul, 2002; Paul & Kamila, 2018). Settlements and rice paddy fields protected by earthen embankments were located behind the natural habitat zone at a distance ranging from 1 to 3 km inland. How it is able to identify the habitat-friendly options of the degraded shorelines will be decided by using the temporal satellite images (Landsat 8 and Google earth) and existing knowledge on habitat dynamics. However, the mechanisms for promoting and managing the coastal squeeze problems may be grouped as: (i) physical alteration of the environment; (ii) land acquisition for conservation and protection on private land; (iii)

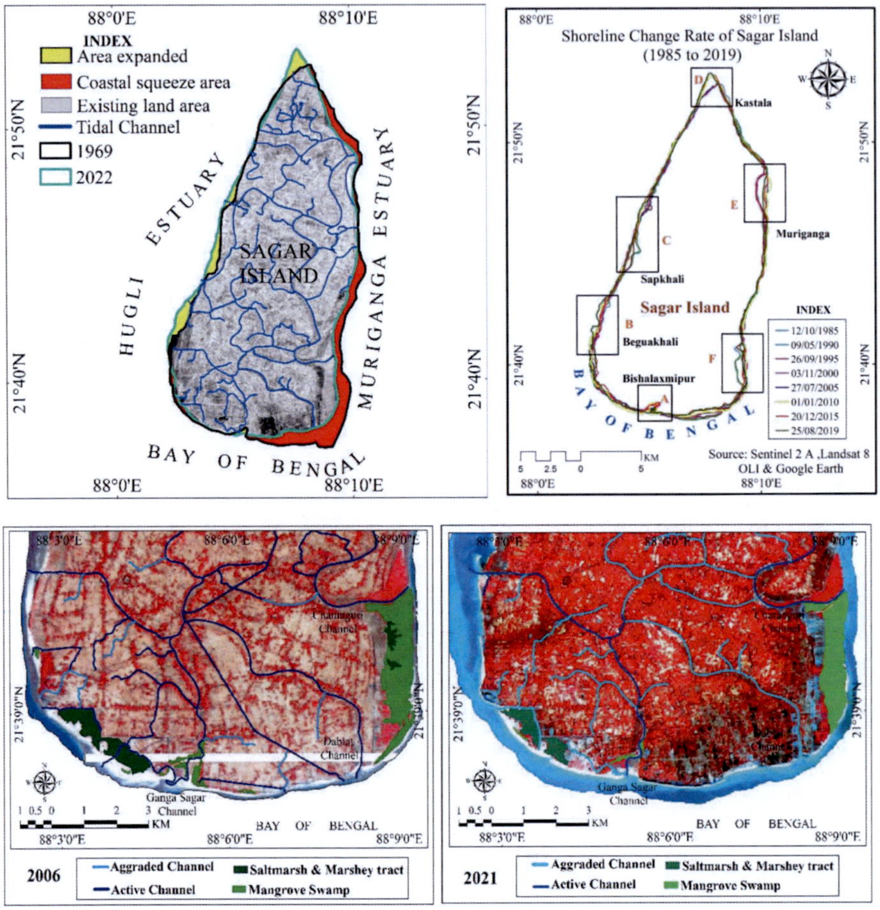

Fig. 26.2 Coastal squeeze on the east and south of the island: Shoreline change rates in the island sections; and temporal changes of coastal wetland habitats (2006–2021)

regulatory or policy interventions; and (iv) financial incentives for managing coastal squeeze like other areas of the tropical coast in Australia (Leo et al., 2019). Villages and shore fringe embankments are exposed on the intertidal region due to such coastal squeeze problems despite the relocation or realignment of the embankments and the rehabilitation of the displaced people of the damaged settlement on the island (Paul, 2002; Paul & Paul, 2022). Raising public awareness about coastal squeeze problems can thus be categorised as another mechanism for managing them in this environment.

Table 26.1 Wetland changes along the shorelines under intertidal zones (2006–2021)

Year	Mangrove Swamp (sq. km)	Saltmarsh and Marshy Land (sq. km)
2006	4.8	1.82
2021	2.78	0.93
Net area loss	2.02	0.89

26.4.1.1 Physical Alteration of the Environment in the Island

The methods of physical alteration can modify the physical and biological make-up of the intertidal areas. Modifications may be suggested, such as the removal of sea walls and sea dykes, the recovery and restoration of the habitats and hydrology, or the implementation of the approaches of living shorelines. Section-wise modification is required for the restoration of intertidal habitats along the island's shorelines by allowing the region to interact with tidal and marine exposures for supplying and accumulating sediments for slope, hydrology, and environment adjustment (Table 26.1).

26.4.1.2 Managed Retreat

By using this method of infrastructural realignment, the intertidal region will get sufficient space for the spread of new habitats. It is proven that the managed retreat of development will be more cost-effective than armouring the alluvial coast as an alternative approach against the impacts of sea level rise. Hence, we have to identify the suitable zones where people will live once climate change reworks the coastal landscapes in the region. Already people are confronting with the impacts of advancing sea and landward migration of the high tide lines and low water lines in several places (e.g., Beguakhali, Boatkhali, Shibpur-Dublat, Mandirtala, and Muriganga villages) of the island. The previously restricted areas of the coast under sea walls and embankments are affected by the encroachment of tide waters by this realignment method, and the habitats will be restored very soon by the supply of marine and tidal sediments in the intertidal region.

26.4.1.3 Intertidal Habitat Restoration in the Areas of Coastal Squeeze

Healthy habitats ensure coastal resilience, and therefore an attempt to restore the degraded coastal habitat is considered a key objective in managing the coastal squeeze. Several national and international frameworks are included in the Ramsar Convention, EU Habitat Directive, and National Oceanic and Atmospheric Administration (NOAA), particularly due to the habitat value of vegetation and the benefits of species used in conservation projects (Atkinson, 2003). Similarly, the

Table 26.2 Sections of the shorelines with present intertidal habitat condition in Sagar Island

Coastal sections	Intertidal habitat condition	Resource use conflicts	Defence structures
Southern shoreline fringed with the Bay of Bengal	Mangroves, saltmarshes, tidal mudflats, sand dunes, tidal creeks, sea beaches, sandy tracts, lowered mud banks by removal of sands, sand spits	Dry fish processing platforms, Gangasagar temple ground, recreational and beach tourism, subsistence marine fishing	Aila Bund in Topobon, realigned earthen embankments, lost ground infrastructures of displaced settlements, coastal squeeze
Eastern shoreline fringed with the Muriganga estuary section	Mangroves, saltmarshes, tidal mudflats, sand spits, tidal creeks, bank failure materials, grassland and meadows, terrestrial vegetations exposed	Estuarine and marine fishing, rice-paddy cultivation, bettle vines, orchards, aquaculture firm ponds	Earthen embankments, Aila Bunds at Muriganga, roads, remains of settlement structures, coastal squeeze
Western shoreline fringed with the Hugli estuary section	Mangroves, saltmarshes, meadows and grasslands, tidal mudflats, tidal creeks, sandy beds and bars, dunes, erosive banks and tidal flats	Estuarine fishing, marine fishing, fishing harbours, rice paddy farming, aquaculture ponds, ferry services	Earthen embankments, bank wall structures at Maindirtala and Beguakhali, other bank protection structures, coastal expand and coastal squeeze

focal theme of Environment Day (5 June 2022) in India was habitat restoration. The Forest Department of South 24 Parganas District, including the Tiger Reserve of India, Sundarbans, adopted the Mangrove Plantation Mission to restore the lost habitats in the region. Promotion of dune development in sandy tracts, plantation of long-rooted grasses and heaths or creepers, mangrove plantation in clay zones, and sediment supply into the intertidal region should be considered in favoured sites following defence breaching to restore intertidal habitats (Table 26.2).

26.4.1.4 Methods of Living Shorelines

The method of living shoreline stabilises the shore face and maintains the natural land-sea interface for critical coastal processes in the intertidal zones by allowing the exchange of sediments and allowing the immerged habitats to keep pace with rising seas through landward migration of the tidal zone and vertical accretion of sediments (Bilkovic et al., 2016). The living shoreline is particularly effective along the extent of three coastal sections of the island, in which the land–sea interface is keeping pace with the landward migration of tidal zones and rising seas, but a low erosion rate is recorded due to other forms of human impacts (Table 26.1). The dry fish processing activities, aquaculture ponds with low-height embankments, and tourism and recreation infrastructure behind the restored habitats have restricted the space within the ecological buffers.

26.4.2 Acquisition of Land for Conservation and Protection on Private Land

To ensure the presence of natural ecosystems and landward migration of wide buffers, acquisition of private lands is needed along the intertidal shores, considering the fixed long-term legal agreements for conservation and protection of the coast.

26.4.2.1 Land Acquisition

Private lands on the island are exposed in many sections of the intertidal zone due to erosion and landward migration of the low tide shoreline. Because of the shortage of open land on the backshores, realignment of defence structures was not possible and could not prevent the migration of the intertidal zone in the region. Private lands are located close to the backshores, and as a result of the absence of a buffer, high-cost sea walls are constructed along the high tide line. Thus, high tide is blocked by the sea wall, but the low tide shorelines have migrated landward, making the coastal sections susceptible to coastal squeeze problems. Thus, private land acquisition is also a method to restore habitat and for realignment of the sea walls to the inland areas in the land–sea interface. The local government wanted to purchase the private land for the construction of sea walls or sea dykes after the devastation of the shorelines during and after the previous storms in the Sundarbans (e.g., Aila, 2009; Bulbul, 2019; Amphan, 2020; Yaas, 2021).

Private land acquisition should be completed for resilient redevelopment of disaster-affected areas and habitat reduction zones by securing lands through a fixed long-term legal agreement for managing habitat squeeze problems on the Sundarban coast (Fig. 26.3).

26.4.2.2 Conservation of Habitats and Short-Term Agreements

Every landowner should restrict the use of their own property in order to protect its natural values behind the shorelines of their settlement sites. The landowner in this case retains ownership, rights, and privileges on their property, but their uses are restricted by the conservation easements and conservation covenants (Fitzsmons & Carr, 2014; The Nature Conservancy, 2017). Such development restrictions will allow migrating coastal habitats to adjust to the rapid rate of coastal erosion and sea level rise impacts (NOAA, 2012). Sea walls or sea dykes should be removed in the near future because they restrict the landward migration of intertidal habitats by modifying slopes and act as defences against coastal processes. Thus, short-term agreements with incentive payments to the landowners will motivate the restriction of property uses behind the shorelines in the vulnerable tracts.

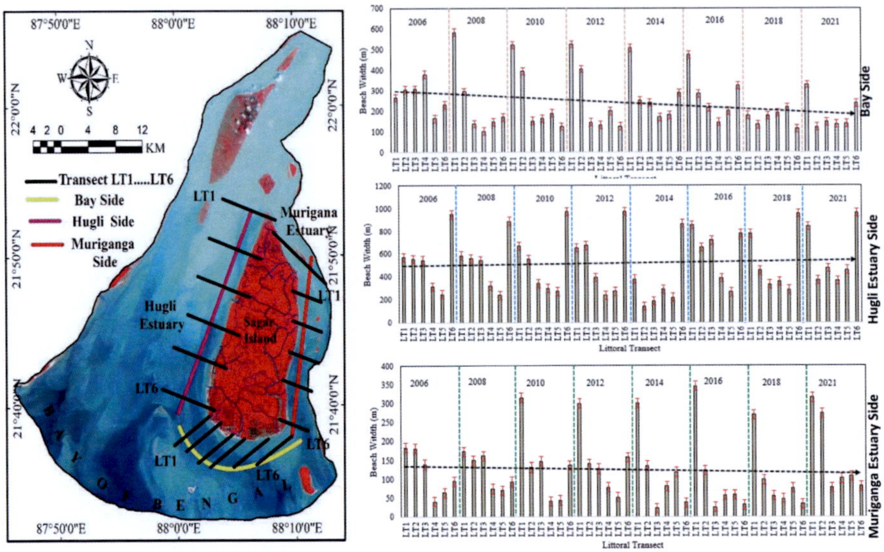

Fig. 26.3 Changing widths of the intertidal habitats along the shoreline section of Sagar Islands

26.4.3 *Regulatory or Policy Interventions*

The coastal regulatory zones (CRZ) and policy interventions use environmental laws in the planning and development of private and public lands along the coast for the conservation of habitats. Thus, for natural landward migration of the shorelines, the adjacent landward buffers with the CRZ rules will protect the natural ecosystems in the coastal zones.

26.4.3.1 Transfer of Development Rights

Along the zoned hazard area of the coast, the landowners may be allowed by the local authority to trade their development rights from a hazardous area to a receiving area (NOAA, 2012). By using this method of Transfer of Development Rights (TDR), the restoration and conservation of mangroves are possible. For the removal of development and reduction of risk in hazardous areas, the land owners can develop the land by planting mangroves in exchange for trading property provided with additional credits from the authority. Thus, the TDR to landowners will provide environmental protection and habitat restoration in the management of coastal squeeze problems.

26.4.3.2 Coastal Setbacks and Policies for Erosion Allowance

The predefined setbacks can regulate all types of coastal development (United Nations Development Programme, 2017). If an 8 m erosion rate per year is recorded along the sections of the coast the total setbacks of infrastructures should be 400 m inland for 50 years' time span to provide space for adjustment of landscapes and hydrology in response to sea level rise and landward migration of intertidal habitats. The sea level rise adaptation overlay zones could be an additional layer in the existing regulation zones, particularly in flood-induced areas, to set the rules for setback.

26.4.4 Financial Incentives for Managing Coastal Squeeze

The carbon mitigation benefits of planting mangroves will provide opportunities for the restoration of coastal wetlands in the Sundarbans. Thus, the NGOs, individuals, experts, and government organisations are monitoring the carbon benefits of the saltmarshes, mangroves, and tidal mudflats to add value to the coastal protection benefits provided by the existing mangroves' newly planted mangroves. Incentives for TDR to landowners may be provided through the implementation of resilience bonds in managing coastal squeeze problems.

To save the coastal wetland habitats and Sundarban communities in the face of sea level rise impacts, a direct attempt is needed to raise money as a parcel tax from the people of Kolkata for emitting pollutants. Credits can be awarded to the coastal communities in protection of the open space coastal habitats by reducing their flood insurance premium. Presently, global carbon markets allow for the purchase and exchange of carbon for the conservation of coastal wetlands. Blue carbon is the term applied to the storage of carbon by marine and coastal ecosystems (Mcleod et al., 2011; Blue Carbon Initiative, 2018). The landowners can be benefited by planting mangroves and saltmarshes on their own lands adjacent to the coast for the restoration, conservation, and management of the coastal habitats. Thus, the coastal wetland and habitats can be incorporated into carbon markets for managing the coastal squeeze problems.

26.5 Effects of Coastal Squeeze in Sagar Islands

The temporal image analysis proves that salt marshes, mangroves, tidal creeks, and tidal flats are lost due to the coastal squeeze problems faced by Sagar Island, Mousuni Island, and Patibunia Island of the Hugli estuary section. Tidal drainage channels on seaward-facing islands are degraded due to the isolation of the natural system by embankment structures and the aggradation process aided by landward migration of intertidal habitat.

During the previous decades (1980s–2020s), sea level rise and climate change impacts eroded a significant amount of land fringed with shorelines on the island's seaward sides and along the estuary fringes (Paul & Paul, 2022). Repeated realignments of embankments were practised after the devastating effects of each cyclonic storm, without the practice of managed retreat by the local administration in the region. The coastal squeeze problems are not understood by the local authorities and people, and thus the coastal vulnerability has increased in the island system in response to erosion and sea level rise impacts (Table 26.3).

26.6 Discussion and Conclusions

Sagar Island was historically reclaimed (1810–1942) by the construction of sea dykes and earthen embankments in the Hugli estuary to avoid the encroachment of tide waters into the island's interior areas other than the selective openings of the large tidal rivers or creeks across the island. However, a large extent of buffers was present in front of the protective embankments all along the island fringes in the near past (during the 1970s), as per the survey of India's toposheets of the island (SOI Toposheets, 1969).

Repeated embankment damages, land erosion, and loss of intertidal habitats with mangroves and saltmarshes have occurred in the region since the events of cyclone landfalls (1988–89, 2009, 2019, 2020, and 2021) and associated sea level rise impacts. As high tide lines were blocked by vertical to inclined sea walls or sea dykes, and low tide shorelines migrated significantly inland in response to sea level rise and episodic high energy phases with storm surges and tidal surges, repeated embankment damages invited salt water encroachment and inundation. On the other side, intertidal habitats were destroyed by the advancing sea and high rates of land erosion along the shores. Thus, the intertidal habitats are squeezed and embankments are realigned further inland as per the available landward buffers on the island. Presently, the island communities and island fringe habitats are suffering from coastal squeeze, thereby producing problems of landscape adjustment at the land–sea interface in response to sea level rise and climate change effects.

The study is preparing a sustainable framework to manage the coastal squeeze problems for the estuarine island to the southwestern parts of the Indian Sundarbans. The management framework highlighted future options for physical changes to the coastal fringe environment, land acquisition methods on private land for conservation and protection of the natural value of the intertidal region, regulatory methods and policy interventions, and financial incentives for landowners to manage coastal habitat squeeze. However, in response to climate change discourses and sea level rise threats, local authorities should consider coastal habitat squeeze management methods in the near future to ensure coastal resilience. The database is prepared for the analysis of coastal squeeze and wetland losses using temporal satellite images in the present study. Managing the coastal squeeze will provide an appropriate model for application in other parts of the Indian Sundarbans. The study also highlighted the coastal squeeze areas along the Muriganga shoreline and Bay of Bengal shoreline, though the western boundary of the island shore showed a minimum rate of squeeze with some expanded area.

Table 26.3 Common adaptation solutions in the estuarine island

Management Issue	Major Programme and Proponent	Funding Sources	Coastal Sections
A. Physical alteration of the shoreline environment			
1. Managed retreat (sea wall/ natural hydrology)	Local government, irrigation department, Sagar-Bakkhali development board	Public funding (Aila embankment funding)	Mandirtala, Muriganga, Topobon, Beguakhali
2. Restoration and assisted recovery (mangrove plantation and recovery of existing marshes and wetlands)	Local government, forest departments, and NGOs	Funding for ecosystem services, public funding	Sagar South, Muriganga, Maya-Goalini, Sapkhali, Chemaguri, Ganga Sagar creek
3. Living shorelines (maintaining natural land water interface)	Local government, agencies, landowners, and NGOs	Public funding, other funding from private sources	Ganga Sagar, Boat Khali, Shibpur-Dublat, Sagar South
B. Land acquisition and protection			
4. Land acquisition (purchase of private land by the government)	Local government, landowners, agencies, and NGOs	Public funding, payment for ecosystem services	Beguakhali, Boatkhali, Shibpur, Dublat, Muriganga, Mandirtala, Kochubaria
5. Land covenants (legal agreements to protect natural values of the land)	Landowners and NGOs	Public funding, insurance premium reductions	Boatkhali, Beguakhali, Muriganga, Mandirtala
6. Conservation of intertidal habitats (coastal developments restriction in the private land)	Local government, government agencies, and NGOs	Public funding, payment for ecosystem services, flood insurance premium reduction	Gangasagar, Topobon, Boatkhali, Beguakhali
C. Regulatory or policy interventions			
7. Coastal setback and erosion allowances (CRZ-1A, distance inland to regulate development types)	Local administration, Sagar-Bakkhali Development Board, irrigation department, Ministry of Environment and Forests (MoEF), and forest department	Public funding	Beguakhali, Sagar south, Gangasagar, Topobon, Boatkhali, Shibpur-Dublat

(continued)

Table 26.3 (continued)

Management Issue	Major Programme and Proponent	Funding Sources	Coastal Sections
8. Sea level rise adaptation overlay zones (500–800 m inland setback rules including conservation easements, freeboard, building elevations, and development density)	MOEF and local government	Public funding	Beguakhali, Sagar south, Gangasagar, Topobon, Boatkhali, Shibpur-Dublat
9. Transferrable development rights (landowners with holdings in a zoned hazard area to trade their development rights)	Local government, local authority	Public funding	Beguakhali, Sagar south, Gangasagar, Topobon, Boatkhali, Shibpur-Dublat, Chemaguri, Muriganga, Kochubaria
D. Adaptation solutions for coastal habitat squeeze with financial mechanisms			
10. Public funding			
(i) Government buyouts			
(ii) Pre- and post-disaster funding			
(iii) Environmental levees			
(iv) Real estate transfer tax			
(v) Integrated coastal zone management (ICZM) funding			
(vi) Environmental spending			
11. Private funding			
(i) Payment for ecosystem services (e.g., blue carbon)			
(ii) Environmental (green and blue) bonds			
(iii) Insurance premium			

References

Aila. (2009). *Cyclonic disturbances over North Indian Ocean during 2009: A report* (Cyclone Warning No. 5/2010) (pp. 1–109). Cyclone Warning Division. https://rsmcnewdelhi.imd.gov.in/download.php?path=uploads/report/27/27_4e34f3_rsmc-2009.pdf

Amphan. (2020). *Super cyclonic storm "AMPHAN" over the southeast Bay of Bengal, (16th–21st May 2020): Summary* (pp. 1–57). Regional Specialised Meteorological Centre tropical Cyclones, Meteorological Department. https://mausam.imd.gov.in/Forecast/marquee_data/indian111.pdf

Atkinson, P. W. (2003). Can we recreate or restore intertidal habitats for shorebirds? *Bulletin-Wader Study Group, 100*, 67–72.

Bilkovic, D. M., Mitchell, M., Mason, P., & Duhring, K. (2016). The role of living shorelines as estuarine habitat conservation strategies. *Coastal Management, 44*(3), 161–174. https://doi.org/10.1080/08920753.2016.1160201

Blue Carbon Initiative. (2018). *What is blue carbon?* Available online at: http://thebluecarboninitiative.org/

Bulbul. (2019). *Very severe cyclonic storm 'Bulbul' (Pronounced as Bul bul) over northwest Bay of Bengal: Cyclone warning and post-landfall outlook for West Bengal Coast: Red message* (IMD, BULLETIN NO.: 35 (BOB/04/2019)). https://mausam.imd.gov.in/Forecast/marquee_data/indian0906.pdf

Fitzsimons, J. A., & Carr, C. B. (2014). Conservation covenants on private land: Issues with measuring and achieving biodiversity outcomes in Australia. *Environmental Management, 54*(3), 606–616.

Leo, K. L., Gillies, C. L., Fitzsimons, J. A., Hale, L. Z., & Beck, M. W. (2019). Coastal habitat squeeze: A review of adaptation solutions for saltmarsh, mangrove and beach habitats. *Ocean & Coastal Management, 175*, 180–190. https://doi.org/10.1016/j.ocecoaman.2019.03.019

Mcleod, E., Chmura, G. L., Bouillon, S., Salm, R., Björk, M., Duarte, C. M., Lovelock, C. E., Schlesinger, W. H., & Silliman, B. R. (2011). A blueprint for blue carbon: Toward an improved understanding of the role of vegetated coastal habitats in sequestering CO2. *Frontiers in Ecology and the Environment, 9*, 552–560.

NOAA. (2012). *Transfer of development rights program, case study: Collier county, Florida's TDR program*. Available online at: http://coastalmanagement.noaa.gov/initiatives/shoreline_ppr_tdr.html

Paul, A. K. (1997). Coastal erosion in West Bengal. *MAEER, MIT Pune Journal, IV*(15k &16k) (special issue on coastal environmental management), 66–84.

Paul, A. K. (2000). Cyclonic storm and their impacts on West Bengal coast. In G. V. Rajamanickam & J. T. Michael (Eds.), *Quaternary sea level variation, Shoreline development and coastal environments* (pp. 8–31). New Academic Publishers.

Paul, A. K. (2002). *Coastal geomorphology and environment* (pp. 1–342). ACB Publication.

Paul, A. K. (2022, August). Dynamic behaviour of the estuaries in response to the phenomenon of global warming in the coastal ecosystems of West Bengal and Odisha, India. In *Transforming coastal zone for sustainable food and income security: Proceedings of the international symposium of ISCAR on Coastal Agriculture, March 16–19, 2021* (pp. 907–931). Springer International Publishing.

Paul, A. K. & Kamila, A. (2018). *Studies on coastal morphometry with total station survey in shore face of Sagar Island to assess the impacts of sea level rise 38th INCA International Congress Hyderabad* (pp. 460–469).

Paul, A. K., & Paul, A. (2022). Adjustment of the coastal communities in response to climate variability and sea level rise in the Sundarban, West Bengal, India. In *Climate change, disaster and adaptations: Contextualising human responses to ecological change* (pp. 201–217). Springer International Publishing.

Paul, A. K., Kamila, A., & Ray, R. (2018). Natural threats and impacts to mangroves within the coastal fringing forests of India. In *Threats to Mangrove Forests: Hazards, vulnerability, and management* (pp. 105–140). Springer.

SOI Toposheet. (1969). *Survey of India 79 C/1 and 79 C/2, West Bengal, Hooghly Estuary Kakdwip Sagar Island 1924 and 1969 maps*. https://onlinemaps.surveyofindia.gov.in/FreeMapSpecification.aspx

The Nature Conservancy. (2017). *Conservation easements*. Available online at: https://www.nature.org/about-us/private-lands-conservation/conservation-easements/allabout-conservation-easements.xml

United Nations Environment Programme. (2017). *Coastal ecosystem-based adaptation: Managed realignment and coastal set-backs*. Available online at: http://www.unep.org/coastal-eba/content/managed-realignment-and-coastal-set-backs

Yass. (2021). *The severe cyclonic storm „Yaas" (Pronounced as „Yass") Over Northwest Bay of Bengal–(Cyclone Warning for Odisha – West Bengal Coasts)* (Tropical Cyclone Advisory Bulletin No. 13). IMD, Regional Specialised Meteorological Centre-Tropical Cyclones. https://mausam.imd.gov.in/imd_latest/contents/cyclone.php

Chapter 27
Assessment of Tourism Carrying Capacity for the Sustainable Tourism Development of South Andaman, India

Swagata Bera and Dipanjan Das Majumdar

27.1 Introduction

Tourism is a fast-growing industry and a valuable sector, contributing significantly to the local economy and society, making it an important economic, social, and cultural activity in many societies, both in developed and developing countries (Telfer & Sharpley, 2007). The World Travel and Tourism Council calculated that tourism generated ₹13.2 lakh crore (US$170 billion), or 5.8% of India's GDP, and supported 32.1 million jobs in 2021. India was ranked 54th out of 117 countries in the Travel and Tourism Development Index for 2021. The Andaman and Nicobar Islands have a lot to offer tourists because of their beautiful scenery, unique natural landforms, wide range of plants and animals, rich history, and native tribes. The number of visitor arrivals at Andaman and Nicobar Islands (ANI) is gradually increasing each year. Tourism has a lot of benefits, but it also has some negative effects on both the physical and social environments. The tourism operations in protected regions must be properly planned, managed, and monitored to ensure their long-term viability. Considering the vulnerable ecosystem and limited carrying capacity of the islands, the Andaman and Nicobar (A&N) administration aims to promote sustainable tourism. The goal of the A&N Administration is to develop sustainable tourism on the islands in such a way that it generates employment and contributes to the economic growth of this Union Territory without harmful effects in the natural balance. For the long-term development of tourism in the Andaman and Nicobar Islands, considerable focus must be placed on tourist management. The

S. Bera (✉)
Department of Geography, Dum Dum Motijheel College, Kolkata, West Bengal, India
e-mail: sbswagatabera@gmail.com

D. D. Majumdar
Department of Geography, Netaji Satabarshiki Mahavidyalaya,
Ashoknagar, West Bengal, India

A. K. Paul, A. Paul (eds.), *Crisis on the Coast and Hinterland*,
https://doi.org/10.1007/978-3-031-42231-7_27

limit or threshold that is set to show if the growth of tourism activities is still within the limits of sustainability is used to figure out the carrying capacity. According to Nghi et al. (2007), a natural, environmental, and socioeconomic system has the highest bearing capacity for tourism when it can accommodate the maximum number of visitors without compromising the system's ability to sustain long-term growth, and when visitor satisfaction levels are maintained during this period. The concept of carrying capacity is intended to exemplify the need to maintain development and activities at a level that is both ecologically and socially sustainable and activities beyond which environmental degradation occurs (Getz, 1982). The concept of Tourism Carrying Capacity (TCC) occupies a key position with regard to sustainable tourism (Tribe et al., 2000). Tourism has a detrimental effect on the socio-cultural landscape and ecology of the Andaman and Nicobar Islands (Reddy, 2004). Despite the drawbacks of the TCC concept, it has been accepted as an appropriate management technique since it enables resource preservation (Queiroz et al., 2014). TCC is a key means for growing and maintaining coastal tourism in ANI sustainably. To this purpose, the use and assessment of TCC for sustainable ecotourism planning may give an overall framework that local communities, planners, and decision-makers may find useful. The management of beaches is essential for preserving their quality and ensuring that they continue to draw tourists to the Andaman and Nicobar Islands, where coastal tourism is a significant economic component of the Union territory. The primary purpose of this study is to evaluate the recreational carrying capacity of various tourist beaches in the South Andaman Districts. This will be accomplished by determining the island's physical carrying capacity (PCC), real carrying capacity (RCC), and effective carrying capacity (ECC). Three major research topics relating to different forms of tourist capacity were formulated to help accomplish this. First, how many people do you think South Andaman can safely accommodate? Has the maximum number of visitors been achieved, or is the beach still manageable? Thirdly, how many people are allowed on the beach at one time?

27.2 Study Area

The Indian Union Territory of Andaman and Nicobar Islands, which is situated in the Bay of Bengal and Andaman Sea, is made up of six districts, with South Andaman being one of them. In the Union Territory of the Andaman and Nicobar Islands, it is the second-largest district. It is located between longitudes 92° and 94° East and latitudes 10° 30′ and 12° 20′ north. The southernmost island in the South Andaman District is Little Andaman, while the northernmost island is South Andaman. The Ten Degree Channel divides Little Andaman Island from Nicobar District, while Jarawa Creek divides South Andaman Island from the North and Middle Andaman Districts. South Andaman district has a total area of 2980 km^2 and the district consists of three tehsils, named as Port Blair, Ferrargunj, and Little Andaman. On the south-easterly coast of South Andaman Island, the administrative center of the Union Territory of the ANI (Port Blair) is located. For this research, we

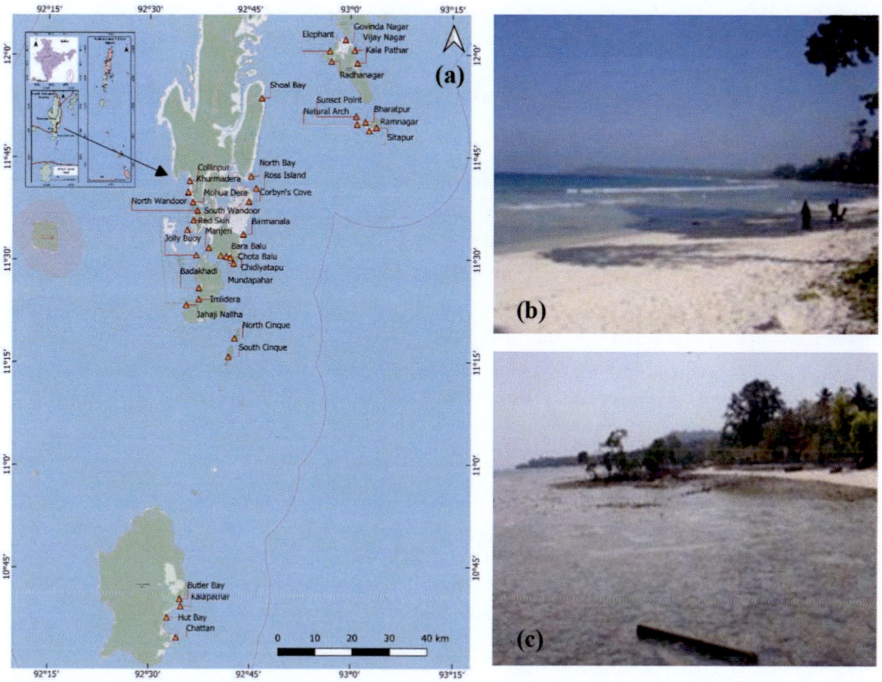

Fig. 27.1 (**a**) Map of the study area indicating the locations of the many beaches that were studied, (**b**) sandy beaches of the Neil Island, and (**c**) coralline beaches of Havelock Island, South Andaman

collected information from 36 distinct beaches spread over a number of islands in the South Andaman District (Fig. 27.1). 15 beaches from South Andaman Island, five from Neil Island, five from Havelock Island, four from Little Andaman Island, three from Rutland Island, two from Cinque Island, and one from each of Ross, Jolly Buoy, and Red Skin Island were selected for this study. The remaining 22 beaches in the South Andaman District are dispersed across the other islands in the district, all of which are linked to Port Blair by boat, with the exception of the 14 beaches on South Andaman Island.

27.3 Methodology

The methodology developed by Nghi et al. (2007) was used to estimate the tourism carrying capacity of various beaches in the South Andaman District. Several authors (Ceballos-Lascurain, 1996; Nghi et al., 2007; Segrado et al., 2008; Zacarias et al., 2011; Bera et al., 2015; Sridhar et al., 2017; Faiz & Komalasari, 2020; Sobhani et al., 2022) from around the world use this methodology to evaluate the tourism capabilities of various tourist destinations. The same methodology was employed in this investigation, albeit with a few modifications. The TCC consists of the following levels.

27.3.1 Physical Carrying Capacity (PCC)

PCC refers to the maximum number of visitors who can physically fit into or onto an area at a given time (Eq. 27.1):

$$\mathrm{PCC} = A / \mathrm{Au} \times \mathrm{Rf} \tag{27.1}$$

where, PCC stands for physical carrying capacity; *A* refers to the available area for tourist use; Au refers to the area required per tourist; and Rf indicates the factor of rotation (number of visits per day). Along with comprehensive field verifications, the Google Earth Pro is used to calculate the total beach area (*A*) of selected beaches in the South Andaman District. Therefore, the rotation factor for each of these thirty-six beaches has been determined independently. The optimal area required per visitor (Au) is 5 m^2, and the rotation factor (Rf) is determined by (Eq. 27.2):

$$\mathrm{Rf} = \text{Open period} / \text{Average time of visit} \tag{27.2}$$

27.3.2 Real Carrying Capacity (RCC)

The number of tourists who can legally visit a certain location at a time after the correction factors (CF) obtained from the site's unique characteristics have been added to the PCC (Eq. 27.3):

$$\mathrm{RCC} = \mathrm{PCC} \times \left(\mathrm{Cf}_1 \times \mathrm{Cf}_2 \times \mathrm{Cf}_3 \times \mathrm{Cf}_4 \times \cdots \mathrm{Cf}_n\right) \tag{27.3}$$

where, RCC stands for real carrying capacity, PCC refer to the physical carrying capacity, and Cf refers to the correction factors. The correction factors are calculated by taking into account the biophysical, environmental, ecological, social, and management aspects of the study area. These are the elements that have a negative impact on tourism activities, as measured by the limiting threshold used to determine a factor's impact level. Excessive sunshine, rainfall, cyclones, beach quality, infrastructure, and temporary closures were the corrective factors that restricted tourism activity and reduced visitor happiness. Consequently, there are correction factors, sometimes referred to as limiting factors, that make tourism hard to do. Correction factors are calculated using the following Eq. 27.4:

$$\mathrm{Cf}_x = 1 - \mathrm{Lm}_x / \mathrm{Tm}_x \tag{27.4}$$

where, Cf_x refers to the correction factors of variable *x*, Lm_x refers to the limiting magnitude of variable *x*, and Tm_x refers to the total magnitude of variable *x*. Here is how to figure out correction factors for different limiting variables.

27.3.3 Excessive Sunshine (Cf_1)

ANI has seen the highest temperatures during the dry summer months (March to May). So, the study area could be limited by the fact that it gets too much sunshine in the summer. The limiting magnitude for this parameter was calculated to be 92 days × 4 h = 368 h of excessive sunlight each year. The total magnitude was calculated based on the number of days in a year, that is, 365 days × 12 h = 4380 h. So, the following Eq. 27.5 was chosen as the correction factor for too much sunshine:

$$\text{Cf}_1 = 1 - \text{Lm}_x / \text{Tm}_x \text{ or } 1 - 368\text{h} / 4380\text{h} = 0.916 \quad (27.5)$$

27.3.4 Rainfall (Cf_2)

The intensity of high rainfall for 4 months restricted visitors to the Andaman group of islands. As a result, access is restricted during the 4 months with the most rain. 122 days was chosen as the limiting magnitude for this parameter. The total magnitude is the sum of all the days in a year (365 days). This is how the correction factor for rainfall was found in Eq. 27.6:

$$\text{Cf}_2 = 1 - \text{Lm}_x / \text{Tm}_x \text{ or } 1 - 122\text{days} / 365\text{days} = 0.666 \quad (27.6)$$

27.3.5 Cyclone (Cf_3)

Cyclone season affects island-to-island ferries and tourism operations. Limiting magnitude was 61 days, total magnitude was 365 days, and the cyclone correction factor (Eq. 27.7) was derived by:

$$\text{Cf}_3 = 1 - \text{Lm}_x / \text{Tm}_x \text{ or } 1 - 61\text{days} / 365\text{days} = 0.8328 \quad (27.7)$$

27.3.6 Beach Quality (Cf_4)

The earlier researchers suggested the metrics were used to assess the beach's quality (Nghi et al., 2007). The results of the evaluation of beach quality and the correction factor for beach quality, which were developed independently for the chosen beaches, are shown in Table 27.1.

Table 27.1 Beach Quality Assessment Matrix for the various beaches in the South Andaman District

Sl. No.	Island	Beach	Material Sand/rock	Slope	Tide	Beach length	Litter	Beach color	Quality of water	Beach quality	Correction factor (Cf_4)
1	South Andaman	Corbyn's Cove	+	+	+	–	–	–	–	4/7	0.429
2		Barmanala	–	+	+	+	–	–	–	4/7	0.429
3		Chidiyatapu	+	+	+	–	–	–	–	4/7	0.429
4		Mundapahar	+	+	+	–	+	–	+	2/7	0.714
5		Chota Balu	+	+	+	–	+	–	–	3/7	0.571
6		Bara Balu	+	+	+	+	+	–	–	2/7	0.714
7		Manjery	–	+	+	–	+	–	–	4/7	0.429
8		North Bay	+	+	+	–	–	+	+	2/7	0.714
9		Soal Bay	+	+	+	+	+	–	–	2/7	0.714
10		Collinpur	+	+	+	–	+	–	+	2/7	0.714
11		Khurmadera	+	+	+	+	+	–	+	1/7	0.858
12		Mohua Dera	+	+	+	+	+	+	+	0/7	0
13		North Wandoor	+	–	+	+	+	+	+	1/7	0.858
14		South Wandoor	+	+	+	+	–	+	+	1/7	0.858
15	Havelock	Radhanagar	+	+	+	+	+	+	+	0/7	0
16		Elephant	+	+	+	+	+	+	+	0/7	0
17		Govinda Nagar	+	+	+	–	–	+	+	2/7	0.714
18		Kala Pathar	+	+	+	+	+	+	+	0/7	0
19		Vijay Nagar	+	+	+	+	–	+	+	1/7	0.858

20	Neil	Bharatpur	+	+	+	+	+	+	+	0/7	0
21		Ramnagar	–	+	+	+	+	+	+	1/7	0.858
22		Sitapur	+	+	–	+	–	–	+	3/7	0.571
23		Sunset Point	+	+	+	+	+	+	+	0/7	0
24		Natural Arch	–	+	–	–	+	+	+	3/7	0.571
25	Little Andaman	Butler Bay	+	+	+	+	+	+	+	0/7	0
26		Hut Bay	+	+	+	+	+	+	+	0/7	0
27		Kalapathar	+	+	–	–	+	+	+	2/7	0.714
28		Chattan	+	+	–	+	+	+	+	1/7	0.858
29	Rutland	Jahaji Nallah	+	–	–	+	+	+	+	2/7	0.714
30		Imlidera	+	+	+	+	+	+	+	0/7	0
31		Badakhadi	+	+	+	+	+	+	+	0/7	0
32	Cinque	North Cinque	+	+	–	+	+	+	+	1/7	0.858
33		South Cinque	+	+	–	+	+	+	+	1/7	0.858
34	Ross	Ross Island	+	–	–	–	+	–	+	4/7	0.429
35	Jolly Buoy	Jolly Buoy	+	+	+	+	+	+	+	0/7	0
36	Red Skin	Red Skin	+	+	+	+	+	–	+	1/7	0.858

Key: Notation: "+" Good or suitable quality; "–" Low or unsuitable quality

27.3.7 Infrastructure Quality (Cf_5)

Tourist destinations' infrastructure includes the standard of their lodgings as well as the cleanliness and convenience of their public restrooms, motels, waiting areas, locker rooms, transportation options, etc. Infrastructural development is crucial to the development of tourism (Malik & Bhat, 2015). During the field study, questionnaires were distributed to tourists to collect their feedback on this criterion. The findings of the tourist assessment were used to estimate the limiting factor for this parameter.

27.3.8 Temporary Closure (Cf_6)

Jolly Buoy Island and Red Skin Island are closed for 6 months out of the year in order to conserve the corals, unlike the other Andaman Islands. Jolly Buoy Island is accessible from November through April. These months do not experience either an exceptionally hot or cold climate. Jolly Buoy Island is likewise closed on Mondays. Tourists may only visit Red Skin Island between May and October, and it is closed on Mondays. The monthly and weekly closures, respectively, have an impact on tourism activity on these two islands. The temporary closure of beaches is another issue that limits beach use. As a result, correction factors for both weekly and monthly closure have been computed using Eq. 27.8 shown below:

$$\mathrm{Cf}_6 = \tfrac{1}{2}\left(\text{Monthly closure} + \text{Weekly closure}\right)\text{or } \tfrac{1}{2}\left(0.496 + 0.856\right) = 0.676 \tag{27.8}$$

27.3.9 Effective Carrying Capacity (ECC)

ECC refers to the highest number of visitors that a location can safely accommodate with its current level of management capability (MC):

$$\mathrm{ECC} = \left(\mathrm{RCC} \times \mathrm{MC}\right) \tag{27.9}$$

where, ECC refers to the effective carrying capacity, RCC refers to the real carrying capacity, and MC refers to the management capacity (Eq. 27.9). Management capacity (MC) is defined as the whole set of conditions necessary for beach tourist management to carry out its tasks and achieve its goals. The MC is very difficult to quantify. The MC was determined based on the available facilities, amenities, legislation, equipment, staffing levels, and budget. During fieldwork, MC was assessed using a perception study of beachgoers.

27.4 Results and Discussion

The estimate of the tourist carrying capacity for each chosen beach serves as the starting point for estimating the tourism carrying capacity for the whole South Andaman District. In order to discover existing data sets and determine the nature, general features, and history of the beach, we first used our methodology and conducted a detailed examination of the region. In order to analyze the physical carrying capacity of the beach an area that is appropriate for tourism, area per user, visit duration, etc., must be evaluated and analyzed (Bera et al., 2015). PCC can be conceptualized as the maximum number of participants a place can accommodate for one activity on a given day, irrespective of other events. Using the 5 m^2 of area that is allocated for each person, the PCC of the beaches is determined. Depending on how cozy beachgoers are and, in some situations, how much room is provided for each individual, this parameter may fluctuate (Khodkar, 2019).

The South Andaman Islands' total useable beach area is 1,520,590 m^2 during the peak tourism season. The total PCC for the beaches of the South Andaman District is 1,023,188 visitors per day. The PCC associated with Hut Bay was the greatest, and the PCC linked with Ross Island was the lowest in comparison to the TCCs of these analyzed beaches. The Little Andaman Island's and Hut Bay's beaches received the highest PCC because they gave guests access to a sizable beach space. The Ross Islands, on the other hand, provide visitors with a comparatively small quantity of beach space. Mohua Dera Beach on South Andaman Island, Radhanagar Beach on Havelock Island, Sunset Point Beach on Neil Island, Jahaji Nallah in Rutland, and South Cinque Beach on Cinque Island all have higher PCC values than the other beaches on these islands. The rotation factor is taken into account when determining PCC for selected beaches since it has an impact on how many visitors may visit a certain beach. Tourists that stay in hotels near the beach spend the majority of their days there, only venturing inland for quick lunches. As a result, the physical boundaries of the beaches, which were previously assigned rotation factors 1 and 2, limit the number of people that may visit them. North Bay Beach on South Andaman Island, Elephant Beach on Havelock Island, and a few other beaches on Rutland, Cinque, Ross, Red Skin, and Jolly Buoy Islands all have low rotation factors as a result of their limited visiting seasons, which significantly affects the PCC of these beaches.

The PCC numbers are theoretical, and the RCC was calculated to check the PCC's extreme value by using some correction factors (Bera et al., 2015). Table 27.2 presents the summary table of various correction factors. Figure 27.2 shows that the RCC has been estimated based on six limiting variables. This study considered correction factors such as excessive sunshine, rainfall, cyclones, beach quality, infrastructural quality, and temporary closure. All of the correction variables that RCC looks at are the same on all of the islands, except for the quality of the beaches and infrastructure. Intense sunshine, among three other climatic correction factors, has not had much of an influence on beach tourism at some beaches (North Bay, Elephant, Jahaji Nallah, Imlidera, Badakhadi, North Cinque Island, South Cinque

Table 27.2 Real carrying capacity (RCC) is derived using a variety of correction factors from physical carrying capacity (PCC). Estimation of the Effective Carrying Capacity after taking the Management Capacity and RCC of different beaches into account

Sl. No.	Island	Beach	Physical Carrying Capacity (PCC)	Correction factors						Real Carrying Capacity (RCC)	Management capacity (MC)	Effective Carrying Capacity (ECC)
				(Cf_1)	(Cf_2)	(Cf3)	(Cf_4)	(Cf_5)	(Cf_6)			
1	South Andaman	Corbyn's Cove	11,400	0.916	0.666	0.833	0.429	0.923	N.A.	2294	0.867	1988
2		Barmanala	18,928	0.916	0.666	0.833	0.429	0.231	N.A.	952	0.267	254
3		Chidiyatapu	8000	0.916	0.666	0.833	0.429	0.462	N.A.	805	0.533	429
4		Mundapahar	6400	0.916	0.666	0.833	0.714	0.308	N.A.	715	0.467	333
5		Chota Balu	4784	0.916	0.666	0.833	0.571	0.154	N.A.	214	0.133	28
6		Bara Balu	25,200	0.916	0.666	0.833	0.714	0.154	N.A.	1407	0.133	188
7		Manjery	2400	0.916	0.666	0.833	0.429	0.231	N.A.	121	0.200	24
8		North Bay	3360	N.A.	0.666	0.833	0.714	0.615	N.A.	819	0.867	710
9		Soal Bay	19,600	0.916	0.666	0.833	0.714	0.231	N.A.	1792	0.133	239
10		Collinpur	10,200	0.916	0.666	0.833	0.714	0.385	N.A.	1554	0.200	311
11		Khurmadera	15,600	0.916	0.666	0.833	0.858	0.385	N.A.	2856	0.133	381
12		Mohua Dera	43,200	0.916	0.666	0.833	N.A.	0.154	N.A.	3377	0.133	450
13		North Wandoor	20,400	0.916	0.666	0.833	0.858	0.462	N.A.	4482	0.267	1195
14		South Wandoor	23,760	0.916	0.666	0.833	0.858	0.538	N.A.	6090	0.467	2842
15	Havelock	Radhanagar	80,000	0.916	0.666	0.833	N.A.	0.846	N.A.	34,400	0.800	27,520
16		Elephant	8160	N.A.	0.666	0.833	N.A.	0.615	N.A.	2786	0.800	2229
17		Govinda Nagar	12,000	0.916	0.666	0.833	0.714	0.615	N.A.	2679	0.533	1429
18		Kala Pathar	21,600	0.916	0.666	0.833	N.A.	0.385	N.A.	4222	0.333	1407
19		Vijay Nagar	24,000	0.916	0.666	0.833	0.858	0.385	N.A.	4025	0.400	1610

				(Cf_1)	(Cf_2)	(Cf3)	(Cf_4)	(Cf_5)	(Cf_6)			
20	Neil	Bharatpur	19,200	0.916	0.666	0.833	N.A.	0.533	N.A.	5254	0.333	1751
21		Ramnagar	17,600	0.916	0.666	0.833	0.858	0.154	N.A.	1181	0.133	157
22		Sitapur	27,600	0.916	0.666	0.833	0.571	0.231	N.A.	1848	0.267	493
23		Sunset Point	38,000	0.916	0.666	0.833	N.A.	0.462	N.A.	8913	0.267	2377
24		Natural Arch	33,600	0.916	0.666	0.833	0.571	0.231	N.A.	2250	0.200	450
25	Little Andaman	Butler Bay	100,000	0.916	0.666	0.833	N.A.	0.462	N.A.	23,454	0.267	6254
26		Hut Bay	312,000	0.916	0.666	0.833	N.A.	0.462	N.A.	73,177	0.200	14,635
27		Kalapathar	12,720	0.916	0.666	0.833	0.714	0.385	N.A.	1775	0.200	355
28		Chattan	19,200	0.916	0.666	0.833	0.858	0.231	N.A.	1932	0.200	386
29	Rutland	Jahaji Nallah	28,800	N.A.	0.666	0.833	0.714	0.231	N.A.	2633	0.200	527
30		Imlidera	8424	N.A.	0.666	0.833	N.A.	0.231	N.A.	1078	0.200	216
31		Badakhadi	8640	N.A.	0.666	0.833	N.A.	0.231	N.A.	1106	0.200	221
32	Cinque	North Cinque	10,212	N.A.	0.666	0.833	0.858	0.231	N.A.	1122	0.333	374
33		South Cinque	15,000	N.A.	0.666	0.833	0.858	0.231	N.A.	1648	0.333	549
34	Ross	Ross Island	960	N.A.	0.666	0.833	0.429	0.154	N.A.	35	0.467	16
35	Jolly Buoy	Jolly Buoy	5760	N.A.	N.A.	0.918	N.A.	0.385	0.676	1375	0.733	1008
36	Red Skin	Red Skin	2400	N.A.	0.666	0.915	0.858	0.385	0.676	140	0.733	103
	Total		1,019,108							204,508		73,441

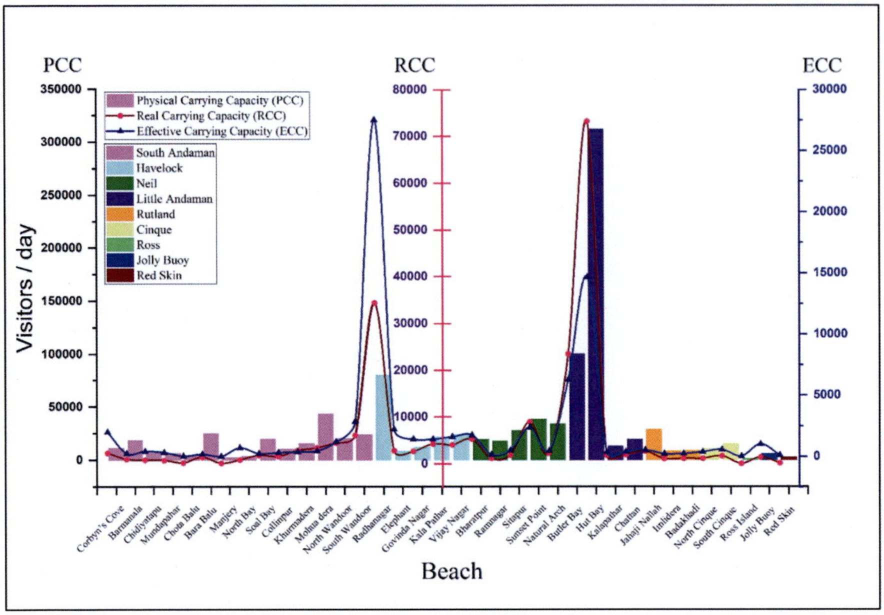

Fig. 27.2 Calculated Tourism Carrying Capacity (PCC, RCC, and ECC) presented graphically for the beaches of South Andaman District

Island, Ross Island, Red Skin Island, and Jolly Buoy Island) since some beaches only allow people at particular times of day. Due to the fact that Jolly Buoy Island is temporarily shut down for the duration of the rainy season, the effects of the rainfall are rather muted on this island. Cyclone factors have the least effect on changing the PCCs of Jolly Buoy and Red Skin Island.

The magnificent beaches of South Andaman are a massive attraction for tourists due to the wide variety of beach-related activities that can be enjoyed there, including snorkeling, scuba diving, game fishing, sea walking, sunbathing, riding in glass-bottom boats, swimming, and a variety of other water sports. One of the most crucial factors that can have an effect on the standard of beach tourism is the condition of the beaches themselves, since poor-quality beaches can have a psychological effect on visitors and lower their level of satisfaction with their beach vacation. This correction factor is assessed by geological criteria such as the material of the beaches, slope, tide, beach length, beach color, litter, and quality of water through a matrix table presented in Table 27.1 (Nghi et al., 2007). Radhanagar, Kala Pathar, and Elephant on Havelock Island; Bharatpur and Sun Set Point on Neil Island; Mohua Dera on South Andaman Island; Butler Bay and Hut Bay on Little Andaman Island; Imlidera and Badakhadi on Rutland Island; and Jolly Buoy Island are all good-quality beaches that support tourism. Corbyn's Cove, Barmanala, Chidiyatapu, and Manjeri on South Andaman Island, as well as Ross Island Beach, are low-quality beaches in the study region. The RCC values for each beach are presented in Table 27.2 and Fig. 27.2. The maximum number of persons that should be permitted in the region is represented by the total RCC of the South Andaman District, or 204,508 visitors per day.

In terms of acceptability, ECC is preferred to the other two categories of carrying capacity (Bera et al., 2015). It helps make plans for beach tourism by figuring out how many people are the best fit for the current situation and management skills. One of the most essential elements that might have a direct influence on the standard of beach tourism is the existing management capability of the area. We evaluated the MC for different beaches by looking at the kinds of tourist attractions, amenities, and safety and security measures they provide (Sridhar et al., 2017). It has been noticed that most of the beaches on South Andaman Island, Neil Island, Little Andaman, Rultand Island, and Cinque Island have a low management capacity (<0.5). This indicates that infrastructure spending on things like transportation, lodging, washrooms, rest rooms, and locker facilities, as well as ensuring the safety of tourists, should be a top concern. The other beaches, like Carbyns Cove, Chidyatapu, North Bay, Radhanagar, Elephant, Govinda Nagar, Red Skin, and Jolly Buoy, which have an MC score of more than 0.5, have enough facilities for tourists. The South Andaman District's effective carrying capacity is 73,441 tourists per day. According to the findings, it appears that each level represents an updated capacity level for the level that came before it. Based on this research, it is clear that the PCC is always greater than the RCC, while the RCC is always greater than the ECC (Sridhar et al., 2017). In 2019, the number of visitors to ANI reached 505,398. It was the highest number of visitors before the COVID-19 pandemic. But the available tourist accommodation in these islands is much lower than the ECC for the region and needs to be raised due to the overall quality of the environment. According to the Andaman and Nicobar Tourism Department, there are 182 hotels without a star rating, 23 lodges, and 22 homestays in the South Andaman District.

Reddy (2007) has pointed out that any attempt to develop tourism on a large scale on the island would have harmful effects on the island's ecology. When there is a large influx of tourists, fresh water is the most urgent issue on the islands. In order to better manage solid and liquid waste, the hospitality industry has to encourage the creation of environmentally friendly housing units that may be clustered on a single site. As Reddy (2007) pointed out, there are additional challenges associated with the growing tourist industry on these islands. The cost of transporting staples like rice, wheat, vegetables, fruits, milk, etc., from the mainland to the islands makes these items more expensive. Simple consumables see price increases due to transportation costs. A larger influx of tourists would put a strain on the area's ability to provide basic necessities like food, water, and shelter. Disposal of trash on these islands, which are kilometers from the mainland, will be difficult. The lack of actions being taken at the panchayat level or the municipal level to dispose of rubbish is a major cause for concern. In order to make the islands self-sufficient in terms of production, it is vital that sustainable practices be implemented, which calls for a comprehensive strategy. The construction of the accommodation units must be encouraged, which are environment-friendly and must be concentrated on a particular site in order to easily manage solid and liquid waste generated from the accommodation sector (Raheem et al., 2017). The tourism carrying capacity of

South Andaman is well conserved to support a higher level of yearly tourist infiltration with enhanced services and management facilities in the future when the overall ECC as well as the annual visitor flow in the Andaman Islands are taken into account.

27.5 Conclusion

Assessment of tourism carrying capacity is one of the most valuable tools for coastal region management. It has the potential to promote balanced regional development, not only through tourism benefits but also through environmental consequences. In this study, carrying capacity analysis is carried out to ensure sustainable usage of natural beaches in various islands of the South Andaman District. Restriction of beach access can alter carrying capacity by as much as 50%. The significance of site-specific correction variables is highlighted by the comparisons between PCC and RCC. Excessive sunlight, one of the most noticeable climatic features of the study region, has the largest impact on reducing carrying capacity. The combination of the true carrying capacity of the study region with its management capability yields the effective carrying capacity (ECC), which is the capacity that should actually be used.

South Andaman District would be able to more swiftly realize its full tourism carrying capacity (TCC) if suitable infrastructure and managerial facilities are put in place. When developing and putting into action a plan for sustainable ecotourism in the Union Territory of Andaman and Nicobar Islands, it is essential for the various stakeholders in the territory to bear in mind the carrying capacity for visitors that the islands already have. To support the effective operation of tourism-related activities, essential public infrastructure, including an appropriate number of hotels and lodges, public washrooms, water supply, power, rubbish management, and communication lines, must be prioritized. It is imperative that every beach provide visitors with a variety of water activities, especially extreme water sports, but ensuring their safety is the first priority. This research on carrying capacity may provide a wide range of conclusions, all of which may inform future strategies for tourism growth.

References

Bera, S., Majumdar, D. D., & Paul, A. K. (2015). Estimation of tourism carrying capacity for Neil Island, South Andaman, India. *Journal of Coastal Sciences, 2*, 46–53.

Ceballos-Lascuráin, H. (1996). *Tourism, ecotourism, and protected areas: The state of nature-based tourism around the world and guidelines for its development*. IV World Congress on National Parks and Protected Areas.

Faiz, S. A., & Komalasari, R. I. (2020). The assessment of tourism carrying capacity in Lombok Island. *IOP Conference Series: Earth and Environmental Science, 592*, 012002. https://doi.org/10.1088/1755-1315/592/1/012002

Getz, D. (1982). A rationale and methodology for assessing capacity to absorb tourism. *Ontario Geography, 19*, 92–101.
Khodkar, G. (2019). *Beach carrying capacity assessment: Case study for sustainable use of Kusadası beaches*. Retrieved from https://acikbilim.yok.gov.tr/handle/20.500.12812/232175
Malik, M. I., & Bhat, M. S. (2015). Sustainability of tourism development in Kashmir - Is paradise lost? *Tourism Management Perspectives, 16*, 11–21. https://doi.org/10.1016/j.tmp.2015.05.006
Nghi, T., Lan, N. T., Thai, N. D., Mai, D., & Thanh, D. X. (2007). Tourism carrying capacity assessment for Phong Nha-Ke Bang and Dong Hoi. *VNU Journal of Science, Earth Sciences, 23*, 80–87.
Queiroz, R. E., Ventura, M. A., Guerreiro, J. A., & Cunha, R. T. D. (2014). Carrying capacity of hiking trails in Natura 2000 sites: A case study from North Atlantic Islands (Azores, Portugal). *Journal of Integrated Coastal Zone Management, 14*(2), 233–242.
Raheem, P., Vincy, M. V., Brilliant, R., & Jude, E. (2017). Estimation of beach carrying capacity for Lakshadweep Islands South India. *Global Journal of Current Research, 5*(3), 120–127.
Reddy, M. V. (2004). *Community-based sustainable ecotourism: The case of the Andaman and Nicobar Islands*. Report submitted to the Royal Geographic Society. April/May, 2004.
Reddy, S. (2007). Mega tourism in Andaman and Nicobar Islands: Some concerns. *Journal Human Ecology, 21*(3), 231–239.
Segrado, R., Muñoz, A. P., & Arroyo, L. (2008). Medición de la capacidad de carga turística de Cozumel. *El Periplo Sustentable, 13*, 33–61.
Sobhani, P., Esmaeilzadeh, H., Sadeghi, S. M. M., & Marcu, M. V. (2022). Estimation of ecotourism carrying capacity for sustainable development of protected areas in Iran. *International Journal of Environmental Research and Public Health, 19*, 1059. https://doi.org/10.3390/ijerph19031059
Sridhar, R., Yuvaraj, E., Sachithanandam, V., Mageswaran, T., Purvaja, R., & Ramesh, R. (2017). Tourism carrying capacity for beaches of South Andaman Island, India. In *Tourism - From empirical research towards practical application*. InTech. https://doi.org/10.5772/62724
Telfer, D. J., & Sharpley, R. (2007). *Tourism and development in the developing world*. Routledge.
Tribe, J., Font, X., Griffiths, N., Vickery, R., & Yale, K. (2000). *Environmental management for rural tourism and recreation*. Cassell.
Zacarias, D. A., Williams, A. T., & Newton, A. (2011). Recreation carrying capacity estimations to support beach management at Praia de Faro, Portugal. *Applied Geography, 31*(3), 1075–1081. https://doi.org/10.1016/j.apgeog.2011.01.020

Chapter 28
Tourism Climate Index (TCI) for Assessing the Favourable Period for Tourism Recreation Activities with the Application of Geospatial Techniques

Farhin Sultana and Ashis Kumar Paul

28.1 Introduction

The land and sea interface of coastal areas are influenced by maritime climate throughout the year with seasonal and local control on its parametric variability (Rutty et al., 2020; Scott et al., 2016; Yu et al., 2021). On the basis of the climatic comfort and plenty of natural resources people are gathered in the coastal areas. Furthermore, natural beauties attract people in the coastal areas. Therefore, the favourable weather condition has played an important role in the intensification of the coastal tourism and it is important to evaluate interrelationship between the climatic conditions and coastal tourism. It is also necessary to understand the favourable period of visiting in a particular coastal area for refreshment and recreation of the visitors or tourists. The weather condition is ever changing in the coastal environment, which influences the tourist's flow through making their decision about preferences of holiday destination sites. Mieczkowski (1985) first analysed the relationship between the climate and tourism through introducing the Tourism Climate Index (TCI). There are some benefits of TCI indexing, as it can promote international tourism and provide information about the better use of the charming weather condition of a region. However, the changing nature of global and regional climate and weather conditions effects on the tourism industry and reduces the attractive potentiality of the tourist as well as the destination sites (Kūle et al., 2013; Zhong et al., 2019, Gao et al., 2022). As a result, the regional level, country level, and global economy may be affected by the climate change and reducing tourism. Furthermore, the increasing temperature and depleting groundwater level caused water scarcity in the coastal areas with other parts of the territory. The high temperature and rainfall also develop many diseases like malaria and dengue, and that is

F. Sultana · A. K. Paul (✉)
Department of Geography, Vidyasagar University, Midnapore, West Bengal, India
e-mail: akpauleastcoast@gmail.com

A. K. Paul, A. Paul (eds.), *Crisis on the Coast and Hinterland*,
https://doi.org/10.1007/978-3-031-42231-7_28

dangerous for human health (Alonso-Pérez et al., 2021). So, it is important to analyse the individual effects of the climatic parameter on the tourists by evaluating TCI in this study (Olya & Alipour, 2015).

Climate is not only affecting the well-being of the tourists but also influences tourists' entertainment like outdoor activities, relaxation, and their happiness. The different climatic parameters that affect the human health are temperature, precipitation, humidity, and wind speed. Temperature and humidity are the main factors that affect the coastal environment. The scorching heat and the excessive humidity are very uncomfortable for the tourist. Sometime, the excessive sunshine may stimulate human blood circulation and that may create serious illness. So, the ideal climatic condition should be satisfying the tourist's physical and mental health. The dry and stormy wind in the summer is very harmful for the tourists to employ their outdoor activities. Excessive rainfall can also create problems for tourists' activities. Therefore, in our natural environment climate is the most effective and influential parameter for coastal tourism.

28.2 Assessment Method of the TCI Index

The assessment of favourable climatic conditions for tourist flow into the destinations has needed to assess the parametric climatic data. Therefore, the climatic data of tourism destinations has been collected for 35 years (1979–2014). The meteorological data of minimum and maximum temperature, precipitation, sunshine hours, relative humidity (mean of relative humidity and mean of minimum of relative humidity), and wind speed are collected from NOAA-based selective stations at the coordinate of 21.69 N and 88.12 E for the sites of Bakkhali, Freserganj, and Benubon, 21.69 N and 88.43 E coordinated for Henry's Island, 21.69 N and 87.81 E for Gangasagar, Nayachar, and Mandirtala, and 21.6999 N and 87.52 E coordinated for Mandarmani, Dadanpatrabar, and Rasulpur tourism areas (the climatic data for the destinations of Dakshin Purushottampur, Boatkhali, and Beguakhali are not available for analysis of TCI values). The estimation of TCI depends on the analysis of the aforementioned meteorological variables of the tourism destinations. The TCI is estimated based on Eq. 28.1 (Mieczkowski, 1985; Ma et al., 2020; Roshan et al., 2016):

$$\mathrm{TCI} = 4\mathrm{Cld} + \mathrm{Cla} + 2R + 2S + W \tag{28.1}$$

However, the individual variables are weighted according to their relative importance for tourist well-being. Where Cld = day time comfort index and composed of maximum temperature and minimum daily relative humidity; Cla = daily comfort index and composed of minimum temperature and mean of minimum of relative humidity; P = Precipitation in mm; S = daily sunshine hours; and W = Wind speed in km/hour. Therefore, the day-time comfort index carries 40% weight in the TCI equation, the daily comfort index carries only 10% weight, precipitation, and sunshine hours assigned a weight of 20% in each, and wind speed weighted as 10% and develops the above formula. Furthermore, Eq. 28.1 is simplified through numerical transformations of the above climatic parameters and simplified into a 5-scaled optimal rating for every variable. Therefore, Eq. 4.1 can be rewritten as Eq. 28.2:

$$\mathrm{TCI} = 2\{(4X5) + 5 + (2X5) + 5\} = 100 \tag{28.2}$$

Depending on Eq. 28.2, the entire meteorological parameters of a 35 year period are analysed, and the monthly climatic comfort condition of each destination site is determined (Table 28.1). Therefore, the parameter-based rating systems are also shown in Table 28.2. Finally, the monthly favourable period for tourists' recreation activities in the coastal region is laid out with the help of GIS software (Arc GIS). The indicator shares the 20% weight in the TCI formula. In TCI-Equation the precipitation rate is obtained from this table (Table 28.2). Therefore, in this rating system, when the precipitation amount is increasing, the rating system will decrease, and this rating system is adopted in the study from Mieczkowski (1985).

28.3 Results and Discussion

28.3.1 Analysis of the TCI Value of Each Month in the Coastal Destinations

The month-wise TCI results of the month of January reveal (Fig. 28.1a) that the sites of Mandarmani (81.83), Dadanpatrabar (81.83), Rasulpur (81.83), Nayachar Island (78.79), Mandirtala (78.79), and Gangasagar (78.79) are the most favourable destinations for the tourists' recreational activities for the very comfortable climatic conditions. However, the TCI values for the remaining tourism destinations of Benubon (71.77), Bakkhali (71.77), Henry's Island (74.36), and Freserganj (71.77) provide the very good climatic conditions for supporting the tourists' recreational activities. Similarly, the TCI of February shows that (Fig. 28.1b) the destinations of Nayachar Island (80.44), Mandirtala (80.44), and Gangasagar (80.44) have

Table 28.1 Tourism climate index numerical values and descriptive category of climatic comfort

Numerical values of indices	Descriptive category
90–100	Ideal
80–89	Excellent
70–79	Very good
60–69	Good
50–59	Acceptable
40–49	Marginal
30–39	Unfavourable
20–29	Very unfavourable
10–2019	Extreme unfavourable
0–9	Impossible

Table 28.2 Precipitation variable rating system; Sunshine variable; and Wind speed variable for the assessment of the Tourism Climate Index (TCI)

Precipitation variable	Mean monthly precipitation (mm)	Rates	Mean monthly h/min of sunshine per day	Wind speed (km/h)	Normal system
5	0.0–14.9	5	10 h or more	< 2.88	5
4.5	15.0–29.9	4.5	9 h to 9 h 59 min	2.88–5.75	4.5
4	30.0–44.9	4	8 h to 8 h 59 min	5.76–9.03	4
3.5	45.0–59.9	3.5	7 h to 7 h 59 min	9.04–12.23	3.5
3	60.0–74.9	3	6 h to 6 h 59 min	12.24–19.79	3
2.5	75.0–89.9	2.5	5 h to 5 h 59 min	19.80–24.29	2.5
2	90.0–104.9	2	4 h to 4 h 59 min	24.30–28.79	2
1.5	105.0–119.9	1.5	3 h to 3 h 59 min	28.80–38.52	1
1	120.0–134.9	1	2 h to 2 h 59 min	>38.52	0
0.5	135.0–149.9	0.5	1 h to 1 h 59 min		
0	Above 150	0	Less than 1 h		

excellent climatic conditions and provide the most favourable destinations in the study area. However, the remaining tourists' destinations of Mandarmani (70.56), Dadanpatrabar (70.56), Rasulpur (70.56), Benubon (77.08), Bakkhali (77.08), Henry's Island (78.63), and Fregerganj (77.08) get the very good climatic conditions for the tourists' recreational activities. In the month of March, the temperature becomes hot and the climatic condition is a little unfavourable for the tourists. After assessing the TCI value of the month of March it shows that (Fig. 28.1c) the destinations of Benubon (86.33), Bakkhali (86.33), Henry's Island (83.16), and Fregerganj (86.33) get excellent climatic conditions for tourists' arrival and recreational activities. The destinations of Nayachar Island (54.87), Mandirtala (54.87), and Gangasagar (54.87) get good climatic conditions. The remaining coastal destinations of Mandarmani (54.72), Dadanpatrabar (54.72), and Rasulpur (54.72) are favourable for tourism activities due to the acceptable climatic condition.

Considering all the climatic parameters, the TCI is assessed and the result of the month of April shows that (Fig. 28.1d) the destinations of Benubon (72.36), Bakkhali (72.36), and Fregerganj (72.36) have very good climatic conditions. However, Henry's Island (64.16) shows a good climatic condition. The remaining tourism destinations of Mandarmani (48.66), Dadanpatrabar (48.66), Rasulpur (48.66), Nayachar Island (52.89), Mandirtala (52.89), and Gangasagar (52.89) have acceptable conditions for the tourists due to the scorching heat and the high humidity that may be uncomfortable for the tourists. In the month of May (Fig. 28.1e) after analysis of the TCI it shows that in Benubon (62.80), Bakkhali (62.80), and Freserganj (62.80), the climatic condition is good. However, Nayachar Island (49.19), Mandirtala (49.19), Gangasagar (49.19), and Henry's Island (53.30) have acceptable climatic conditions. The remaining coastal destinations of Mandarmani (43.80), Dadanpatrabar (43.80), and Rasulpur (43.80) have marginal climatic condition due to the hot wind during this month that is unfavourable for the tourists'

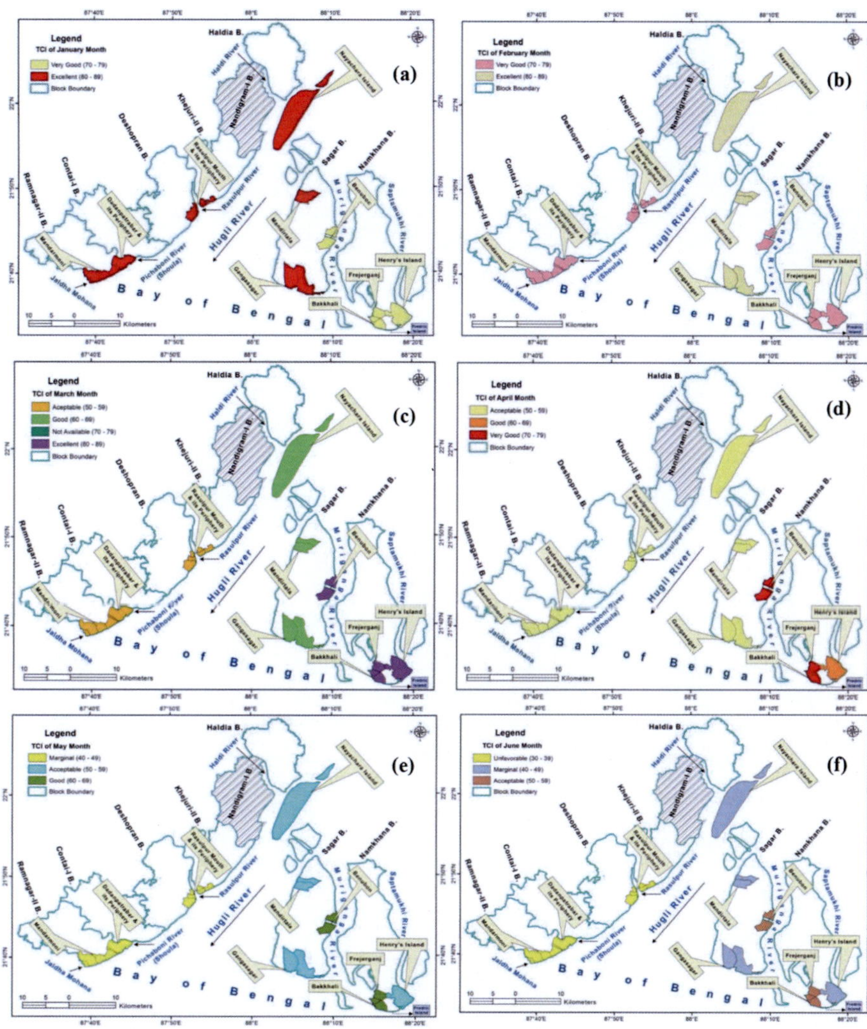

Fig. 28.1 Tourism Climate Index maps for the month of (**a**) January; (**b**) February; (**c**) March; (**d**) April; (**e**) May; and (**f**) June in the ecotourism destination sites

activities. In the month of June, the monsoonal rainy season starts. The TCI of the month of June shows that (Fig. 28.1) Benubon (60.25), Bakkhali (60.25), and Freserganj (60.25) have the acceptable climatic conditions for the tourists. Nayachar Island (42.01), Mandirtala (42.01), Gangasagar (42.01), and Henry's Island (53.77) have acceptable climatic conditions for the tourists. Mandarmani (39.52), Dadanpatrabar (39.52), and Rasulpur (39.52) have unfavourable climatic conditions for the tourists (Fig. 28.1f).

In the month of July, the climatic conditions are the same (Fig. 28.2a) as it is in the month of June. Benubon (71.86), Bakkhali (71.86), Henry's Island (74.97), and Freserganj (71.86) have the very good climatic conditions. Whereas, Nayachar Island (62.46), Mandirtala (62.46), and Gangasagar (62.46) have the good climatic conditions. Mandarmani (49.41), Dadanpatrabar (49.41), and Rasulpur (49.41) have a marginal climatic condition for the tourists. In the month of August, the TCI shows that (Fig. 28.2b) Benubon (72.61), Bakkhali (72.61), Henry's Island (73.55), and Freserganj (72.61) have very good climatic conditions. Nayachar Island (64.95), Mandirtala (64.95), and Gangasagar (64.95) have good climatic conditions. Mandarmani (54.8), Dadanpatrabar (54.8), and Rasulpur (54.8) have acceptable climatic conditions for the tourists. The TCI for the month of September reveals that (Fig. 28.2c) Benubon (70.75), Bakkhali (70.75), Henry's Island (71.5), and Freserganj (70.75) have very good climatic conditions. Nayachar Island (62.29),

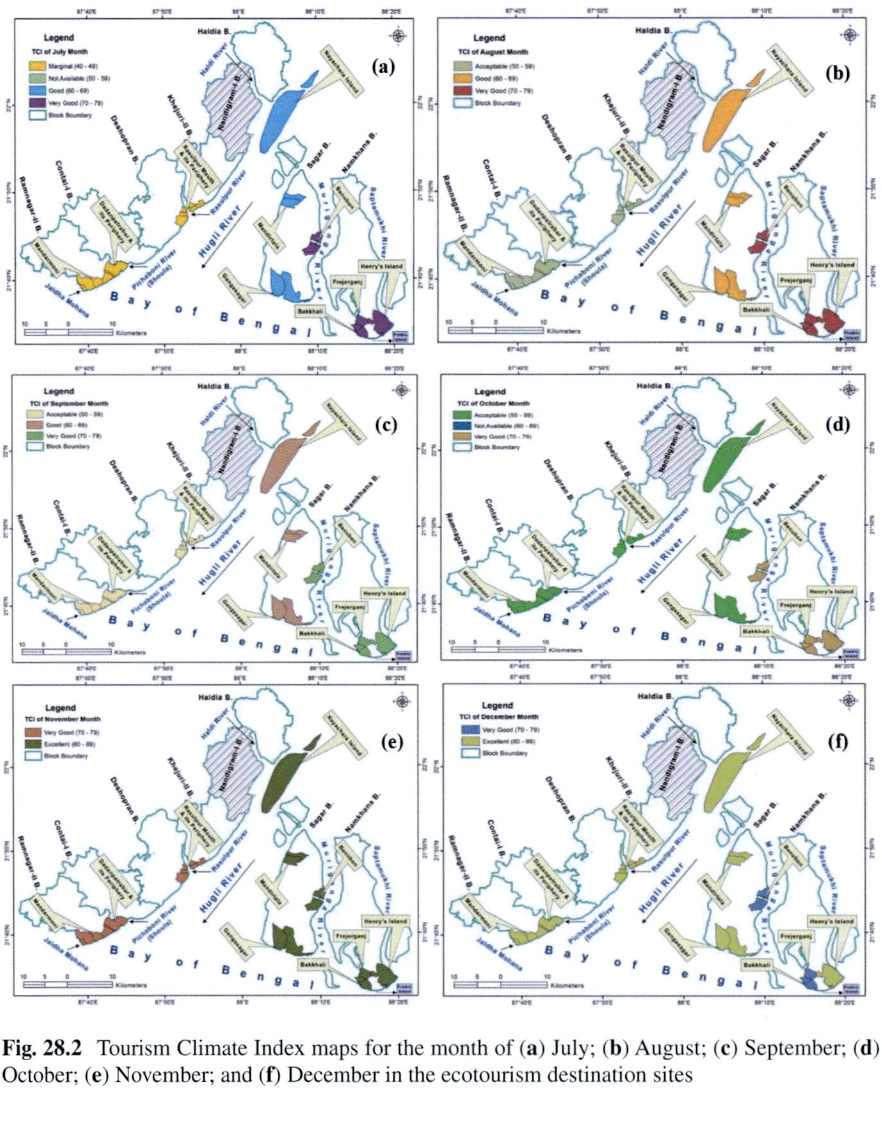

Fig. 28.2 Tourism Climate Index maps for the month of (**a**) July; (**b**) August; (**c**) September; (**d**) October; (**e**) November; and (**f**) December in the ecotourism destination sites

Mandirtala (62.29), and Gangasagar (62.29) have good climatic conditions for the tourists. Mandarmani (50.22), Dadanpatrabar (50.22), and Rasulpur (50.22) have acceptable climatic conditions for the tourists in the coastal destination sites.

In the month of October, the TCI shows that (Fig. 28.2d) the climatic conditions for the coastal destinations are favourable. Benubon (75.30), Bakkhali (75.30), Henry's Island (75.5), and Freserganj (75.30) have the very good climatic conditions. Mandarmani (54.80), Dadanpatrabar (54.80), Rasulpur (54.80), Nayachar Island (60.43), Mandirtala (60.43), and Gangasagar (60.43) have the acceptable climatic conditions for the tourists. The winter season starts in the month of November and the climatic conditions are very comfortable for the coastal tourists. The TCI of that month shows that (Fig. 28.2e) Nayachar Island (82.21), Mandirtala (82.21), Gangasagar (82.21), Benubon (78.69), Bakkhali (78.69), Henry's Island (80.16), and Freserganj (78.69) get excellent climatic conditions for the tourists. Mandarmani (77.13), Dadanpatrabar (77.13), and Rasulpur (77.13) have very good climatic conditions for the tourists in the coastal destination sites of the study area. The climatic condition in the month of December is very much comfortable. Therefore, the TCI result of that month reveals that (Fig. 28.2f) Mandarmani (83.83), Dadanpatrabar (83.83), Rasulpur (83.83), Nayachar Island (81.64), Mandirtala (81.64), Gangasagar (81.64), and Henry's Island (76.66) have an excellent climatic conditions. However, the remaining destination sites of Benubon (74.5), Bakkhali (74.5), and Freserganj (74.5) are perceived as good climatic conditions for tourists' recreational activities.

The overall climatic condition of the 12 months of the entire study area shows that the month of June is the most unfavourable condition for tourism activities in the coastal destinations. However, to some extent the tourism activities are acceptable in few sites. In few destinations during the month of May and July, the tourism activities are accepted with the support of the recreational activities by the tourists. But the marginal TCI prevails for some destinations due to the hot and humid condition coupled with the occurrences of low pressures and cyclonic behaviour of the coastal region in the destinations. However, the months of March, April, August, September, and October are acceptable for playing the recreational activities with good quality of TCI. Only the months of January, February, November, and December are the most favourable months for playing the safe recreational activities in coastal tourism destinations. These months are most favourable for the tourists' movements in the destinations. During this time the tourism pressure is maximum in the coastal destinations.

In the year 2014–2015, the survey result suggests that the month of January 2015 (7768121) is the peak month for domestic tourists, and the lowest tourist pressure is seen during the months of June 2014 and May 2014 (NCAER, 2015). However, in the case of the foreign tourists, the tourist pressure is the same all over the year (Table 28.3). However, during these months all the hotels and restaurants are able to accommodate the tourists and provide the local foods or marine foods at their optimum level. The result also reveals that the hotel and restaurants of the tourist's destinations earn a maximum profit during the favourable climatic condition of this period (Table 28.3).

Table 28.3 Category of climatic condition of the destination sites

Month	Coastal destination sites	Category of climatic conditions
January	Mandarmani, Dadanpatrabar and its periphery, Rasulpur mouth and its periphery, Nayachar Island, Mandirtala, and Gangasagar	Excellent
	Benubon, Henry's Island, Bakkhali, Fraserganj	Very good
February	Nayachar Island, Mandirtala, Gangasagar	Excellent
	Mandarmani, Dadanpatrabar and its periphery, Rasulpur mouth and its periphery, Benubon, Bakkhali, and Fraserganj	Very good
March	Benubon, Bakkhali, Freserganj, Henry's Island	Excellent
	Nayachar Island, Mandirtala, Gangasagar	Good
	Mandarmani, Dadanpatrabar and its periphery, Rasulpur mouth and its periphery	Acceptable
April	Benubon, Freserganj, Bakkhali	Very good
	Henry's Island	Good
	Mandarmani, Dadanpatrabar and its periphery, Rasulpur mouth and its periphery, Nayachar Island	Acceptable
May	Benubon, Freserganj, Bakkhali	Good
	Nayachar Island, Mandirtala, Gangasagar, Henry's Island	Acceptable
	Mandarmani, Dadanpatrabar and its periphery, Rasulpur mouth and its periphery	Marginal
June	Benubon, Freserganj, Bakkhali	Acceptable
	Nayachar Island, Mandirtala, Gangasagar, Henry's Island	Marginal
	Mandarmani, Dadanpatrabar and its periphery, Rasulpur mouth and its periphery	Unfavourable
July	Benubon, Freserganj, Bakkhali, Henry's Island	Very good
	Nayachar Island, Mandirtala, Gangasagar	Good
	Mandarmani, Dadanpatrabar and its periphery, Rasulpur mouth and its periphery	Marginal
August	Benubon, Freserganj, Bakkhali, Henry's Island	Very good
	Nayachar Island, Mandirtala, Gangasagar	Good
	Mandarmani, Dadanpatrabar and its periphery, Rasulpur mouth and its periphery	Acceptable
September	Benubon, Freserganj, Bakkhali, Henry's Island	Very good
	Nayachar Island, Mandirtala, Gangasagar,	Good
	Mandarmani, Dadanpatrabar and its periphery, Rasulpur mouth and its periphery	Acceptable
October	Benubon, Freserganj, Bakkhali, Henry's Island	Very good
	Mandarmani, Dadanpatrabar and its periphery, Rasulpur mouth and its periphery, Nayachar Island, Mandirtala	Acceptable
November	Nayachar Island, Mandirtala, Gangasagar, Benubon, Freserganj, Bakkhali, Henry's Island	Excellent
	Mandarmani, Dadanpatrabar and its periphery, Rasulpur mouth and its periphery	Very good
December	Mandarmani, Dadanpatrabar and its periphery, Rasulpur mouth and its periphery, Nayachar Island, Mandirtala, Gangasagar, Henry's Island	Excellent
	Benubon, Freserganj, Bakkhali	Very good

28.4 Conclusion

TCI is the most ideal index that analyses the relationship between the climatic condition and its effect on the human well-being and their activities. January, February, November, and December are the most favourable months for the tourists arriving in the all coastal tourism destination sites of the study area. In the month of March, Mandarmani, Dadanpatrabar, Rasulpur, Nayachar, Mandirtala, and Gangasagar have acceptable climatic conditions; and Benubon, Bakkhali, Henry's Island, and Freserganj have the excellent climatic conditions for the travelling tourists. In the month of April, Mandarmani, Dadanpatrabar, and Rasulpur have marginal climatic comfort; Nayachar, Mandirtala, and Gangasagar on the other hand have acceptable conditions; and Benubon, Bakkhali, Henry's Island, and Freserganj have very good climatic comfort conditions for the tourists. In the month of May, Mandarmani, Dadanpatrabar, Rasulpur, Nayachar, Mandirtala, and Gangasagar have marginal climatic comfort and Benubon, Bakkhali, Henry's Island, and Freserganj have good climatic comfort in the coastal destinations. In the month of June Mandarmani, Dadanpatrabar, and Rasulpur have the unfavourable climatic condition for the tourists, in Nayachar, Mandirtala, and Gangasagar the climatic condition is marginal, and in Benubon, Bakkhali, Henry's Island, and Freserganj the climatic comfort is good. In the month of July, the climatic comfort for the coastal tourism destination sites Mandarmani, Dadanpatrabar, and Rasulpur is marginal, in Nayachar, Mandirtala, and Gangasagar it is good, and in Benubon, Bakkhali, Henry's Island, and Freserganj the climatic comfort condition is very good. In the month of August, September, and October, the climatic comfort condition for Mandarmani, Dadanpatrabar, and Rasulpur is acceptable, in Nayachar, Mandirtala, and Gangasagar it is good, and in Benubon, Bakkhali, Henry's Island, and Freserganj the climatic comfort condition is very good.

References

Alonso-Pérez, S., López-Solano, J., Rodríguez-Mayor, L., & Márquez-Martinón, J. M. (2021). Evaluation of the tourism climate index in the Canary Islands. *Sustainability, 13*(13), 7042.

Gao, C., Liu, J., Zhang, S., Zhu, H., & Zhang, X. (2022). The Coastal Tourism Climate Index (CTCI): Development, validation, and application for Chinese coastal cities. *Sustainability, 14*(3), 1425.

Kūle, L., Haller, I., Varjopuro, R., & Alberth, J. (2013). *Climate change impacts on coastal tourism in the Baltic Sea Region*. Coastline Reports, 91.

Ma, S., Craig, C. A., & Feng, S. (2020). The Camping Climate Index (CCI): The development, validation, and application of a camping-sector tourism climate index. *Tourism Management, 80*, 104105.

Mieczkowski, Z. (1985). The tourism climatic index: a method of evaluating world climates for tourism. *Canadian Geographer/Le Géographe Canadien, 29*(3), 220–233.

National Council of Applied Economic Research (NCAER). (2015). *India: Regional tourism satellite, accounts, 2015–16 West Bengal*, The Ministry of Tourism, Government of India, Assessed 2019.

Olya, H. G., & Alipour, H. (2015). Risk assessment of precipitation and the tourism climate index. *Tourism Management, 50*, 73–80.

Roshan, G., Yousefi, R., & Fitchett, J. M. (2016). Long-term trends in tourism climate index scores for 40 stations across Iran: The role of climate change and influence on tourism sustainability. *International Journal of Biometeorology, 60*, 33–52.

Rutty, M., Scott, D., Matthews, L., Burrowes, R., Trotman, A., Mahon, R., & Charles, A. (2020). An inter-comparison of the holiday climate index (HCI: Beach) and the tourism climate index (TCI) to explain Canadian tourism arrivals to the Caribbean. *Atmosphere, 11*(4), 412.

Scott, D., Rutty, M., Amelung, B., & Tang, M. (2016). An inter-comparison of the holiday climate index (HCI) and the tourism climate index (TCI) in Europe. *Atmosphere, 7*(6), 80.

Yu, D. D., Rutty, M., Scott, D., & Li, S. (2021). A comparison of the holiday climate index: Beach and the tourism climate index across coastal destinations in China. *International Journal of Biometeorology, 65*, 741–748.

Zhong, L., Yu, H., & Zeng, Y. (2019). Impact of climate change on Tibet tourism based on tourism climate index. *Journal of Geographical Sciences, 29*, 2085–2100.

Chapter 29
Assessment of Diversity in Landscape Ecology in Parts of the Purba Medinipur Coastal District, West Bengal, with Geospatial Technology

Amrit Kamila, Jatisankar Bandyopadhyay, and Ashis Kumar Paul

29.1 Introduction

Landscape ecology is the study of how landscape structure influences the profusion and allocation of organisms at the multiplicity of scales of the landscape. So, landscape ecology focuses on three characteristics of the landscape such as structure, function, and modification. Structure stands for the spatial associations between the typical ecosystems or rudiments, whereas function deals with the relations among the spatial features, and modification comprised the change in the arrangement and purpose of the ecological mosaic over time (Troll, 1968; Godron & Forman, 1983). Two systems collaborate to create landscapes, using edge-specific topographical methods and precise perturbation of factor conditions. Landscapes vary greatly in aerial range, with scales restricted to small areas of a few metres or hundreds of metres being finer than landscapes. Because of the area's geomorphology, the multifaceted landforms and parental materials present are comparatively invariable over a landscape (Neef, 1984; Zonneveld, 1989). Tropical marine ecosystems are frequently dynamic and spatially heterogeneous seascapes in which different habitat types (e.g., coral reef, seagrass, open water, mangrove, and sand) are linked by a variety of biological, physical, and chemical processes. Water movements, including tides and currents, facilitate the exchange of nutrients, chemical pollutants, pathogens, sediments, and organisms among components of the seascape (Paul, 2002; Woodroffe, 2002). Landscape ecology has traditionally been inadequate to the lessons of terrestrial systems; however, the scientific questions and techniques

A. Kamila (✉) · J. Bandyopadhyay
Department of Remote Sensing & GIS, Vidyasagar University,
Midnapore, West Bengal, India
e-mail: amritkamila90@gmail.com

A. K. Paul
Department of Geography, Vidyasagar University, Midnapore, West Bengal, India

A. K. Paul, A. Paul (eds.), *Crisis on the Coast and Hinterland*,
https://doi.org/10.1007/978-3-031-42231-7_29

are uniformly applicable to maritime and coastal sub-systems. The mutual association between spatial outline and ecological procedure and the overarching consequence of scale on this connection were being investigated in some maritime settings as the common discipline of landscape ecology was growing through the latter two decades of the last century. As with all mechanisms of the biosphere, these associations are decisive for the successful supervision of marine and coastal sub-systems (Naveh, 1995; Wiens, 1997). However, comparatively recent progress in geographic information systems, remote sensing, and computer technologies has commenced dealing with these issues and is now authorising the appraisal of patterns and processes in oceans. This present work intends to emphasise research that is adapting the conception of landscape ecology to respond to ecological questions within coastal sub-systems, to deal with the exclusive challenges expressed in these landscapes, and to motivate an exchange of thoughts and clarification of universal problems.

29.2 Geographical Setup of the Study Area

The studied coast is a large coastal region in the state's far south-west corner. A part of the District of Purba Medinipur, West Bengal, along the Bay of Bengal includes the coastal plain. This rising coastal plain is made up of sand and mud sedimented by the fluvial and aeolian processes and is also a mid-eastern division of the Kanthi Coastal Plain, it covers an area of about 29,439 ha, or 294.39 km^2 (Fig. 29.1). Geologically, the area is characterised by typical Holocene alluvial deposits originating from the Subarnarekha and Ganges River networks. The nature of this geomorphic part is mainly characterised by sand dunes and marshes parallel to the coast.

29.3 Materials and Methods

The present study deals with Survey of India (SOI) toposheets, satellite images of Landsat 5 and Sentinel-2 Multi-Spectral Instrument (MSI), and Google Earth images with different temporal phases. The study has also used climactic data from the India Meteorological Department (IMD), Kolkata; a geological map from the Geological Survey of India (GSI) official portal; and a soil map from the National Bureau of Soil Survey and Land Use Planning (NBSS & LUP) for the assessment of different parameters, and simultaneously repeated field surveys have been conducted with pre-designed questioners for validation of the research results. The radiometric and geometric errors of satellite imagery are methodically corrected, and spectral radiance to surface spectral reflectance is converted through the gain bias method.

A high-resolution Digital Elevation Model (DEM) has been prepared through the bootstrap iterations for determining the precision statistics stochastically (Sharma et al., 2010) using the Advanced Spaceborne Thermal Emission and Reflection

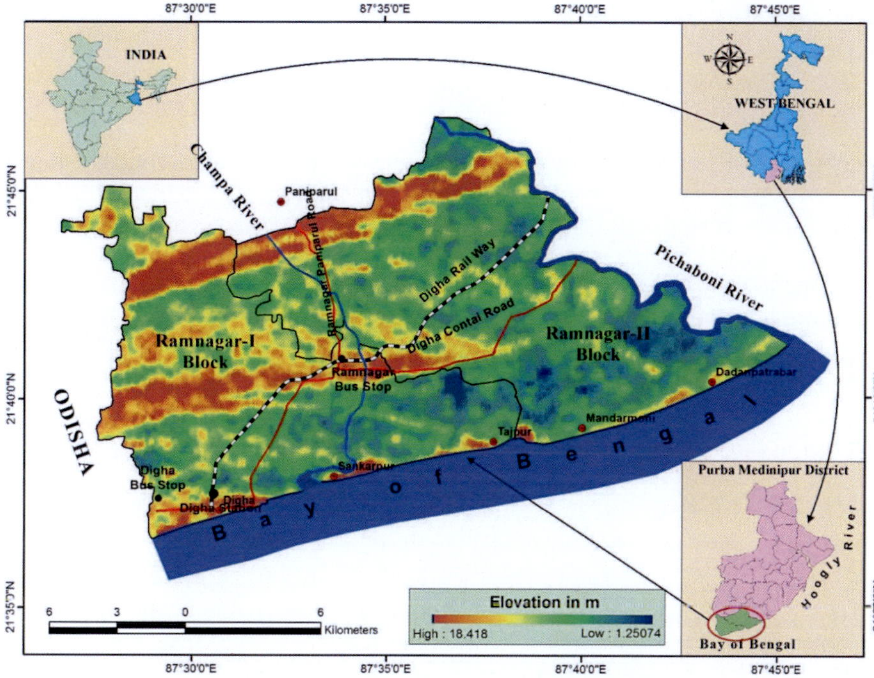

Fig. 29.1 The study areas with landscape ecological components depict the beach ridge chenier swale topography of Kanthi Coastal Plain along the northern fringe of the Bay of Bengal coast

Radiometer (ASTER) Global Digital Elevation Model (GDEM) (Version-215 m, 2014), Shuttle Radar Topography Mission (SRTM: 1 arc-second, 30 m, 2014) DEM, Google Earth elevation, in situ 2000 Ground Control Points (GCP), and Total Station survey data. This model is very significant for identifying the landscape characteristics and evolutionary stages of coastal tracts. High-resolution DEM, existing radiocarbon dating records, Optically Stimulated Luminescence (OSL) dating records, and geomorphological and stratigraphic signatures have been considered for the validation of entire processes of landscape evolution. Image classification techniques, the Shannon-Weiner Diversity Index, Sorenson's Coefficient, and Hierarchical Cluster Analysis (HCA) techniques have been used to estimate, to explore, and to analyse the species richness, evenness, and spatial diversity of coastal habitats in the study area.

Human observational studies are also conducted through repeated field studies to better understand the relationship between human activities (resource uses and land use and land cover changes), coastal hazards, and coastal morphodynamics. At the same time, biological processes have been estimated through phyto-geomorphological mapping. Several quantitative methods incorporate the link between spatial patterns and ecological processes at broad spatial and temporal scales. This linkage of time, space, and environmental change can assist managers in applying plans to solve environmental problems in the coastal landscapes.

29.4 Results and Discussions

29.4.1 Topographic Character and Landform Order

The physiography of the alluvial coast is very important in coastal morphodynamics in response to outer environmental impacts with a significant change in local boundary conditions (Bhandari & Das, 1998; Paul, 2002; Maiti & Bhattacharya, 2009). To predict and establish the evolution of the Chenier coastal plain, the chronology of the coastal evolution is explained using available dating records of different landform units in the existing literature, estimation of present-day wave hydrodynamics and energy level, and estimation of the sediment budget of the near shores (Kamila et al., 2021a). In the wide valley flat surface between the Ramnagar-Deuli beach ridge section and the Digha-Junput beach ridge section, there are three bifurcated ridges in the form of narrow and low-height ridges (Fig. 29.2a, Table 29.1). The three barriers are separated by linear depressions running parallel to the present ridge lines and represent the linear tidal basins of that time. To the east, the wider flats of tidal basins are characterised by the location of younger natural levees and older natural levees and some depressed wetlands (Paul, 2002; Kamila et al., 2020).

The energy of longshore current is calculated and estimated as being highest for the Contai-Paniparul beach ridge chenier, Ramnagar-Deuli beach ridge chenier, and Digha-Junput beach ridge chenier after consideration with the volume of sediment estimation under modern sea-face energy levels. The Chenier Plain is the result of a combination of fluvio-marine deposition of sediment into coastal areas, the occurrence of strong longshore currents at the sea face, the activities of repeated coastal storms, tectonic impacts, and past sea level fluctuations.

However, the shorter beach ridge cheniers are shaped under weaker longshore current energy in the east-east north direction, parallel to the present-day shoreline (Table 29.2). On the other hand, it is also observed that the wide, shallow flats in between landward and seaward beach ridge cheniers were formed by the finer sediments (swale topography) deposited under the lagoonal setting behind the barrier

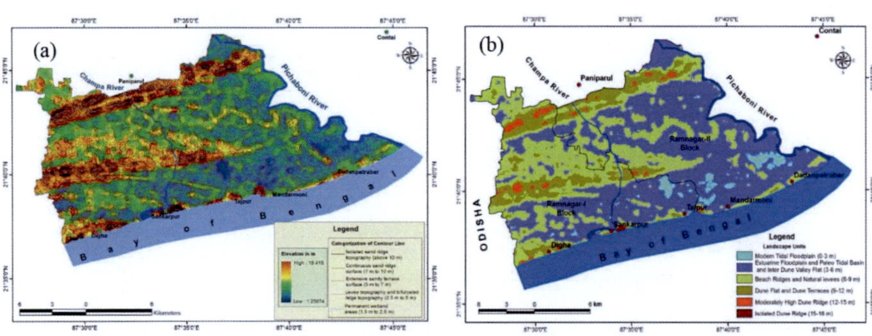

Fig. 29.2 (**a**) The spatial elevation changes depicted by the contour patterns; and (**b**) Landscape habitat fragments in the coastal plain

Table 29.1 Chronology of coastal formations in the study areas of Ramnagar-I and II Administrative Blocks (Kamila et al., 2021a)

Landscapes	Elevation	Pattern	Soil material	Tonal contrast	Existing dating age	Probable age	Ancient processes involved	Sea level indicators
First category of landscapes	Above 10 m	Shore parallel isolated sand ridge	Oxidised sands	Brownish	7000 years BP	Early Holocene	Windblown deposits and erosion by tidal waves	Sea-level steel stand
Second category of landscapes	7–10 m	Shore parallel sand ridge topography	Oxidised sands	Brownish to yellowish	5760 ± 160 years BP	Middle Holocene	Windblown and wind-tidal	Sea-level steel stand
Third category of landscapes	5–7 m	Extensive sandy tract on the side of ridges	Sandy	Greyish	2900 ± 160 years BP	Late Holocene	Over wash reactivated deposits	Transgressive seas
Fourth category of landscapes	2.5–5.0 m	Bifurcated ridges and crenulated levees	Sandy and loamy	Dark grey	2000 ± 100 years BP	Recent to sub-recent	Wave-induced currents and fluvial currents	Regressive seas

Table 29.2 The assessment of hydro-morphodynamics of the past landforms based on modern available data (Maiti, 2013; Kamila et al., 2021a)

Stages	Area of beach ridges in m^2	Average height of beach ridges in m	Volume of depositional sediment in m^3	Duration of deposition in year	Volume of deposition per year in m^3	Energy concentration (e) per year	Area of beach ridges in km^2
1st stage	6,63,96,200	10.38	689,192,556	1240	5,55,800.45	14.98	66.39
2nd stage	1,22,61,700	9.01	110,477,917	1430	77,257.28	2.08	12.26
3rd stage	3,35,20,800	10.25	343,588,200	1430	2,40,271.45	6.48	33.52
4th stage	7,00,6760	9.21	645,322,59	300	2,15,107.53	5.80	7.07
5th stage	90,82,150	8.40	762,900,60	300	23,54,300.20	6.85	9.08
6th stage	64,82,380	9.24	59,897,191	300	1,99,657.30	5.38	6.48
7th stage	1,79,75,600	9.61	172,745,516	1200	1,43,954.60	3.88	17.97

bar systems and the supply of finer sediments by Hugli River mouth discharges into the Late Holocene tidal basin. Most recently, the beach ridge cheniers and shoreline are disconnected by older distributary channels and act as tidal inlet mouths along the beach surface and have been modified by modern coastal processes (Paul, 2002; Kamila et al., 2021a).

29.4.2 Landscape Ecological Diversity

The native vegetation in coastal areas plays a significant role in stabilising the surface against wind erosion and providing a habitat for wildlife. So, the protection of coastal vegetation is important for the long-term protection of beachfront properties. It is not sufficient to describe the patterns of species turnover at an ecotone; one must think about the underlying causes of that turnover, how species are responding to the environment, and the relative distributions of these species along the gradient.

29.4.3 Species Categorisation

The regional species zonation map is prepared with the concern signature of the plant over the entire study area by classifying floral species on a small grid using the Normalized Difference Vegetation Index (NDVI). On the other hand, the

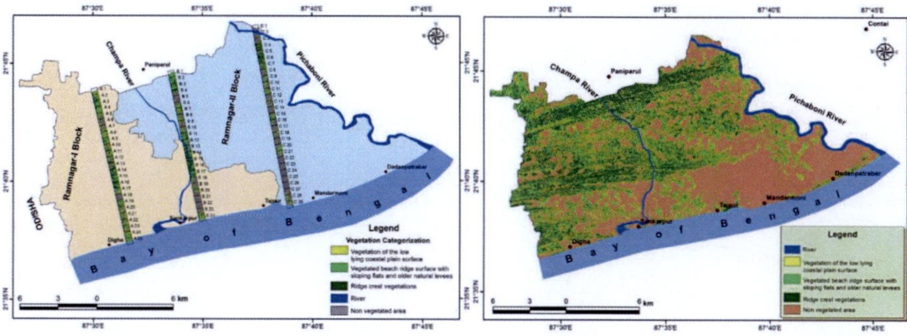

Fig. 29.3 Grid-wise vegetation classification (Left) and overall scenario of vegetation types

classification map of the sampling site can reflect the actual scenario of the plant community in and around the study area. However, the sentinel data offers a higher resolution than other open-source images, so the vegetation is well distributed in this region (Fig. 29.3).

The vegetation is widely distributed and extended over the beach ridge surface, the beach ridge fringed coastal plain, the swale topography, and the inner coastal plain with abandoned creeks and older natural levee bank surfaces. The maximum diversity of vegetation is found in transect B (4.29) in comparison to the other two transects, whereas transect A shows a very high Shannon Diversity Index (4.00), while transect A denotes a very high evenness character compared to the other two transects. The Sorenson's Coefficient is 0.7692, 0.6885, and 0.7731 for AB, BC, and AC transects, respectively, which indicates that the presence of species communities is common in the AC transect when the estimated value is close to 1 (Table 29.3).

The Hierarchical Cluster Analysis result shows that the vegetation belongs to relatively homogeneous groups of species communities. Therefore, the B transect has revealed more heterogeneous characteristics than the other two transects in the present study. Accordingly, the A and C transects denote the consistency of the clustering habit of the species (Kamila et al., 2021b).

29.4.4 *Identification of Micro Landscape Units*

The study deals with a high-resolution DEM to understand the micro-landscape units of the entire study area. The DEM is categorised into six units based on the altitudinal variations and morphometric characteristics such as isolated dune ridge (15–18 m); moderately high dune ridge (12–15 m); dune flat and dune terraces (9–12 m); beach ridge and natural levees (6–9 m); estuarine flood plain and paleo-tidal basin; interdune valley flat (3–6 m), and modern tidal flood plain (0–3 m) to understand the morphological setup (Fig. 29.2b). After considering the elevation of each intersection, a longitudinal profile is constructed from sea to land to estimate

Table 29.3 Estimation of species diversity through Shannon Weiner Diversity Index and Sorenson's Coefficient

Transect	The maximum diversity (H_{max})	Evenness (EH)	Shannon Diversity Index (H)	Sorenson's Coefficient
Transect – A	4.2484	0.9430	4.0067	AB = 0.7692
Transect – B	4.2904	0.9169	3.9341	BC = 0.6885
Transect – C	3.8918	0.9173	3.5700	AC = 0.7731

grid-based micro-terrain units. Each grid represents a different type of terrain unit (Kamila et al., 2021c).

29.4.5 Linkage Between Morphological Units and Coastal Habitats

The current study area also contains several types of habitat units, such as large trees, small trees, shrubs, heaths, and grasslands, which are closely connected and interact with each other, so that this integrated ecosystem creates a large coastal ecosystem. The transact method identifies five plant community types (e.g., grasses, heaths, shrubs, small trees, and large trees) with 106 species and 25 types of micro-landscape ecological units in the current study area (Kamila et al., 2021c). In Table 29.4, it is very clear that small tree habitation is much more frequent than other habitation of floral species, whereas heathland habitation occupies a very small part of the land. Other habitation zones are lying in between the land cover of these two classes. The beach ridge and natural levees, on the other hand, occupy a larger percentage of land in each habitation zone than other micro morphological units. As a result, micro-morphological units, primarily beach ridges and natural levees, are extremely important in terms of floral species diversity and abundance (Fig. 29.4, Table 29.4).

The maximum variability of floral species exists in the beach ridge surface, beach ridge fringed coastal plain surface, inner coastal plain surface with a narrow beach ridge segment, beach ridge separated by a swale valley, inner coastal plain surface with an older levee bank, inner coastal plain with abandoned creeks, and the inner coastal plain with beach ridge remnant surface. Because the dune furrows, dune valleys, sloping flats of the coastal sand dunes, and ancient beach ridge topography retain sufficient soil moisture content in this sensitive area.

Due to the tidal inundations, the wetlands of the coastal belt support potential zones for the growth and extension of floral habitats. Sediment recycling and nutrient recycling are progressively improved in the micro zones due to the wide-ranging growth of plant communities in hot and humid tropical environments. Finally, the

Table 29.4 Percentage of the area occupied by the plant ecology in different morphological units

Sl. No.	Morphological units	Occupied area of plant ecology (Total area 146.24 km^2)				
		Grass land area in percentage (%)	Heath land area in percentage (%)	Scrub land area in percentage (%)	Small tree area in percentage (%)	Large tree area in percentage (%)
1	Isolated dune ridge (15–18 m)			0.03	0.04	0.14
2	Moderately high dune ridge (12–15 m)	0.13	0.07	0.19	1.01	1.75
3	Dune flat and dune terraces (9–12 m)	2.13	0.57	5.32	7.87	7.96
4	Beach ridge and natural levees (6–9 m)	2.02	1.07	16.99	20.46	10.73
5	Estuarine flood plain and Paleo tidal basin and inter dune valley flat (3–6 m)	3.04	1.04	5.86	4.43	5.29
6	Modern tidal flood plain (0–3 m)	1.66	0.14	0.01	0.02	0
	Total area in percentage (%)	8.97	2.89	28.58	33.79	25.84

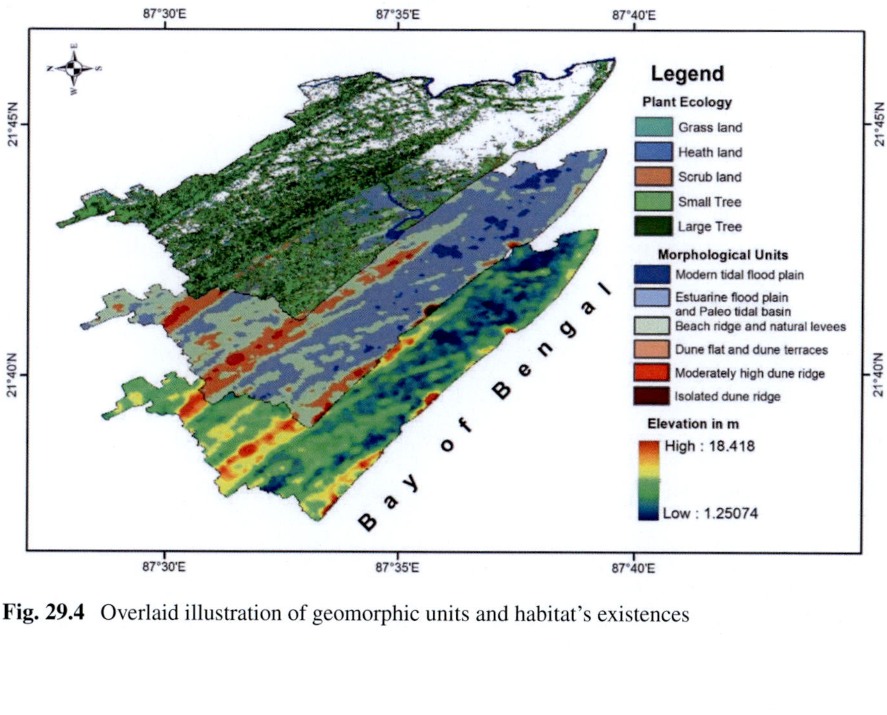

Fig. 29.4 Overlaid illustration of geomorphic units and habitat's existences

spatial allocations of the diverse plant variability zones are incorporated into the microtopographic units of the coastal landscape.

29.5 Conclusions

The coastal plain of alluvium surface with beach ridge chenier and swales was formed during the Early Holocene, Middle Holocene, and Late Holocene periods ranging from 7000 YBP to 500 YBP and the sub-recent stage. There are seven stages of coastal chenier formations on the studied coast (Largely Contai-Paniparul beach ridge chenier, Ramnagar-Deuli beach ridge chenier, and Digha-Junput beach ridge chenier). On the other hand, it is also observed that the wide, shallow flats in between landward and seaward beach ridge cheniers were formed by the finer sediments (swale topography) deposited under the lagoonal setting behind the barrier bar systems and the supply of finer sediments by Hugli River mouth discharges into the Late Holocene tidal basin. According to the hierarchical cluster analysis, the Shannon diversity index, and Sorenson's coefficient, the species diversity is very high in the beach ridge fringed coastal plain, swale topography, beach ridge surface, and inner coastal plain with abandoned creeks and older natural levee bank surfaces. These diverse vegetation zones have high soil moisture contents and subsurface groundwater sources, and they are weakly susceptible to infrastructure development. However, dense orchards have grown up as a result of plantations established by people who have lived there since the beginning of time.

The study reveals that there is a significant connection between the flora and nature of coastal alluvium and influences from the sea and the tides. Backwaters, tidal channels, and tidal flood plains are known to deposit finer alluvial sediments during seasonal and periodic tidal floods in the coastal belt. Gradually, they become very important sediment sinks, temporary floodplain reservoirs, physical buffers, and important bio-shields against advancing seas in coastal lowlands. They also have the very substantial chemical and biological functions of retaining pollutants and filtering water and moisture, making them ideal breeding grounds for fish and resting places for other animals (particularly avifauna).

References

Bhandari, G. N., & Das, S. C. (1998). A study of beach erosion for appropriate protection of Digha Coast. In *Coastal Zone problems, proceedings of national workshop* (pp. 51–60). Jadavpur University.

Godron, M., & Forman, R. T. T. (1983). Landscape modification and changing ecological characteristics. In *Disturbance and ecosystems* (pp. 12–28). Springer.

Kamila, A., Bandyopadhyay, J., & Paul, A. K. (2020). An assessment of geomorphic evolution and some erosion affected areas of Digha-Sankarpur coastal tract, West Bengal, India. *Journal of Coastal Conservation, 24*(5), 1–14.

Kamila, A., Paul, A. K., & Bandyopadhyay, J. (2021a). Exploration of chronological development of coastal landscape: A review on geological and geomorphological history of Subarnarekha chenier delta region, West Bengal, India. *Regional Studies in Marine Science, 44*, 101726.
Kamila, A., Paul, A. K., & Bandyopadhyay, J. (2021b). Assessment of species diversity and topographic variability on beach dune complex of Digha Sankarpur coastal tract, Purba Medinipur district, West Bengal, India. *International Journal of Ecology and Environmental Sciences, 3*(1), 87–97.
Kamila, A., Bandyopadhyay, J., & Paul, A. K. (2021c). Assessment of landscape ecological connectivity for sustainable management of Digha–Shankarpur Coastal Tract, West Bengal, India. *Journal of the Indian Society of Remote Sensing, 49*(11), 2701–2719.
Maiti, S. (2013). Interpretation of coastal morphodynamics of Subarnarekha estuary using integrated cartographic and field techniques. *Current Science, 104*(12), 1709–1714.
Maiti, S., & Bhattacharya, A. K. (2009). Shoreline change analysis and its application to prediction: A remote sensing and statistics-based approach. *Marine Geology, 257*(1–4), 11–23.
Naveh, Z. (1995). Interactions of landscapes and cultures. *Landscape and Urban Planning, 32*(1), 43–54.
Neef, E. (1984). Applied landscape research. *Applied Geography and Development, 24*, 38–58.
Paul, A. K. (2002). *Coastal geomorphology and environment: Sundarban Coastal Plain, Kanthi Coastal Plain, Subarnarekha Delta Plain*. ACB publications.
Sharma, A., Tiwari, K. N., & Bhadoria, P. B. S. (2010). Vertical accuracy of digital elevation model from Shuttle Radar Topographic Mission–a case study. *Geocarto International, 25*(4), 257–267.
Troll, C. (1968). Landschaftsökologie. In *Pflanzensoziologie und Landschaftsökologie* (pp. 1–21). Springer.
Wiens, J. A. (1997). Metapopulation dynamics and landscape ecology. In *Metapopulation biology* (pp. 43–62). Academic.
Woodroffe, C. D. (2002). *Coasts: Form, process and evolution*. Cambridge University Press.
Zonneveld, I. S. (1989). The land unit—a fundamental concept in landscape ecology, and its applications. *Landscape Ecology, 3*(2), 67–86.

Chapter 30
Overwash Vulnerabilities in Chilika Lagoon With Barrier Spit Morphology of Odisha Coast, an Assessment With Remote Sensing Approach

Sk. Majharul Islam and Ashis Kumar Paul

30.1 Introduction

Spits are the favourable place for washovers. During the high energy phase sediment-laden sea water crosses the beach berm and deposits sediments on the beaches. Sometimes these waters overtop a dune and enter the lagoon directly. Washover deposits are causing loss of property and damage to infrastructure, damages protective barriers or dunes as they are lowered by washover, blocks drainage channels and burry salt marsh and other coastal vegetations. Coastal areas that experience regular washover must account for long-term management plans. The study of washover deposits also helps to assess energy, environments, and timings of paleo-storms. Chilika lagoon is situated in the East Coast of India about 50 km southwest from Bhubaneswar city in Odisha. It is the largest coastal lagoon in India with 1135 km^2 water cover area (Paul, 2002, 2014, 2022; Paul et al., 2014). The lagoon has significant ecological, economic, as well as social importance in the state. Different controlling agents like riverine input, sea input through tidal inlet, wave, wind, storm surge, etc., control the lagoon system.

Washovers are easily identifiable features in high-resolution (preferably <1 m) aerial photographs or satellite images. In this study, the washovers are identified and measured from satellite images as well as in ground. Spits are zoned according to their morphology and environment. Zone-wise washover vulnerability has been calculated following the methodology of Garcia et al. (2010). However, the earlier researchers have developed an integrated methodology for evaluation of overwash vulnerability and applied to the barrier regional scale for the assessment of

S. M. Islam (✉)
Bangabasi College, Rajkumar Chakraborty Sarani, Kolkata, West Bengal, India
e-mail: skmajharulislam@gmail.com

A. K. Paul
Department of Geography, Vidyasagar University, Midnapore, West Bengal, India

A. K. Paul, A. Paul (eds.), *Crisis on the Coast and Hinterland*,
https://doi.org/10.1007/978-3-031-42231-7_30

breaching hazard and extension in barrier island systems in the world wise regions (Garcia et al., 2010; Rodrigues et al., 2012; Bergillos et al., 2016; Plomaritis et al., 2018). The majority of the research on washover is confined in Portugal, Spain, or Netherlands. There is no such study on washover vulnerability at Chilika lagoon spit. This study will produce washover vulnerability hazard zones of Chilika lagoon spits.

30.2 About the Study Area

The Chilika lagoon is situated in the western part of the Mahanadi Delta in Odisha coast along the Bay of Bengal shoreline. Geomorphologically, this area is under a typical coastal lagoon system combined with spits, inlets, and protected waterbody. The spit plays a vital role to maintain lagoon water quality and ecology, which retain the livelihood of the people living in lagoon islands and lagoon fringe societies. There is some hilly area on the northwestern part of the lagoon and the elevation of these hills ranges from 800 to 946 m. As the low elevated part of Chilika spits is lying within 1–2 m in height above mean sea level, thus it is prone to washover impacts. Some part of the spit is flat, but some parts of the spit are covered with vegetated dune elevated up to 15 m in height (Fig. 30.1a) in the region. The depth of the spit back lagoon is also ranging from 1.70 to 3.80 m at present.

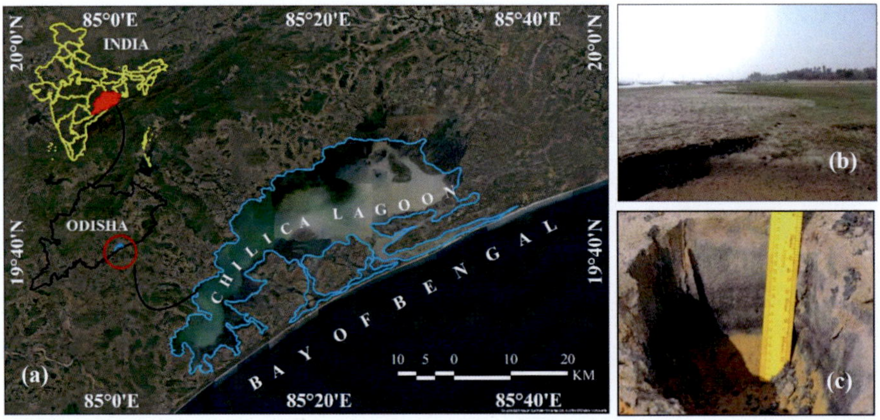

Fig. 30.1 (**a**) The study area of Chilika lagoon fringe spits; (**b**) Washover deposits at Mirzapur, Odisha; and (**c**) Placer deposits at the base of washover fan at Mirzapur, Odisha, after cyclone Phailin

30.3 Materials and Method

30.3.1 *Washover Vulnerability Index*

There are three types of indices used in this study to detect washover vulnerabilities at Chilika lagoon spits such as Overwashed Shoreline Ratio (OSR), Maximum Overwash Intrusion Recurrence (MOIR), and Complete Barrier Overwash (CBO). The methods of the above indices were suggested by Garcia et al. (2010) in their research work. In this case, it is possible to find out how vulnerable any part of a barrier bar is by only three simple parameters such as opening of the washover fan, length of the washover intrusion, and length and width of the barrier bar. The Chilika lagoon spit has been divided into six distinct zones according to the morphological and environmental differences (Paul et al., 2014). These three indices have been applied to each of the six spit zones of the lagoon.

30.3.2 *Overwashed Shoreline Ratio (OSR)*

The percentage of a spit affected by overwash is known from the Overwashed Shoreline Ratio (OSR). There are two parameters required to calculate it, the length of the spit and the sum of the openings of the washover. Among the washover data of the study period for any one spit, the highest value is expressed by OSRm. The OSRm is used as a parameter of the washover vulnerability index (Eq. 30.1):

$$\text{OSR} = \left(\varsigma \text{washover opening} / \text{shoreline extent}\right) \times 100 \tag{30.1}$$

30.3.3 *Maximum Overwash Intrusion Recurrence (MOIR)*

MOIR indicates the maximum intrusion history of overwash and represents the recurrence of maximum overwash intrusion. Maximum washover intrusion length at 5, 10, 15 years recurrence interval can be calculated from MOIR. In this study, MOI 5 (the maximum overwash intrusion length with a 5-year return period) has been used as a parameter of the washover vulnerability index (Eq. 30.2):

$$\begin{aligned}\text{MOIR} = N + 1/\varsigma\ &\text{number of active washovers} \\ &\text{with maximum intrusion greater than } X\end{aligned} \tag{30.2}$$

where N = number of years of the analysed period; X = the inferior limit of each overwash intrusion class; Return period (T) or recurrence interval corresponds to the time interval between occurrences of events with magnitude equal or greater than a defined value; and $T(x)$ = period of observation/number of events with magnitude equal or greater than x.

30.3.4 Complete Barrier Overwash (CBO)

Complete barrier overwash refers to the amount of overwash entering the lake over the spit. To figure it out, the length of the overwash intrusion and the width of the spit need to be known. It is calculated by dividing the length of the overwash intrusion by the width of the barrier. Main washover vulnerability index uses CBO 2/3 criteria. It means, what percentage washover fan has exceeded 2/3 width of the spit (Eq. 30.3):

$$\text{CBO} = \text{overwash intrusion} / \text{barrier width} \tag{30.3}$$

30.4 Results and Discussion

30.4.1 Dynamic Variation in Wave and Tide

According to the study of Chandramohan and Nayak (1994), the significant wave height varies between 0.5 and 2.2 m during June to September, 0.2–2.4 m during October to January, and 0.2–2.2 m during February to May. During the cyclone Phailin, the maximum significant wave height of 7.3 m and maximum wave height of 13.5 m was measured at 50 m water depth when the cyclone is only 70 km away from Gopalpur (Amrutha et al., 2014). According to the tide gauge reading of Gopalpur port, this area is under semi-diurnal micro-tidal (tidal range is less than 2 m) environment. Micro-tidal zone is favourable for barrier bar formation and also lagoon formation. Tides are semi-diurnal with an average tidal range between 2.39 and 0.85 m in height during the occurrence of neap tide and spring tides, respectively (Mohanty et al., 2012). According to Gopalpur Ports Limited, the High Astronomical Tide (HAT) level reaches 2.2 m above Mean Sea Level (MSL) in a year.

30.4.2 Longshore Drifts

This area is prone to significant wave breakers during the southwest monsoon when longshore drifts occur parallel to the shoreline. The longshore current of this area changes direction seasonally. During March to October the longshore current flows toward the northeast, but during November to February the longshore current flows toward the southwest direction. However, the net longshore transport in this region is guided by northerly direction (Mishra et al., 2011; Mohanty et al., 2012). The northward longshore transport during the southwest monsoon is the main cause of the northward transport of littoral drift materials. During the period of monsoon month, the coastal current measurement at Gopalpur coast in 2008 showed the fastest flow of current recorded as 74.5 cm/s on 3 August (Mishra et al., 2011). Sediments washed over from the sea and deposited in the form of fan or lobe in wide areas of the back spit lagoon. Sediments are varying in different types of grain sizes ranging from fine sediments to course sediments over washover deposits. Normally fine riverine sediments and coarse marine sediments deposit within the washover fan, but sometimes shells, woods, pebbles, and boulders can be found in unsorted manner within the sediment deposits (Fig. 30.1a, b).

30.4.3 Spit Dynamism

There are two major groups of spits in the Chilika lagoon. The older spits (Puri to Satapada section) are originated from the Mahanadi Delta and extended towards the southwest shoreline along the Bay of Bengal. The new set of spits has been developed from southwest of the lagoon and extended towards northeast direction along the shoreline. The old spit is broader than new spits. The length of the old spit is about 40 km, whereas the new spit is 70 km long and the width of the new spit is ranging from 90 to 1230 m, whereas the width of old spit ranges from 1500 to 3500 m. Northeastward longshore drift nourishes the new spits of Chilika lagoon. There are some anthropogenic factors which reshaped the new spits in the recent past. From the British period new spits had been cut several times to increase the entry of sea water to the lagoon. Then the artificial opening has been shifting northeastward during the course of time, particularly by the cyclones. This northeastward shifting rate of the last artificial opening during September 2000 to September 2022 was about 300 m/year. Though this east northeastward sifting mainly occurs during the cyclones. A certain cyclone shifts the Spit inlet about a kilometre during a single storm. The Chilika lagoon spit has been categorized for their morphological and local difference in six separate zones such as Palur hills to Naba beach (New spit), Naba beach to New mouth inlet of the year 2000 (New spit), New mouth inlet of the year 2000 to western side of present inlet in the year 2020 (New spit), Eastern side of present inlet in the year 2020 to the end point of outer channel (New spit), Satapada to Sipakuda (Old spit), and Mirzapur to the end point of outer channel (Old spit) (Table 30.1).

Table 30.1 Morphological features of the Chilika lagoon spits

Zone	Location	Length, width, and average elevation (m)	Significant geomorphic characteristics
1	Palur hills to Naba beach (New spit)	19,621, 1200, and 8	Wide spit, frontal dune condition with fairly dense casuarina plantation
2	Naba beach to new mouth inlet of the year 2000 (New spit)	27,427, 300, and 8	Narrow spit, presence of elevated frontal dune with casuarina plantation
3	New mouth inlet of the year 2000 to western side of present inlet in the year 2020 (New spit)	6809, 200, and 2	Sandy flat surface with less than 1 m elevation
4	Eastern side of present inlet in the year 2020 to the end point of outer channel (New spit)	10,602, 150, and 2	Sandy flat surface with less than 1 m elevation
5	Satapada to Sipakuda (Old spit)	9079, 2500, and 5	Sheltered area
6	Mirzapur to the end point of outer channel (Old spit)	19,721, 3000, and 8	Presence of old dunes with vegetation cover

Table 30.2 Recent cyclones in the northern Bay of Bengal, which affected Chilika lagoon

Name of the storm	Date of landfall	Landfall location	Maximum wind speed (km/h)	Storm surge at Gopalpur coast (m)
Odisha Super Cyclone	1999/10/29	Between Paradip and Puri, Odisha	260	6
Phailin	2013/10/12	Gopalpur, Odisha	215	6
Hudhud	2014/10/12	Visakhapatnam, Andhra Pradesh	185	2
Titli	2018/10/10	Palasa, Andhra Pradesh	150	1
Fani	2019/05/03	Puri, Odisha	215	2
Bulbul	2019/11/09	Sagar Island, West Bengal	140	2
Amphan	2020/05/20	Bakkhali, West Bengal	240	5
Yaas	2021/05/26	Dhamra, Odisha	140	4

30.4.4 Cyclone Landfalls

Cyclone-induced storm surge is the main agent of the washover process in this area. There are eight cyclones which affected Chilika lagoon since the last super cyclone in Odisha in 1999 (Table 30.2). During the Orissa super cyclone, the maximum significant wave height measured was 8.44 m on 28 October 1999 (Rajesh et al., 2005). However, maximum significant wave height and a maximum wave height at Gopalpur were 7.3 m and 13.5 m, respectively, during cyclone Phailin in 2013 (Amrutha et al., 2014). Maximum significant wave height and a maximum wave height were 5.8 m and 9.2 m, respectively, on 12 October 2014 during cyclone Hudhud at Gopalpur which was standing about 20 km southwest from the lagoon (Table 30.2).

30.4.5 Formation and Structure of Washover Fan

Washover deposits have been used for paleo-storm analysis. Deposition of coarse material cover the area in the form of a washover fan during the periods of repeated storm surge activities. Many paleo super cyclones have been identified using sedimentological analysis of ancient washover deposits. Washover deposits help to increase the lagoon filling process and it also helps to form islands in the lagoon. The washover fan of the Chilika lagoon spit is mainly made of fine sands. There are some coarse sand and shell fragments and biotic materials in the washover deposits. Some washover deposits have a layer of black placer deposits. The width of the washover fans are varying from 150 to 300 m, the intrusion of the washover fans are ranging between 100 and 200 m, and the depth of the washover was about 0.2–0.5 m. The washover fans are normally oriented southwest to northeast direction (Fig. 30.2a, b).

30.4.6 Distribution of Washover Fan

Washover deposits are distributed over the spits of Chilika lagoon. Maximum washover deposits are small in size (i.e., < 500 m^2) and observed between beach berm and fore dune area. The frequency of the washover fan can be observed up to 20/km in the middle part of the new spit of Chilika lagoon. After cyclone Phailin, 624 washover points have been identified of which 140 washover lobes have an area of over 500 m^2 which are included in this study (Fig. 30.2a, b). Average area of these washover lobes was 13,078 m^2. The frequency of the washover fan was highest in the middle part of the new spit (Fig. 30.2b).

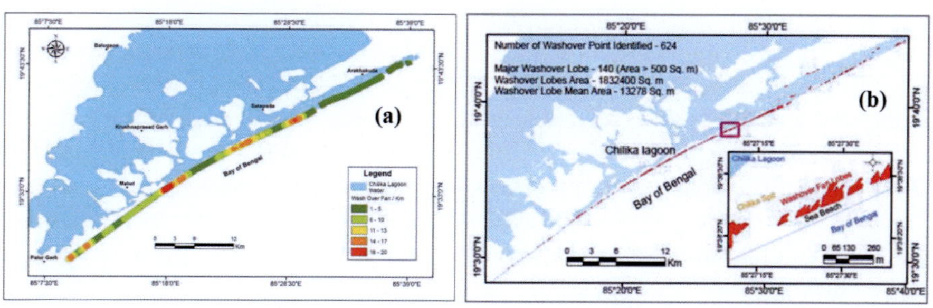

Fig. 30.2 (**a**) Frequency of washover fan in the study area after cyclone Phailin; and (**b**) Distribution of washover fan over the study area after the cyclone Phailin

30.4.7 Integration of Indices

The values obtained from OSRm, MOI 5, and CBO 2/3 indices are divided into four categories. The highest values obtained from individual indices are given four and the lowest values are given one. These values are then added back to obtain the vulnerability value for each spit.

The value of the vulnerability index will be ranging between 1 and 12. For example, an overwash vulnerability index value of 1–3 of a spit indicates minimum vulnerability, a value of 4–6 indicates moderate vulnerability, 7–9 indicates high vulnerability, and 10–12 indicates extreme vulnerability (Table 30.3 and 30.4).

According to the index value, both sides of the Chilika inlet of new spits (Zone 3 and 4) are extremely vulnerable zone to washover (Table 30.4 and Fig. 30.3) activities and Naba beach to new mouth inlet of the year 2000 zone of new spit (Zone 2) is highly vulnerable to washover. Palur hills to Naba beach zone of new spit (Zone 1) is covered with vegetated dunes, thus according to the study the vulnerability of this zone is low. Zones 5 and 6 in the old spit are relatively less vulnerable to washover according to the present study. Zone 5 is sheltered by Zone 2 and there are no significant washovers identified during the study period. Presently, the forest department of Odisha is planting trees, especially casuarina in the newly formed spits to arrest sand particles which increase the elevation of the frontal part of the spit and which also dissipate energy and stabilize the spits temporarily in the region.

30.5 Conclusion

The overwash vulnerability assessment is a major environmental issue along the barrier beach shoreline of Chilika Lagoon. The younger and older spits are grouped into six sections of shoreline for the estimation of overwash intensity for the region. In the method of shoreline vulnerability assessment, the overwash vulnerability has three components: (i) Overwashed Shoreline Ratio (OSR), (ii) Maximum Overwash Intrusion Recurrence (MOIR), and (iii) Complete Barrier Overwash (CBO). The younger sand spits have been extended from the Rushikulya river mouth to Brahmagiri point on the Puri coast from two different directions along the Bay of Bengal shoreline, and they are acting as a barrier against the open marine environment and lagoonal settings of the protected lake. The new mouth spit of Chilika has two inlets, that is, an active inlet and a dying inlet. During the stronger cyclonic storm surges, overwash sand fan deposits occur and extend across the spit and barrier bar into the lagoonal water body by transporting sediments from the narrow beaches and spits. The present assessment of the overwash vulnerability shows that the moderate and extreme levels of vulnerability ranking are reflected towards the younger spits from Naba Beach to the new mouth and old mouth points. However, the ranking of low vulnerability represents the older spits in the northeastern and southwestern corners of Chilika Lagoon.

Table 30.3 Washover vulnerability index

Zone	Location	Length, width, and average elevation (m)	OSR (max)	OSR Score	CBO (average)	CBO (2/3%)	CBO Score	MOIR (5 Years)	Moir Score	Vulnerability
1	Palur hills to Naba beach (New spit)	19,621, 1200, and 8	7.94	1	0.07	0	1	134 m	1	3
2	Naba beach to new mouth inlet of the year 2000 (New spit)	27,427, 300, and 8	56.32	3	0.39	6	1	255 m	2	6
3	New mouth inlet of the year 2000 to western side of present inlet in the year 2020 (New spit)	6809, 200, and 2	43.91	2	0.89	92	4	545 m	4	10
4	Eastern side of present inlet in the year 2020 to the end point of outer channel (New spit)	10,602, 150, and 2	33.36	2	0.92	83	4	545 m	4	10
5	Satapada to Sipakuda (Old spit)	9079, 2500, and 5	0	1	0	0	1	0	1	3
6	Mirzapur to the end point of outer channel (Old spit)	19,721, 3000, and 8	8.49	1	0.04	0	1	85 m	1	3

Table 30.4 Final scores of the washover vulnerability classification after integration of the OSRM, MOI 5, and CBO 2/3

Zone of the Spit	Vulnerability Classes			
	Low = 1-3	Medium = 4-6	High = 7-9	Extreme = 10-12
Zone-1				
Zone-2				
Zone-3				
Zone-4				
Zone-5				
Zone-6				

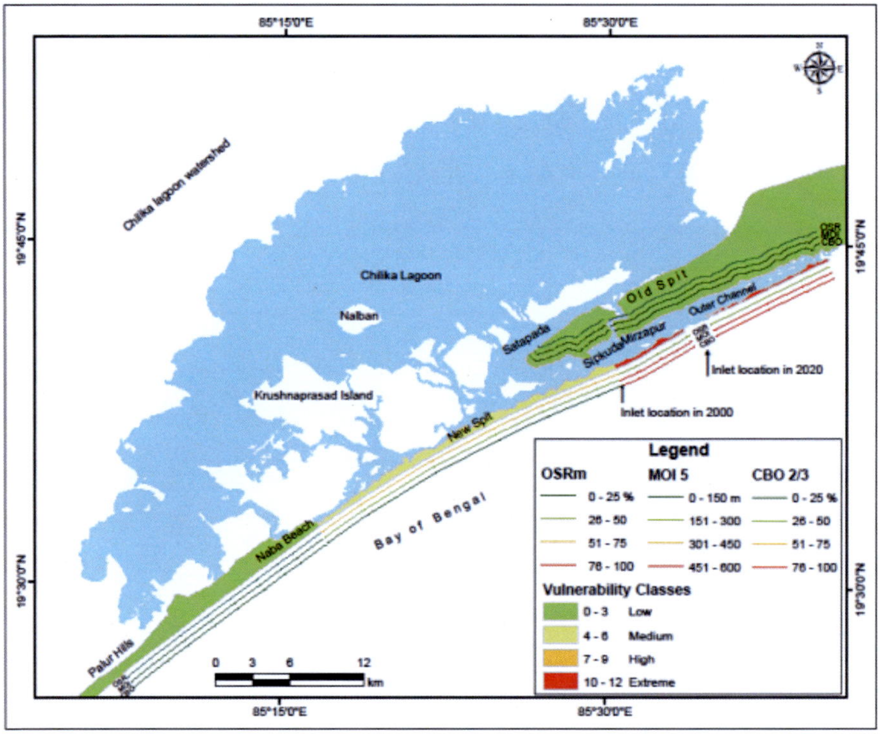

Fig. 30.3 Washover vulnerability map of Chilika spit

Significant transport of sediments from the sand dunes, back spit surfaces, and sea beaches took place during the occurrences of storm surges, with features of overwash fan lob deposits into the inner part of the lagoon. The study also represented the two categories of overwash fans found in the Chilika lagoon, such as (i) older overwash fans and (ii) younger overwash fans. The increasing rate of overwash vulnerability indicates the impact of sea level rise and the intensity of cyclone-induced storm surges in a regional setting.

References

Amrutha, M. M., Sanil Kumar, V., Anoop, T. R., Nair, B., Nherakkol, A., & Jeyakumar, C. (2014, September). Waves off Gopalpur, northern Bay of Bengal during cyclone Phailin. In *Annales geophysicae* (Vol. 32, No. 9, pp. 1073–1083). Copernicus GmbH.

Bergillos, R. J., Masselink, G., McCall, R. T., & Ortega-Sánchez, M. (2016). Modelling overwash vulnerability along mixed sand-gravel coasts with XBeach-G: Case study of Playa Granada, southern Spain. *Coastal Engineering Proceedings, 1*(35), 13.

Chandramohan, P., & Nayak, B. U. (1994). A study for the improvement of the Chilka lake tidal inlet, east coast of India. *Journal of Coastal Research*, Autumn, 1994, 10(4), 909–918.

Garcia, T., Ferreira, O., Matias, A., & Dias, J. A. (2010). Overwash vulnerability assessment based on long-term washover evolution. *Natural Hazards, 54*(2), 225–244.

Mishra, P., Patra, S. K., Murthy, R., Mohanty, P. K., & Panda, U. S. (2011). Interaction of monsoonal wave, current and tide near Gopalpur, east coast of India, and their impact on beach profile: A case study. *Natural Hazards, 59*(2), 1145–1159.

Mohanty, P. K., Patra, S. K., Bramha, S., Seth, B., Pradhan, U., Behera, B., ... & Panda, U. S. (2012). Impact of groins on beach morphology: A case study near Gopalpur Port, east coast of India. *Journal of Coastal Research*, 28(1), 132–142.

Paul, A. K. (2002). *Coastal geomorphology and environment* (pp. 1–342). ACB Publication.

Paul, A. K. (2014). Morphology of estuaries and tidal inlets: An emphasis on Hugli, Subarnarekha and Chilika systems along the Bay of Bengal shoreline. In *Proceedings of the national conference on Modern Trends in Coastal and Estuarine Studies*, Tilak Maharashtra Vidyapeeth, Pune, 6–7 Feb 2014.

Paul, A. K. (2022, August). Dynamic behaviour of the estuaries in response to the phenomenon of global warming in the coastal ecosystems of West Bengal and Odisha, India. In *Transforming coastal zone for sustainable food and income security: Proceedings of the international symposium of ISCAR on Coastal Agriculture, March 16–19, 2021* (pp. 907–931). Springer International Publishing.

Paul, A. K., Islam, S. M., & Jana, S. (2014). An assessment of physiographic habitats, geomorphology and evolution of Chilika Lagoon (Odisha, India) using geospatial technology. In C. Finkl & C. Makowski (Eds.), *Remote sensing and modeling. Coastal research library* (Vol. 9). Springer. https://doi.org/10.1007/978-3-319-06326-3_6

Plomaritis, T. A., Ferreira, Ó., & Costas, S. (2018). Regional assessment of storm related overwash and breaching hazards on coastal barriers. *Coastal Engineering, 134*, 124–133.

Rajesh, G., Joseph, K. J., Harikrishnan, M., & Premkumar, K. (2005). Observations on extreme meteorological and oceanographic parameters in Indian seas. *Current Science, 88*(8), 1279–1282.

Rodrigues, B. A., Matias, A., & Ferreira, Ó. (2012). Overwash hazard assessment. *Geologica Acta, 4*, 427–437. https://doi.org/10.1344/105.000001743; www.geologica-acta.com

Chapter 31
Changing Livelihood Security Index Along the Coastal Belt of Purba Medinipur District, an Assessment Using Spatial Information Systems

Jasmin Parvin and Ashis Kumar Paul

31.1 Introduction

Livelihood security refers to the accessibility and sustainability of living resources such as food, drinking water facilities, medical care, educational opportunities, and the housing condition of an individual or household (Serrat & Serrat, 2017; Patidar, 2019). The leading indicators described in livelihood security are economic security, health security, habitat security, educational security, and food security. Everyone is adapting to natural adversity to keep their livelihood alive. West Bengal's coast has a 220 km coastline, and seven million people live along the Bay of Bengal coast (ICZMP-WB, 2012). The coastal zones of Purba Medinipur, South 24 Parganas, and North 24 Parganas districts jointly hold 12% of total population of the state of West Bengal (Paul, 2002; Paul & Paul, 2022). Purba Medinipur district is a part of the lower Indo-Gangetic plain and eastern alluvial coastal plain. The Bay of Bengal lies toward south of Purba Medinipur district along a 65.5 km coastline. The lives and livelihoods of the coastal people are endangered by adversity of emerging marine hazards and obstruction from climatic stress. As the distance from the sea increases, there is a change in livelihood options as well as adaptive strategies to deal with a changing climate variability. Vulnerability reflects on the livelihood calendar of villagers and their socioeconomic conditions.

The livelihood of the people in the study area has been affected by climate variability and its associated hazards. Villagers have been trying to cope with or adapt to such variability and adversity. The villagers are constantly shifting their livelihood calendar with the seasonal shifting of natural calamities and trying to improve the resilience capacity or flexibility of the livelihood calendar to adjust to natural adversity through various adaptation methods. However, many are trying to change

J. Parvin (✉) · A. K. Paul
Department of Geography, Vidyasagar University, Midnapore, West Bengal, India
e-mail: jasuparvin@gmail.com

A. K. Paul, A. Paul (eds.), *Crisis on the Coast and Hinterland*,
https://doi.org/10.1007/978-3-031-42231-7_31

their traditional livelihoods to improve their standard of living. Their adaptive strategies to vulnerability are successful or not, reflecting only on their livelihood outcomes, such as food security, income security, health security, social security, and educational security. However, livelihood diversification may reduce the vulnerability of livelihoods.

31.2 Study Area

Purba Medinipur has twenty-five coastal administrative blocks, among them five blocks are touching the sea along the Bay of Bengal shoreline, such as Ramnagar-I, Ramnagar-II, Contai-I, Deshopran, and Khejuri-II (Paul, 2002). The study is conducted in three administrative blocks: Ramnagar-I, Ramnagar-II, and Contai-I. These coastal blocks of the Purba Medinipur district are more vulnerable than other blocks. This area is vulnerable as it is exposed to extreme climatic events in the previous decades as evidenced from the frequency and intensity of cyclone landfalls (Fig. 31.1).

The region is densely populated, and most of the population is dependent on the natural resources of the region. Natural disasters of any kind make it difficult to access natural resources, which endanger the livelihoods of the people in the region, which is already reflected in their socio-economic condition and changing

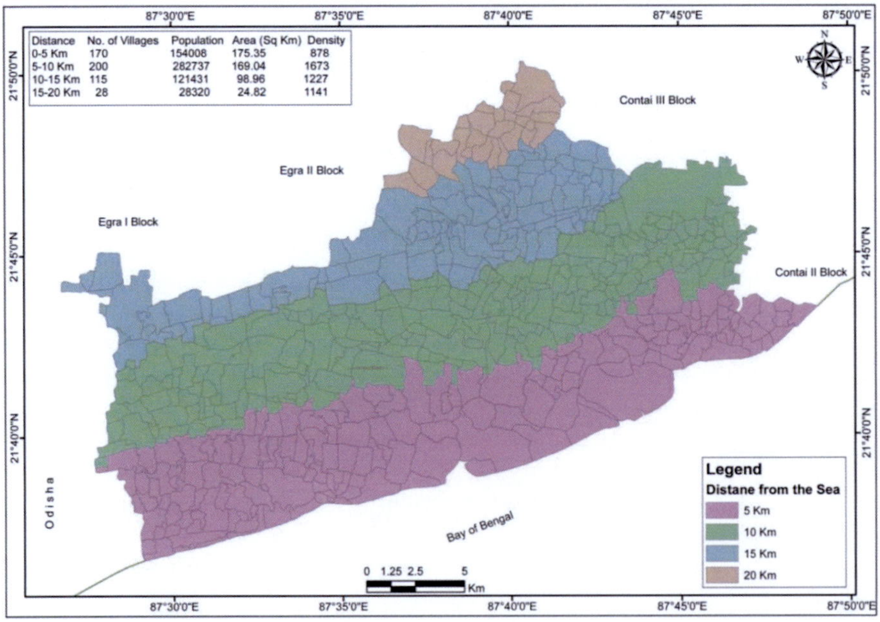

Fig. 31.1 Shore parallel coastal zones along the coastal belt of Purba Medinipur district (Ramnagar-I, II and Contai-I block) at a different distance from the shoreline

livelihood patterns. The study area has been divided into several distance zones of 5 km in width and lying parallel to the shoreline, that is, Zone I (0–5 km), Zone II (5–10 km), Zone III (10–15 km), and Zone IV (15–20 km), to understand the relationship between distance and different livelihood indicators (Fig. 31.1). As livelihood diversity varies with the distance from the coast (Fig. 31.1; Table 31.1), the impact of climatic hazards on various livelihoods also varies, which shows the difference in their standard of living.

31.3 Assessment of the Livelihood Security Index

The livelihood of the people in the coastal areas of Purba Medinipur is generally associated with fishing, farming, tourism, and transportation. Natural calamities of the past have produced damages to the livelihoods such as on agriculture and fishing activities in the coastal zones. Cyclones have caused extensive damage to the tourism industry in the previous decades. The main goal of this work is to understand the security status of the livelihoods of the people. Livelihood security refers to the accessibility and sustainability of the resources that every human being needs to survive, such as food, drinking water, health, access to education, housing, etc. (Serrat, 2017). Livelihood security refers to the adverse conditions faced for their livelihood and how each family fulfils their basic needs by overcoming those adversities. To learn about the status of livelihood security in the coastal villages of Purba Medinipur, five hundred households were surveyed in fifty villages in Ramnagar-I, Ramnagar-II, and Contai-I blocks. The Livelihood Security Index (LSI) has been developed using the five-point scale method to understand the livelihood security status of the people. The scores of economic, food, habitat, and education are indicators of each household; the index is formed by aggregating those scores (CARE, 2004; Mishra & Debata, 2021). Each indicator has been standardized using the method taken from the Human Development Index to calculate the Life Expectancy Index, which has been followed by the work of Rahaman and Aktar (2010). The Household Livelihood Security Index (HLSI) is formed by aggregating these standardized values using the following Eqs. 31.1, 31.2:

$$\text{HLSI} = \sum j = 1 Z_{ind} j \quad (31.1)$$

$$Z_{ind} j = \text{Indicator value} - \text{minimum value} / \text{maximum value} - \text{minimum value} \quad (31.2)$$

$$LSI = \sum i6 = 1\, W_i.\text{HLSI} / \sum i6 = 1 \;\; W_i \cdots \quad (31.3)$$

here, HLSI refers to the Household Livelihood Security Index, J refers to the number of indicators used in the index; LSI refers to the Livelihood Security Index, and W_i refers to the Weight determined by the number of indicators used in each index.

31.4 Results and Discussion

31.4.1 Economic Security

Economic security refers to a stable income source that enables a household to meet its basic needs (ICRC, 2015). Every family has different expectations because their needs and wants are different. Five sub-indicators have been taken into account to understand household economic security, such as monthly household income, assets, active population, current loan, and value of the land used by them. Household wise monthly income depends on the type of their occupations. Fishermen in coastal areas have a monthly income of around 9000 rupees. In the middle portion of the study area, along and within beach ridges, people are primarily engaged in farming or wage labor. Their monthly income ranges from ten thousand to fifteen thousand rupees. While some cultivators share the land for shrimp farms, their monthly income is forty thousand to fifty thousand. However, economic inequality is very high in this region. The economic situation of fishermen in Zone I is poor because of low income and having large family members, a lack of education, and being more vulnerable to climatic hazards. Family size affects the active population. Most of the households living along the shoreline have a family size of ten to twelve persons, and the number of earning members is one or two. In comparison, most of the surveyed houses in the distant villages from the shore have four to five family members. If the number of family members is excessively higher than the number of earning members, then there is pressure on the earning members, and per capita income will be low. A loan acts as a positive indicator in the economic security index, meaning that the loan helps to manage their livelihood in terms of security. The loan helps maintain the living standards of the people if their livelihood is affected by any adverse circumstances. Most of the villagers in Zone II, Zone III, and Zone IV get benefits from loans from cooperative societies as they have cultivable land, which helps them improve their standard of living.

Land value also plays a vital role in securing their livelihood. Land value is also high in the shore fringe areas because investors want to build hotels in this area for tourism activities. There is a lot of diversity of livelihoods in the coastal areas, but they are more susceptible to disruption due to natural calamities. People in zone I and zone II are more economically vulnerable than those in zones III and IV (Fig 31.2a). Seasonality influences livelihood strategy, resulting in income fluctuation that directly affects livelihood security (Devereux et al., 2012). The income of the people varies in different seasons. The villagers living away from the coast are ahead in terms of economic security, as most of them have a stable source of income. Besides rice-paddy farming and fish farming, many people have been involved in other activities. Many are cultivating agricultural land and working as hotel workers, shopkeepers, businesses, wage laborers, and other activities, fishermen are working as wage laborers when the lean period starts in their occupation. On the other hand, those involved in tourism do not want to be engaged in other work in the off-season because they know the demand for tourism will return after a few months.

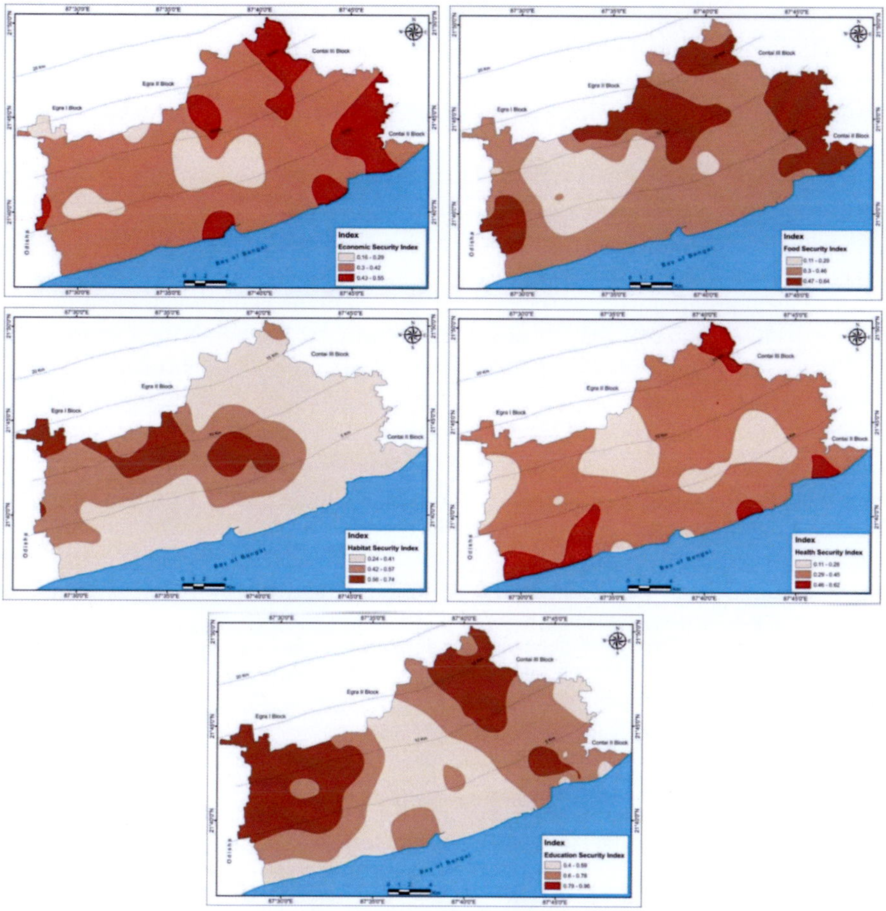

Fig. 31.2 (**a**) Economic security status along the coastal belt of Purba Medinipur District (Ramnagar-I, II and Contai-I block) with (**b**) Food security status, (**c**) Habitat security status, (**d**) Heath security status, and (**e**) Education security status

Diversification of livelihoods increases household income and increases economic security. Livelihood diversification is more prevalent along the shore fringe in tourism-dominated areas than in distant villages.

31.4.2 Food Security

According to FAO (2003), food security refers to the situation where all people have economically and physically sufficient access to the right amount of safe, nutritious food to meet their dietary needs, leading an active and healthy life. The quality and nutritional value of food depend on several factors, such as the nature of the food in their diet and its quantity, and whether the diet contains an adequate amount of carbohydrates, proteins, fats, vitamins, etc. Another important factor is how much food is stored in the house, which also affects food security. Food security has three sub-indicators: food frequency, dietary diversity, and food storage.

People in coastal areas have different food diets. Food security depends on the nutritional value and quality of the foods included in their food list. The presence of carbohydrates, protein, and fat in their diet is high, but there is a lack of vitamins and minerals on the diet chart. On the other hand, the villagers in Zone III and Zone IV try to include carbohydrates, protein, fat, and vitamins in their diet. Food stocking is a positive indicator that ensures food security by providing food in adverse conditions. Most farmers in the village distant from the coast have farmland, so they have grains in their homes all year round; they survive eating them during the disaster. On the other hand, villagers living near the coast have no such opportunities and find themselves in trouble during disasters. During the cyclone, they mostly rely on government relief care units. In coastal villages, the quality and nutritional value of the food in people's diets are low, particularly in the fishermen's community. The impact of a cyclone or other natural disaster on agriculture is greater in Zone II than in Zone III and Zone IV. As a result, middle-class families are in trouble, which disrupts food security (Fig. 31.2b).

31.4.3 Habitat Security

Accommodation is essential for survival. Habitat security refers to all family members living in a healthy environment with sufficient sunlight and an airy place that provides a safe and secure life (Mishra & Debata, 2021). Habitat security has three sub-indicators: house structure, sanitary facilities, and drinking water facilities. Most of the houses in the coastal villages are made of the materials of mud. The walls are made of mud or bricks, and the roof is made of tarpaulin, asbestos, tin, straw, etc. The number of rooms inside the house is very small; about 10–12 people live in two or three rooms. In many cases, cattle are kept inside the house because they cannot be accommodated elsewhere. Most houses along the shore fringe are damaged regularly by storm surges or broken down by stormy winds twice or three times a year. The sanitation facility is also not good; the toilet is often left uncovered. *Pucca* houses can be seen in areas far away from the coast. The overall home environment in Zone III and Zone IV is far better than in Zone I and Zone II (Fig. 31.2c). Drinking water accessibility with distance from the sea shows that drinking water access is better in villages near the sea. During the summer season, drinking water became reddish because of excessive iron content. Some villagers in the shore fringe area face drinking water salinity problems due to chloride mixing.

31.4.4 Health Security

Coastal areas are densely populated, and the people have built their livelihoods based on natural resources and have to work hard to earn a living. Fishermen need a lot of physical labor to go fishing in trawlers, so they suffer from sickness for about 10–12 days a month and cannot go to work. Diseases are more prevalent in the fishing community because its members are not health conscious, spend more time in unsafe and unhygienic environments, drink alcohol, and have poor lifestyles. The health security index has two sub-parameters: one is the accessibility of hospitals, and the other is being unable to work. Health problems are more prevalent among the villagers engaged in labor-intensive work, such as day laborers, fishermen, wage laborers, etc. (Fig. 31.2d). During and after cyclones, their physical and mental health is affected, and their mental health is also disrupted during evacuation processes.

31.4.5 Education Security Index

Educational qualifications determine the nature of a livelihood. This indicator affects all other indicators of the livelihood security index. This study used two sub-indicators to calculate the educational security index, that is, male literacy rate and female literacy rate. Educational qualification in the study area reveals that the educational status differs in Zone-I from Zone-IV. The literacy rate is low among the villagers in Zone I and Zone II (Fig. 31.2e). The present generation is more interested in vocational education, such as hotel management, vocational training, hospitality management, etc., than formal education. Although women are skilled in technical knowledge, most of them are uneducated. A study shows that male and female literacy is much higher in villages located away from the coast than in villages near the sea (Table 31.1).

31.4.6 Determination of Overall Livelihood Security Index (LSI)

The Livelihood Security Index (LSI) is a composite index (Singh & Hiremath, 2010). Each indicator is interrelated, and the rank of any indicator affects the value of the LSI as with education, economic conditions affect people's income, health, food, shelter, and standard of living. The overall livelihood security index helps to identify which places and people are more vulnerable and whether their livelihood strategies are successful or not. Villages with low livelihood diversity are more susceptible to hazards.

Table 31.1 Village-wise ranking of economic security index, habitat security index, education security index, health security index, food security index along the coastal belt of Purba Medinipur district (Ramnagar-I, II, and Contai-I block)

Village Name	Distance from the coast in km	Economic Security Index	Rank	Habitat Security Index	Rank	Education Security Index	Rank	Health Security Index	Rank	Food Security Index	Rank	LSI	Rank
Tedubi	20.7	0.444	11	0.451	17	0.819	9	0.570	2	0.457	17	0.548	1
Rani Basan	13.1	0.473	7	0.383	23	0.917	4	0.440	11	0.500	8	0.543	2
Natdighi	15.3	0.463	8	0.374	27	0.875	6	0.465	8	0.475	11	0.530	3
Baguran Jalpai	1.9	0.491	4	0.351	31	0.772	13	0.500	6	0.502	7	0.523	4
Kiyakuli	6.6	0.504	1	0.589	7	0.900	5	0.210	45	0.409	29	0.522	5
Mukundapur	6.1	0.308	38	0.475	14	0.900	5	0.315	30	0.608	2	0.521	6
Purusottompur	2.3	0.385	21	0.333	33	0.917	4	0.360	20	0.595	4	0.518	7
Badhia	12.3	0.294	42	0.604	6	0.900	5	0.317	29	0.420	28	0.507	8
Kulberia	17.3	0.423	12	0.317	37	0.900	5	0.365	19	0.528	5	0.507	9
Karanji	6.3	0.495	3	0.352	30	0.629	24	0.418	12	0.604	3	0.499	10
Ahammadpur	8.4	0.256	46	0.710	2	0.576	29	0.313	31	0.640	1	0.499	11
Shikharbar	9.6	0.395	19	0.618	5	0.955	2	0.252	41	0.254	44	0.495	12
Majna	13.8	0.406	16	0.377	25	0.861	7	0.450	9	0.356	35	0.490	13
Tajpur	12.9	0.500	2	0.376	26	0.750	14	0.325	26	0.385	33	0.467	14
Palta beria	18.9	0.451	9	0.284	43	0.667	19	0.565	4	0.428	24	0.479	15
Depar Sasanbar	10.5	0.267	45	0.464	15	0.778	11	0.412	13	0.471	12	0.478	16
Panipiya	3.1	0.484	6	0.390	21	0.778	12	0.295	33	0.428	25	0.475	17
Uttar Shimulia	5.8	0.318	37	0.561	9	0.921	3	0.250	42	0.325	39	0.474	18
Dulalpur	13.1	0.343	32	0.382	24	0.742	15	0.330	25	0.489	10	0.457	19
Paschim Chachuraput	4.6	0.486	5	0.358	29	0.583	27	0.404	15	0.447	20	0.456	20
Jaldah	1.7	0.398	18	0.400	20	0.655	21	0.322	28	0.407	30	0.436	21
Kadua	10.2	0.299	40	0.733	1	0.588	25	0.107	47	0.465	14	0.439	22

Duttapur	0.7	0.376	26	0.274	47	0.520	33	0.617	1	0.393	32	0.436	23
Dahadaya	2.9	0.375	27	0.422	18	0.675	18	0.350	22	0.357	34	0.436	24
Ghatua	8.9	0.384	22	0.350	32	0.780	10	0.240	43	0.425	27	0.436	25
Mankaraiput	2.2	0.412	15	0.319	36	0.542	31	0.440	11	0.460	15	0.434	26
Srirampur	10.9	0.360	31	0.385	22	0.632	23	0.325	26	0.466	13	0.434	27
Gadadharpur	0.4	0.402	17	0.253	48	0.515	35	0.567	3	0.430	23	0.433	28
Santeswarpur	10.2	0.369	29	0.507	13	0.822	8	0.178	46	0.283	41	0.432	29
Dakhin purusottompur	0.5	0.445	10	0.275	45	0.580	28	0.360	20	0.426	26	0.417	30
Islampur	6.6	0.271	44	0.627	4	0.515	34	0.292	34	0.433	22	0.428	31
Ragu sardarbar jalpai	2.7	0.380	25	0.325	35	0.585	26	0.370	18	0.438	21	0.419	32
Purba Mukundapur	1.5	0.370	28	0.285	41	0.585	26	0.493	7	0.355	36	0.418	33
Saiyadpur	5.0	0.383	23	0.374	27	0.660	20	0.442	10	0.240	46	0.420	34
Biramput	2.9	0.295	41	0.310	38	0.676	17	0.410	14	0.395	31	0.417	35
Shyamraibar Jalpai	2.0	0.305	39	0.332	34	0.562	30	0.345	23	0.528	6	0.414	36
Ramchandra nagar	6.7	0.331	33	0.667	3	0.484	39	0.300	32	0.282	42	0.413	37
Dadanpatrabar	1.0	0.366	30	0.274	46	0.400	44	0.535	5	0.460	16	0.407	38
Bhuianjibar	9.5	0.413	14	0.556	11	0.465	41	0.275	37	0.325	38	0.407	39
Danda belbani	11.2	0.422	13	0.402	19	0.439	42	0.265	38	0.498	9	0.405	40
Haldia	6.9	0.324	35	0.454	16	0.730	16	0.380	17	0.123	48	0.402	41

(continued)

Table 31.1 (continued)

Village Name	Distance from the coast in km	Economic Security Index	Rank	Habitat Security Index	Rank	Education Security Index	Rank	Health Security Index	Rank	Food Security Index	Rank	LSI	Rank
Hirapur	4.7	0.272	43	0.285	40	0.962	1	0.353	21	0.128	47	0.399	42
Dakshin Shitala	6.3	0.214	49	0.558	10	0.644	22	0.278	36	0.283	40	0.396	43
Uttar Shitala	7.2	0.231	48	0.576	8	0.503	36	0.398	16	0.265	43	0.395	44
Mandarmani	0.5	0.328	34	0.291	39	0.520	33	0.255	40	0.450	18	0.369	45
Silampur	0.5	0.382	24	0.281	44	0.430	43	0.335	24	0.450	18	0.376	46
Sonamuhi	0.7	0.320	36	0.284	42	0.490	38	0.325	27	0.450	19	0.374	47
Lalpur	10.0	0.170	50	0.552	12	0.541	32	0.233	44	0.250	45	0.349	48
Jamura Shyampur	0.2	0.385	20	0.274	47	0.480	40	0.256	39	0.350	37	0.349	49
Karonji	5.9	0.243	47	0.363	28	0.493	37	0.288	35	0.282	42	0.334	50

According to the rank of LSI, the livelihood security of the villages in the south-central part of the study area like Jamra Shyampur, Paschim Gadadharpur, Purba Mukundpur, Mandarmani, and Duttapur is low because of the high climatic exposure and low economic viability. The diversity of livelihoods in these villages is also low, and villages are affected mainly by inundation during cyclones or spring tides, saline water encroachment, and coastal erosion. Due to their economic backwardness, they spend less on food, education, health, and habitat, so their standard of living is very low. On the other hand, LSI values are higher in regions where livelihoods are diverse and the hazard effect is less. The level of livelihood security is much higher in villages like Tedubi, Ranibasan, Kulberia, etc., on the northeastern side of the study area. Tedubi has good education and health in the economic security index and ranks first based on the livelihood security index among the studied villages. Karonji has low economic, food, and health security, making it the last based on the livelihood security index among the studied villages (Table 31.1).

In study areas, LSI values are high in only thirteen villages, whereas thirty-five villages have low to moderate livelihood security. Most of the villages located in shore fringe areas are affected by climatic hazards, but they are trying to improve their livelihoods and quality of life through various adaptation processes (Fig. 31.3). LSI values are lower in areas where climate variability and hazard effects are more significant, such as cyclone inundation, coastal erosion, flooding, and storm surge activities. On the other hand, LSI values are higher in areas where livelihood depends on secondary and tertiary activities. Involvement in multiple jobs has increased income, improved quality of life, secured livelihoods, and increased resilience capacity to deal with climatic hazards (Fig. 31.3). People in Zone I and Zone II are more vulnerable than those in Zone III and Zone IV because of the differences in their livelihoods, incomes, and climatic hazards exposure. Villagers in Zone III and Zone IV have been modifying or altering their livelihoods through various adaptive strategies. On the other hand, people who live in Zone-I have more occupational opportunities. Still, their livelihoods become vulnerable due to climate risks, as exposure to climatic hazards is high in low-lying shore fringe areas.

31.5 Conclusions

The study reveals that Livelihood Security Index (LSI) values are lower in those villages where climate variability and hazard effects are more significant, such as cyclones, inundation, coastal erosion, flooding associated with storm surges, salt water encroachment, etc.; especially farmers and fishermen communities have low livelihood security, and their adaptability to climate vulnerability has not increased as much as is required. The percentage of households with a low livelihood security index is higher in studied villages located between 0 and 5 km from the coast. Multiple livelihood options are only available in the region under 10 km from the shoreline, but the options are gradually decreasing away from the coast. Due to the impact of coastal hazards, particularly climatic hazards, the livelihood of villagers

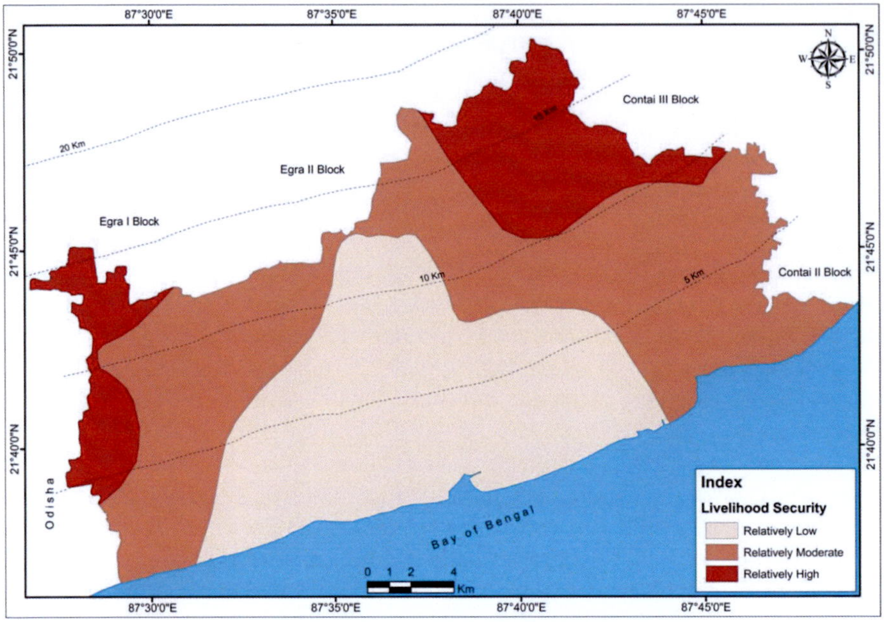

Fig. 31.3 Overall livelihood security status in Ramnagar-I, II, and Contai-I block

becomes vulnerable. Finally, the livelihood security index value shows better socio-economic conditions for the local people who are living in the landward parts of the coasts, ranging from 15 to 20 km from the shoreline.

References

CARE. (2004). *Measuring livelihood impacts: A review of livelihoods indicators, Livelihood Monitoring Unit (LMU) Rural Livelihoods Program CARE Bangladesh*. TANGO International, Inc.

Devereux, S., Sabates-Wheeler, R., & Longhurst, R. (2012). *Seasonality, rural livelihoods and development, seasonality rural livelihoods and development* (p. 352). ISBN 9781849713252, by Routledge.

FAO. (2003). *Food security: Conceptualizing the linkages/commodity policy and projections service*. Commodities and Trade Division Food and Agriculture organization of the United Nations.

ICRC. (2015, June 18). *What is economic security?* International Committee of the Red Cross. 18 June 2015. Link: https://www.icrc.org/en/document/introduction-economic-security

ICZMP, West Bengal. (2012). *Coastal zones, ICZM project, SPMU, IESWM, Dept. of Environment, Govt. of West Bengal*. https://www.iczmpwb.org/main/coastal_zones.php

Mishra, A., & Debata, B. (2021). Livelihood security among rural poor: Evaluating the impact of rural livelihood mission in Odisha, India. *Cogent Economics & Finance, 9*(1), 1978705.

Patidar, H. (2019). Livelihood security in rural India: Reflections from some selected indicators. *Forum for Development Studies, 46*(1), 147–185. https://doi.org/10.1080/08039410.2018.1519517

Paul, A. K. (2002). *Coastal geomorphology and environment* (pp. 1–342). ACB Publication.

Paul, A. K., & Paul, A. (2022). Adjustment of the coastal communities in response to climate variability and sea level rise in the Sundarban, West Bengal, India. In *Climate change, disaster and adaptations: Contextualising human responses to ecological change* (pp. 201–217). Springer International Publishing.

Rahman, S., & Akter, S. (2010). *Determinants of livelihood security in poor settlements in Bangladesh: International Working Paper Series* (Vol. 10(01), p. 21). University of Plymouth-UK.

Serrat, O. (2017). The sustainable livelihoods approach. In *Knowledge solutions*. Springer. https://doi.org/10.1007/978-981-10-0983-9_5

Serrat, O., & Serrat, O. (2017). The sustainable livelihoods approach. *Knowledge solutions: Tools, methods, and approaches to drive organizational performance*, 21–26.

Singh, P. K., & Hiremath, B. N. (2010). Sustainable livelihood security index in a developing country: A tool for development planning. *Ecological Indicators, 10*(2), 442–451.

Chapter 32
Urban Sprawling and Its Emerging Consequences in Response to Climate Variability: A Study in Coastal Urban Areas of Digha, Contai, and Haldia Using Remote Sensing Approach

Dipankar Mondal, Subrata Jana, and Ashis Kumar Paul

32.1 Introduction

Coastal cities are always vulnerable in the face of climate change and associated hazards (O'Brien et al., 2006; Stojanov et al., 2016). The urban areas of the West Bengal coast, particularly in Purba Medinipur district, are facing similar kinds of problems, particularly in the last two decades (Mondal, 2021). The haphazard urbanization and urban sprawling over the immature low-lying tidal flats and dune landscapes insist on land degradation (Mondal et al., 2022). The sea surface temperature (SST) is gradually increasing due to the effects of global warming, which further increases cyclone vulnerability in the coastal areas of the Bay of Bengal (Nayak & Bhaskaran, 2014; Srinivas et al., 2017). Moreover, the increasing sea level promotes an increase in tidal range and wave height, which creates land inundation and puts pressure on the livelihoods of the coastal people (Khan et al., 2012; Ahammad et al., 2013). The northern head of the Bay of Bengal coast is severely affected by sea level rise, tidal inundation, storm surge, shoreline erosion, and coastal flooding (ALi, 1999; Paul, 2002; Hazra et al., 2002). Despite all these kinds of shortcomings, people are encroaching on the coastal areas of the Purba Medinipur district, and the population pressure is tremendously increasing in that area only because of the plenty of resource exploration options and resultant livelihood practices (Mondal et al., 2021). Heavy downpours during cyclone landfalls in the coastal plains (Thomalla & Schmuck, 2004; Ali et al., 2007; Hoarau et al., 2012; Sahoo & Bhaskaran, 2018) are prone to saltwater flooding due to storm surges and tidal

D. Mondal (✉) · A. K. Paul
Department of Geography, Vidyasagar University, Midnapore, West Bengal, India
e-mail: dipankargeo@gmail.com

S. Jana
Department of Geography, Belda College, Paschim Medinipur, West Bengal, India

A. K. Paul, A. Paul (eds.), *Crisis on the Coast and Hinterland*,
https://doi.org/10.1007/978-3-031-42231-7_32

waves, primarily during the Highest Astronomical Tide (HAT) phase of the monsoon months (Paul, 2002; Chittibabu et al., 2004). The wetlands, lowlands, sand dune habitats, and agricultural lands are occupied for urban sprawling (Jelgersma et al., 1993; Jana & Paul, 2019). Moreover, the Intergovernmental Panel on Climate Change (IPCC) in its working report (Nicholls, 2004; Cazenave & Cozannet, 2014) predicted the areas of Indian coastal urbanization in the scenario of global warming induced sea-level rise impacts. In this concern, the urbanized areas of the coastal plain of Medinipur littoral tract, West Bengal, have been selected for understanding the environmental conflict related to the uncontrolled sprawling of urban areas over the wetlands, sand dunes, lowlands, and shore fringes, and possible adjustments with emerging impacts of coastal chaos due to climate variability at present. The current study areas include urban areas such as Digha and Contai in the Kanthi (Contai) coastal plain and Haldia township area in the Hugli-Haldi estuarine floodplain.

32.2 Study Area

The shore fringe resort town of Digha is located and extended over the shore along parallel sand dunes and beach ridges fringed with the Bay of Bengal and separated by the Digha estuary and Jaldah estuary into three coastal plain sectors (the Digha sector, the Sankarpur-Tajpur sector, and the Mandarmani sector). The Contai municipality town is widely extended over the coastal sand dune ridge and Holocene tidal basin of the supralittoral tract with an area of 14.35 km^2 (Fig. 32.1). Haldia municipality town is located on the supralittoral tract of the Hugli-Haldi estuarine floodplain surface with area encompassing 99.97 km^2. The littoral or intertidal fringes can produce impacts over the supralittoral tracts (above the intertidal littoral fringes) during tidal waves, cyclone surges, and during the seasonal high sea levels (June to November) along the region. The areas of urban centres have emerged during 1958 (Contai), 1990 (Digha), and 1997 (Haldia) in the littoral tract (BADP, 2014; LUDCP, 2015; Mondal, 2021) to fulfil functions like residential and trade centre development, expanding resort centres for coastal tourism processes, and expansion of port area development with emerging institutional hubs under the coastal district of Purba Medinipur of West Bengal. These three urban centres have remained in diverse physical and socio-economic settings.

32.3 Database and Methodology

In the study, Landsat 5 and 8 and the Shuttle Radar Topography Mission (SRTM) data sets for different periods were collected from the United States Geological Survey (USGS). The SRTM-based digital elevation model (DEM) data of 30 m resolution was considered for terrain analysis and the mapping of waterlogged areas in the present study. The SRTM image was resampled to 10 m spatial resolution after subpixel

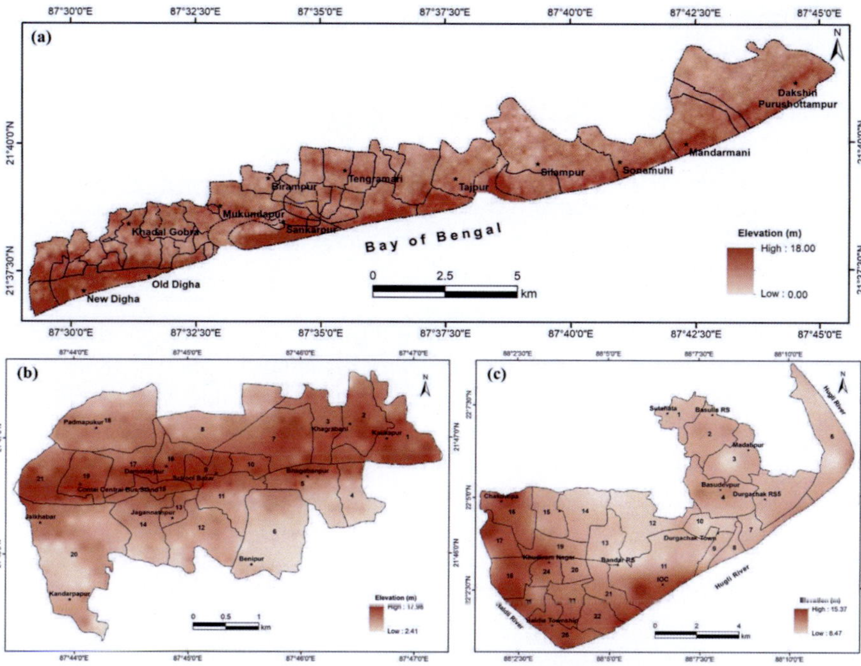

Fig. 32.1 The DEM depicts the physiographic characters of (**a**) Digha-Mandarmoni resort town, (**b**) Contai municipality town, and (**c**) Haidai municipality town (West Bengal)

analysis (Mokarrama & Hojati, 2018). The vertical accuracy of the resampled SRTM data was validated with randomly taken 500 differential global positioning system (DGPS)-based elevations, which divulge ±0.12 m root mean square error (RMSE). Moreover, the Google Earth image was also considered for accurate detection of the different morphological features, which are finally processed into ArcGIS 10.1 software to get the final output. The supervised classification techniques, Spectral Angle Mapper (SAM) and Support Vector Machine (SVM) algorithms, are used in the four different Landsat images of the selected sites from 1991, 2001, 2011, to 2018 for the classification of urban sprawl mapping (Pal and Mather 2005; Bouaziz et al. 2017). Based on the classified images, the built-up areas were extracted from the selected images for three distinct urban sites (Machineni et al., 2019).

32.4 Results and Discussion

32.4.1 Physical Setup of the Urban Centres

The urban centres of Digha, Contai, and Haldia remained over the coastal plain alluviums dominated by sand and clay materials. The surface elevation of the coastal plain ranges from 0 to 18 m above the mean sea level (MSL), making it susceptible to flooding during cyclonic storm surges and heavy rains. Digha is located on the shorefront beach ridge sand dune surface, Contai is located on the inland dune ridge surface, and Haldia is located on the estuarine floodplain surface. The maximum elevations of the three urban centres are 18 m (Digha), 17.98 m (Contai), and 15.37 m (Haldia) (Fig. 32.1). The shallow groundwater aquifers of Digha, Contai, and Haldia are liable to saltwater encroachment as the over-extraction of groundwater is continued in indiscriminate ways for the municipality's water supply. Thus, the environmental setup of the three urban areas is very sensitive to the erosion and inundations on the alluvial coasts.

Over time, the tourism centres were extended into different sections (Digha, Sankarpur-Tajpur, and Mandarmani) from 1950 to 2018 over the beach ridges and sand dunes. The entire digha sankarpur development authority (DSDA) area is situated within the low-lying tracts under beach ridges and wetlands. The elevation of the area varies from 0 m (MSL) to 18 m. The higher elevation is observed along the shorefront dune ridges in the Digha, Sankarpur, and Mandarmani coastal stretches, whereas the lower elevation is found in the wetland areas in the inner parts. Geomorphologically, the coastal stretch is situated within the distinctive geomorphic setup of a sandy beach, a beach ridge with dunes, swale topography, a mature swamp, a recent tidal basin, and swampy tracts (Fig. 32.2a). Among those geomorphic features, most of the area is occupied by the recent tidal basin (25.17 km^2), ancient tidal basin (15.62 km^2), and beach ridge with dune surface (15.21 km^2).

Contai is a municipality town in the Purba Medinipur district. The urban core extends over the beach ridge sand dunes complex of the Contai- Paniparul 1 sand ridge, which lies landward at an 8–10 km distance from the current Bay of Bengal shoreline. The elevation of the entire area varies from 2.41 to 17.98 m above the MSL (Fig. 32.2b). The beach ridge with sand dunes (6.28 km^2) and Holocene tidal basin (5.40 km^2) have taken up the majority of the area, with the other geomorphic features being the ancient tidal basin (1.83 km^2) and palaeo channel with levee (0.76 km^2).

The municipality town, or port town, of Haldia is situated on the estuarine bank margins of the rivers Hugli and Haldi in Purba Medinipur district. The elevation of this area varies from 8.47 to 15.37 m (Fig. 32.2c). The higher elevation is found in the bank margin areas at the confluence of the rivers Hugli and Haldi. Lower elevated areas are still found in the inner low-lying areas. The six major diversified geomorphological features of ancient tidal deposit, Holocene fluvial deposit, mature swamp, mudflat, paleo channel with levee, and swamp are observed within the area (Fig. 32.2c). The Holocene fluvial deposit (61.22 km^2), Paleo channel with levee (15.25 km^2), and Ancient tidal basin (10.21 km^2) have formed the majority of the area. The mature swamp, swamp, and mudflat surfaces have occupied areas of 6.62 km^2, 2.86 km^2, and 1.66 km^2, respectively. The urban infrastructures extend across the tidal flats, wetlands, and low-lying alluvial surfaces of the estuarine floodplain of basement clay and silty sands. Over-extraction of groundwater at an alarming rate is reflected in the rapid fall of groundwater depth in the region. Currently, massive dumping of urban wastes has filled up the wetlands and lowlands along the river estuaries' bank margins. The vast expanse of river bodies and

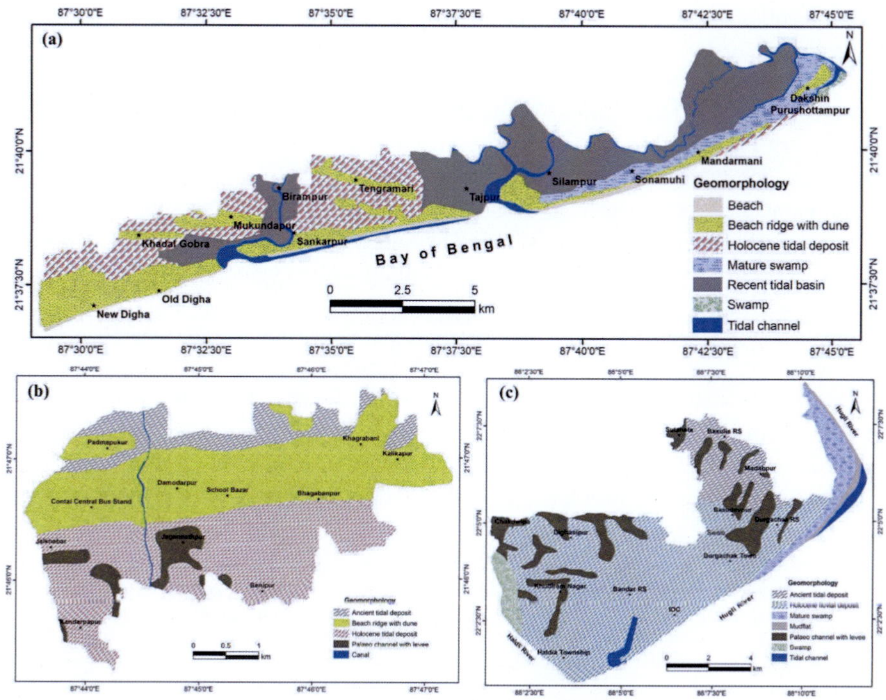

Fig. 32.2 Morphogenetic regions of the urban landscapes; (**a**) Digha-Mandarmani (**b**) Contai, and (**c**) Haldia

extensive saltmarshes of Nayachar Island and the Haldi estuary banks serve as an effective filter for pollutants emitted by industries and port areas in Haldia. The spread of the town may further occupy wetlands and reduce the pollutant trapping capacity of the open marshes. Restoration of wetlands at the urban fringes will control the level of pollution in the inner township areas.

32.4.2 *Trend of Urbanization and Urban Sprawling*

The superimposed built-up areas of four periods in Digha (Fig. 32.3a) depict the concentration of the built-up areas aligned along the shorefront dune ridge in a linear pattern. The urbanization and growth of built-up areas are influenced by tourism activities. The most compact pattern is found in the New Digha and Old Digha tourist destinations. The growth of built-up areas in the total area is very negligible compared to the other areas. But, it is more significant in terms of the existence and risk of this kind of urban expansion. However, the trend of urban growth is increasing from 0.83% (1991) to 1.07% (2001), and from 1.88% (2011) to 3.74% (2018) of the total study area (67.04 km^2) of Digha, respectively (Fig. 32.3a; Table 32.1).

After the 2000s, the growth of built-up areas has significantly increased in the forms of hotels, homestays, market areas, car parking sites, roads, and other supporting accommodations in that area. As a result, such a drastic change is found within the built-up areas of the Digha area. With such infrastructural development, the natural landscape of the dune ridges is destroyed, and sand is transferred to the low, elevated parts for flattening of the land to prepare suitable land for infrastructural development. In other shorefront areas like Mandarmani, built-up structures have existed along the High Tide Line (HTL). The higher rate of shoreline retreat pushed the hotel structures into the interior part of the area after wetland filling. Recently (2018), several built-up structures have been constructed and observed in the far interior low-lying areas of the Mandarmani, Sankarpur, and Tajpur coastal sectors. Moreover, new tourist spots have appeared (e.g., Baguran Jalpai) in recent times, with the most investment in the tourism sector for the best possible earnings. The entire shoreline (Udaypur to Tajpur) is already concretized to protect the shoreline retreat, which shows as the built-up area (Fig. 32.3a).

In the Contai municipality area, the urban growth is based on residential, commercial, and administrative activities. The town of Contai is the main gateway to Digha and other tourism areas of the coastal zones. All kinds of transport and communication and accommodating supply systems in the Digha-Mandarmani tourism

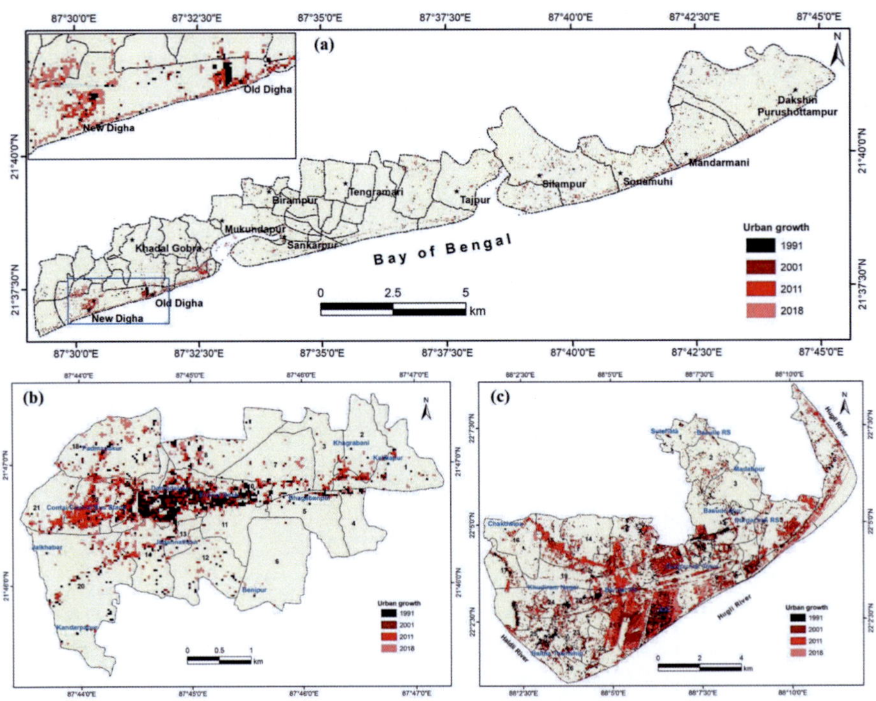

Fig. 32.3 Urban sprawling at (a) DSDA area, (b) Contai Municipality area, and (c) Haldia Municipality area from 1991 to 2018

Table 32.1 Urban sprawling at different urban centres during 1991–2018

	Built-up area of total area (67.04 km²) at Digha		Built-up area of total area (14.35 km²) at Contai		Built-up area of total area (99.97 km²) at Haldia	
Year	km²	%	km²	%	km²	%
1991	0.56	0.83	0.61	4.25	4.72	4.72
2001	0.71	1.07	0.84	5.86	13.25	13.25
2011	1.26	1.88	1.40	9.75	22.71	22.71
2018	2.51	3.74	2.21	15.39	29.36	29.37

area are provided through the Contai urban centre. Such important, diversified urban activities promote the faster growth of that urban centre. As per the result of the land use and land cover (LULC) change, the extracted built-up areas and their superimposition clearly indicate (Fig. 32.3b) that the urban centre is mainly concentrated in the central part of the municipality town (during 1991) over the dune ridge. After the 2000s, built-up areas and settlements expanded in the east and west, along the dune ridge and its fringe areas, with some settlements also built on both sides of the Contai-Digha main road. In the later phase (during 2018), most of the settlements were built in the low-lying areas, occupying the agricultural land and wetland filling in the northern and southern parts of the main town. But, very interestingly, the built-up areas have not grown at a significant level on the eastern side, despite the more suitable land over the dune ridge (part of ward numbers 1, 2, 3, and 7). The major problem in that area is the encroachment of saltwater in the subsurface and groundwater aquifers (CGWB, 2014). The land under the built-up area is growing likewise, by 4.25% (1991), 5.86% (2001), 9.75% (2011), and 15.39% (2018), to the total area (67.04 km²) of the Contai municipality (Fig. 32.3b; Table 32.1). In the Contai urban centre, too, the urban expansion rate has significantly increased since the 2000s, which intensified during 2011–2018.

The most dramatic change in the LULC pattern and in the built-up area took place in the Haldia municipality centre during the years 1991 to 2018. The town initially developed as a port-industry base; afterwards, complete urban infrastructures developed, coupling with residential and market complexes. In 1991, the urban built-up areas were found in some scattered form, mainly in the three sites (the IOC, Durgachak, and Township areas) around the port (Fig. 32.3c). In similar ways (as observed in the Digha and Contai), after the 2000s, the built-up areas significantly increased in the central part, western and eastern levee side areas. The brick-field or brick kiln areas that were extracted as built-up areas have remained on the river Hugli's eastern levee. The settlements and some industries are built in the roadside areas on the north-western side of the study area. After 2010, the settlements were mostly built-up in the low-lying areas in the central part through land-filling. Among the three sites, the highest percentage (29.37%) of total land remains under the built-up area. The growth rate of urban areas is also highest in comparison to the other two urban areas. Since 1991, the area under the built-up portion of the total municipal area has been 4.72% (1991), 13.25% (2001), 22.71% (2011), and 29.37% (2018) (Fig. 32.3c; Table 32.1). After degrading the natural landscape, vegetation cover, and agricultural land, such urban expansion over the most risk-prone areas of river banks, low-lying areas, and wetlands will create misery for urban dwellers in the future.

32.4.3 Nature of Climate Variability

32.4.3.1 Impact of Tidal Fluctuation

In the global and regional scenario of sea level rise (SLR), the local sea level is rising at the rate of 2.2–3.1 mm/y in the northern Bay of Bengal (Kusche et al., 2016). With a mean tidal range of 4.38 m the entire coastal region experiences meso-tide (2–4 m) variation (Church et al. 2006). At the Haldia gauge station, the analysis result of the available data (from 1971 to 2014) shows a positive (rising) trend (0.73) of tide level with the highest (1750 mm) and lowest (6918 mm) during 2013 and 1972, respectively. The tide level in Digha reflects the open coastal environment of the Bay of Bengal, with a higher trend (0.82) of increasing tide level between 1977 and 2012, with the highest (5344 mm) and lowest (1655 mm) tide levels in 1999 and 1992, respectively. The overall result clearly indicates the increasing trend of tide levels all along the Medinipur coastal stretch.

The increasing tide level creates a severe impact on the shoreline and interior parts of the coastal zone. During high tide, the Mandarmani coastal zone has been severely affected, and tidewater has entered into the existing hotels at the seafront positions. Waves are repeatedly acting landward and eroding the shorefront sand dunes and other coastal landforms. The saline seawater encroached into the far elevated coastal landforms and the natural coastal habitats, which are significantly affected by saline water. The life-supporting occupations of the coastal stakeholders are severely affected due to land erosion and saltwater inundation, and that will be tremendously increased in the near future.

32.4.3.2 Cyclonic Events and Associated Risk

During 1999, the entire coastal part of Purba Medinipur district was tremendously affected by the Odisha super cyclone (Kalsi, 2006). The Bengal coast has been affected by the cyclones that provoked the coastal parts of Odisha, West Bengal, and Bangladesh. The month-wise events of Depression (D), Cyclonic Storm (CS), and Severe Cyclonic Storm (SCS) in the Bay of Bengal during 1891–2018 show that the maximum number of cyclonic events have been observed in the pre-monsoon (May), monsoon (July–September), and in the later phase of the monsoon (October–November) (Fig. 32.4). However, the SCS mainly occurs in the months of May and October–November. The overall trend of D, CS, and SCS during 1891–2018 reveals that the trends of D (−0.12) and CS (−0.48) have been decreasing, whereas the trend of SCS has increased at a significant level (0.11) (Fig. 32.5). The cyclone

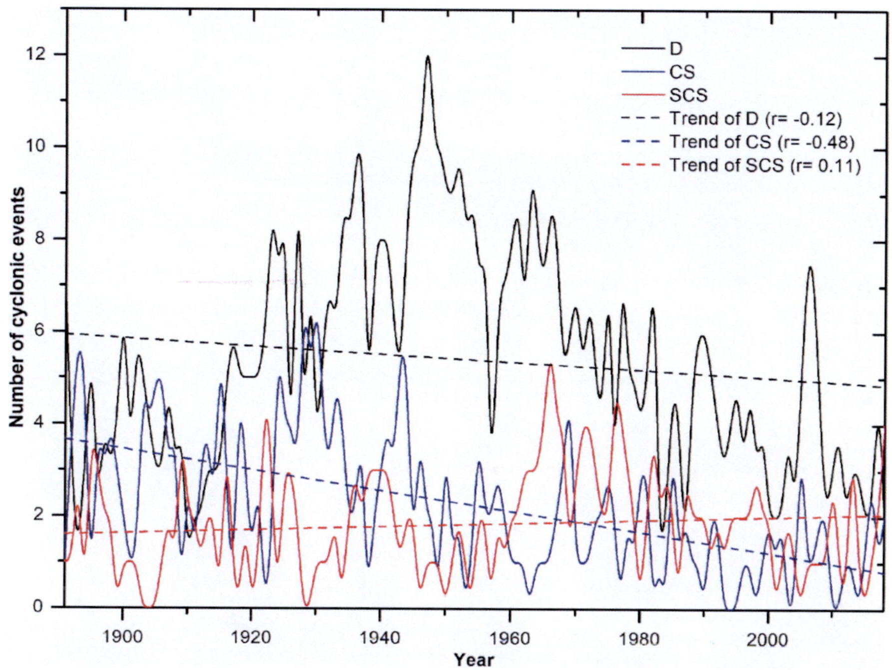

Fig. 32.4 Year-wise cyclonic events of depression (D), cyclonic storm (CS), and severe cyclonic storm (SCS) and their trends in the Bay of Bengal during 1891–2018 (showing the vulnerabilities)

tracks on the Odisha and West Bengal coasts over a temporal span of 20 years indicate that the highest number (259) of cyclonic events have been observed during 1931–1950. Although the number of cyclones was at a low (45) from 2011 to 2018, these only occurred in the last 8 years. The number of cyclonic events has been decreasing since the 1970s when compared to previous periods (from 1891 to 1970).

32.4.4 Inundation and Waterlogging Problems

32.4.4.1 Tidal Inundation at Digha

The tidal range in the meso-tide-dominated environment at the Digha coast faced a severe problem in terms of coastal inundation. Concerning the 4.5 m tidal amplitude, the inundation zones have been demarcated for the Digha coastal stretch (Fig. 32.5a). The tidewater inundation zone shows that about 48% land area (32.45 km^2) of the total area (67.04 km^2) will remain under tidewater. However, about 29.62 km^2 of area remains below 4.5 m elevation, where lots of settlements (mainly hotels and resorts) are built-up as a tourist support system after landfilling. These areas are mostly liable for tidal inundation and are fully inundated during cyclonic storm surges. About

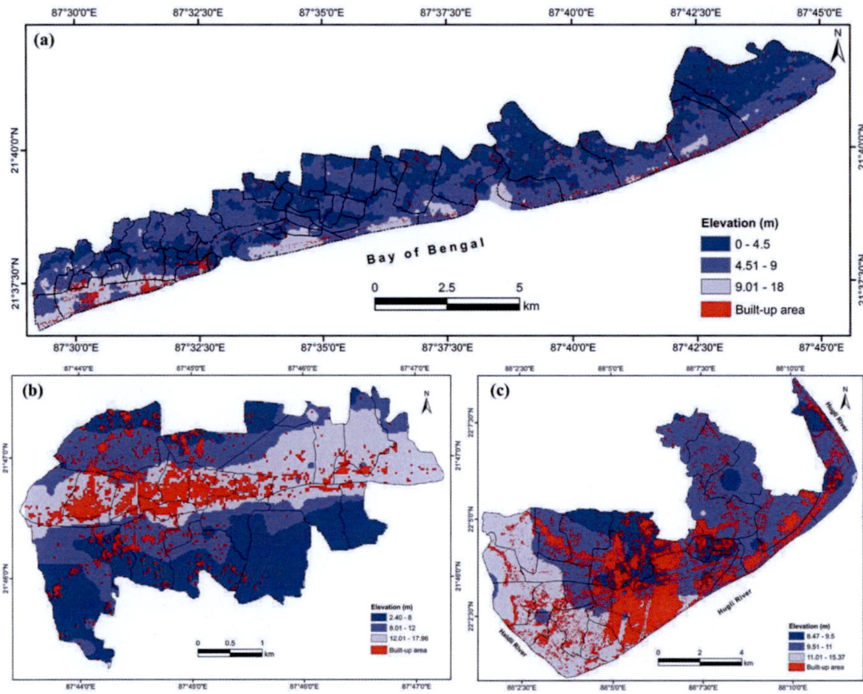

Fig. 32.5 Water logging and saltwater inundation prone areas at (**a**) Digha urban centre, and water logged areas at (**b**) Contai and (**c**) Haldia urban centres

31.08 km^2 remain within the 4.5–9 m elevation zone of the dune fringe areas. The built-up areas have been formed after flattening the dune landscape. Only 6.34 km^2 of land remains within the 9–18 m elevation of the shorefront dune ridges, but most of these ridges have been degraded for the construction of hotels, resorts, and roads in the DSDA area. Different types of LULC classes have existed within the entire inundation zone. Mainly, agricultural land and vegetation areas are affected by this kind of saltwater inundation, which will intensify further in the near future. However, it will be beneficial for the aquaculture practices in the wetland areas. Moreover, the recent trend of urban expansion in low-lying areas (after lowland filling) can severely endure the predicted tidewater inundation level.

32.4.4.2 Water-Logging at Contai

The Contai municipality area has not suffered from the waterlogging problem due to its elevated dune landscape. However, the urban areas have been extended over the lowland areas in recent times. During monsoonal storm rainfall events, these lowland settlements are prone to waterlogging and its associated problems. The lower elevation is observed on both sides of the sand dunes, and the high slope has been observed at the dunes' fringes. Micro-level contour pattern and terrain characteristics also confer the relative existence of the dune ridge and lowlands. The heavy rainfall (average 1200 mm) in the southwest monsoon months is responsible for waterlogging in low-lying areas. Rainwater is unable to drain in the mildly sloppy surfaces, causing waterlogging in the natural depressions on both sides of the dune. There are not any significant canals or any natural drainage systems to drain such a volume of rainwater. Waterlogging can occur in Costa Rica's urban outskirts during short-term rainfall events, particularly during cyclone landfalls. In the year 1997, during the landfall of the Contai Cyclone, 495 mm of rainfall took place in a 48 h period (Paul, 2002). If such an incidence takes place in the future with the climate change process, then the 4.44 m elevated areas may be affected by large-scale waterlogging in the Contai municipality areas, which are estimated to be 22% (3.21 km^2) of the total area (14.35 km^2). The estimated waterlogged areas (Fig. 32.5b) are mainly used for agriculture practices and newly constructed settlement areas. Furthermore, recent urban expansion trends over low-lying areas may increase future risk and vulnerability for residents.

32.4.4.3 Tidal Inundation at Haldia

The Haldia urban centre has existed in the estuary fringe problematic zone. Therefore, tidewater encroachment into the interior and land inundation will create a severe impact in the faster-growing urban-industrial areas. The elevated natural levee has already been embraced by urban-industrial infrastructures. Recently (2019), in the low-lying interior part, a new ward has been incorporated with the municipality. Furthermore, an exaggerated rate of lowland filling for urban infrastructure construction has been observed in the municipality area. The existing tidal channels have also been degraded by landfilling, therefore, the entire land area in the interior part of the municipality has remained risk-prone and vulnerable. The micro-terrain morphometry reveals a bowl-shaped lower elevation landscape in the municipality's centre (Fig. 32.5c). Considering the elevation differences and estuarine tidal environment, the 0.98 m depth of ponding has been considered over the 8.47 m elevated land surface (Fig. 32.5c). The rainwater-logging and saltwater inundations will affect about 6.69 km^2 (about 17%) of the total municipality area (99.97 km^2). Within the estimated inundation zones, high-density urban infrastructures, agricultural land, and vegetation have existed. Therefore, the inundation zone will create more risk and vulnerability in the future.

32.5 Conclusions

Coastal urbanization across the landscapes dominated by unconsolidated alluviums of beach ridges, sand dunes, tidal wetlands, floodplains, and swales provided risks of flood-related inundations. During previous cyclone attacks, the urban areas of

Digha-Mandarmoni resort town, Contai, and Haldia municipality towns experienced salt water inundation and rain-affected waterlogging due to storm surges, tidal surges, and storm rainfalls. Based on the DEMs and classification of inundations and waterlogging areas, inundation vulnerabilities have been estimated for the three coastal urban centres of West Bengal. In all three urban built-up zones, drainage outlets, low lying areas, and tidal flat habitats are occupied by urban sprawling in sensitive coastal plain environments. Overwithdrawal of groundwater in the population concentrated areas of unconsolidated alluvium surfaces are directly responsible for the lowering trends of groundwater tables and saltwater encroachments into the coastal aquifers. In this case, Haldia and Digha urban centres are more affected than Contai municipality. The emerging consequences of such urban sprawl have proven that coastal urban centres will suffer from physical vulnerabilities in response to sea level rise and climate change threats.

References

Ahammad, R., Nandy, P., & Husnain, P. (2013). Unlocking ecosystem-based adaptation opportunities in coastal Bangladesh. *Journal of Coastal Conservation, 17*(4), 833–840.

Ali, A. (1999). Climate change impacts and adaptation assessment in Bangladesh. *Climate Research, 12*(2–3), 109–116.

Ali, M. M., Jagadeesh, P. V., & Jain, S. (2007). Effects of eddies on Bay of Bengal cyclone intensity. *Eos, Transactions American Geophysical Union, 88*(8), 93–95.

BADP. (2014). *Beachfront area development plan for Digha Shankarpur Area*. Prepared by I-Win Advisory Services Limited under the Digha-Shankarpur Development Authority, Digha, West Bengal (pp. 1–309). Accessed on 12 Jan 2023. Accessed from http://www.iczmpwb.org/main/pdf/pea/Digha%20Shankarpur%20Integrated%20Beachfront%20Area%20Development%20Plan.pdf

Bouaziz, M., Eisold, S., & Guermazi, E. (2017). Semiautomatic approach for land cover classification: A remote sensing study for arid climate in southeastern Tunisia. *Euro-Mediterranean Journal for Environmental Integration, 2*(1), 1–7.

Cazenave, A., & Cozannet, G. L. (2014). Sea level rise and its coastal impacts. *Earth's Future, 2*(2), 15–34.

CGWB. (2014). *Report on status of ground water quality in Coastal Aquifers of India, Central Ground Water Board, Ministry of Water Resources, River Development & Ganga Rejuvenation, Government of India, Faridabad*. Accessed on 29 Dec 2022. Accessed from https://cgwb.gov.in/WQ/Costal%20Report.pdf

Chittibabu, P., Dube, S. K., Macnabb, J. B., Murty, T. S., Rao, A. D., Mohanty, U. C., & Sinha, P. C. (2004). Mitigation of flooding and cyclone hazard in Orissa, India. *Natural Hazards, 31*(2), 455–485.

Church, J. A., White, N. J., & Hunter, J. R. (2006). Sea-level rise at tropical Pacific and Indian Ocean islands. *Global and Planetary Change, 53*(3), 155–168.

Hazra, S., Ghosh, T., DasGupta, R., & Sen, G. (2002). Sea level and associated changes in the Sundarbans. *Science and Culture, 68*(9/12), 309–321.

Hoarau, K., Bernard, J., & Chalonge, L. (2012). Intense tropical cyclone activities in the northern Indian Ocean. *International Journal of Climatology, 32*(13), 1935–1945.

Jana, S., & Paul, A. K. (2019). Assessment of morphogenetic sedimentary depositional environments of different morphological surfaces of middle-lower and deltaic courses of Subarnarekha River. *Journal of Coastal Sciences, 6*(1), 1–11.

Jelgersma, S., Van der Zijp, M., & Brinkman, R. (1993). Sea level rise and the coastal lowlands in the developing world. *Journal of Coastal Research, 9*(4), 958–972.

Kalsi, S. R. (2006). Orissa super cyclone-A synopsis. *Mausam, 57*(1), 1.

Khan, A. S., Ramachandran, A., Usha, N., Punitha, S., & Selvam, V. (2012). Predicted impact of the sea-level rise at Vellar–Coleroon estuarine region of Tamil Nadu coast in India: Mainstreaming adaptation as a coastal zone management option. *Ocean & Coastal Management, 69*, 327–339.

Kusche, J., Uebbing, B., Rietbroek, R., Shum, C. K., & Khan, Z. H. (2016). Sea level budget in the Bay of Bengal (2002–2014) from GRACE and altimetry. *Journal of Geophysical Research: Oceans, 121*(2), 1194–1217.

LUDCP. (2015). *Land use and development control plan*. Digha Sankarpur Development Authority, Purba Medinipur, West Bengal (pp. 1–28). Accessed on 11 Jan 2023. Accessed from https://www.wbhidcoltd.com/upload_file/report_publication/report11.pdf

Machineni, N., Sinha, V. S., Singh, P., & Reddy, N. T. (2019). The impact of distributed landuse information in hydrodynamic model application in storm surge inundation. *Estuarine, Coastal and Shelf Science, 231*, 106466. https://doi.org/10.1016/j.ecss.2019.106466

Mokarrama, M., & Hojati, M. (2018). Landform classification using a sub-pixel spatial attraction model to increase spatial resolution of digital elevation model (DEM). *The Egyptian Journal of Remote Sensing and Space Science, 21*(1), 111–120.

Mondal, D. (2021). *Coastal urbanization and population pressure with related vulnerabilities and environmental conflicts-A case study at Medinipur littoral tract, West Bengal* [Doctoral thesis submitted to Vidyasagar University].

Mondal, D., Mandal, M., Jana, S., & Paul, A. K. (2021). Estimation of relationships between land surface temperature and land use diversity in the Contai municipality area, West Bengal. *Indian Journal of Geography, 17–18*, 47–57.

Mondal, D., Jana, S., & Paul, A. K. (2022). Status of land conversion and urban sprawling over coastal tract of West Bengal: A study on Haldia municipality area. *Indian Journal of Geography, 19*, 1–4.

Nayak, S., & Bhaskaran, P. K. (2014). Coastal vulnerability due to extreme waves at Kalpakkam based on historical tropical cyclones in the Bay of Bengal. *International Journal of Climatology, 34*(5), 1460–1471.

Nicholls, R. J. (2004). Coastal flooding and wetland loss in the 21st century: Changes under the SRES climate and socio-economic scenarios. *Global Environmental Change, 14*(1), 69–86.

O'Brien, G., O'keefe, P., Rose, J., & Wisner, B. (2006). Climate change and disaster management. *Disasters, 30*(1), 64–80.

Pal, M., & Mather, P. M. (2005). Support vector machines for classification in remote sensing. *International Journal of Remote Sensing, 26*(5), 1007–1011.

Paul, A. K. (2002). *Coastal geomorphology and environment: Sundarban coastal plain, Kanthi coastal plain, Subarnarekha delta plain*. ACB Publications.

Sahoo, B., & Bhaskaran, P. K. (2018). A comprehensive data set for tropical cyclone storm surge-induced inundation for the east coast of India. *International Journal of Climatology, 38*(1):403–419.

Srinivas, C. V., Mohan, G. M., Rao, D. V., Baskaran, R., & Venkatraman, B. (2017). Numerical simulations with WRF to study the impact of sea surface temperature on the evolution of tropical cyclones over Bay of Bengal. In Mohapatra et al. (Eds.), *Tropical cyclone activity over the north Indian Ocean* (pp. 259–271). Springer.

Stojanov, R., Kelman, I., Ullah, A. A., Duží, B., Procházka, D., & Blahůtová, K. K. (2016). Local expert perceptions of migration as a climate change adaptation in Bangladesh. *Sustainability, 8*(12), 1223.

Thomalla, F., & Schmuck, H. (2004). 'We all knew that a cyclone was coming': Disaster preparedness and the cyclone of 1999 in Orissa, India. *Disasters, 28*(4), 373–387.

Index

A. K. Paul, A. Paul (eds.), *Crisis on the Coast and Hinterland*,
https://doi.org/10.1007/978-3-031-42231-7

The manufacturer's authorised representative in the EU is Springer Nature Customer Service Centre GmbH, Europaplatz 3, 69115 Heidelberg, Germany. If you have any concerns regarding our products, please contact ProductSafety@springernature.com

Printed and bound by CPI Group (UK) Ltd, Croydon, CR0 4YY
30/09/2026
02224059-0003